Lecture Notes in Mechanical Engineering

Lecture Notes in Mechanical Engineering (LNME) publishes the latest developments in Mechanical Engineering—quickly, informally and with high quality. Original research reported in proceedings and post-proceedings represents the core of LNME. Volumes published in LNME embrace all aspects, subfields and new challenges of mechanical engineering.

To submit a proposal or request further information, please contact the Springer Editor of your location:

Europe, USA, Africa: Leontina Di Cecco at Leontina.dicecco@springer.com
China: Ella Zhang at ella.zhang@springer.com
India: Priya Vyas at priya.vyas@springer.com
Rest of Asia, Australia, New Zealand: Swati Meherishi at swati.meherishi@springer.com

Topics in the series include:

- Engineering Design
- Machinery and Machine Elements
- Mechanical Structures and Stress Analysis
- Automotive Engineering
- Engine Technology
- Aerospace Technology and Astronautics
- Nanotechnology and Microengineering
- Control, Robotics, Mechatronics
- MEMS
- Theoretical and Applied Mechanics
- Dynamical Systems, Control
- Fluid Mechanics
- Engineering Thermodynamics, Heat and Mass Transfer
- Manufacturing
- Precision Engineering, Instrumentation, Measurement
- Materials Engineering
- Tribology and Surface Technology

Indexed by SCOPUS, EI Compendex, and INSPEC

All books published in the series are evaluated by Web of Science for the Conference Proceedings Citation Index (CPCI)

To submit a proposal for a monograph, please check our Springer Tracts in Mechanical Engineering at https://link.springer.com/bookseries/11693

Cucuk Nur Rosyidi · Pringgo Widyo Laksono ·
Wakhid Ahmad Jauhari · Muhammad Hisjam
Editors

Proceedings of the 6th Asia Pacific Conference on Manufacturing Systems and 4th International Manufacturing Engineering Conference

APCOMS-IMEC 2022, Surakarta, Indonesia

2022 APCOMS IMEC

Springer

Editors
Cucuk Nur Rosyidi
Faculty of Engineering
Universitas Sebelas Maret
Surakarta, Central Java, Indonesia

Pringgo Widyo Laksono
Faculty of Engineering
Universitas Sebelas Maret
Surakarta, Central Java, Indonesia

Wakhid Ahmad Jauhari
Faculty of Engineering
Universitas Sebelas Maret
Surakarta, Central Java, Indonesia

Muhammad Hisjam
Faculty of Engineering
Universitas Sebelas Maret
Surakarta, Central Java, Indonesia

ISSN 2195-4356 ISSN 2195-4364 (electronic)
Lecture Notes in Mechanical Engineering
ISBN 978-981-99-1244-5 ISBN 978-981-99-1245-2 (eBook)
https://doi.org/10.1007/978-981-99-1245-2

This Springer imprint is published by the registered company Springer Nature Singapore Pte Ltd.
The registered company address is: 152 Beach Road, #21-01/04 Gateway East, Singapore 189721, Singapore

Organizing Committee

Chairman of the Organizing Committee

Cucuk Nur Rosyidi, Universitas Sebelas Maret, Indonesia

Hononary Chair

Yuserrie Zainuddin, Universiti Malaysia Pahang, Malaysia
Kamal Zuhairi Zamli, Universiti Malaysia Pahang, Malaysia
Zamberi Bin Jamaludin, Universiti Teknikal Malaysia Melaka, Malaysia
Seri Rahayu Binti Kamat, Universiti Teknikal Malaysia Melaka, Malaysia

The Conference Co-chairs

Eko Pujianto, Universitas Sebelas Maret, Indonesia

Steering Committee

Secretary
Retno Wulan Damayanti, Universitas Sebelas Maret, Indonesia

Publication
Pringgo Widyo Laksono, Universitas Sebelas Maret, Indonesia
Wakhid Ahmad Jauhari, Universitas Sebelas Maret, Indonesia

Technical Program
Bambang Suhardi, Universitas Sebelas Maret, Indonesia
Muh Hisjam, Universitas Sebelas Maret, Indonesia

Organizing Committee

Coordinator
Dewi Fajar Setyorini, Universitas Sebelas Maret, Indonesia

Secretary
Aprillia Dewi Brias Alma, Universitas Sebelas Maret, Indonesia
Putri Dwi Larasati, Universitas Sebelas Maret, Indonesia
Dwi Sulistyo Widya Habsari, Universitas Sebelas Maret, Indonesia
Annisa Puja Ramadhandyna, Universitas Sebelas Maret, Indonesia

Event
Devina Nur Affifah, Universitas Sebelas Maret, Indonesia
S. Febryanti Valentina, Universitas Sebelas Maret, Indonesia
Jessica Paleta, Universitas Sebelas Maret, Indonesia
Erina Annatasya Octaviani, Universitas Sebelas Maret, Indonesia
Hanin Tsimar Andriyani, Universitas Sebelas Maret, Indonesia
Nauval Hernandoko, Universitas Sebelas Maret, Indonesia
Rahmat Herpradipto, Universitas Sebelas Maret, Indonesia

Publication and Documentation
Hafsah Qonita, Universitas Sebelas Maret, Indonesia
Imam Bagus Wibisono, Universitas Sebelas Maret, Indonesia
N. Aldi Pascagama, Universitas Sebelas Maret, Indonesia
Ilham Fairuzaman, Universitas Sebelas Maret, Indonesia
Dhea Naomi Kenlaksita, Universitas Sebelas Maret, Indonesia

Sponsorship
Nathania Angelica, Universitas Sebelas Maret, Indonesia
H. Shafa Keysa Rinjani, Universitas Sebelas Maret, Indonesia
N. R. Rizki Ananda Putra, Universitas Sebelas Maret, Indonesia

Logistic
Ahnaf Rafif Fadlurrohman, Universitas Sebelas Maret, Indonesia
Yoshan Ardhi Pratama, Universitas Sebelas Maret, Indonesia
Cristin Angel, Universitas Sebelas Maret, Indonesia
Agus Kurnia Akbar, Universitas Sebelas Maret, Indonesia
S. Elisa Melodian Charista, Universitas Sebelas Maret, Indonesia
Fadhila Diah Ayu, Universitas Sebelas Maret, Indonesia
Issacian Mutiara Paska, Universitas Sebelas Maret, Indonesia
Sekar Zaneta Amirulputri, Universitas Sebelas Maret, Indonesia

Preface

We are immensely pleased to welcome all delegates and distinguished guests to the 6th Asia Pacific Conference on Manufacturing System and the 5th International Manufacturing Engineering Conference, held in Surakarta, Indonesia. Due to the global pandemic of COVID-19, we conduct this conference in hybrid mode. However, we believe that the objective of the conference as the forum to share and discuss the ideas and research on industrial engineering still can be achieved. The theme of APCOMS-iMEC 2022 is "Post Pandemic Recovery: Industrial Engineering Perspective". This theme is selected since in the last two years Indonesia and the global world have been facing pandemic COVID-19. The pandemic has made a tremendous impact on quality of life for individuals and community. It also disrupted all aspects including the industrial operations. It takes a lot of efforts to restore it to the previous normal situation. A continuous innovation on technology and commercialization is needed in responding the needs of individuals, society, and industries toward the new normal phase.

In this conference, the committee received 77 papers from Indonesia, Germany, Malaysia, Thailand, and Philippines. After rigorous review process, we accepted 62 papers and 1 paper was withdrawn. All accepted and presented papers will be forwarded for consideration to be published in the *Lecture Notes in Mechanical Engineering*, a leading periodical from Springer. As the high numbers of papers presented from various contributors, we hope this triggers the collaboration among the researches who participated in this event. International engagement and cooperation are very important in solving the global problem which is in line with the vision of this conference.

We especially thank to all collaborating universities: Universitas Sebelas Maret, Institut Teknologi Bandung, Universiti Malaysia Pahang, Universiti Teknikal Malaysia Melaka, and Kasetsart University. We hope the collaboration can be extended in other areas such as research, faculty, and student exchange and other academia activities. We also thank to all keynote speakers, participants, sponsors, and partners for their contribution and all committees for their contributions and great effort to organize this conference.

Surakarta, Indonesia
October 2022

Cucuk Nur Rosyidi
Pringgo Widyo Laksono
Wakhid Ahmad Jauhari
Muhammad Hisjam

Contents

Manufacturing Process

Development and Design of IoT-Based Poultry Weighing Tools 3
Rayinda P. Soesanto, Muhammad Mi'radj Isnaini, Afrin F. Rizana, Lutfia Septiningrum, and Fandi Achmad

Effect of Machining Parameters on Surface Quality of Aluminium Puncher for Microchannel Fabrication Using Micro Cutting Process 11
Mohd Saifuldin Mohd Mokhtar and Ahmad Razlan Yusoff

Wear Behaviour of Tungsten Carbide in End Milling Process of Aluminium Alloy 6061-T6 with Minimal Quantity of Tri-hybrid Nanofluids 21
W. Safiei, M. M. Rahman, A. R. Yusoff, and M. Y. Ali

Experimental and Simulation Approach; Investigation Effect of Axial Feed Rate to the Cutting Force in Dimple Milling Using Ball End Mill Tool 33
M. A. H. Shaharudin and A. Razlan Yusoff

Assembly Sequence Optimization to Minimize Reorientation and Tool Changes Using Genetic Algorithm 51
William, Fariz Muharram Hasby, Wisnu Aribowo, and Sukoyo

Surface Roughness Analysis on Sheet Metals During AWJ Machining Using Low Pressure 59
Nazrin Mokhtar, Mebrahitom Asmelash, Kushendarsyah Saptaji, and Azmir Azhari

Manufacturing System

Design of Intervention for Operator in the Inspection Department of PT. XY Using Pahl and Beitz Method 67
Rahmaniyah Dwi Astuti, Bambang Suhardi, Pringgo Widyo Laksono, Amirul Wika Rosida, and Muhammad Junus Atthariq

Supply Chain Network Design for e-Groceries Using Clustering and Linear Programming 87
A. D. P. Prady, G. M. A. Alaudin, T. P. Achmad, and R. T. Cahyono

A Closed-Loop Supply Chain Inventory Model with Remanufacturing and Cleaner Production System 97
Wakhid Ahmad Jauhari, Anda Iviana Juniani, Wiwik Sulistyowati, and Anik Dwiastuti

Analysis of Cardboard Sheets Supplier Selection Using Fuzzy AHP and TOPSIS Methods at PT. XYZ 107
Santika Sari and Dewi Maharani

Investigation of Monitoring Electrical Consumption on Load Using IoT 115
Muhammad Zuhairi Azharudin and Norazlianie Sazali

Investigation of IoT Devices for Fire Alarm Detection Systems Using Raspberry Pi 123
Nik Nurjihan Nisrina Nik Zaid and Norazlianie Sazali

Electric Motorcycle Adoption Research: A Bibliometric Analysis 131
Yuniaristanto, Wahyudi Sutopo, Muhammad Hisjam, and Hendro Wicaksono

Design of Halal Traceability System Model for Edge Sewing Rectangular Veil Products at PT. YKI and PT. KPS Bandung 139
Gina Rahayu Wardiani, Mohammad Mi'radj Isnaini, and Dradjad Irianto

The Design of Cultural Tourism Potential Indicators at Rembang Regency Using SECI Method 151
Mia Amelia, Lutfia Septiningrum, and Augustina Asih Rumanti

Packaging Design for Kelincidunia Cassava Chips Using Ergonomic Function Deployment (EFD) in UD. Wahyu Pendawa Lestari 161
Niken Utami Tyastuti, Bambang Suhardi, Rahmaniyah Dwi Astuti, and Annisa Syahliantina

Six Sigma Application on Software Improvement Process 183
Tatinia A. R. Amalia, Mohammad Mi'radj Isnaini, and Dradjad Irianto

Multi-skilled Operator Assignment Model for Long Cycle Time and Low-Volume Assembly Line: A Case Study of Aircraft Assembly Line 195
Anas Ma'ruf and Shinta Uli Agnes Monica Gultom

Analysis of Employment Increasing for Potential Tourism Industry in Indonesia: An Exploration Method Approach 205
Lutfia Septiningrum, Fandi Achmad, Augustina Asih Rumanti, and Iwan Inrawan Wiratmadja

Development of an Optimization Model for Home Health Care Routing and Scheduling Problem 213
Annisa Puja Ramadhandyna, Pringgo Widyo Laksono, and Cucuk Nur Rosyidi

Development of a Sustainable Supplier Selection Model and Order Allocation Considering Transportation Alternatives and Quantity Discounts 223
Rosifa Sahara Prasiwi, Cucuk Nur Rosyidi, and Pringgo Widyo Laksono

Optimization Model for Lot Sizing and Scheduling Problem Considering Multiple Suppliers, Quantity Discounts, Carriers, and Production Time Constraints 233
Hani Aninda Intan Permatasari, Cucuk Nur Rosyidi, and Pringgo Widyo Laksono

An EOQ Model for Perishable Product Considering Carbon Emissions Resulting from Refrigerated Truck and Cold Storage 247
Hanin Tsimar Andriyani, Cucuk Nur Rosyidi, and Pringgo Widyo Laksono

Web Design: Develop the Decision Support System for Judgement Additional Order (JAO) Using Waterfall Method 257
Yoshan Ardhi Pratama, Pringgo Widyo Laksono, and Eko Liquiddanu

Optimization Model for Perishable Product with Different Payment Schemes and Carbon Tax Regulation 271
Shafa Keysa Rinjani Hananta, Cucuk Nur Rosyidi, and Wakhid Ahmad Jauhari

Segmentation of Potential Sustainable Tourism Based on Support System Facility Perspective 285
Riska Yanu Fa'rifah, Fandi Achmad, Lutfia Septiningrum, and Iwan Inrawan Wiratmadja

Study of Learning Factories Concept for Production Laboratories in Bandung Polytechnic of State 293
Fachri Koeshardono, Sutrimo, Aris Suryadi, Muhammad Syahrain Basirisa, Lili Awalia, and Mugo Muchjayadi

Applying Importance Performance Analysis (IPA) to Measure the Customer Satisfaction on Food Safety Practices in Restaurant 301
Gusti Fauza, Aisyah Rizky Millenia, Asri Nursiwi, Dwiningtyas Padmaningrum, Setyaningrum Ariviani, and Hari Prasetyo

Natural-Based Tourism Potential Identification from Indonesia in Post Pandemic Situation ... 317
Rayinda P. Soesanto, Afrin F. Rizana, and Artamevia S. Rizaldi

Development of Framework for Flexible Job Shop Scheduling Based on Digital Twin to Tackle Disturbing Events ... 325
Muhammad Usman Nisar, Anas Ma'ruf, and Abdul Hakim Halim

Online Transportation Order Matching Considering Passenger Priority Using Recency Frequency Monetary Method ... 331
Andreas Handojo, Nyoman Pujawan, Budi Santosa, and Moses Laksono Singgih

Manufacturing Losses Analysis of Ring Spinning Machine Based on Overall Equipment Effectiveness Evaluation: A Textile Case Study ... 339
Hasna Khairunnisa, Annas Toat, Ahmad Darmawi, and Agus Ardiyanto

Improving the Efficiency of the Coal Car Body Assembly Process Using Work Standards at PT INKA ... 349
Farhana Mazaya Putri and Bambang Suhardi

The Sustainability-Oriented Total Cost of Ownership Model of the Electric Motorcycle ... 365
Yuniaristanto, Sayyidah Maulidatul Afraah, Wahyudi Sutopo, and Muhammad Hisjam

Supplier Selection and Order Allocation: Systematic Literature Review ... 375
Enty Nur Hayati, Cucuk Nur Rosyidi, Wakhid Ahmad Jauhari, Pringgo Widyo Laksono, and Retno Wulan Damayanti

A Literature Review on the Energy Efficiency Issue in Operating Room Planning and Scheduling ... 383
Ahmad Kholid Alghofari, Cucuk Nur Rosyidi, Wakhid Ahmad Jauhari, and Pringgo Widyo Laksono

Economic Order Quantity Model for Perishable Product by Considering Donation Factor ... 391
Cucuk Nur Rosyidi and Syania Hilda Andini

Optimization of the Automotive Manufacturing Plant Using Systematic Layout Planning (SLP) with Modular Layout ... 401
Muhammed Nafis Bin Osman Zahid and Muhammad Zulhilmi Bin Shamsul Amran

Inventory Model for Reusable Items with Uniform Demand Arrival 411
Yosi Agustina Hidayat

Introduction of AHP Decision Model for Prioritizing Critical Risk of WMSD in the Automotive Working Environment 425
Fazilah Abdul Aziz, Nik Mohd Zuki Nik Mohamed, and Ahmad Nasser Mohd Rose

Design Thinking Process for Solving Supplier Selection and Risk Management Problem: A Case Study 437
Tiara Priscilla Nandyasari and Yosef Daryanto

Decision Support System for Tuition Subsidy in XYZ Senior High School Student with Families Affected by Pandemic 447
Irsyad Thalib Abdillah, Mia Amelia, Dana Azizah, Nurul Fitri Dwi Andini, Wali Hakim Pradana, Amelia Kurniawati, and Afrin Fauzya Rizana

Cost Analysis for Warranted Products Protected by Lemon Laws Considering Failure Interaction 457
Fakhri I. Alifin, Andi Cakravastia, and Bermawi P. Iskandar

Literature Review of Lean Manufacturing, Six Sigma and Lean Six Sigma Benefit 473
M. A. K. Tengtarto, M. L. Singgih, N. Siswanto, and P. Suwignjo

Design Process of Whippletree-Like Mechanism for Versatile Gripping in Arm Prosthesis 485
Wildan Trusaji, Dradjad Irianto, Ardianto Satriawan, Casi Setianingsih, Erza Nurtriandari, and Rachmat Zulkarnain Goesasi

Factors that Affect the Job Performance of the Health Workers in Indonesia: A Configurational Analysis 495
Dicka Korintus Kurnianto, Amanda Syifa Ariqoh, Ferdinanda Pascha Hasian, Imanuella Indah Nugroho, Josyika Nisfullaili, and Retno Wulan Damayanti

Best-Fit Heuristic Algorithm for Multi-stack Rectangular Packing Problem of NC Router Machine 509
Irfansyah Ali, Wisnu Aribowo, and Sukoyo

Cost Minimization Model for Raw Materials Procurement Planning at an Indonesian Aluminium Producer 517
Audira Hutami, Praditya Ajidarma, and Dradjad Irianto

Measuring Customer Satisfaction on Educational Management Services in University 527
Cindy Saraswati, Hari Prasetyo, and Gusti Fauza

Perception of Employees on Lean Six Sigma Practices and Corporate Sustainability in Automotive Manufacturing Industry 539
Nur Syahirah Mohamad Azlan and Noraini Mohd Razali

Cognitive-Affective Model of Perceived User Satisfaction (CAMPUS) on the Design of E-Learning Interface: A Literature Review 549
Melati Salma and Ari Widyanti

PSO-LFDE Algorithm on Constrained Real-Parameter Optimisation Test Functions 559
Nafrizuan Mat Yahya and Nur Iffah Mohamed Azmi

A Condition-Based Maintenance Lease Contract with Continuously Monitored Equipment 569
H. Husniah, R. Wangsaputra, and B. P. Iskandar

Designing Drug Inventory Management System Design Using ABC-VED and Probabilistic Model to Minimize Total Inventory Cost in Public Health Service 577
Inaya Izzati, Ari Yanuar Ridwan, and Hardian Kokoh Pambudi

Elaborate on Sustainable Production and Consumption in Mocaf Agro-Industry 593
Dessy Srikandi, Muhammad Hisjam, and Eko Pujiyanto

Manufacturing Automation

Deep Learning Implementation for Rami Fiber Image Classification Using Convolutional Neural Network Algorithm 607
Anisa, Muhammad Nihri Jundullah, Latifu Itsnaini Khoirun Nafiah, Loviana Ajeng Sari, and Pringgo Widyo Laksono

Control and Electrical System Design Consideration for a 3D Concrete Printer 617
Fadlika Septiansyah, Sonki Prasetya, and Djoko Nursanto

Comparative Study on the Automated Warehouse with Cartesian Robot Using PLC and Arduino 627
Thitisak Aussawarangkul, S. A. Che Ghani, Pakanun Wattanasinbumrung, and Worapong Sawangsri

Materials

Study on Plasticity of Cu–Ni Alloy at Atomic Scale: Insight from Molecular Dynamics Simulation 639
M. Zaenudin, Ujiburrahman, Ade Sunardi, Yasya Khalif Perdana Saleh, and Adhes Gamayel

Deformation Behaviour of a High Carbon Co–Cr–Mo Medium-Entropy Alloy via Thermal Oxidation 649
Rosdi Daud, H. Mas Ayu, R. Norizah, and A. K. Nasution

Mechanical Performance Evaluation of Al_2O_3–ZrO_2–Ti Composite Materials 657
Nur Aziera, Mebrahitom Asmelash, Azmir Azhari, and Kushendarsyah Saptaji

Author Index 665

Manufacturing Process

Development and Design of IoT-Based Poultry Weighing Tools

Rayinda P. Soesanto, Muhammad Mi'radj Isnaini, Afrin F. Rizana, Lutfia Septiningrum, and Fandi Achmad

Abstract The increasing popularity of IoT implementation should drive the livestock sector to take the IoT benefits to obtain the best livestock, including in poultry farming. However, in Indonesia, many activities in poultry farming are still carried out without the utilization of smart technology, for example, in the weighing activity. The weighing process in poultry often increases the potential to produce inaccurate results. This will affect the profits obtained by poultry farmers because the weighing results determine the profits obtained. Therefore, this study aims to design and develop an IoT-based tool to weigh poultry in chicken coop farms to observe and control the condition of chicken coop needs. The method used in this research is identification and field observation to determine the farmer's needs. Then the design of the tool is carried out, and verification of the system and design of the tool is carried out to stakeholders. The results of this study show that IoT scales can measure weight accurately, and the IoT system runs as expected. This study also proposes a business process for weighing activities using IoT-based weighing equipment.

Keywords IoT · Livestock · Poultry farm · Smart technology

R. P. Soesanto · A. F. Rizana · F. Achmad (✉)
Telkom University, Bandung 40257, Indonesia
e-mail: fandiac.achmad@gmail.com

M. M. Isnaini
Bandung Institute of Technology, Bandung 40132, Indonesia

L. Septiningrum
University of Yogyakarta, DI Yogyakarta 55281, Indonesia

C. N. Rosyidi et al. (eds.), *Proceedings of the 6th Asia Pacific Conference on Manufacturing Systems and 4th International Manufacturing Engineering Conference*,
Lecture Notes in Mechanical Engineering,
https://doi.org/10.1007/978-981-99-1245-2_1

1 Introduction

The Internet of Things (IoT) has become a revolutionary technology that links the physical and digital worlds in order to solve real-world problems [1]. The utilization of smart technology to support the process in the organization has become common in various domains such as healthcare, smart grid, and agriculture [1, 2]. By providing smart solutions for precision farming, greenhouse management, and livestock monitoring, this domain has begun to revolutionize the agriculture industry [1]. IoT enables the vision of smart agriculture by collecting, analyzing, and processing data in real-time, allowing for better overall farm management and assisting farmers in making more informed decisions [1, 3]. In addition, it has already been recognized that enabling IoT technologies in agricultural industry as well as livestock farming systems is essential for achieving sustainable growth [4]. The increasing popularity of IoT implementation drives studies in the agriculture field also take the IoT benefits to obtain the best livestock [5]. Several studies have proposed IoT technologies to support several processes in poultry farming, such as environmental monitoring systems to monitor the living environment and average weight of the poultry [2, 5–7], predict the weather and disease of poultry to support decision making [7], monitoring animal's health [1, 8] and so on. The implementation of IoT in the various aspects of livestock processes shows that IoT has great potential for improving processes in this field.

The problem that occurs in poultry farming in Indonesia is that the current process of weighing the poultry often has the potential to produce inaccurate results since the weighing process is often carried out manually. The weighing activities tend to be viewed as simple activities, but the error in this activity turns out to have a significant effect on the company. It is because the profits obtained by the company are largely determined by the weight of the poultry. Through the existence of IoT, there is a approach to be implemented to minimize the manual work by performing automation to various activities performed in a poultry farm [9], including weighing activities. Thus, the objective of this study is to design and develop an IoT-based tool for weighing poultry on farms. The result of this study is expected to produce IoT-based weighing tools that can measure the weight of the poultry in the harvesting period accurately as well as provide a real-time database and produce an online receipt for the third party and customer.

2 Methodology

In this study, a design will be carried out to build a system that can observe and control the condition of chicken coop needs. The study began by identifying the problems contained in the chicken drum, then making in-depth observations related to what was done. Furthermore, by identifying the needs of chicken drums and with users related to the design of the chicken drum needs system. This system is designed to

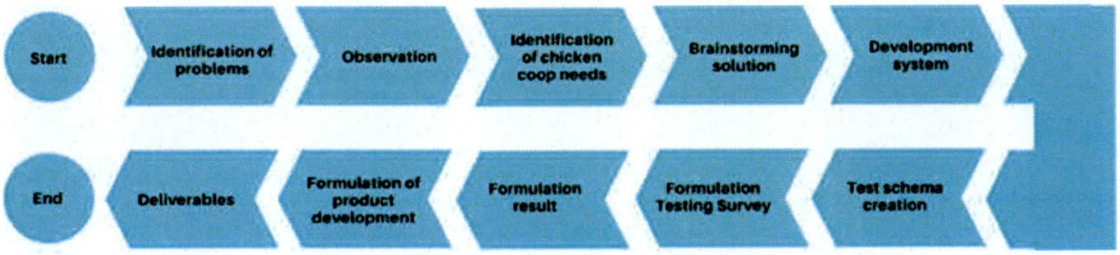

Fig. 1 Methodology

use several sensors that are connected to the internet, where the function is to connect the data recorded in the system and send chicken drums to the user and control it when conditions are not in accordance with the conditions they should be. Next, the data from these sensors is sent to the API and displayed on the smart technology designed (Fig. 1).

3 Result and Discussion

The first step is to identify the needs for the IoT system. To identify the needs, observation is used in this study to acquire the requirement for the IoT system. After the observation is done, the next step is to generate the solution through brainstorming. The communication process is centered on learning from the present business processes and investigating the issues that current users are currently experiencing. Following data processing and analysis, the researchers concluded based on the process' functional and non-functional requirements. The output from the brainstorming is the mockup and the proposed business process. Figure 2 shows the mockup scale design for this study.

The mockup includes two different designs: sitting scales with one point and sitting scales with four points. After the mockup is developed, the next step is to design the architecture of the technology. Figure 3 shows the architecture of the IoT.

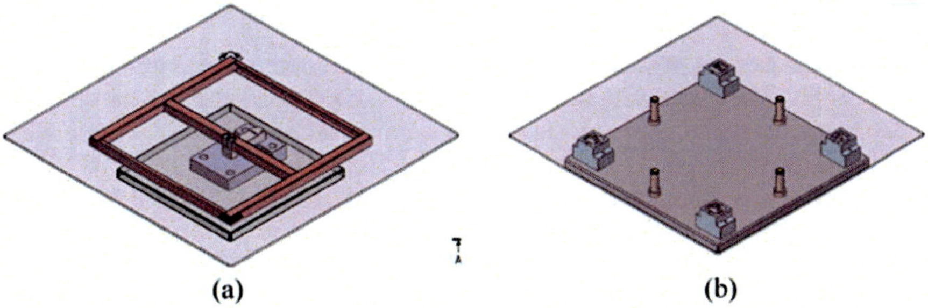

Fig. 2 Scales mockup, **a** one point flat scales, **b** four point flat scales

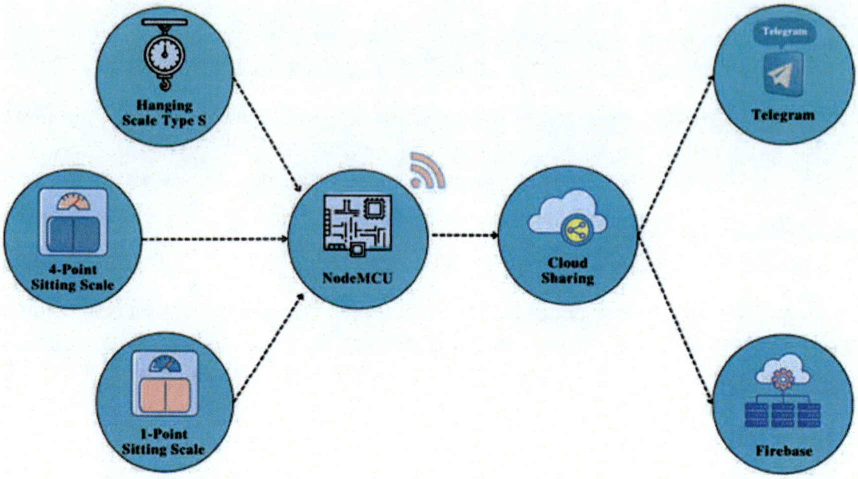

Fig. 3 IoT architecture

From Fig. 3, the picture depicts some of the hardware included in IoT-based scaling. In this study, the sensor nodes and gateway are built using open-source microcontroller boards. The microcontroller handles the readings of the poultry measurement. Attached to the scales are the loadcell sensor to measure the weight and the OLED display to display the weight result. NodeMCU serves as a tool that combines the relay module and weight sensor to send data over the internet. NodeMCU has a Wi-Fi Module so that it can operate on the internet network. The loadcell sensor will give accurate information about the poultry weight and send it to the NodeMCU. The NodeMCU will then relay the information through the internet using Firebase API, so the data and message will be sent to the smartphone using a messaging application. Figure 4 shows the flow diagram of the system.

From Fig. 4, it is known that the result will be displayed on the smartphone using telegram apps. The NodeMCU will send the measurement information to the user via Firebase API, and the user will see the result in the telegram apps while the result will also be shown on the OLED screen of the IoT Scale. Figure 5 depicts the user's interactions with the device.

The farmer can also use a smartphone with messaging apps and an Internet connection to retrieve the sensor data from the chicken house. The result from the telegram apps is shown in Fig. 6. From Fig. 6, it is known that the IoT system can measure the weight with the command from the user by using the command "/readings" in the telegram apps. The command will be sent through the NodeMCU via Firebase API and the NodeMCU will start to measure the poultry weight and display the result in the OLED screen and pass the value to the Telegram apps via Firebase API.

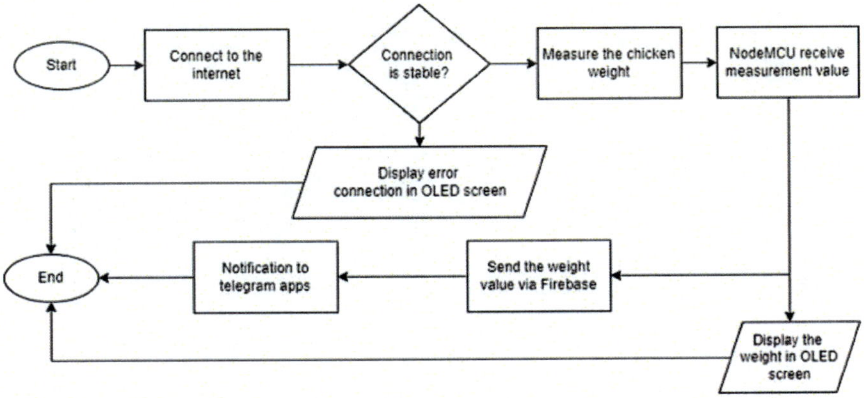

Fig. 4 Flow diagram for weighting poultry

Fig. 5 Use case diagram for poultry IoT

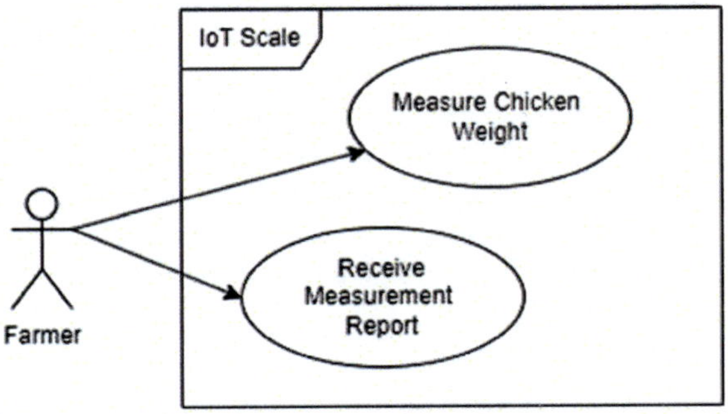

Fig. 6 Measurement result in telegram

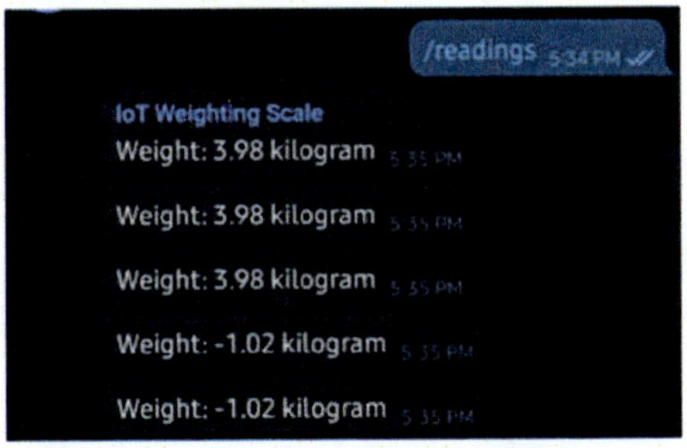

The result of the IoT scale is shown in Fig. 7. After the development phase is done, the next step is to verify the system with the stakeholder. From the testing phase, it is known that the scale can measure the weight accurately, and the IoT system is running as expected.

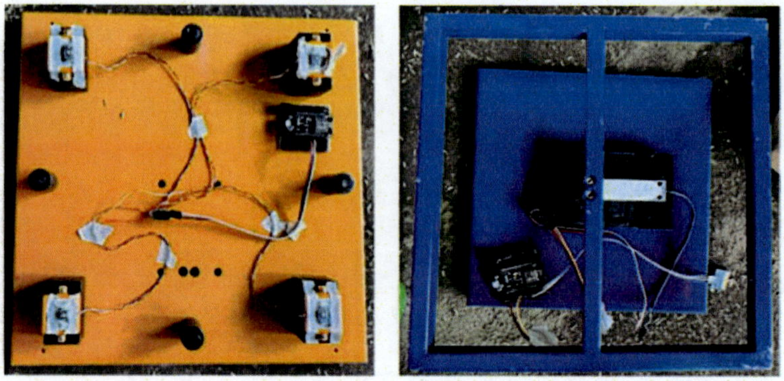

(a) (b)

Fig. 7 IoT scales result

From the testing result, the suggested model is affordable for the farmer. The usage of open-source boards and sensors lowers the cost of the hardware. The value of the weight is constantly measured from the Telegram apps, and the trigger is sent immediately when the measurement by NodeMCU is done to the farmer. The scales are designed by using the Internet as media. In the testing phase, there are some areas in the poultry that cannot be accessed by the Internet, so it is necessary to relay the internet connection through the entire poultry in order to make sure that the IoT scale can be accessed. Figure 8 shows the proposed business process for weighing poultry.

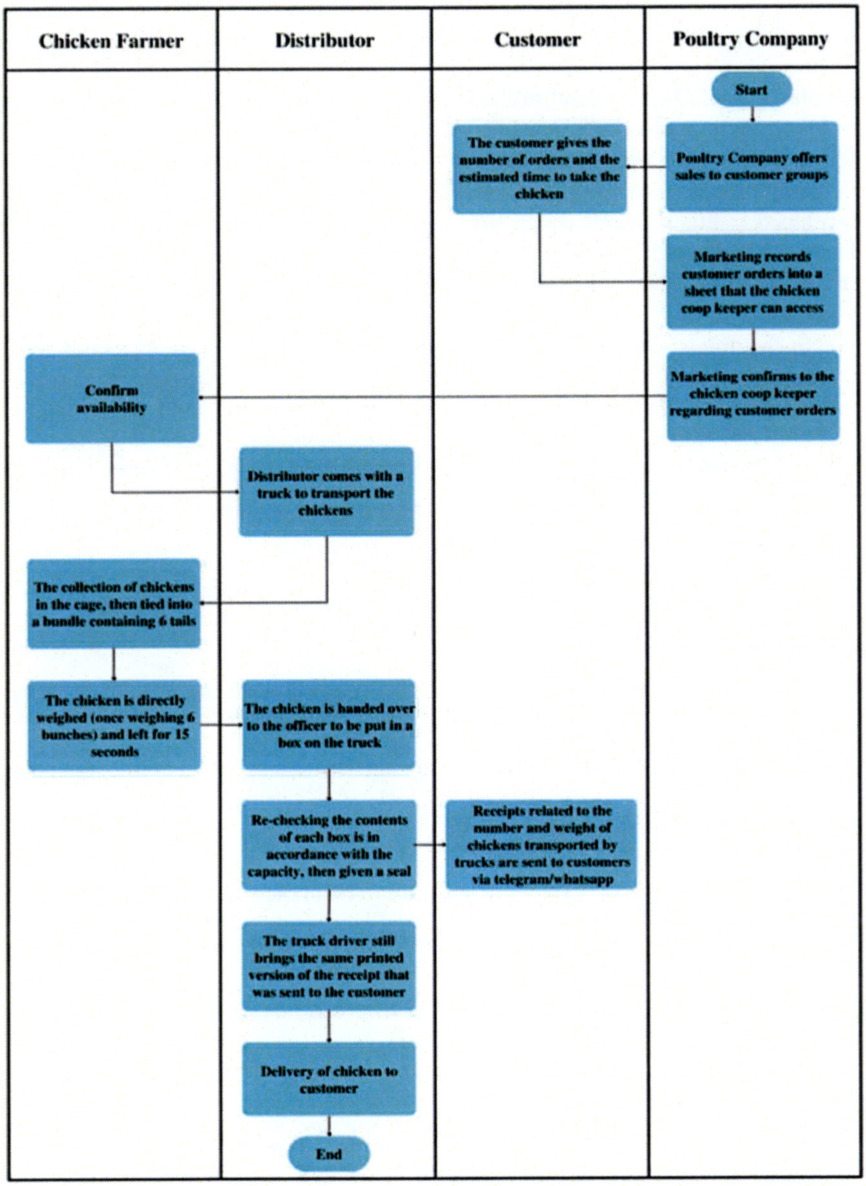

Fig. 8 Proposed business process for weighting poultry using IoT

4 Conclusion

In today's era, IoT-based solutions are essential to obtain better results from farming in every sector, including in poultry farming. Thus, this study is attempted to design and develop an IoT-based tool to facilitate obtaining more accurate results of weighing process in poultry farm. An IoT-based weighing device was successfully developed in this research which is equipped with the ability to provide real-time databases and generate online receipts for third parties and customers, as well as propose their business processes. It is expected that by designing this tool, it can help poultry farmers to obtain accurate results regarding poultry weight that will be sold to customers. Other than that, the tool designed hopefully could minimize the potential for fraud by irresponsible parties who often provide inappropriate reports regarding the weight of poultry.

Acknowledgements This research was funded by the "Pemulihan Ekonomi" research scheme of LPPM ITB for the fiscal year 2022.

References

1. Farooq MS, Sohail OO, Abid A, Rasheed S (2022) A survey on the role of IoT in agriculture for the implementation of smart livestock environment. IEEE Access 10:9483–9505. https://doi.org/10.1109/ACCESS.2022.3142848
2. Lokman Ali M, Ab Rahman M, Shahidah Afifi Md Taujuddin N (2020) Smart chicken farm monitoring system. Evol Electr Electron Eng 1(1):317–325 [Online]. Available: http://publisher.uthm.edu.my/proceeding/index.php/eeee
3. Köksal, Tekinerdogan B (2019) Architecture design approach for IoT-based farm management information systems. Springer US, vol 20, no 5
4. Symeonaki E, Arvanitis KG, Loukatos D, Piromalis D (2021) Enabling IoT wireless technologies in sustainable livestock farming toward agriculture 4.0. In: IoT-based intelligent modelling for environmental and ecological engineering, Springer, Berlin, pp 213–232
5. Manshor N, Rahiman ARA, Yazed MK (2019) IoT based poultry house monitoring. In: 2019 2nd International Conference on Communication Engineering and Technology, ICCET 2019, pp 72–75
6. Zhang J, Kong F, Zhai Z, Han S, Wu J, Zhu M (2016) Design and development of IOT monitoring equipment for open livestock environment. Int J Simul Syst Sci Technol 17(26):2–7. https://doi.org/10.5013/IJSSST.a.17.26.23
7. Akhund TMNU, Snigdha SR, Reza MS, Newaz NT, Saifuzzaman M, Rashel MR (2021) Self-powered IoT-based design for multi-purpose smart poultry farm. Smart Innov Syst Technol 196:43–51. https://doi.org/10.1007/978-981-15-7062-9_5
8. Andonovic I, Michie C, Cousin P, Janati A, Pham C, Diop M (2018) Precision livestock farming technologies. In: 2018 global internet things summit. GIoTS 2018
9. Sitaram KA, Ankush KR, Anant KN, Raghunath BR (2018) IoT based smart management of poultry farm and electricity generation. In: 2018 IEEE international conference on computational intelligence and computing research, ICCIC 2018, pp 1–4. https://doi.org/10.1109/ICCIC.2018.8782308

Effect of Machining Parameters on Surface Quality of Aluminium Puncher for Microchannel Fabrication Using Micro Cutting Process

Mohd Saifuldin Mohd Mokhtar and Ahmad Razlan Yusoff

Abstract Precision and surface finish plays an important role in modern industry. Both of these elements are more challenging to achieve in micro-sized machining due to their extremely small size and ease of damage. The quality of surface roughness and accuracy is directly influenced by the machining parameters used. The present work involves a detailed experimental analysis of surface quality for microchannel fabrication. This experiment was formed using the Response Surface Methodology (RSM) method by using three main machining parameters namely spindle speed (r/min), feed rate (mm/min), and depth of cut (μm) to produce an Aluminium microchannel puncher. The results of the study show that the best surface roughness is obtained by the combination of the following parameters; spindle 14,000 r/min, feed rate of 90 mm/min, and depth of cut 50 μm. This study concludes that the surface roughness of Al6061 steel is greatly influenced by two main parameters, namely depth of cut and feed rate.

Keywords Surface quality · Micro machining · RSM

1 Introduction

Microchannels can be produced through several methods. Among the commonly used ones are etching [1], photolithography [2], wire electrical discharge machining [3], laser ablation process [4], and micro-machining [5]. Among the following methods, micro-milling has several advantages such as the production of complex

M. S. M. Mokhtar (✉) · A. R. Yusoff
Faculty of Manufacturing and Mechatronic Engineering Technology, Universiti Malaysia Pahang, Pekan Campus, 26600 Pekan, Pahang, Malaysia
e-mail: mdsaifuldin@uctati.edu.my

A. R. Yusoff
Center for Advanced Industrial Technology, University Malaysia Pahang, Pekan Campus, 26600 Pekan, Pahang, Malaysia

C. N. Rosyidi et al. (eds.), *Proceedings of the 6th Asia Pacific Conference on Manufacturing Systems and 4th International Manufacturing Engineering Conference*, Lecture Notes in Mechanical Engineering,
https://doi.org/10.1007/978-981-99-1245-2_2

microchannel shapes with high precision and efficiency at low cost [5]. However, machining micro-sized features are very challenging, especially to get the required accuracy and surface quality. The main factors that affect the roughness of the machined surface quality are the type of cutting tool, the material used, and the cutting condition. The correct cutting parameters must be chosen to achieve the desired level of surface quality, which determines product quality.

Many researchers believed that the three basic factors that affect surface roughness are depth of cut, feed rate, and cutting speed [5–7, 10]. Studies by [8] have found that the surface roughness of the channel base was dependent on cavities, which are controlled by the cutting conditions. A low feed rate will also contribute to an increase in surface roughness. For a combination of machining parameters; rpm 10,000–20,000, feed rate 0.5–1.0 μm/z, and depth of cut 125–150 mm, there is a large cavity at the base of the channel that affects the quality of the surface. Optimizing cutting parameters such as cut depth, spindle speed, and feed rate can reduce the level of surface roughness in micromachining. The optimization of machining parameters was done by Bagaber and Yusoff [9] on the dry turning of stainless steel. They used multi-responses based on response surface methodology with Box-Behnken design (BBD) in the optimization process. As a result, the surface roughness was decreased by 13.89% and the minimum power consumption by 6.78%. The cutting parameters for the micromachining of CuNi18Zn20 samples were optimized by Abeni et al. in 2021[5]. The Analysis of Variance (ANOVA) technique was used to analyze the relationship between process parameters and outputs. Next, a multifocal 3D microscope was used to analyze the burrs' extension. Finally, an ideal parameter set was chosen by taking into account the amount of time required to fabricate each prototype. They have concluded that feed per tooth, depth of cut, and their interplay have a significant impact on surface roughness (Sa). Surface roughness (Sa) will also increase as feed rate and feed per tooth increase.

One of the biggest challenges in micromachining is to get high-quality results. Therefore, the selection of the correct cutting parameters at the initial stage is very important. Optimization and modeling of these parameters will produce better machining results. The objective of this study is to examine the effect of machining parameters on surface quality during the machining of Aluminium (Al6061) microchannel puncher. The Response Surface Methodology (RSM) has been employed to construct a systematic experiment that enable the analysis and exploration of variables influences, aiming to propose the optimal options for machining.

Table 1 Factor and level

Factor/condition	Level
Spindle speed (r/min)	8000, 14000, 20000
Feed rate (mm/min)	30, 90, 150
Depth of cut (μm)	20, 50, 80
Cutting condition	Dry cutting

2 Experiment

2.1 *Design of Experiment (DOE)*

Different factor levels, which are listed in Table 1, were used in the experiments to better understand how different parameters, such as spindle speed, depth of cut and feed rate, affect the surface quality of an aluminium microchannel. In order to prevent any tool wear effects, each set of experiments was carried out with a brand-new end mill. The experiments were designed and analyzed by using Response Surface Methodology (RSM). Central Composite Design (CCD) is a common statistical tool for optimizing process variables. It is often used to extract the regression model equations and operational conditions from the related experiments. The RSM process involves the design of experiments, the study of output responses, the generation of a regression model equation and the evaluation of the significance of determining factors based on the responses. This present study consists of three (3) factors as independent variables and one (1) response as a dependent variable as shown in Table 3.

2.2 *Microchannel Machining*

In this experiment, uncoated four flutes solid carbide end mills with diameters of 2.0 mm and 4.0 mm were used. This experiment involved producing microchannels puncher on an aluminium substrate Al6061 alloy steel. The chemical composition of the Al6061 alloy steel is presented in Table 2. To conduct the experiments, a precision 5-axis DMG CNC milling machine was employed. Figure 1a shows the design of the microchannel puncher utilized in this experiment, featuring a channel height of 100 μm and a width of 200 μm. The CATIA software was employed for designing the microchannel features and generating the CNC programming. The workpiece size was 40 mm x 40 mm x 15 mm. Before commencing the actual machining process, the workpiece surface was flattened using face milling tool to ensure a perfectly flat surface. For microchannel machining process, a Ø4.0 mm end mill tool was used during the roughing process, except for fabricating the channel and circular features, which were machined using a Ø2.0 mm end mill tool.

Table 2 The chemical composition of Al6061 alloy steel

Element	Content (%)
Aluminium, Al	97.9
Magnesium, Mg	1
Silicon, Si	0.60
Copper, Cu	0.28
Chromium, Cr	0.20

Table 3 Design of experiment and response output

Run	Spindle speed, (r/min)	Feed rate, (mm/min)	Depth of cut, (μm)	Surface quality, (μm)
1	20,000	150	80	1.8
2	8000	90	50	1.9
3	14,000	90	50	1.7
4	20,000	30	80	1.7
5	14,000	90	20	2.3
6	8000	150	80	1.9
7	20,000	30	20	2.4
8	8000	30	20	3.0
9	14,000	90	50	1.9
10	14,000	90	50	1.7
11	8000	30	80	1.7
12	20,000	90	50	1.8
13	14,000	90	50	1.5
14	8000	150	20	2.2
15	14,000	90	50	1.6
16	14,000	150	50	1.7
17	14,000	30	50	1.6
18	14,000	90	50	1.7
19	20,000	150	20	2.3
20	14,000	90	80	1.7

2.3 *Surface Roughness Measurement*

The surface roughness value were visually examined using a 3D Confocal Microscope (LEXT OLS5000 by Olympus). Three sidewall surfaces of microchannels were randomly selected to measure surface roughness. The microchannel sidewall surfaces are specifically the measured area for roughness values, as shown in Fig. 1b. The microchannels were divided into two sub-parts using wire electrical

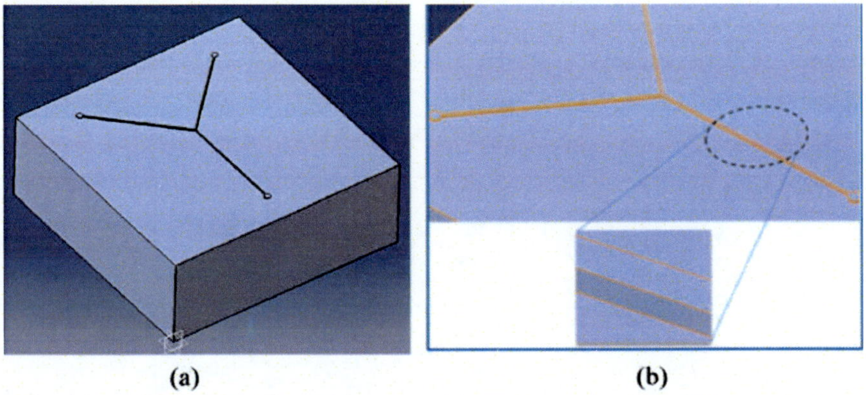

Fig. 1 a The microchannel puncher design. b The microchannel puncher sidewall surface

discharge machining (WEDM) in order to measure the surface roughness of the sidewall surfaces. Table 3 shows surface roughness measurements of the microchannel sidewall surfaces.

3 Discussion

To determine the adequacy and significance of the quadratic model, analysis of variance (ANOVA) was required. Table 4 presented details on the P-value for the model source, each model term, and interactions. The significant model terms were indicated with a p-value less than 0.05.

The surface quality was significantly affected by model terms C, BC, and C^2. The lack of fit of the model for surface quality was not significant, according to the p-value for the term in Table 4 which was more than 0.05. There is a 40.77% chance that lack of fit F-value this large could occur due to noise. In other words, the model was suitable for the following experiment because it fit the responses data that had been collected.

The fit statistic is shown in Table 5. The fitness of the model for surface quality was expressed by the R2 value (0.9210) indicating that 92.1% of the variability in the response could be explained by the model. Meanwhile, the coefficient of determination (adjusted R2) was calculated to be 0.8498 indicating that only 15.02% of the total variation was not included in the model. A negative Predicted R^2 implies that the overall mean may be a better predictor of response than the current model. In some cases, a higher order model may also predict better. Adeq Precision measures the signal-to-noise ratio. A ratio greater than 4 is desirable. The value ratio of 12.752 indicates an adequate signal. This model can be used to navigate the design space. The model developed was useful for predicting the best machining parameters that will reduce surface roughness in microchannel machining.

Table 4 ANOVA for surface quality

Source	Sum of squares	DF	Mean square	F-value	P-value	
Model	2.31	9	0.2568	12.95	0.0002	Significant
A—spindle speed	0.0490	1	0.0490	2.47	0.1471	
B—feed rate	0.0250	1	0.0250	1.26	0.2878	
C-DOC	1.16	1	1.16	58.28	< 0.0001	
AB	0.0450	1	0.0450	2.27	0.1629	
AC	0.0200	1	0.0200	1.01	0.3390	
BC	0.1800	1	0.1800	9.07	0.0131	
A^2	0.0736	1	0.0736	3.71	0.0829	
B^2	0.0036	1	0.0036	0.1833	0.6776	
C^2	0.2705	1	0.2705	13.64	0.0042	
Residual	0.1984	10	0.01984			
Lack of fit	0.1100	5	0.0220	1.25	0.4077	Not significant
Pure error	0.0883	5	0.0177			
COR total	2.51	19				

Table 5 The fit statistic

Std. Dev.	0.1408	R^2	0.9210
Mean	1.90	Adjusted R^2	0.8498
C.V.%	7.39	Predicted R^2	0.1166
		Adeq precision	12.7523

Multiple linear regression was used to model the relationship between variables (spindle speed, depth of cut, and feed rate) and output measurements (surface quality). The following equation, Eq. (1) in terms of coded factors presents the surface quality obtained.

$$1.68455 + -0.07 * A + -0.05 * B + -0.34 * C + 0.075 * AB + 0.05 * AC + 0.15 * BC + 0.163636 * A + -0.0363636 * B + 0.313636 * C \tag{1}$$

where; A—RPM, B—feed, and C—DOC, respectively.

The residual plots (Fig. 2a) showed linear behavior and an extrapolate straight line, demonstrating the accuracy of the model prediction. The plots (Fig. 2b) demonstrate the correlation between the predicted and observed (actual) values of the model response. A smaller residual indicates a closer match between the predicted and actual values. When plotting projected against actual values, the point will be on or very near a straight line.

Identifying the connections between variables and responses is made easier by looking at the measured responses in three dimensions (3D) plots. The relationship

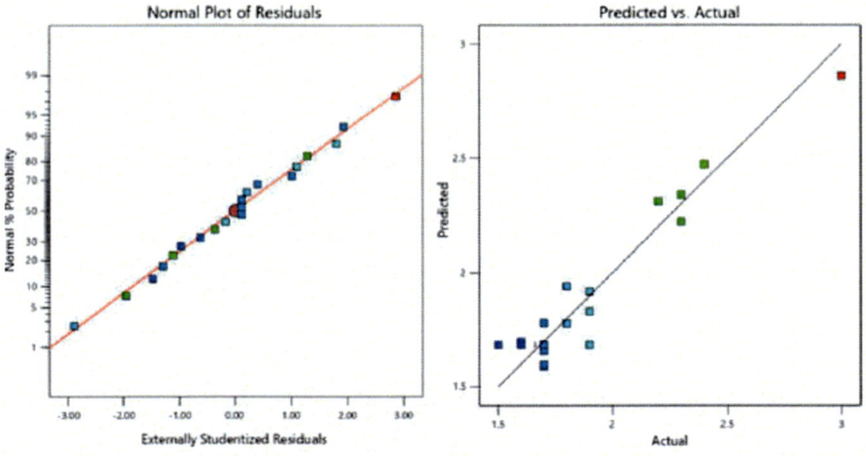

(a) (b)

Fig. 2 **a** Normal probability plot of residuals; **b** plot of predicted values versus actual

between the input factor and surface quality is shown in a 3D surface plot in Fig. 3, which was created based on the Eq. (1). The plots demonstrate that the lowest surface roughness value, indicated by the blue area, was observed at a feed rate of 90 mm/min and a depth of cut of 50 μm (Fig. 3a), while the spindle speed was set at its moderate value of 14,000 r/min (Fig. 3b).

Optimization works to find a combination of factors simultaneously to achieve the requirements that have been set in the initial stage. Design expert software is used to form a mathematical model in implementing the numerical optimization process. Figure 4 shows the 3D Optimization Criteria Setting for surface roughness used in this study. The lower bound setting used is 0.01, while the upper bound setting is 1.92.

This shows that the best range of surface roughness to achieve is 1 μm. This suggests that the minimum value of the setting parameter is the optimal range that needs to be achieved. The suggested optimum conditions are obtained from the combination of parameters; 15,904 r/min spindle speed, 30 mm/min feed rate, and 78.8 μm depth of cut. The predicted response for surface roughness is 1.5 μm (Fig. 5).

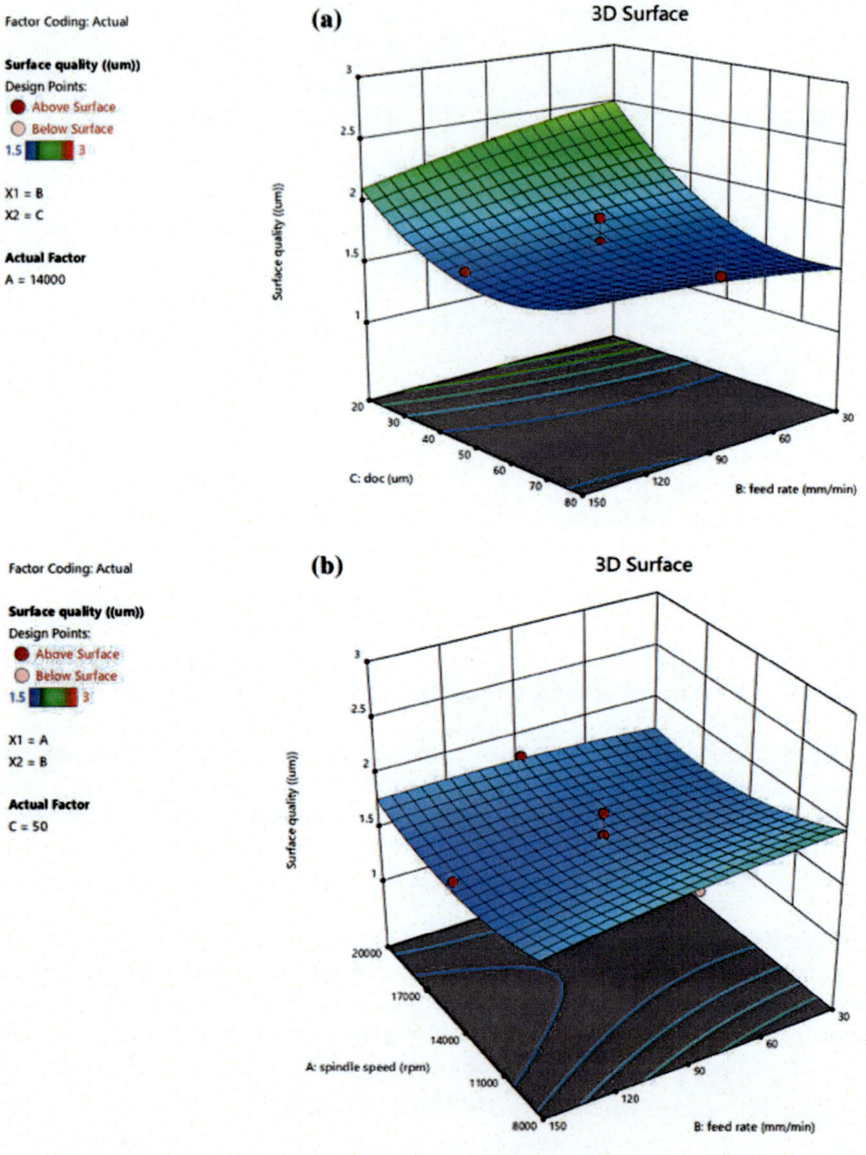

Fig. 3 3D Surface graph of surface quality, **a** effect of depth of cut and feed rate; **b** effect of spindle speed and feed rate

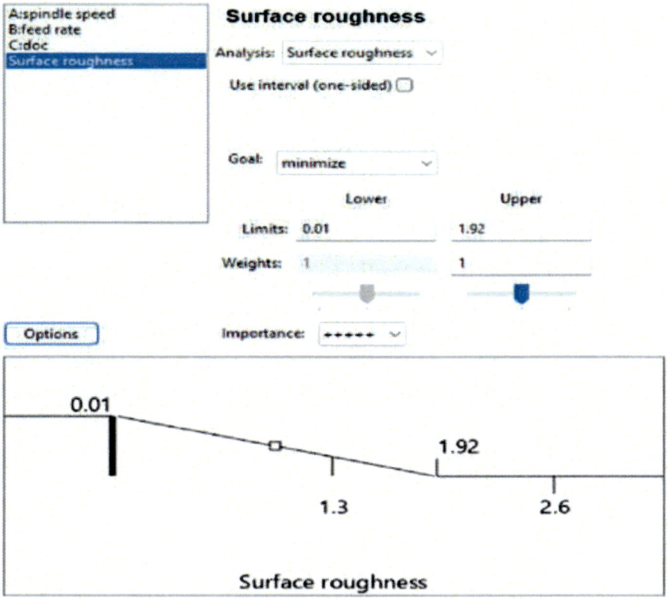

Fig. 4 3D optimization criteria setting for surface roughness

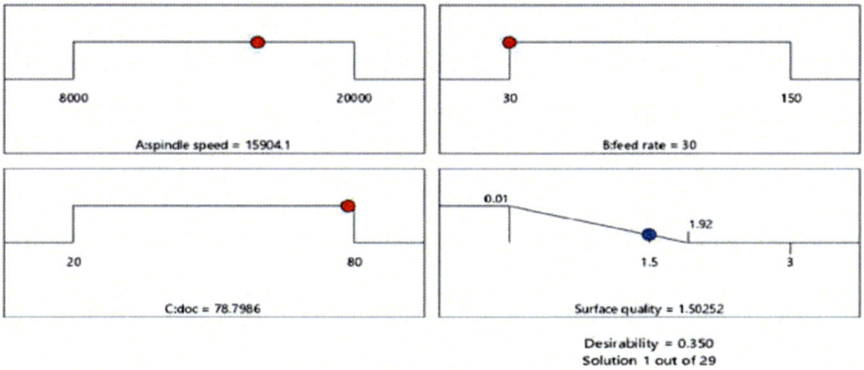

Fig. 5 Suggested optimum conditions and predicted response

4 Conclusion

The effect of using multiple machining parameters on the surface quality of aluminium (Al6061) puncher during microchannel machining was investigated using Response Surface Methodology (RSM). Analysis of Variance (ANOVA) identified the optimal cutting parameters that yield minimal surface effects on the surface of the lateral channels. The analysis results revealed that the depth of cut and feed rate are the primary factors contributing to surface roughness. The best surface roughness was achieved through the experimental process with the following parameters combination; spindle speed of 14,000 r/min, feed rate of 90 mm/min, and depth of cut of 50 μm. Similarly, the suggested optimum condition is achieved through the combination of parameters; spindle speed of 15,904 r/min, feed rate of 30 mm/min, and depth of cut of 78.8 μm.

However, it is crucial to validate the suggested optimal condition and predicted response through experimental validation. This validation is essential for comparing the results of this current study with the optimization outcomes suggested by the Design Expert software. The surface roughness model can be predicted using forecasted values obtained from the regression equation. Overall, this experiment successfully achieved its objective of identifying the optimal combination of machining parameters to produce the finest surface roughness on the sidewall of the aluminium puncher microchannel surface. The experimental data obtained and the regression equation demonstrate good agreement that can be utilized to predict the surface roughness of aluminium microchannels in future applications.

Acknowledgements The authors would like to thank Universiti Malaysia Pahang for funding this research project through the UMP Post Graduate Research Scheme (Grant no. PGRS2003176).

References

1. Lee DK, Kwon JY, Cho YH (2019) Fabrication of microfluidic channels with various cross-sectional shapes using anisotropic etching of Si and self-alignment. Appl Phys A Mater Sci Process 125(5):1–7
2. Wang GJ, Hsueh CC, Hsu SH, Hung HS (2007) Fabrication of PLGA microvessel scaffolds with circular microchannels using soft lithography. J Micromechan Microeng 17(10)
3. Sarma P, Patowari PK (2018) Fabrication of metallic micromixers using WEDM and EDM for application in microfluidic devices and circuitries. Micro Nanosyst 11:137–147
4. Gao K, Liu J, Fan Y, Zhang Y (2019) Ultra-low-cost fabrication of polymer-based microfluidic devices with diode laser ablation. Biomed Microdev 21(4)
5. Abeni A, Metelli A, Cappellini C, Attanasio A (2021) Experimental optimization of process parameters in CuNi18Zn20 micromachining. Micromachines 12(11)
6. Khan K, Varghese A, Dixit P, Joshi SS (2019) Effect of tool path complexity on top burrs in micromilling. Procedia Manuf 34:432–439
7. Yousuff CM, Danish M, Ho ETW, Basha IHK, Hamid NHB (2017) Study on the optimum cutting parameters of an aluminum mold for effective bonding strength of a PDMS microfluidic device. Micromachines 8(8)
8. Wang T, Wu X, Zhang G, Chen Y, Xu B, Ruan S (2020) Study on surface roughness and top burr of micro-milled Zr-based bulk metallic glass in shear dominant zone. Int J Adv Manuf Technol 107(9–10)
9. Bagaber SA, Yusoff AR (2017) Multi-responses optimization in dry turning of a stainless steel as a key factor in minimum energy. Int J Adv Manuf Technol 96(1–4)
10. Khalil K, Mohd A, Mohamad COC, Faizul Y, Zainal Ariffin S (2021) The optimization of machining parameters on surface roughness for AISI D3 steel. J Phys Conf Ser 1874(1)

Wear Behaviour of Tungsten Carbide in End Milling Process of Aluminium Alloy 6061-T6 with Minimal Quantity of Tri-hybrid Nanofluids

W. Safiei, M. M. Rahman, A. R. Yusoff, and M. Y. Ali

Abstract Nowadays, using nanotechnology in science and industry improves the yield of different processes. The machining process using hybrid nanofluids requires further research to better understand the mechanism of tool wear and the fundamental aspects are not yet ventured. In machining, tool wear is common problems that exist for quite some time. In addition, milling process of Aluminium Alloy was challenged due to a strong adhesion particularly in higher temperature. Deposition of chips material during the process at the tool edge may induce several tool failures such as build-up edge, chipping and flaking. Eventually, tool life, manufacturing cost and product quality were the factors that normally effects by tool wear. However, the severity of tool wear can be reduced by applying a cutting fluid to the tool-workpiece interface. This paper intends to discover the effects of tri-hybrid nanofluids in end milling process of Aluminium Alloy 6061-T6 mainly on wear conditions of uncoated and double-layered PVD coated inserts. In this research works, three different nanoparticles SiO_2-Al_2O_3–ZrO_2 were dispersed in 60:40 of deionized water and ethylene glycol. The concentration was prepared between 0.06 and 0.12 wt.%. The MQL system with assisted air pressure was employed to deliver newly developed tri-hybrid nanofluids. During metal cutting process, the metal working fluid was supplied intermittently based on flow rate setting in the MQL system to the cutting zone with a very minimal quantity. A single insert was used and changed for every 100 mm of cutting length at different machining parameters. The effects on wear mechanisms were closely examined at the flank area using scanning electron microscope. Through comprehensive investigation, the wear mechanisms consist of

W. Safiei (✉) · A. R. Yusoff
Faculty of Manufacturing and Mechatronic Engineering Technology, Universiti Malaysia Pahang, 26300 Pekan, Pahang, Malaysia
e-mail: wahaizads@ump.edu.my

M. M. Rahman
College of Engineering, Universiti Malaysia Pahang, 26300 Gambang, Pahang, Malaysia

M. Y. Ali
Mechanical Engineering Programme Area, Faculty of Engineering, Universiti Teknologi Brunei, Tungku Highway, Gadong 1410, BE, Brunei Darussalam

C. N. Rosyidi et al. (eds.), *Proceedings of the 6th Asia Pacific Conference on Manufacturing Systems and 4th International Manufacturing Engineering Conference*, Lecture Notes in Mechanical Engineering,
https://doi.org/10.1007/978-981-99-1245-2_3

attrition, flaking, abrasion and coating delamination. Other phenomenon such as thermal crack was observed in the wear region. The tool failures have a relationship with machining parameters and cutting tool condition itself. It can be concluded that, coating delamination and abrasion quite severe for coated inserts. While, uncoated tools were severe with attrition mode of failures. At extreme machining condition, higher temperature and friction forces at the tool-workpiece interface have a significant effect on the tool failures. For further investigation, the effects of tri-hybrid nanofluids on wear behaviour of tungsten carbide inserts can be examined for other machining process with different workpiece material.

Keywords End milling process · Tri-hybrid nanofluids · Flank wear · Attrition · Abrasion · Flaking · Delamination · Thermal crack

1 Introduction

In metal cutting process, lubrication is important aspect that has noticeable effects on the machining performance. Applying lubricants during machining can reduce cutting temperature and mechanical forces resulting outstanding process performance. Despite that condition, tool wear is a natural and inevitable phenomenon for all cutting conditions as the process requires shear stress in removing unwanted material. Nevertheless, the severity of tool failures in machining should be mitigated to prolong the tool life. The attempts to address tool failures can be achieved by obtaining optimum machining condition based on the metal cutting process parameters. Moreover, the wear phenomenon can be enhanced by deeply understanding the correlation between the effects of cutting process, the fundamental of material sciences and the changes of physical properties of the machined surface material during process.

Conventionally, metal working fluids with flooded technique were employed to offer tribology characteristic particularly on the tool-workpiece interface. However, the large quantity of conventional metal-cutting fluid has potential of exposing harmful chemical elements in long hours resulting skin and lung disease. In addition, flooded lubrication technique requires more working fluid and additional cost is needed to treat the fluid with biocide for [1] reuse purposes which led to a higher manufacturing cost [2]. For the last few years, near dry machining approach seems promising in enhancing the metal cutting process. Such idea was well recognized with the application of minimum quantity lubricant (MQL) and the enhancement of thermal-physical properties of cutting fluid with the introduction of single type nanofluid. The limitation of thermal conductivity of single type nanofluid in transferring the heat has encouraged another milestone discovery of hybrid nanofluids as promising and innovative metal working fluids [1].

Machining Aluminium Alloy is a real challenge due to its mechanical properties. It possesses superior higher strength to weight ratio and strong resistance to corrosion [3]. This material has been widely used in aerospace, automotive, construction and weapon industries. However, at higher temperature, there are high tendency of Aluminium chip adhesions on the tool which may lead to other wear problems such as built-up edge and fracture [4, 5].

In order to improve the cutting tool performance, a multi-layer surface coating material was deposited with a right combination of etching durations and deposition temperatures [6]. It was crucial to obtain an optimum condition between these two factors as higher etching duration and deposition temperature reduces the coating micro-hardness. At low hardness, deposition materials were easily peeling-off in elevated temperature and higher shear stress. Zheng et al. [7] improved Ti(C, N) based cutting inserts with addition of carbon in controlled condition with the aim to reduce cutting temperature and alleviate the wear effects when removing Aluminium Alloy in dry turning process. Sharif and Rahim [8] investigated the failure modes of TiAlN-coated and uncoated when drilling Titanium Alloy. It was observed that the dominant failure modes for both cutting inserts were non-uniform flank wear, chipping and catastrophic failure.

Therefore, to reduce cutting temperature and improve the process effectively, the capability of cutting fluid to transfer the heat from the heat source in end milling process must be enhanced. For that reason, water-based tri-hybrid nanofluids were developed with a combination of SiO_2–Al_2O_3–ZrO_2 nanoparticles by considering fluid's stability and its thermal conductivity. The objective of this study is to investigate the wear phenomenon upon machining Aluminium Alloy 6061-T6 with tri-hybrid nanofluids are delivered at the machining zone with a very minimal quantity. This is a novelty aspect in this investigation where the effects of wear conditions of both uncoated and dual- layered coated were examined particularly at the flank area upon machining Aluminium Alloy 6061-T6 with the present of newly developed tri-hybrid nanofluids. In addition, the behaviour and interactions between machining process parameters, workpiece material, coated materials and variety of tool failures were presented in this paper.

2 Methodology

In this experiment, a newly developed tri-hybrid nanofluids was supplied at the cutting zone with a minimal quantity. The nanofluids were prepared at low concentration between 0.08 and 0.12 wt% to obtain homogenous solution. In order to enhance fluid's stability, dispersion agent CTAB was added with 1:3 ratio over nanoparticles concentration. A newly cutting fluid was delivered at the cutting zone using minimum quantity lubricant method. A combination of nanofluid and MQL in machining process promotes green and sustainable manufacturing. Uncoated and PVD double- layered coated tungsten carbide inserts were used to mill through slot of 100 mm Aluminium Alloy 6061-T6. In order to differentiate the effects of process parameters including the usage of cutting fluid on the cutting tool, a single inserts was used and changed for every 100 mm of cutting length. The cutting insert was mounted on a 10 mm diameter tool holder to ensure smooth cutting when conducting the experiments. A

Table 1 Inserts used in the study (Ceratizit Co.)

CERATIZ IT code number	Type	Composition	Grain size (um)	Hardness
CTW4615	Uncoated	Co 6.0%	1	HV301630
CTPP235	PVD TiAlTaN	Co 10.5%, mixed carbide 2.0%	1–2	HV301400

total 64 inserts of uncoated and coated cutting inserts were examined under SEM at different magnifications upon machined the workpiece. Cutting speed, feed rate and depth of cut were the machining inputs applied. Other factors involved were MQL Flow rate and nanofluids concentration. The range of cutting speed was between 9000 and 12,000 rpm which consider high in machining (Table 1).

3 Wear Mechanisms

3.1 Attrition and Micro-attrition

High squeezing and shearing force induced various tool failures including attrition mainly on the flank side. Rough attrition was severe on uncoated tungsten carbide rather than coated tools. However, attrition and its plowing were observed on the PVD TiAlTaN tungsten carbide as can be seen in Fig. 1. Continue shearing forces occur between two surfaces materials resulting minor dislodge of the cutting tool material.

3.2 Flaking and Micro-flaking

Flaking was one of wear mechanisms and severely obvious on the tool edge under high magnifications. It seems like a peeling-off material located along the wear area as shown in Fig. 1. The formation of flaking is due to rapid shearing and rubbing process between surface material and the cutting tool. Micro-flaking was also observed of double-layered PVD TiAlTaN tungsten carbide. The rough surface that consists of flaking and attrition has remarkable effects on the machined surface quality.

3.3 Flank Wear

SEM micrographs for uncoated tungsten carbide for a combination of parameters with cutting speed at 10,000 RPM, MQL flow rate of 1.2 ml/min and concentration

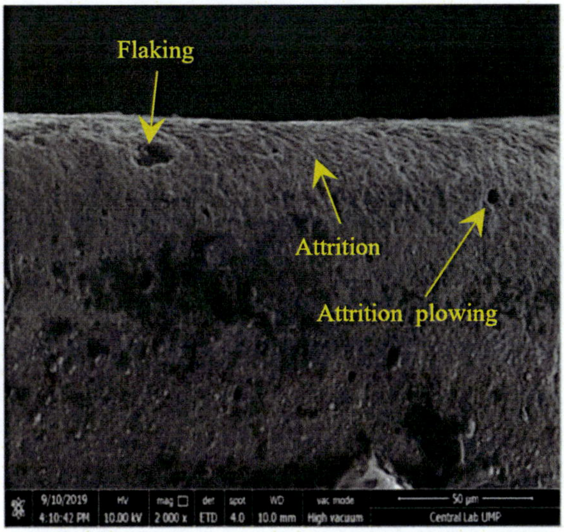

Fig. 1 Flaking and attrition observed on the tool edge of double-layered PVD TiAlTaN after machining with parameters at 8000 rpm, 150 mm/rev, 0.75 mm, 1.2 mL/min and 0.1 wt.% of cutting speed, feed rate, depth of cut, MQL flow rate and concentration

0.1 wt.% is shown in Fig. 2a. Attrition and abrasion are the main wear mechanisms. Besides attrition, flaking and micro-flaking are observed as well on the PVD TiAlTaN tungsten carbide in Fig. 2b. Figure 3 shows condition of uncoated tungsten carbide with a variety of tool failures upon machining Aluminium Alloy 6061-T6 at different depth of cut and feed rate. Chip adhesion was observed on the tool edge in Fig. 3a. Tool failures were severe at higher feed rate of 200 rev/min with a combination of plowing, attrition, micro-attrition and abrasion as can be seen in Fig. 3b. Najiha and Rahman [9] revealed that attrition plowing and abrasion marks severely observed at the parameters of 5300 rpm and 440 rev/min of cutting speed and feed rate when machined Aluminium Alloy 6061-T6. No severe edge damage was observed. At extreme machining condition, higher temperature and friction at the tool- workpiece interface have a significant effect on the tool failures. Figure 4 shows flank wear conditions of PVD TiAlTaN tungsten carbide at different combination of machining parameters. It is observed that work materials adhered at the tool edge regardless of cutting speed. Attrition is observed with dull surfaces at the tool-chip contact zone. Both abrasion and micro-abrasion marks are observed at the flank vicinity just beside the attrition zone. Micro-abrasion is a result of broken abrasive metal pieces rubbing against the tool surface [4]. A combination of abrasion and attrition is attributed to higher cutting speed, feed rate, depth of cut, moderate MQL flow rate and low nanofluids concentration.

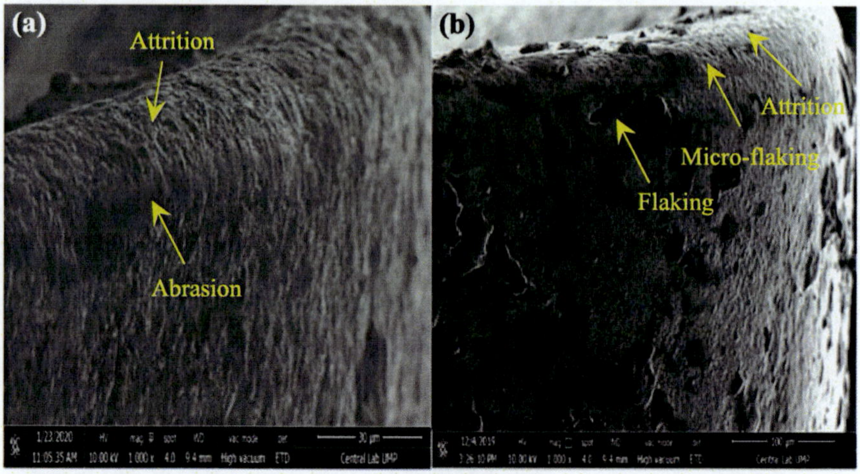

Fig. 2 Flank wear with attrition, abrasion and flaking observation at the tool edge of **a** uncoated tungsten carbide at 10,000 rpm and 1.2 mL/min of MQL flow rate, **b** PVD TiAlTaN tungsten carbide at 9000 rpm of cutting speed and 0.9 mL/min of MQL flow rate

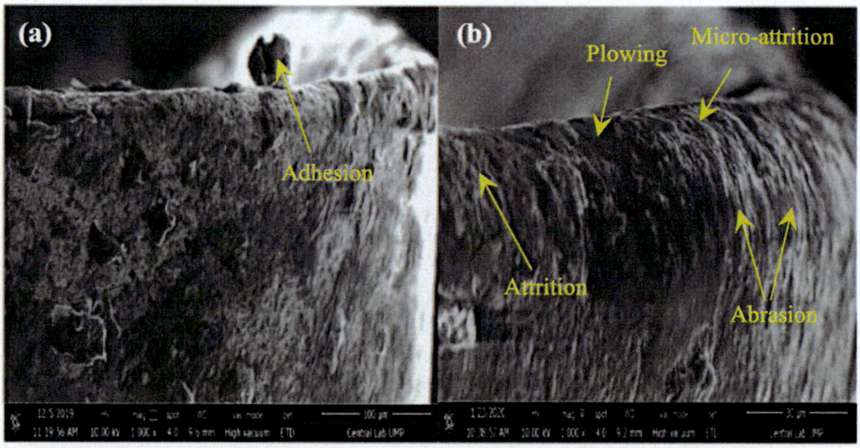

Fig. 3 Flank wear condition of uncoated tungsten carbide at **a** 9000 rpm, 0.5 mm and 100 mm/min, **b** 9000 rpm, 1.0 mm and 200 mm/min of cutting speed, depth of cut and feed rate

3.4 Coating Delamination

Coating surface experienced rapid shrinkage when the material was exposed at higher and lower temperature during the process. Intermittent heating and cooling occur at the friction interface where the heat was generated due to shear stress and the heat dissipated to the surroundings by the flow of hybrid nanofluids. This phenomenon

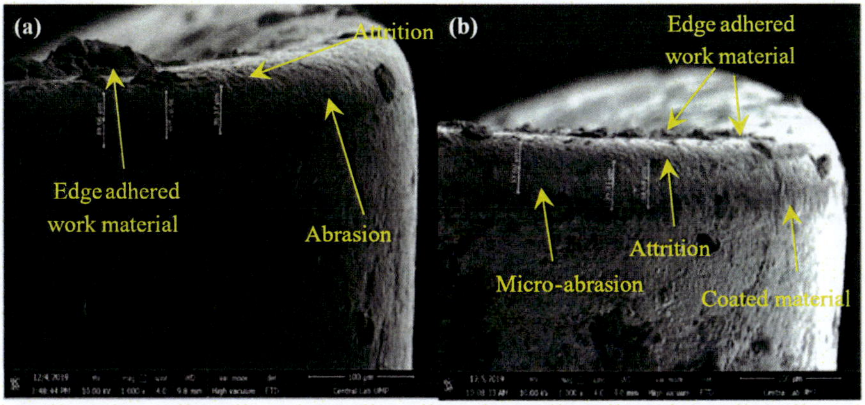

Fig. 4 Flank wear condition of PVD TiAlTaN tungsten carbide at **a** 11,000 rpm, 0.5 mm and 100 mm/min, **b** 9000 rpm, 1 mm and 150 mm/min of cutting speed, depth of cut and feed rate

of cyclic thermal leads to pre-coating and coating delamination near the cutting edge. At initial phase, pre-coating was severely observed just because of intermittent heating and cooling on the material with no application of shearing. However, coating material was fully delaminated just beside the wear area due to cyclic thermal and shear force as shown in Fig. 5. Xichun et al. [10] observed coating delamination of similar PVD TiAlNTiN coated inserts at the cutting-edge vicinity on the rake face.

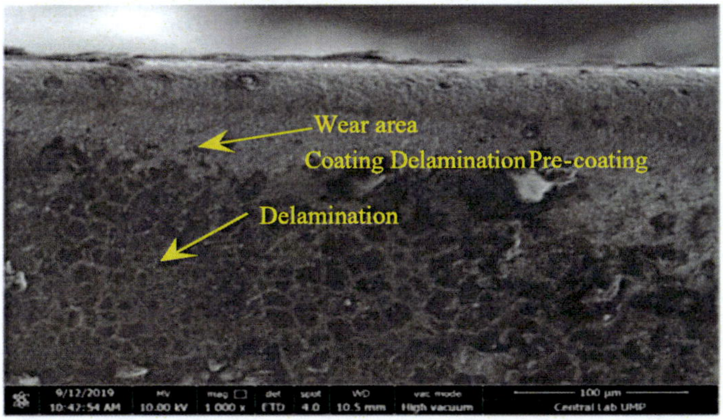

Fig. 5 Coating delamination observed on flank wear vicinity of double-layered PVD TiAlTaN tungsten carbide at 150 mm/min of feed rate and 0.75 mm of MQL flow rate

3.5 Abrasion and Micro-abrasion

Figure 6 shows the formation of abrasion and micro-abrasion marks occurs on PVD TiAlTaN tungsten carbide. Minor dislodging of carbide grains at the boundary of abrasion resulting micro-abrasion marks. Deposited material consists of adhesion and micro-welded are seen on the inserts as a result of materials flowing during the cutting process at a combination cutting parameters of 11,000 rpm (cutting speed), 200 mm/min (feed rate), 1.0 mm (depth of cut), 0.9 mL/min (MQL flow rate) and 0.09 wt.% (concentration). Chenbing et al. [11] revealed that abrasive wear of PVD TiAlN/AlCrN coated carbide was a prominent wear mechanism in both dry and MQL condition in turning process of Titanium Alloy. There was a condition where the adhesive material was peeling off from the flank area during the process leaving some mechanical marks with contour on the surface. In dry machining, abrasive material gradually rubbed off the surface leaving severe abrasion marks on the edge [12].

Figure 7 shows a combination of attrition and abrasion of uncoated tungsten carbide that machined at 8000 and 10,000 rpm of cutting speed. The wear mechanisms were observed under high magnification. Rough attrition was severe on the top edge of the cutting tool. It was due to mechanical shear stress between tool and workpiece interface during the process. Besides, material adhesion was observed on the cutting tool edge that possibly induce other wear mechanisms when rapid tool was used for another series of cutting process.

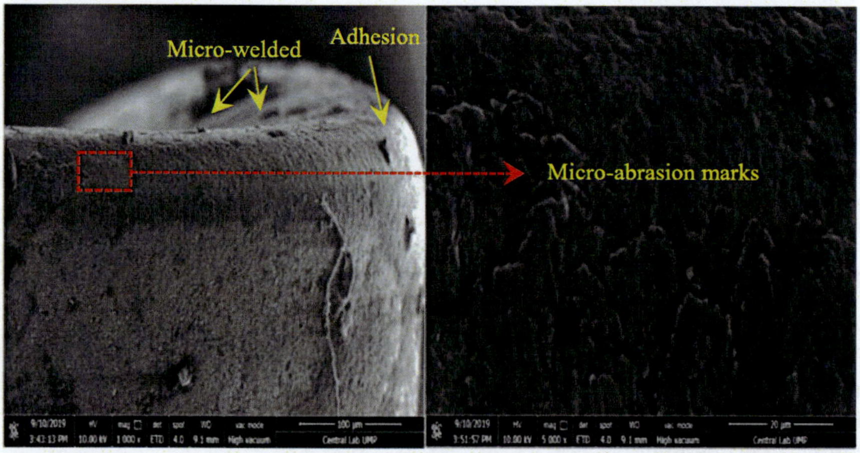

Fig. 6 Observation of abrasion marks and deposited material on double-layered PVD TiAlTaN tungsten carbide

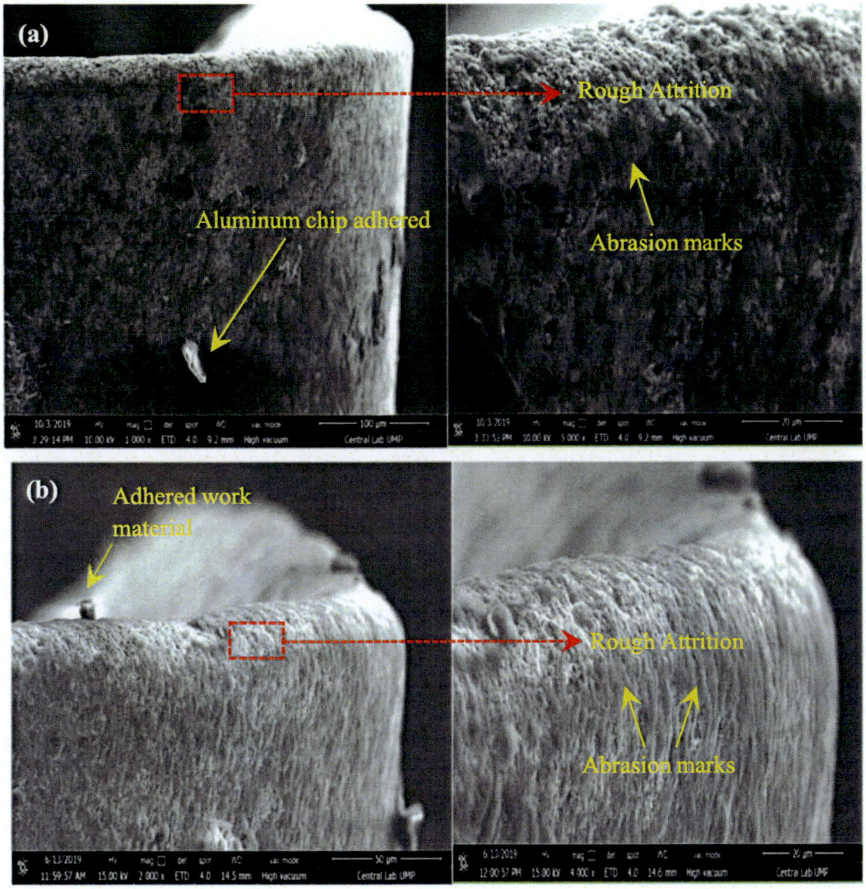

Fig. 7 Attrition wear mechanism observed on uncoated tungsten carbide with **a** 8000 and **b** 10,000 rpm of cutting speed

3.6 Thermal Crack

Figure 8 shows thermal crack occurred on the flank face of PVD TiAlTaN inserts at higher cutting speed and higher speed rate upon removing Aluminum Alloy 6061-T6. In minimum quantity lubricant (MQL) method, nanofluids are delivered to the cutting zone intermittently depending on the pumping stroke setting. Thus, a combination of tri-hybrid nanofluids and MQL method offers thermal cyclic loads due to heating of the friction forces and cooling of the nanofluids resulting expansion and shrinkage of the substrate material that led to thermal crack especially on the outer surface of cutting tool. The thermal crack accelerates the process of material fracture which led to flaking and chipping along the crack line at 250 mm/min of feed rate. However, the thermal crack did not produce catastrophic failures on the cutting

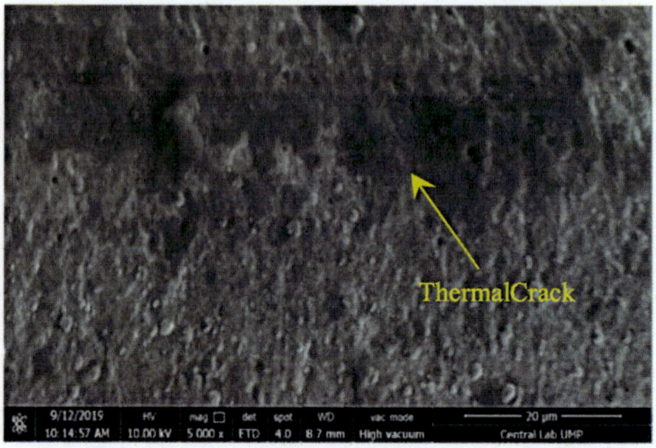

Fig. 8 Thermal crack observation on flank face of PVD TiAlTaN tungsten carbide at 10,000 rpm, 250 mm/min and 0.75 mm of cutting speed, feed rate and depth of cut

tools. Heat concentration at the tools remains instead of being transferred due to low conductors of carbide material in the cutting tool. This phenomenon causes major cyclic thermal stress that has high tendency to generate thermal crack. However, tri-hybrid nanofluids can mitigate the cyclic thermal stress with the capability to provide cooling and lubricating effects on the interface area. As a result, the cause of thermal crack occurrence can be reduced with the combination of optimal cutting solutions and tri-hybrid nanofluids.

A bigger size of worn area occurred on the tool edge when conventional coolant was applied at the cutting zone. Other wear behavior such as built-up edge was not observed and abrasion marks were reduced when graphene oxide of nanofluids applied in the milling process of Titanium Alloy [10, 13]. Thermal and physical properties of cutting fluid play a significant role in enhancing the process of heat transfer and offer excellent lubrication characteristics. Therefore, a new developed tri-hybrid nanolfuids offer a significant role as "roller effect" that may be attributed to the thin oil film is created on the surface which is referred to rolling friction instead of shear force that occur between the workpiece flank surface and the cutting tool, then substantially reduce the friction, cutting force, surface integrity and tool wear. Work material adhesion was not observed particularly at higher nanofluids concentrations and MQL flow rate which are associated with the excellent thermal-physical properties of nanofluids that can dissipate the heat to surrounding, resulting lower cutting temperature especially at tool-chip interface.

4 Conclusion

The presence of coating delamination among coated inserts in the region of the flank wear after machining Aluminium Alloy 6061-T6 was found as a result of high temperatures at the tool- workpiece interface, which contributed to tool wear failures. The process of coated tool failures involved thermal crack, pre-coating delamination and coating delamination. As a result of coating delamination, rough friction was generated during the cutting process. It was also revealed that there were other wear processes present, such as adhesion, attrition and abrasion. Aluminium chip adhesion high probably may induce other tool failures such as chipping, flaking and built-up edge. However, abrasion was a substantial problem for coated inserts, in addition to coating failures. Nevertheless, throughout the machining process, attrition was shown to be more prevalent in uncoated tungsten carbide due to rough surface shear stress in unprecedented condition. At extreme machining condition, higher temperature and friction at the tool-workpiece interface have a significant effect on the tool failures. Therefore, the application of tri-hybrid SiO_2–Al_2O_3–ZrO_2 nanofluid was the first attempted utilizing a combination of three different nanoparticles and MQL method to decrease the cutting temperature and cutting force of Aluminium Alloy 6061-T6 during the end milling process which in turn able to enhance the process and prolong the tool life of the cutting inserts.

Acknowledgements The authors are grateful for the support given by Universiti Malaysia Pahang for providing research grants PDU213204.

References

1. Ranga Babu JA, Kumar KK, Srinivasa Rao S (2017) State-of-art review on hybrid nanofluids. Renew Sustain Energy Rev 77:551–565
2. Muaz M, Choudhury SK (2019) Experimental investigations and multi-objective optimization of MQL-assisted milling process for finishing of AISI 4340 steel. Measurement (Lond) 138:557–569
3. Demir H, Gündüz S (2009) The effects of aging on machinability of 6061 aluminium alloy. Mater Des 30(5):1480–1483
4. Chang H, Li S, Shi R (2017) Design and manufacturing technology of high speed milling cutter for aluminum alloy. Procedia Eng 174:630–637
5. Wakabayashi T, Suda S, Inasaki I, Terasaka K, Musha Y, Toda Y (2007) Tribological action and cutting performance of MQL media in machining of aluminum. CIRP Ann Manuf Technol 56(1):97–100
6. Ling CL et al. (2022) Effect of substrate roughness and PVD deposition temperatures on hardness and wear performance of AlCrN-coated WC-Co. Surf Coat Technol 436
7. Zheng ZP, Lin N, Zhao LB, Li X, He YH (2018) Fabrication and wear mechanism of Ti(C,N)-based cermets tools with designed microstructures used for machining aluminum alloy. Vacuum 156
8. Sharif S, Rahim EA (2007) Performance of coated- and uncoated-carbide tools when drilling titanium alloy-Ti-6Al4V. J Mater Process Technol 185(1–3):72–76
9. Najiha MS, Rahman MM (2016) Experimental investigation of flank wear in end milling of aluminum alloy with water-based TiO_2 nanofluid lubricant in minimum quantity lubrication technique. Int J Adv Manuf Technol 86(9–12):2527–2537
10. Li A, Zhao J, Hou G (2017) Effect of cutting speed on chip formation and wear mechanisms of coated carbide tools when ultra-high-speed face milling titanium alloy Ti-6Al-4V. Adv Mechan Eng
11. Ni C et al (2020) Effects of machining surface and laser beam scanning strategy on machinability of selective laser melted Ti6Al4V alloy in milling. Mater Des 194

12. Ni C et al (2022) Machining performance and wear mechanism of PVD TiAlN/AlCrN coated carbide tool in precision machining of selective laser melted Ti6Al4V alloys under dry and MQL conditions. J Manuf Process 79(March):975–989
13. Hong SY, Markus I, Jeong WC (2001) New cooling approach and tool life improvement in cryogenic machining of titanium alloy Ti-6Al-4V. Int J Mach Tools Manuf 41(15):2245–2260

Experimental and Simulation Approach; Investigation Effect of Axial Feed Rate to the Cutting Force in Dimple Milling Using Ball End Mill Tool

M. A. H. Shaharudin and A. Razlan Yusoff

Abstract A recent study indicated that the dimple has potential to improve the parts performance in a way of minimizing friction on a sliding mechanical component. Despite outnumbered dimple fabrication methods, milling process is considered as the versatile process that could produce various dimple profile with complex shape and the process is extremely dependent on the process variable such as machining parameters and cutting tool condition. The present research work aiming to study the effect of the milling machining parameters which is the axial feed rate and tool diameter to the cutting force by means of experimental work and finite element analysis (FEA). The concave dimple profile is machined at different axial feed rate to a flat Al6061 specimen using ball end mill attached to 3-axis CNC milling machine and the cutting force captured by KISTLER force dynamometer and the results will be compared with the FEA results. Based on the results, shows that cutting force increased directly proportional with the increasing plunging feed rates and tool diameters. The investigation work has successfully characterized the influence of studied milling parameters to the cutting forces and the research work will be continued further on the incline milling technique and micro size ball end mill.

Keywords Dimple fabrication · Milling process · Ball nose end mill · Finite element analysis

M. A. H. Shaharudin (✉) · A. R. Yusoff
Faculty of Manufacturing and Mechatronics Engineering Technology, Universiti Malaysia Pahang, 26060 Pekan, Pahang, Malaysia
e-mail: alihanafiah@ump.edu.my

A. R. Yusoff
e-mail: razlan@ump.edu.my

A. R. Yusoff
Center for Advanced Industrial Technology, Universiti Malaysia Pahang, 26060 Pekan, Pahang, Malaysia

C. N. Rosyidi et al. (eds.), *Proceedings of the 6th Asia Pacific Conference on Manufacturing Systems and 4th International Manufacturing Engineering Conference*, Lecture Notes in Mechanical Engineering,
https://doi.org/10.1007/978-981-99-1245-2_4

1 Introduction

Surface texturing is the process of creating a specific pattern on the surface of a workpiece to trans-form it into a functional surface. Texturing with features or profiles with multiple scales to creating a features pattern could be done ranging from micro scale up to millimeter range. The features pattern such as dimple enables many new applications of the engineered surfaces, where specific dimple shapes can be used to influence the functional properties of a surface especially the tribological characteristics [1]. Based on literature, various dimple shape has been studied but the most common shape that intensively being study is round shape where the comparative study found that the round dimple textured surface is capable to reduce the surface coefficient of friction (COF) up to 30% compared to untextured surface [2].

In general, various processes could be employed to manufacture the dimple structure and the selection of the process depends on a few criterions namely workpiece material, workpiece shape and size, dimple size and shape and required accuracy. As for milling process, it is considered as the most versatile process to machine various micro dimple profile and based on the past research [3–8] on application of milling to fabricate a micro-dimple, most of the research approached are done intensively on round and hemispherical shape dimple using ball end mill tool. It is due to hydrodynamic performance of the hemispherical shape dimple and capability of the ball end mill cutting tool to form the concave profile of the micro-dimple with acceptable tolerance and surface quality.

Meanwhile in milling process, cutting force is one of predominant criterion in evaluating the machining process performance. The cutting force depend greatly on the machining parameters and the tool conditions [9]. The machining parameters basically referred to three relative motions of the milling cutting tool including the rotational speed of the cutting tool, advancement of cutting tool towards work material along the intended path (either axial or radial direction) is also known as feed rate and the penetration motion be-low the workpiece surface to necessary depth so called depth of cut. In fact, the simultaneous action of all these 3 relative motions has resulting the advancement of cutting tool along the intended toolpath generating the intended shape. According to a study done on effect of machining parameters to the cutting force addressed the cutting force directly proportional to the feed rate and depth of cut but inversely proportional to the cutting speed [10].

Meanwhile the condition of the cutting tool refers to the state of wear at the tool cutting edge. As the cutting process started, smooth and sharp tool cutting edge is progressively wear off causing an increasing of roughness at the cutting edge. In turn, rough tool cutting edge will be imprinted to the machined surface and deteriorate its roughness. Based on study done [11] on effect of tool wear to the final surface roughness at micro level shows tool wear directly proportional to the numbers of machined micro-dimple or materials removed volume. Despite effect the surface quality of the machined workpiece, tool wear also causing a significant increase in measured cutting force during machining as well as excessive force will lead to tool breakage which possibly led to decreased productivity.

In milling process, machining a micro profile using small tool diameters is quite challenging where the cutting force become very critical especially when the toolpath consist of plunging or axial feed motion in the z-axis direction. With the low cutting speed right at the center of the tool has causing the rubbing action [12] rather than cutting which increased the cutting force in the z-axis direction. This could lead to excessive cutting force that accelerate the tool wear faster or even worst could break off the cutting tools and ruined the operations economically. As in dimple milling machining, the nature of the dimple profile itself has caused the toolpath to have the plunging or axial feed motion where the BEM tool need to be positioned perpendicular to the workpiece surface and descending to certain depth to form the intended dimple dimension. Due this situation, it is critical to acknowledge the limit of the plunging feed rate or axial feed rate as well as the effect of the plunging feed rate to the cutting force. So, this research work aims to investigate the effect of machining parameters specifically the feed rate to the generated cutting force with different BEM sizes ranging from Ø1.0 to Ø4.0 mm as well as simulate the dimple machining process using DEFORM 3D software to compare the simulated cutting force.

2 Methodology

The present research work consists of two main parts; the experimental work and the FEA simulation works. In order to determine the interaction between tool axial feed rate and the generated cutting force, parametric studies have been conducted to determine the range of appropriate axial feed rate values and an experiment has been carefully setup to manipulating the values. Whereas in the FEA simulation, the setup tries to imitate as close as possible the actual experimental work to ensure the reliability of the simulation results later.

2.1 Experiment Setup

In the experimental work, the tool axial feed rate values will be manipulated during machining the dimple profile (round convex profile) using different size of ball end mill. The axial feed rate will be varying from 10 to 100 mm/min while the total depth of cut will be kept constant at 0.4 mm to maintain the dimple aspect ratio of 0.4 for the smallest tool size. The experiment will be conducted using Makino KE-55 3-axis CNC Milling with a pre-programmed toolpath according to designated dimple position and the force will be captured using KISTLER dynamometer (Fig. 1a). The cutting tool which is 2 flute carbide ball end mill tool size ranging from 1.0 to 4.0 mm in diameter and the workpiece made off Aluminum alloy Al6061-T6 will be used in the experimental works.

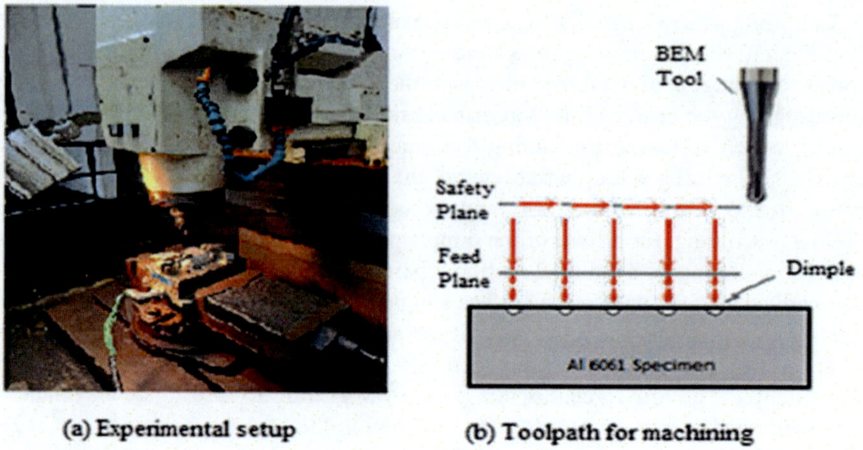

Fig. 1 The experiment setup used to machine micro-dimple and the programmed tool path used the experiments

The dimple profile will be machined using pre-programmed toolpaths (Fig. 1b) and the tool motion begin with positioning motion to the intended position at safety plane with rapid feed followed by descending motion to feed plane and then plunging motion of the Ball End Mill tool to the $z = -0.4$ mm. At this point (plunging motion), the tool machined the dimple profile with different axial feed rate (z-axis) and the BEM tool will dwell about 4 s before the BEM tool retrack to z-axis feed plane with rapid feed.

2.2 *Finite Element Analysis*

In the FEA, the setup is designed to simulate as close as possible the dimple cutting process to validate the friction coefficient used in the simulation with the actual experiment cutting force as well as observing the material deformation in the dimple milling operation. Assuming the friction coefficient are the same regardless the cutting tool size, the FEA simulation will be conducted using the BEM tool Ø1.0 mm with axial feed rate of 50 mm/min. The simulation results then will be compared with the actual experimental work for validation to determine the close match friction coefficient value. The simulation will be conducted using DEFORM-3D software (version 10, Scientific Forming Technologies Corporation, Columbus, Ohio) while the 3D geometry of the BEM tool constructed using CATIA software based on the cutting-edge profile generated by MATLAB software.

The DEFORM 3D simulation process consists of two parts namely pre-processor and post-processor. In the pre-processor parts, the 3D models, meshing, motion and speed parameters, and boundary conditions are assigned as close as possible with the

Table 1 Common materials properties of ball end mill tool and workpiece applied in the DEFORM simulation

Properties	Ball end mill	Test specimen
Materials	Coated tungsten carbide (WC)	Aluminum alloy (AL6061-T6)
Density (kg m^{-3})	14,500	2730
Poisson ratio	0.21	0.33
Elastic module (MPa)	612	68.9
Yield strength (MPa)	–	276
Ultimate tensile strength (MPa)	370	310
% Elongation	–	18
Specific heat (J kg^{-1} K^{-1})	337	896
Thermal conductivity (W m^{-1} K^{-1})	110	187

actual experimental work done. While the post-processor part is where simulation and analysis of results are performed. In any simulation, the crucial element properties for both tool and specimen such as mechanical and thermal properties need to be defined to ensure the accuracy of the FEA simulation. This material's property values govern the physical law and affecting directly to the computational results of the simulation. Due to that, the value of the mechanical and thermal properties for both BEM tool and the workpiece is carefully determined and summarized in Table 1.

2.3 *Pre-processing*

In pre-processing setup, both 3D of ball end mill tool and the workpiece meshed with tetrahedral element type with numbers of element approximately between 40,000 and 50,000 and the size of element ranging from 0.002 to 0.02 mm respectively. Localize meshing methods are also applied to the workpiece where a specific portion (region where the chip formation) on the surface workpiece has a finer mesh size (ratio 1:0.25), rather than the entire workpiece and the summaries of the meshing parameters tabulated in Table 2. The workpiece movement and rotation were restricted in all directions, whereas the ball end mill tool was allowed to be rotated around the tool axis (clockwise) and move in the *Z*-axis direction (axial feed direction) with predetermined axial feed speed to simulate the cutting action as shown in Fig. 2. The initial temperature of both 3D object preset at 30 °C and BEM tool was treated as a rigid object (assumed to be no shape deformation occurred during the cutting operation), and the test specimen was considered as a plastic object (deformed according to the governing physical law).

Table 2 Summary of meshing parameters and boundary condition applied in simulation

	Ball end mill tool	Workpiece
Element type	Tetrahedral	Tetrahedral
No. of element	Approx. 40,000–50,000	Approx. 40,000–50,000
Meshing size	Min. 0.002 mm–Max. 0.05 mm	Min. 0.002 mm–Max. 0.05 mm
Local mesh ratio	None	0.25 at chip formation area
Degree of freedom	2 (rotation about tool-axis and movement-Z-axis)	0
Shape model	Rigid	Plastic
Initial temperature	30 °C	30 °C
Friction model	Shear friction, m = 0.5	Shear friction, m = 0.5

Fig. 2 The position setup of Ball End Mill tool and workpiece in the simulation pre-processing

In the present simulation works, Lagrangian Incremental method was used to simulate the cutting phenomenon between the BEM tool and the aluminum workpiece. For the material model, the workpiece is assigned with Johnson–Cook material model since it's a well-known model that capable to predict the mate-rial movement and deformation accurately. In the Johnson–Cook material model the plastic flow stress during the cutting process is formulated as:

$$\sigma = (A + B\varepsilon^n)\left(1 + C\frac{\dot{\varepsilon}}{\dot{\varepsilon_0}}\right)\left(1 - \left(\frac{1 - T_r}{T_m - T_r}\right)^n\right) \quad (1)$$

where σ, is the materials flow stress, $\dot{\varepsilon}$, is effective plastic strain, $\dot{\varepsilon_0}$, effective strain rate, A, B, C, n and m are material constants that are found by material tests. T is instantaneous temperature; T_r is room temperature and Tm is melting temperature of a given material.

At the interface between the tool-chip is modeled to have contact friction and this friction significantly affects the simulation especially the cutting force since the frictional energy will convert into heat. The friction conditions at the tool-chip interface are modeled differently according to its deformation regions. In the sliding region, the friction is modeled as Coulomb friction while at the sticking region modeled as shear friction [13] as mathematically formulate as follows:

$$\text{Coulomb friction, } F_s = \mu p \tag{2}$$

$$\text{Shear friction, } F_s = mk \tag{3}$$

where F_s frictional force, μ friction coefficient, p interface pressure, m friction factor, k shear yield stress

Even though friction is known to exist at the tool-chip contact interface, the determination of the actual value of friction is difficult, especially in micromachining operations due to micro size tool. As reference or baseline values, the present simulation will use the Coulomb friction model at the tool-chip interface with coefficient value of $\mu = 0.5$ [14, 15] based on previous research works.

The 3D model of BEM tool is designed using CATIA software based on the derived cutting-edge curves plotted using MATLAB software. The BEM model was constructed as two flutes with a 30° helix angle, and a summary of the tool geometry design is tabulated in Table 3. The derivations of the cutting edge are based on the 3D cartesian coordinate system of the intersection curve between the spherical sur-face and the helical surface as shown in Fig. 3 where the center of the spherical, O is the origin. Based on the intersection definition, thus any point on the cutting edge of the BEM at the spherical region can be expressed by the following equations.

$$x = R\sqrt{1 - (c\varphi)^2}.\cos\varphi \tag{4}$$

$$y = R\sqrt{1 - (c\varphi)^2}.\sin\varphi \tag{5}$$

$$z = Rc\varphi \tag{6}$$

$$c = \frac{1}{\tan\varphi} \tag{7}$$

where x, y and z coordinate elements, R tool radius, φ angle parameters, c helix parameters.

Table 3 Summary of tool geometries parameters applied in FEA simulation

Tool parameters	Value
Diameters (mm)	Ø1.0
Rake angle	5°
Clearance angle	9°
Helix angle	30°
Cutting edge radius	5 μm

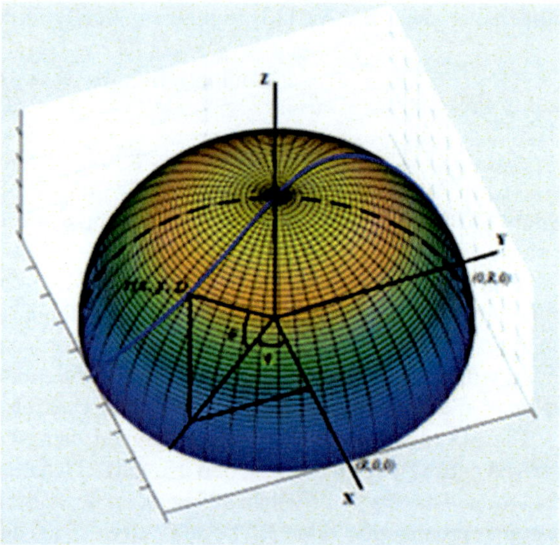

Fig. 3 The 3D cartesian coordinate system applied to the construction of BEM tool cutting edge with helix angle of 30°

2.4 Post Processing

In the post processing parts, the simulation cutting force data on each axis (F_x, F_y, and F_z) will be plotted against machining time to validate or compared with the actual force captured by the load cell in the actual experimental work. In addition, the maximum cutting force, F_{max} for each axial feed rate and tool size will be determined and the maximum cutting force, F_{max} will be plot against the axial feed rate also visualized the correlation between maximum cutting force and axial feed rate for specific tool sizes.

3 Result and Discussion

3.1 Experiment Results

The experimental works has been conducted to investigate the effect of the axial cutting feed rate to the generated cutting force during machining with different BEM tool size Ø1.0, Ø2.0, Ø3.0 and Ø4.0 mm. In the experiments, spindle speed has been preset to constant at 4000 r.p.m, depth of cut of 0.4 mm while the axial feed rate varies from 10 to 100 mm/min. In total 12 different experiment has been conducted and the present sections will represent the experiment results (captured by the dynamometer in the experiment works) in form of plotted graph cutting force against time as shown in Figs. 4, 5, 6, 7, 8, 9, 10, 11, 12, 13, 14, and 15.

Based on the experiment results shows a consistence force pattern for all applied BEM tool size where the highest force captured in z-axis and the force in x and y-axis could be assumed to be equal in term of magnitude. At low axial feed rate, the time taken to reach the intended depth are longer, causing the force is spread over the time resulting in low cutting force in all axis direction especially in the z-axis direction. While the force in x-axis and y-axis maintains almost close to zero for all studied BEM tool sizes. In another hand, at high axial feed rate value, the cutting force are accumulated in short period of time causing the cutting force spike high in all axes and obviously in z-axis directions and this force pattern are consistence with all size of BEM tools.

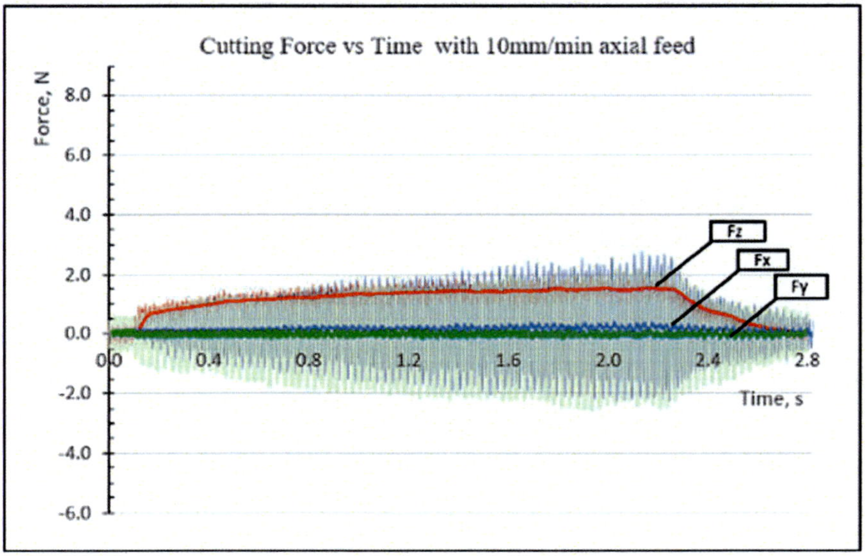

Fig. 4 Captured cutting force with BEM tool Ø1.0 mm and axial feed rate of 10 mm/min

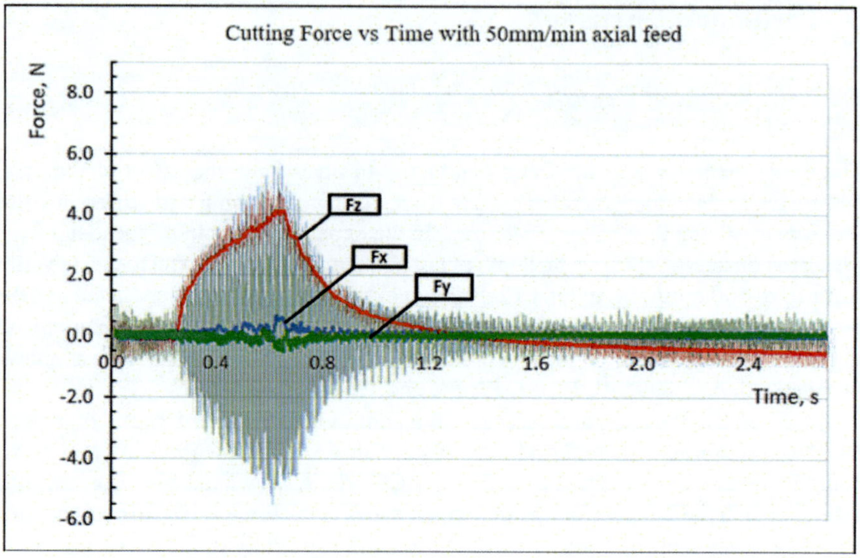

Fig. 5 Captured cutting force with BEM tool Ø1.0 mm and axial feed rate of 100 mm/min

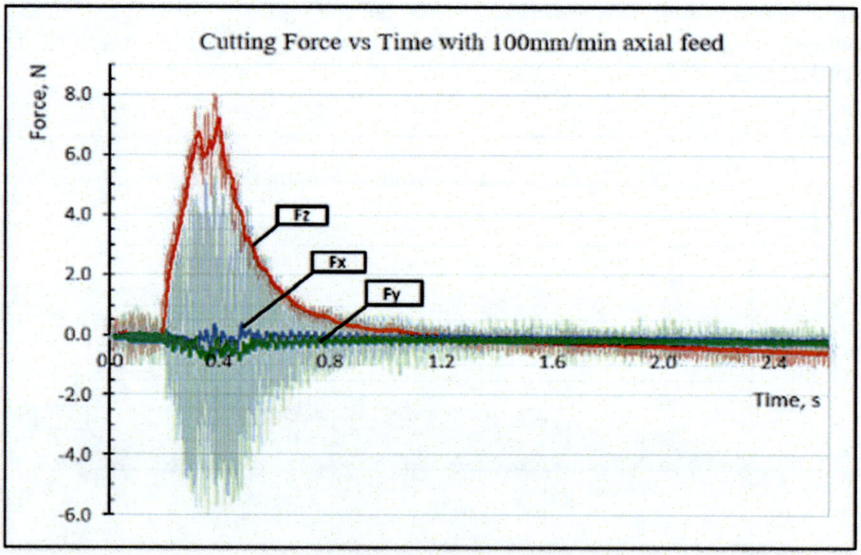

Fig. 6 Captured cutting force with BEM tool Ø1.0 mm and axial feed rate of 50 mm/min

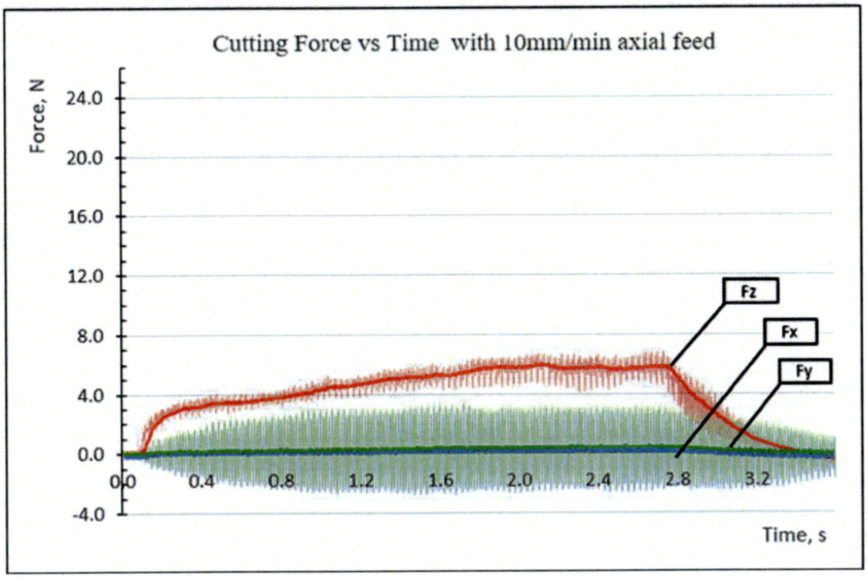

Fig. 7 Captured cutting force with BEM tool Ø2.0 mm and axial feed rate of 10 mm/min

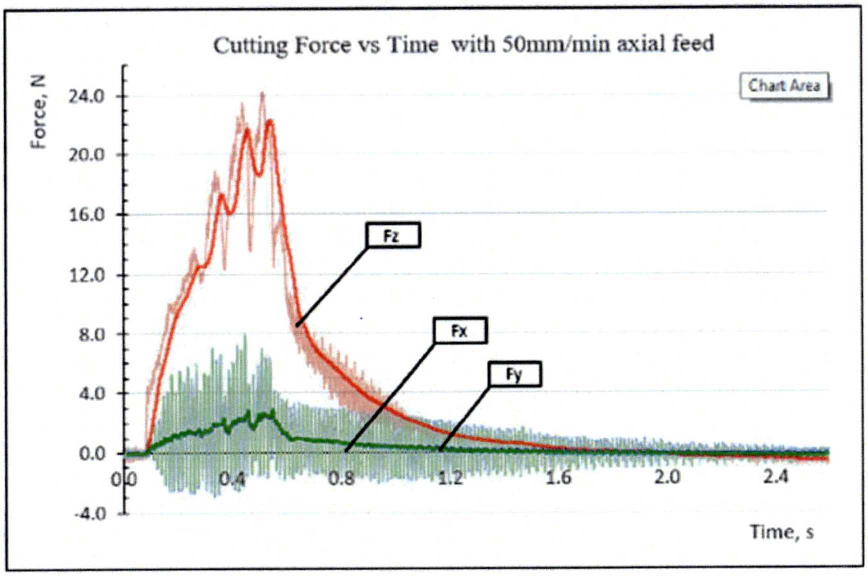

Fig. 8 Captured cutting force with BEM tool Ø2.0 mm and axial feed rate of 50 mm/min

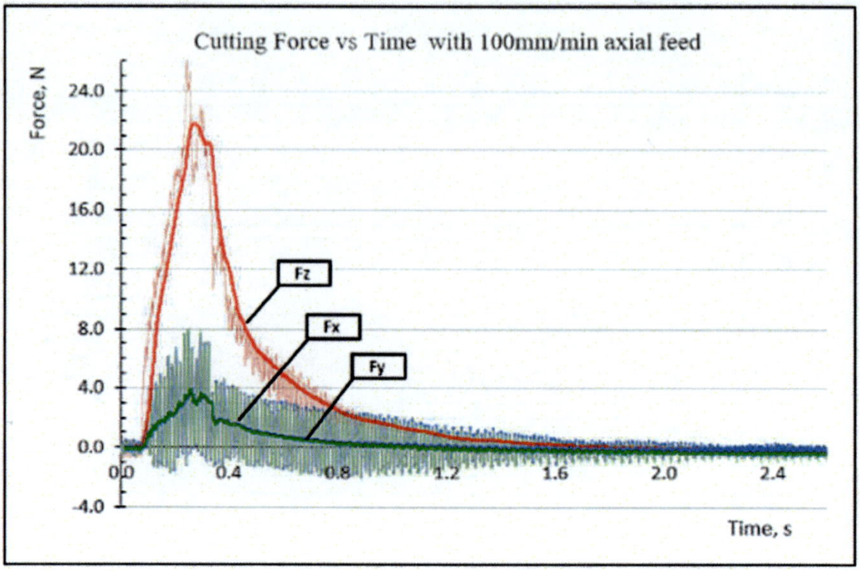

Fig. 9 Captured cutting force with BEM tool Ø3.0 mm and axial feed rate of 10 mm/min

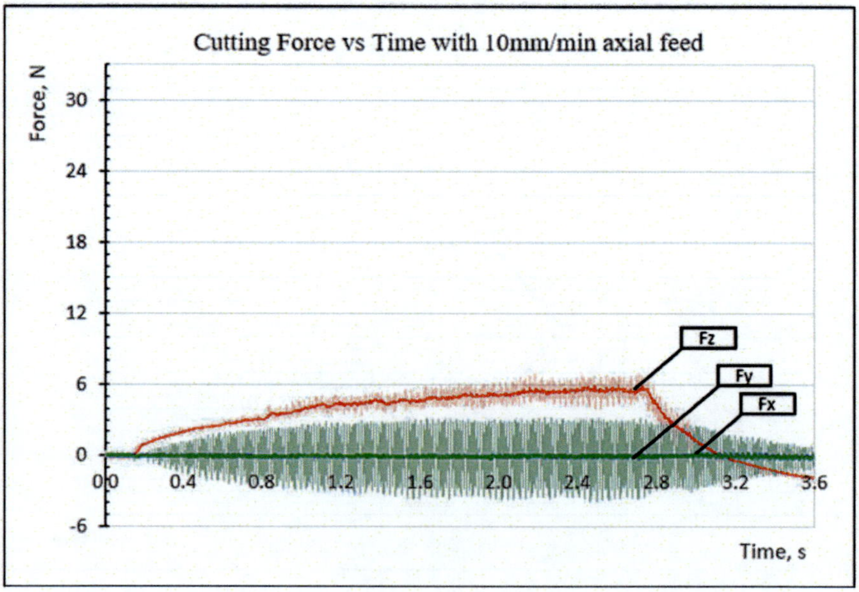

Fig. 10 Captured cutting force with BEM tool Ø3.0 mm and axial feed rate of 100 mm/min

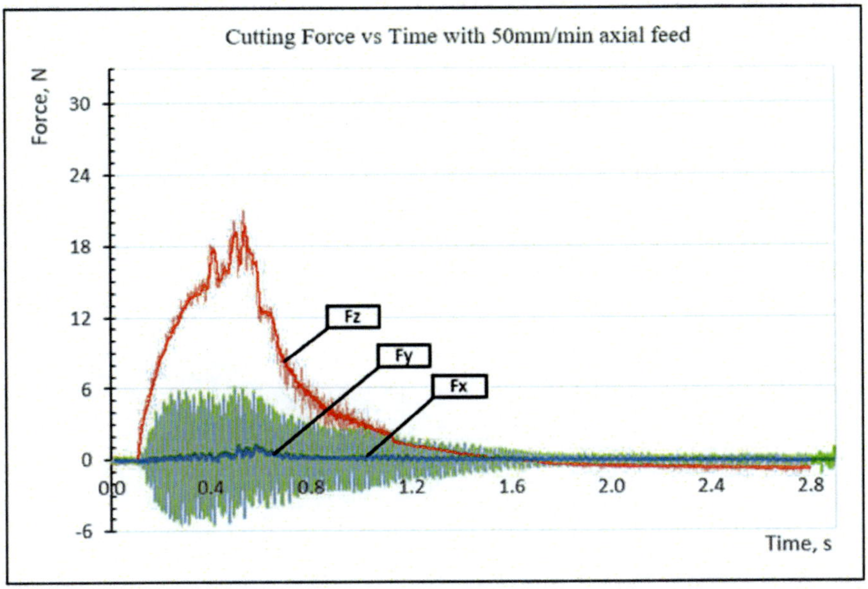

Fig. 11 Captured cutting force with BEM tool Ø2.0 mm and axial feed rate of 100 mm/min

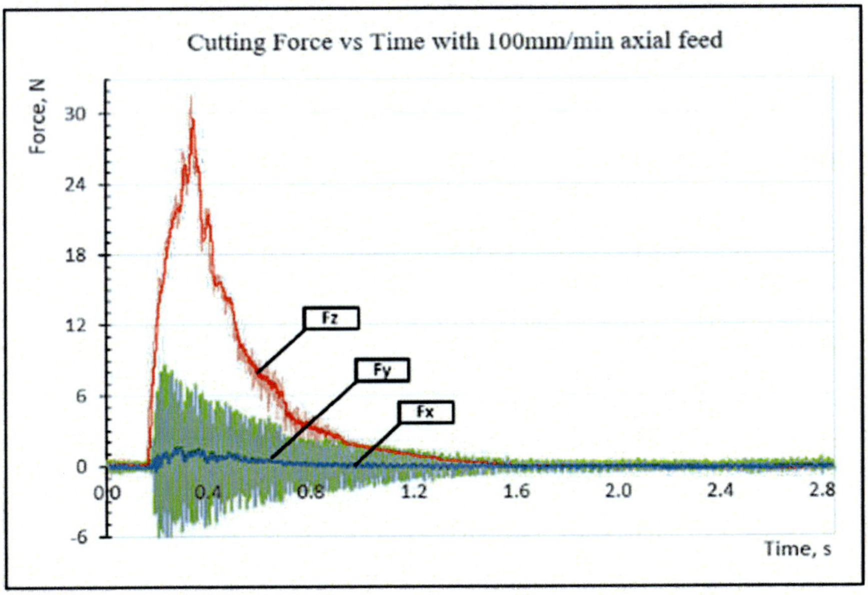

Fig. 12 Captured cutting force with BEM tool Ø3.0 mm and axial feed rate of 50 mm/min

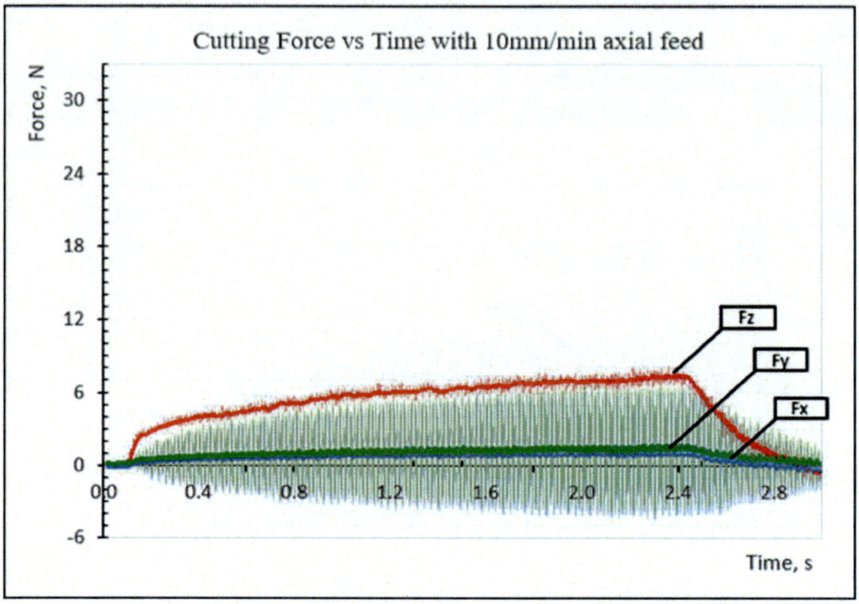

Fig. 13 Captured cutting force with BEM tool Ø4.0 mm and axial feed rate of 10 mm/min

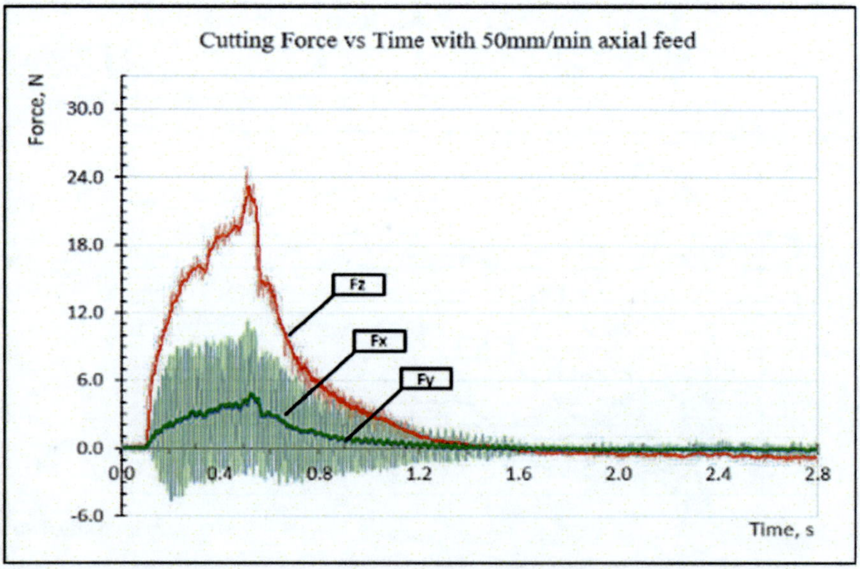

Fig. 14 Captured cutting force with BEM tool Ø4.0 mm and axial feed rate of 50 mm/min

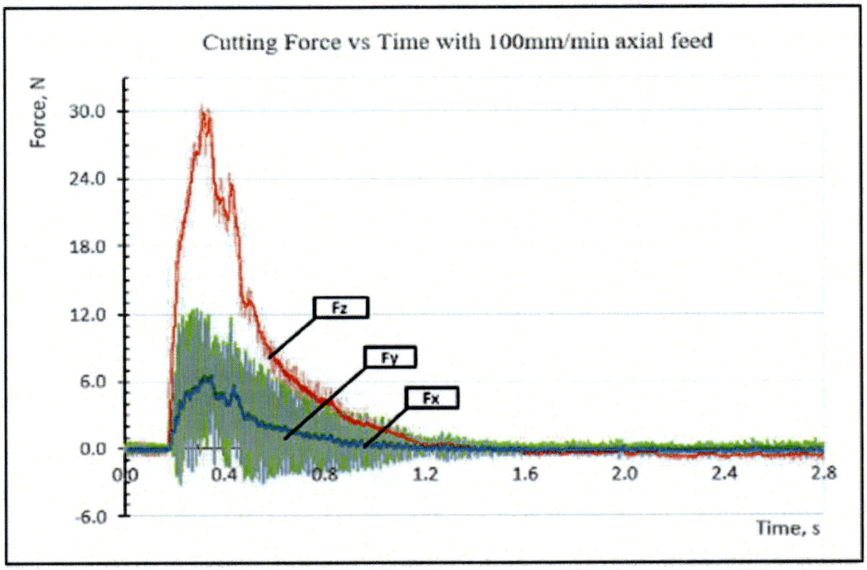

Fig. 15 Captured cutting force with BEM tool Ø4.0 mm and axial feed rate of 100 mm/min

In addition to that, the analysis extended to the maximum cutting force occurrence where the maximum cutting force in z-axis are tabulate in the chart Fig. 16. Based on the chart, shows that highest cutting force, F_z occurred with BEM tool Ø4.0 with 100 mm/min axial cutting speed and the lowest cutting force, F_z occurred with BEM tool Ø1.0 with 10 mm/min axial cutting speed. In respect to the chart, two characteristics related to tool size and the feed can be conclude. First, an increase in tool size will increase the cutting force due to the bigger materials volume to be cut. Second, an increase of the tool axial feed rate will increase the cutting force. In case of dimple machining, the removed material volume is constant so cutting with lower axial feed rate causing shallow tool cutting edge penetration in every tool revolution consequently producing thin chip thickness, longer machining time as well as generating lower cutting force. Meanwhile, cutting the dimple with high axial feed rate will forced the tool cutting edge penetrate deeper in the materials for every tool revolution producing thicker chip thickness, shorter machining time and consequently generating higher cutting force.

3.2 Simulation Results

As mentioned previously, the FEA simulations are conducted to simulate as close as possible the dimple machining process in order to validate the coefficient value used in the FEA simulation. In the FEA simulation, BEM tool Ø1.0 mm were used to

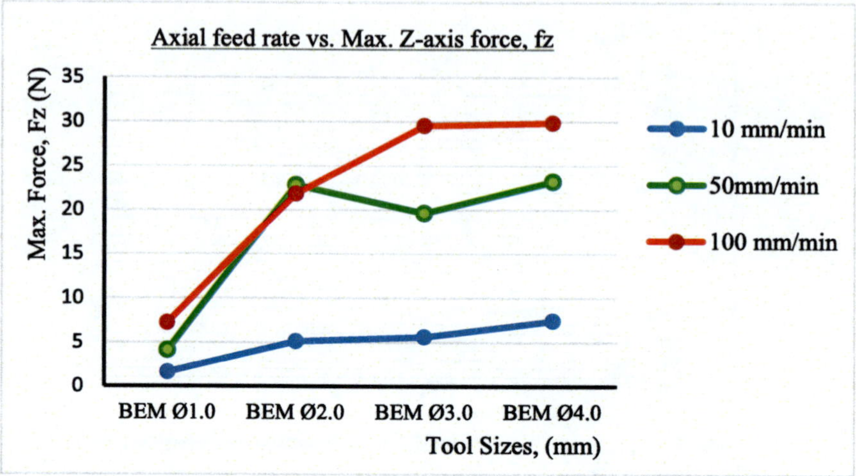

Fig. 16 Maximum cutting force, F_z generated in the experimental work at different axial feed rates and BEM tool sizes

form the dimple profile with axial feed rate of 100 mm/min and at the tool and chip interface, the shear friction model was employed to simulate the cutting process. In the following section will present the FEA simulation results in form of graph simulated cutting force versus machining time. Based on the simulation results graphs (Fig. 17), for all axial feed rate values shows that the maximum cutting force occurred in the z-axis direction while force on the x-axis and y-axis almost equal during simulations. The results also show that the maximum cutting force occurred in z-axis direction and the force in x-axis and y-axis are almost equal.

As comparison with the actual cutting as shown in experiment result Fig. 6, both simulation and experimental results find good agreement with the same cutting condition in term of rank of cutting force as well as the magnitude of the cutting force in all axes. In the rank of cutting force, both simulation and experiment results shows that the maximum cutting force occurred in the z-axis direction with magnitude of 8.0N and 7.6 N in the simulation and experiment respectively. This conclude that the Coulomb friction model with coefficient value of $\mu = 0.5$ capable to predict accurately the cutting force in the dimple cutting operation as being proof by previous research work [14, 15].

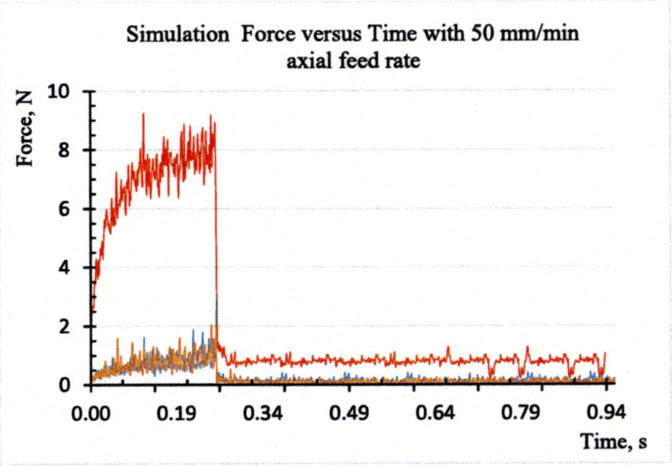

Fig. 17 Maximum cutting force, F_z generated in the experimental work at different axial feed rates

4 Conclusion

In this experiment, the main concern is engaged to analyze the correlation between axial feed rate, (f_r) and tool size to the cutting force. Based on the result gained and the data analysis, proves that the work has successfully characterized the influence of axial feed rate and tool diameters to the generated cutting force using experimental and FEA simulation approach. Other than that, the study concludes:

- The cutting force are increase directly proportional with the increase axial feed rate. and cutting tool sizes.
- The cutting force are increase directly proportional with the increase cutting tool sizes.

Based on the drawn conclusion, the future work should work out on the incline milling method as well as move on into micro level machining operation. One last critical point on micro machining, a serious measure should be taken in term of the tolerance and accuracy during machining especially during the tool and workpiece/ specimen setup to ensure the reliability and repeatability of the gathered data.

Acknowledgements This research paper is part of author doctoral research work and financially support by Universiti Malaysia Pahang (UMP) under Post Graduate Research Scheme (PGRS200324)

References

1. Yu H, Wang X, Zhou F (2010) Geometric shape effects of surface texture on the generation of hydrodynamic pressure between conformal contacting surfaces. Tribol Lett 37(2):123–130
2. Liew KW, Kok CK, Ervina Efzan MN (2016) Effect of EDM dimple geometry on friction reduction under boundary and mixed lubrication. Tribol Int 101:1–9
3. Yan J, Zhang Z, Kuriyagawa T, Gonda H (2010) Fabricating micro-structured surface by using single-crystalline diamond endmill. Int J Adv Manuf Technol 51(9–12):957–964
4. Matsumura T, Takahashi S (2012) Micro dimple milling on cylinder surfaces. J Manuf Process 14(2):135–140
5. Graham E, Park CI, Park SS (2013) Fabrication of micro-dimpled surfaces through micro ball end milling. Int J Precis Eng Manuf 14(9):1637–1646
6. Pratap T, Patra K (2018) Fabrication of micro-textured surfaces using ball-end micro-milling for wettability enhancement of Ti-6Al-4V. J Mater Process Technol 262(June):168–181
7. Denkena B, Grove T, Schmidt C (2018) Machining of micro dimples for friction reduction in cylinder liners. Procedia CIRP 78:318–322
8. Resendiz J, Graham E, Egberts P, Park SS (2015) Directional friction surfaces through asymmetrically shaped dimpled surfaces patterned using inclined flat end milling. Tribol Int 91:67–73
9. Bagaber S, Yusoff A (2018Apr) Multi-responses optimization in dry turning of a stainless steel as a key factor in minimum energy. Int J Adv Manuf Technol 1:96
10. Özlü E, Budak E (2007) Experimental analysis and modeling of orthogonal cutting using material and friction models
11. Dehen S, Segebade E, Gerstenmeyer M, Zanger F, Schulze V (2020) Milling parameter and tool wear dependent surface quality in micro-milling of brass. Procedia CIRP 87:95–100
12. Pratap T, Patra K (2018) Micro ball-end milling; An emerging manufacturing technology for micro-feature patterns. Int J Adv Manuf Technol 94(5–8):2821–2845
13. Zorev NN (1963) Inter-relationship between shear processes occurring along tool face and shear plane in metal cutting. ASME, New York, pp 42–49
14. Li S, Sui J, Ding F, Wu S, Chen W, Wang C (2021) Optimization of milling aluminum alloy 6061–T6 using modified johnson-cook model. Simul Model Pract Theory 111:1023–1030
15. Wang H, Zhang T, Wang S, To S (2020Mar) Characterization of the friction coefficient of aluminum alloy 6061 in ultra-precision machining. Metals (Basel) 3(10):336

Assembly Sequence Optimization to Minimize Reorientation and Tool Changes Using Genetic Algorithm

William, Fariz Muharram Hasby, Wisnu Aribowo, and Sukoyo

Abstract The topic of production planning not only includes capacity planning or production sequence planning, but also topics that overlap with product engineering such as assembly sequence planning. In this paper, we implement assembly sequence optimization for one of the products in a multiproduct flow shop system, specifically by searching for an assembly sequence that involves a minimum number of reorientations and tool changes. As assembly sequence optimization is an NP-Hard combinatorial problem, we use the Genetic Algorithm metaheuristic technique to find the best solution. We also combine the Taguchi Method to find the best meta-parameters for the Genetic Algorithm, such that the resulting solution is better than a solution with arbitrary meta-parameters assigned. The assembly sequence generated reduced the reorientations and equipment changes time by 14.4% (5.74 min from 6.74 min) and reducing the total assembly time by 3.3% (29 min from 30 min).

Keywords Assembly sequence planning · Genetic algorithm · Taguchi method · Reorientation · Tool change

1 Introduction

In the production planning phase, there are many alternatives that can be taken to increase the efficiency of a given production system. Besides designing an optimal job schedule or determining the most efficient use of a production system capacity, altering the assembly sequence for an assembled product can also be implemented such that the total assembly time can be minimized, and productivity can be increased.

Assembly sequence optimization can be prioritized in a multiproduct flow shop environment, where the interaction of multiple routings from multiple products and multiple machines may complicate the effort of finding the optimal production plan.

William · F. M. Hasby (✉) · W. Aribowo · Sukoyo
Department of Industrial Engineering, Institut Teknologi Bandung, Bandung 40132, Indonesia
e-mail: farizmhasby@itb.ac.id

C. N. Rosyidi et al. (eds.), *Proceedings of the 6th Asia Pacific Conference on Manufacturing Systems and 4th International Manufacturing Engineering Conference*, Lecture Notes in Mechanical Engineering,
https://doi.org/10.1007/978-981-99-1245-2_5

In this paper we present a similar situation where a multiproduct make-to-order flow shop company would like to pivot and make one of its bestselling products as a make-to-stock product. As the shop floor capacity will be decreased due to the resources committed to produce the make-to-stock product, it is important to reduce the assembly time for the product to free up capacity that can be used for the other make-to-order products.

Assembly sequence planning is a categorized combinatorics problem with highly constrained feasible solution [1]. It is also an NP-Hard complex problem that will also massively increase computation effort as the number of components involved in the problem increases. In this paper we use genetic algorithm to solve the assembly sequence optimization problem, where the aim is to reduce non-value-added activities such as tool changing and reorienting parts during the assembly [2]. We also try to optimize the results of the genetic algorithm by using experimental design methods, as it has been shown that optimizing the parameters of the genetic algorithm models can lead to better results compared with non-optimized models [3].

2 Literature Research

Assembly sequence planning (ASP) is one of the phases of production planning where an assembly sequence of a given product can be standardized and will be used as long as the product is produced. Assembly sequence will significantly impact the whole of the assembly process, such as the assembly resources, layout, and assembly cost. The effect is more pronounced especially for small to medium enterprises [4].

Chen et al. [5] classified three different methods to solve the assembly sequence planning problem, which are graph-based method, knowledge-based method, and artificial intelligence method. Each of these methods have the capability to solve the problem, but each of these methods have different drawbacks. The Graph-based method can search the entire feasible solution space, but requires a lot of time to generate solutions. Knowledge-based method can mimic expert knowledge by using similar problems to generate a solution, but is hard to find an optimal solution. The artificial intelligence apprach optimized the assembly sequence using a specific objective criteria, but requires large computational resources.

Furtermore, Rashid et al. [1] has summarized the three most used algorithms in artificial intelligence to solve the assembly sequencing problem, which are Genetic Algorithms (GA), Ant Colony Optimization (ACO), and Particle Swarm Optimization (PSO). Other methods that utilize global complete search (Depth First Search, Breath First Search, etc.) are not preferable due to the long time they require to come up with a solution.

Genetic Algorithm is a metaheuristic inspired by the natural selection process for finding suboptimal solutions to a problem. GA process flow can be categorized as GA function and parameter definition, initialization, selection, crossover and mutation [6]. Through the process, the algorithm will select the best individual that can withstand the selection process. As such, GA is suitable for solving combinatoric problems such as the ASP problem described earlier.

Special consideration should be given to the infeasible solutions generated by the GA procedure when the procedure is implemented to solve ASP problems. For example, research by Tseng et al. [7] shows that their use of the GA algorithm generated 253 feasible solutions and 247 infeasible solutions. As the infeasible solutions will consume computing power and possibly also prohibit the algorithm to find a better solution, it is important to restrict the generation of infeasible solutions and their use as a parent for further solutions in the algorithm. To that end, a penalty score could be assigned to the infeasible solutions [2], a check for infeasible solution could be conducted in each subsequent iterations [8], or a modified operator that will only process feasible solutions [9] can also be used in the algorithm flow itself.

The meta parameters of the GA itself can be further optimized so that a better solution can be reached. The combination of these parameters, such as the number of generations, overall population number, crossover probability, and mutation probability, will affect the optimal solution generated by the algorithm. Kucukkoc et al. [3] have shown that Response Surface Treatment method is more sensitive in finding the optimal parameter combinations compared to Taguchi method. Conversely, the Taguchi method is less cumbersome in computing power compared to the response surface treatment.

3 Model and Case Study

To solve the ASP problem, we utilize a model based on previous research outlined above. To elaborate, the model is aiming to reduce the number of tool changes and part reorientation [2] while using GA operator for feasible solutions [9] and using the Taguchi Method to find the optimal parameter combination for this particular case.

Sequence Representation

Assembly sequence is represented by a list of components.

$$S = \{P_{ij}\}, \ i, j \in \{1, \ldots, n\}$$

P_i component i assembled at sequence j
i component index
j sequence index
n total number of components.

Objective Function

Minimize the total assembly time by reducing the assembly reorientation (Z_1, Z_2) and tool changes (Z_3). The reorientation consists of two elements, which are 90°

reorientation and 180° reorientation.

$$Z = w_1 Z_1 + w_2 Z_2 + w_3 Z_3 + C \tag{1}$$

Z Total time to reorient and change tool.
w_1 Time to do 90° reorientation
Z_1 Total number of 90° reorientations to assembly *S*
w_2 Time to do 180° reorientation
Z_2 Total number of 180° reorientations to assembly *S*
w_3 Time to do tool change
Z_3 Total number of tool changes to assembly *S*
C Total value-added time to assembly all components.

Feasibility Criteria

Feasible assembly sequence is checked using the Precedence Relations. The feasibility is checked for each component and for each assembly.

$$U(P_i) = \sum_{\forall Pl} PM(P_i, P_l),\ P_l \notin K \tag{2}$$

$U(P_i)$ Binary number indicating whether a component P_i can be assembled to the current sub-assembly K (0 if the component is not feasible, 1 if the component is feasible).
$PM(P_i, P_l)$ Binary number indicating the precedence relation between component P_i and component P_l (0 if there is no precedence relation; 1 if there is a precedence relation i.e., the component P_i must be assembled before componentP_l).

Then, the total feasibility for the assembly is checked by the following equation:

$$L = \sum_{\forall Pi} U(P_i) = \begin{cases} 0, \textit{sequence is feasible} \\ \geq 1, \textit{sequence is not feasible} \end{cases};\ \forall Pi \in S \tag{3}$$

L Feasibility of assembly sequence.

In the genetic algorithm itself, the assembly sequence solution is represented by an ordered list of length *n*. For example, in Table 1, the assembly sequence is to assemble part, first, part 3, second, and so on.

Table 1 Assembly sequence representation

Assembly sequence							
2	3	5	8	1	6	4	7

Table 2 Taguchi experiment configuration

Factors	Levels				
G	50	300	550	800	1050
P	20	40	60	80	100
C	0.1	0.3	0.5	0.7	0.9
M	0.05	0.1	0.2	0.3	0.5

The fitness function that will be evaluated in the GA is the same as with Eq. (1). The evaluation will take place in each successive generation after the selection, crossover and mutation operator has been applied to the current solution. The parent mating pool is chosen using the Roulette Wheel method, ensuring fitter individuals will be assigned a higher probability to be chosen as a parent. The crossover operator and the mutation operator are adapted from [8] and ensures the feasibility of the generated solution.

For the Taguchi method, four factors with five levels each are used to find the optimal parameter combination. The factors considered are the number of generations (G), number of population (P), crossover probability (C) and mutation probability (M). 25 experiments will be done, and for each experiment there will be 25 replications. The configuration of factors and level involved in this experiment is given in Table 2.

The model will be used to solve an ASP problem involving 30 components, with up to 9 identical parts for each component. The assembly involves 6 different orientations (both directions of each x, y and z axis) and 5 different tools.

4 Results and Discussion

The use of Taguchi Method for the problem can be summarized in Table 3. Table 3 depicts the Signal to Noise Ratio for all the combination of factors and their level, where a small SNR value shows that the particular level for the corresponding factor will minimize the fitness function used in the model. The best parameter combination for this study is 550 for G, 100 for P, 0.9 for C and 0.3 for M.

Using the optimal combination, the model is then run 10 times and the best solution is then chosen as the final result for this case. The assembly sequence is shown in Table 4 with the corresponding fitness value and running time of the model.

Further analysis of the model performance can be seen in Fig. 1, which displays the maximum fitness value, minimum fitness value and average fitness value for each generation in the GA. It can be seen that the solution converged around the 200th generation to the optimal value of 344.6 s.

The GA model has been successful to reduce the total non-value-added time, consisting of reorientation time and tool change time. The solution generated reduced the non-value-added time by 14.4% from the existing assembly sequence (5.74 min

Table 3 Signal to noise ratio result

Levels	G	P	C	M
1	– 54.82	– 54.65	– 54.75	– 54.45
2	– 54.23	– 54.4	– 54.37	– 54.28
3	– 54.1*	– 54.26	– 54.1	– 54.26
4	– 54.12	– 54.07	– 54.25	– 54.15*
5	– 54.15	– 54.05*	– 53.95	– 54.28

* - shows the optimal setting for each corresponding factors as shown by the lowest signal to noise ratio

Table 4 Final assembly sequence result

Fitness value (second)	Running time (second)	Assembly sequence									
344.6	74.521	9	7	22	1	2	3	19	17	4	20
		18	14	8	29	23	12	6	28	27	5
		13	21	10	25	30	24	11	26	15	16

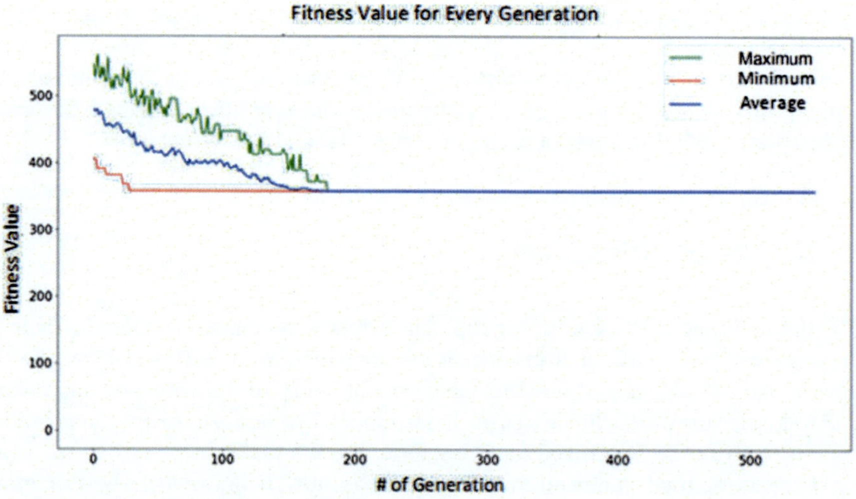

Fig. 1 Performance of GA model

from 6.74 min). The solution will reduce the overall assembly time by 3.3% (29 min from 30 min). Further research can be concentrated on finding an optimal solution for the problem. Since assembly sequence is only generated once for a given product, the tradeoff between time intensive solution methods and the optimal assembly time might be worth pursuing.

References

1. Rashid MFF, Hutabarat W, Tiwari A (2011) A review on assembly sequence planning and assembly line balancing optimisation using soft computing approaches. Int J Adv Manuf Technol 59:335–349
2. Mishra A, Deb S (2016) An intelligent methodology for assembly tools selection and assembly sequence optimization. In: CAD/CAM, robotics and factories of the future. Springer, India, pp 323–333
3. Kucukkoc I, Karaoglan AD, Yaman R (2013) Using response surface design to determine the optimal parameters of genetic algorithm and a case study. Int J Prod Res 51(17):5039–5054
4. Alharbi F, Wang Q (2018) Solving an assembly sequence optimisation problem using the genetic algorithm. In: Proceeding of 2018 international conference on electronics, control, optimization and computer science (ICECOCS), pp 1–5
5. Chen WC, Hsu YY, Hsieh LF, Tai PH (2010) A systematic optimization approach for assembly sequence planning using Taguchi method, DOE, and BPNN. Expert Syst Appl 37(1):716–726
6. Haupt RL, Haupt SE (2004) Practical genetic algoritms, 2nd edn. Wiley, New Jersey
7. Tseng YJ, Kao HT, Huang FY (2010) Integrated assembly and disassembly sequence planning using a GA approach. Int J Prod Res 48(20):5991–6013
8. Choi YK, Lee DM, Cho YB (2009) An approach to multi-criteria assembly sequence planning using genetic algorithms. Int J Adv Manuf 42:180–188
9. Marian RM, Luong LH, Abhary K (2006) A genetic algorithm for the optimisation of assembly sequences. Comput Industr Eng 50(4):503–527

Surface Roughness Analysis on Sheet Metals During AWJ Machining Using Low Pressure

Nazrin Mokhtar, Mebrahitom Asmelash, Kushendarsyah Saptaji, and Azmir Azhari

Abstract The present study discusses the effect of abrasive waterjet machining (AWJ) parameters on surface roughness during cutting of sheets metals of stainless steel 304 and pure copper. A relatively low hydraulic pressure below 150 MPa was used. Several machining parameters were chosen namely pressure, traverse rate, stand-off distance and abrasive mass flowrate. It was found that increasing the pressure leads to an increase in the surface roughness. Varying other parameters did not clearly show any trends on the surface roughness. A higher surface roughness happened during AWJ machining of stainless steel 304 compared to pure copper due to its former higher hardness. The surface roughness at the upper region closer to the top surface is higher than the bottom section regardless of machining parameters. Furthermore, embedded particles are more dominant in pure copper than stainless steels 304. It can be concluded that a low water pressure in AWJ machining process can be used to cut sheet metals successfully with acceptable cutting quality.

Keywords Abrasive waterjet machining · Sheet metals · Surface roughness

1 Introduction

Abrasive waterjet (AWJ) machining process has progressed rapidly and received wide interest amongst researchers in material cutting method [1, 2]. AWJ machining uses a beam of water to increase velocity of the abrasive particles to fast enough to cut through much harder target materials. AWJ technology are used widely in many

N. Mokhtar · M. Asmelash · A. Azhari (✉)
Faculty of Manufacturing and Mechatronic Engineering Technology, Universiti Malaysia Pahang, 26600 Pekan, Pahang, Malaysia
e-mail: azmir@ump.edu.my

K. Saptaji
Department of Mechanical Engineering, Faculty of Engineering and Technology, Sampoerna University, Jakarta, Indonesia

C. N. Rosyidi et al. (eds.), *Proceedings of the 6th Asia Pacific Conference on Manufacturing Systems and 4th International Manufacturing Engineering Conference*, Lecture Notes in Mechanical Engineering,
https://doi.org/10.1007/978-981-99-1245-2_6

fields ranging from automotive, medical and aerospace due to its inherent nature of cold machining process [1]. AWJ is not limited to cutting only but also find it useful in surface treatment, cleaning and coating removal [3, 4].

AWJ machining process is excellent for 2-D machining processes. It is highly promising to cut complex structures or sharp corners of any angles with complicated profiles [5]. Various AWJ processing parameters may affect the quality of cut including hydraulic pressure, traverse rate, abrasive mass flowrate, nozzle size and stand-off distance [6]. The present works aims to examine the effect of several AWJ processing parameters on the surface roughness of very thin sheet metals i.e. stainless steel 304 and pure copper.

2 Methodology

A self-developed low pressure abrasive waterjet machine as shown in Fig. 1a was utilized for the experimentation. A pneumatic-driven liquid pump is used to generate water pressure up to 200 MPa. The machine is numerically controlled by a PC based controller using Mach3Mill CNC software system for the movement of nozzle in three directions (X, Y and Z).

Table 1 lists the processing parameters and their levels. Since, the sheet metals have very low thickness of 1 mm, a relatively small hydraulic pressure below 150 MPa was selected. After the parameters were set accordingly, the workpiece samples were machined with a slot length of 75 mm as illustrated in Fig. 1b.

A surface roughness instrument (Mahr MarSurf PS1) was utilized to assess the quality of the machined surface. The average arithmetic roughness parameters (R_a) was acquired according to EN ISO 4288 standard. It was evaluated over the edge of the machined specimen at three different depths of top, middle and bottom regions which are approximately at the depths of 0.25, 0.5 and 0.75 mm respectively as shown in Fig. 1b. Three different measurements at each region were taken to reduce the variability of surface finish data. Scanning electron microscopy (SEM) was used to evaluate the characteristic of the AWJ machined surfaces.

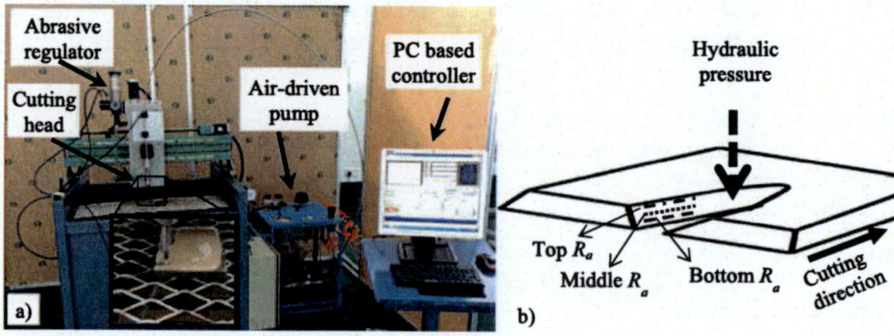

Fig. 1 **a** The self-developed CNC waterjet machine (the inset shows the cutting process), and **b** illustration of AWJ cutting on sheet metals

Table 1 Machining parameters and their levels

No.	Parameters	Levels		
		1	2	3
1	Type of material	Copper	Stainless steel	–
2	Pressure (MPa)	69	103	138
3	Traverse rate (mm/min)	5	10	15
4	Stand-off distance (mm)	3	6	9
5	Abrasive mass flow rate (g/s)	0.15	0.6	0.75

3 Result and Discussion

The effect of material type on the average R_a at different machined depths is shown in Fig. 2a. It can be noticed that in general stainless steel experienced higher surface roughness than the copper. The stainless steel is harder than copper hence it is more resistance to removal of material thus generating a rougher surface [7].

Figure 2b shows the influence of pressure on the average R_a at different machined depths. In general, it can be noted that a higher water pressure leads to an increase in the surface roughness. This is in contrast to previous findings where a higher water pressure resulted in higher abrasives energy thus easily cutting the workpiece causing a smooth surface [8]. However, the range of pressure used in the present study (i.e. 69–139 MPa) is significantly low compared to previous studies above 150 MPa. The workpiece was cut through without experiencing a smoothening action thus producing rough edges [9].

The effect of nozzle traverse speed on the average R_a at different machined depth is shown in Fig. 2c. The surface roughness has shown no significant changes during the machining at different traverse speeds from 5 to 15 mm/min. Theoretically, the increase in the nozzle traverse speed would result in fewer abrasive particle impacting on the cutting target and hence less material would be removed [6]. This would cause rougher surface as traverse speed increases. However, this is not the case in the present study. A probable explanation is due to a small range of traverse speed between 5 to 15 mm/min which is trivial in causing significant changes to the surface roughness. This limited range was chosen to complement the use of low waterjet pressure.

The effect of stand-off distance on the surface roughness are shown in Fig. 2d. The surface roughness shows no clear trend as the stand-off distance varies. It is expected that a higher stand-off distance would increase the jet diameter thus diverging its coherence and reducing its effective kinetic energy [6]. However, this is not the case in the present study which can be attributed to the very thin workpiece of 1 mm. The jet was able to cut through the workpiece smoothly without leaving any significant roughness within the range of stand-off distance.

The effect of the abrasive mass flowrate on the average R_a is presented in Fig. 2e. The surface roughness has shown a slight decrease as the abrasive mass flowrate increases. The increase of abrasive flowrate would increase the penetration energy

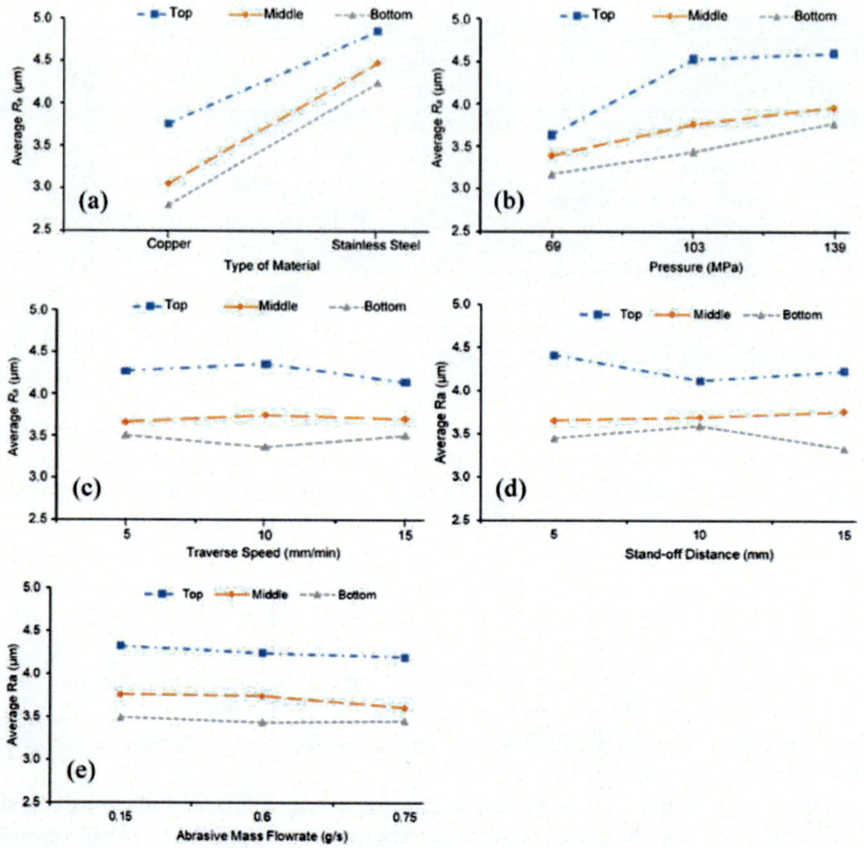

Fig. 2 Effect of **a** type of material, **b** pressure, **c** traverse speed, **d** stand-off distance, and **e** abrasive mass flowrate on the average R_a at different machined depths

[10]. However, beyond a critical value of abrasive mass flowrate may decrease the penetration energy due to an increase in particle fragmentation and interference of the abrasive particle itself within the mixing chamber [6]. In the present case, the amount of abrasive mass flowrates used are very small less than 1 g/s due to the application of low water pressure. As a result, no critical threshold of abrasive mass flowrate was observed thus resulting in very marginal changes to the surface roughness.

Furthermore, it can be noticed from Fig. 2 that the surface roughness at the upper section closer to the top surface is higher than the bottom section regardless of machining parameters. This can be attributed to the fact that the upper section experienced more particle collisions during the jet entry thus roughening the machined surface [11]. Besides, there is a noticeable lack of striation mark observed in the present study due to the very thin specimen of only 1 mm which is less than the depth of the particle decelerating to cause a smooth cut to the work material [12].

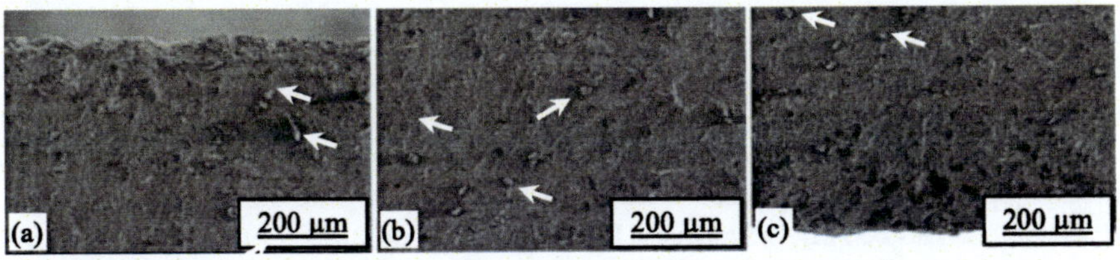

Fig. 3 SEM images of machined surfaces for stainless steel 304 specimen at **a** top, **b** middle and **c** bottom regions (white arrows indicate embedded abrasive particles)

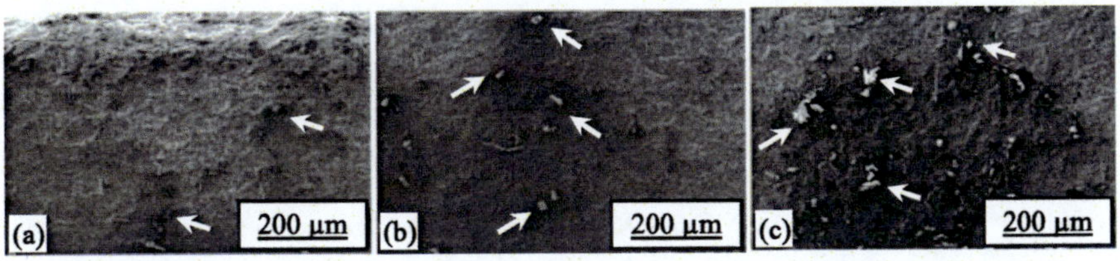

Fig. 4 SEM images of machined surfaces for pure copper specimen at **a** top, **b** middle and **c** bottom regions (white arrows indicate embedded abrasive particles)

Figures 3 and 4 show SEM images for the cutting regions at top, middle and bottom respectively for stainless steel 304 and pure copper specimens. These regions are separated by 0.25 mm depth apart. The stainless steel 304 specimens (see Fig. 3) have shown noticeable fewer and smaller embedded abrasive particle as compared to pure copper (see Fig. 4) which have large embedded abrasive particle. This is caused by the higher ductility and lower hardness of copper thus enabling more and large abrasive particles to be easily embedded into the workpiece material during machining [13]. Also it can be observed that the amount of embedded abrasive particles are noticeably higher from the top to the bottom regions for both specimens. The increasing embedded abrasive particle are caused by the decreasing of abrasive particles velocity and particle colliding which resulted in more embedded abrasive particles in the bottom penetration zone. However, this is more predominant in pure copper, while stainless steel 304 mostly has a consistently sized embedded particle.

4 Conclusion

The effect of abrasive waterjet machining (AWJ) parameters on surface roughness during cutting of stainless steel 304 and pure copper sheet metals was investigated. A low hydraulic pressure below 150 MPa was found sufficient to cut both sheet metals of 1 mm thickness. Pressure has shown a clear effect on the surface roughness with an increasing trend. Other parameters namely traverse rate, standoff distance and abrasive mass flowrate have not shown distinct trends on the surface roughness probably due to the minimal changes in their parametric levels. AWJ machining of stainless steel 304 produced a higher surface roughness compared to pure copper due to its former higher hardness. The surface roughness at the upper section closer to the top surface is higher than the bottom section regardless of machining parameters. Furthermore, embedded particles are more dominant in pure copper than stainless steels 304. It can be concluded that a low water pressure in AWJ machining process can be used to cut sheet metals successfully with acceptable cutting quality.

Acknowledgements Authors would like to gratefully acknowledge the financial support from the Ministry of Higher Education Malaysia through FRGS/1/2019/TK03/UMP/02/25 (RDU1901161).

References

1. Llanto JM, Tolouei-Rad M, Vafadar A, Aamir M (2021) Recent progress trend on abrasive waterjet cutting of metallic materials: a review. Appl Sci 11:3344
2. Ting HY, Asmelash M, Azhari A, Alemu T, Saptaji K (2022) Prediction of surface roughness of titanium alloy in abrasive waterjet machining process. Int J Interact Design Manuf (IJIDeM) 16:281–289
3. Qutaba S, Asmelash M, Saptaji K, Azhari A (2022) A review on peening processes and its effect on surfaces. Int J Adv Manuf Technol 120:4233–4270
4. Alzaghir AF, Nawi MNM, Gebremariam MA, Azhari A (2022) Optimization of Waterjet paint removal operation using artificial neural network. In: Enabling industry 4.0 through advances in mechatronics. Springer, pp 11–20
5. Folkes J (2009) Waterjet—an innovative tool for manufacturing. J Mater Process Technol 209:6181–6189
6. Azmir MA, Ahsan AK (2009) A study of abrasive water jet machining process on glass/epoxy composite laminate. J Mater Process Technol 209:6168–6173
7. Murugan M, Gebremariam MA, Hamedon Z, Azhari A (2018) Performance analysis of abrasive waterjet machining process at low pressure. In: IOP conference series: materials science and engineering. IOP Publishing, p 12051
8. Hashish M, South A (1991) Optimization factors in abrasive—Waterjet machining. J Eng Indust 1
9. Azhari A, Schindler C, Kerscher E, Grad P (2012) Improving surface hardness of austenitic stainless steel using waterjet peening process. Int J Adv Manuf Technol 63:1035–1046
10. Azmir MA, Ahsan AK (2008) Investigation on glass/epoxy composite surfaces machined by abrasive water jet machining. J Mater Process Technol 198:122–128
11. Mazlan NK, Mokhtar N, Gebremariam MA, Azhari A (2022) Prediction of Abrasive Waterjet machining of sheet metals using artificial neural network. In: Enabling Industry 4.0 through advances in mechatronics. Springer, Berlin, pp 43–50
12. Chen FL, Siores E (2001) The effect of cutting jet variation on striation formation in abrasive water jet cutting. Int J Mach Tools Manuf 41:1479–1486
13. Hashish M (1995) Material properties in abrasive-waterjet machining. J Manuf Sci E T ASME 117:578–583

Manufacturing System

Design of Intervention for Operator in the Inspection Department of PT. XY Using Pahl and Beitz Method

Rahmaniyah Dwi Astuti, Bambang Suhardi, Pringgo Widyo Laksono, Amirul Wika Rosida, and Muhammad Junus Atthariq

Abstract Material handling in the textile industry often involves transporting materials such as rolls of cloth in large quantities and sizes and also has a high frequency of repetition. The use of available interventions causes poor working posture for the operator, and often the transported material is dropped in the process of transportation and inspection. To overcome this problem, material handling tools are needed to transport materials safely, have a large carrying capacity to reduce the frequency of repetition, and can be used as a crutch in the inspection process. This study designed the tool's design using the Pahl and Beitz method with an ergonomic approach. The intervention is made in the form of trolleys to assist the fabric transportation process and support during the inspection process. The results show that the tool's design successfully meets the safety factor value through the finite element analysis validation method and was able to reduce the REBA score of the operator's working posture from 13 to 7 when implemented.

Keywords Intervention design · Pahl and Beitz methods · Finite element analysis (FEA) · Safety factor · REBA

1 Introduction

PT XY is one of the manufacturing companies in Central Java that is engaged in textiles or fabric manufacture. Most of the material transportation activities carried out at PT XY are still carried out manually, one of which is the transfer of cloth rollers from the Production Department to the Inspection Department. The choice of humans over machines in carrying out material handling activities is because operators are more flexible in responding to unexpected system changes during production activities. In addition, operators are also more flexible in the division of tasks because

R. D. Astuti (✉) · B. Suhardi · P. W. Laksono · A. W. Rosida · M. J. Atthariq
Industrial Engineering Departement, Universitas Sebelas Maret, 57126 Surakarta, Indonesia
e-mail: rahmaniyahdwi@staff.uns.ac.id

C. N. Rosyidi et al. (eds.), *Proceedings of the 6th Asia Pacific Conference on Manufacturing Systems and 4th International Manufacturing Engineering Conference*, Lecture Notes in Mechanical Engineering,
https://doi.org/10.1007/978-981-99-1245-2_7

it is possible to carry out several functions at once. The activity of moving the cloth rollers aims to ensure that the fabric can be inspected to ensure that the finished product that reaches the customer is a product of good quality.

The activity of moving cloth rollers from the Production Department to the Inspection Department is carried out using a trolley with a capacity of 1 cloth roller. The average weight of a rolled cloth roller is 35 kg. Laying the cloth roller to the inspection table sometimes cannot be done directly. Cloth rollers should be placed on temporary storage racks or placed on the floor in advance to await inspection. Cloth rollers placed on the floor will get dirty easily and can reduce the quality of the fabric. In the Inspection Department itself, it is not possible to add more temporary shelves due to the narrow area. Therefore we need a storage solution that can keep the cloth rollers while waiting for the inspection turn.

If the cloth roller on the inspection table has been inspected, the operator will replace the cloth roller. In each work shift, an operator from the Inspection Department must manually transfer cloth rollers to the inspection table as many as 80–100 cloth rollers, so the frequency of activities carried out by operators is quite high. The more the number of repetitions of the activity performed, the higher the chance for errors or musculoskeletal injuries to the operator [1].

From the existing problems, a solution is needed to solve the problem in material handling activities, in this case, the cloth roller transfer activity. Efforts that can be made by companies to improve performance in material handling activities are by providing training and also providing appropriate work facilities [2]. Therefore, the manufacture of interventions is needed as a means that can reduce the frequency of manual material handling activities carried out by operators as well as a tool that can be used as a support for cloth rollers during inspection activities.

Many approaches can be chosen in the process of designing a tool, one of which is an ergonomic approach because this approach can create tools that are suitable and comfortable for the operator. The ergonomics approach can be done in various ways, one of which is through observation, assessment of work posture, and also the results of interviews with operators and parties involved in a production process. From the results of the work posture assessment that has been carried out, it can be seen that the work posture in material moving activities in the form of cloth rollers at PT XY is still not ergonomic because the overall work posture results in a REBA rating above 11 which mean that the work posture has a very high level of risk, so the level of risk is very high. action on a scale of 4 which means that improvements must be made at the same time to the existing conditions.

The product design method is each procedure, technique, and certain tool that represents a certain number of activities used by the designer in the design process [3]. Product design requires a method that is useful for helping the thinking process and correcting errors that may be caused during the design process. The method used in this study is the rational method developed by Pahl and Beitz. The rational method is a design method that emphasizes a systematic approach. The Pahl and Beitz method is one method that can be used in assisting the product design process. By using the Pahl and Beitz manufacturing methods, the tools created are following the wishes and needs of the operators who use them. Therefore, the manufacture of tools in this study will be carried out based on the steps in the Pahl and Beitz method.

The design of the tool requires validation to prove the feasibility of the product. The tool design validation will use Finite Element Analysis (FEA), where FEA is a simulation carried out to determine technical phenomena that can occur in a product. Simulations can be carried out using Solidworks software which provides a simulation package for finite element analysis [4]. FEA solves problems in an engineering design by dividing the object of analysis into several parts with a certain amount (finite) to perform load testing. The output that will be produced from this FEA is von mises and safety factors.

2 Research Methodology

This section describes in detail the stages of the research carried out. The stages in making the design of this tool generally consist of the initial identification stage and the design stage using the Pahl and Beitz method. Figure 1 shows the flow of the research process.

This research begins with conducting a field study to find out the existing problems. Then a literature study was conducted to solve the problem. Next is to identify existing problems to be the basis for the planning stage and task explanation. The result of the planning stage and task description is the design requirement or design requirements for the manufacture of tools. Next is the planning stage and task description to determine the demand and wish of the design requirements that will produce product design specifications. Next is the product concept planning stage to define demand and wish in the design requirements to get the overall product function and also the sub-functions of the product to be made so that an overview of the components of the tool will be obtained.

At the product concept stage, the solution principle is also determined, where in this study, the solution principle needed is the problem of the choice of material for the tool making which will be described in a morphological chart where each solution principle will be combined and produce several alternative designs. At the product design stage, the three alternatives are made with 3D drawings to produce an overview of the design as consideration for the assessment of existing design alternatives. The best design alternative will be selected and used as a guide at the product detail design stage. The product detail design results are bill of materials, detailed dimensions of the tools, and also the total cost estimate.

Product validation using finite element analysis will be carried out on the design that has been made as evidence that the design is feasible and safe to be applied to the system. In the end, conclusions will be drawn based on the research that has been done.

Fig. 1 Research flowchart

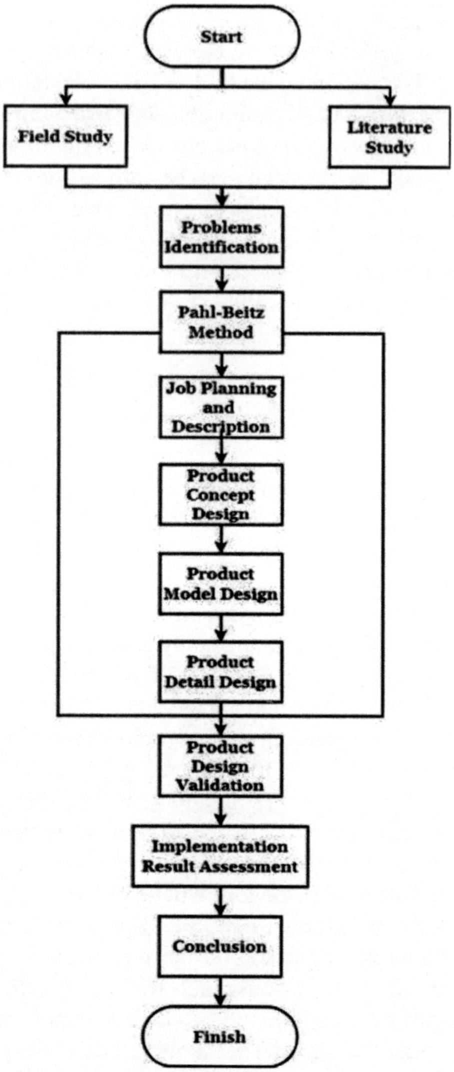

3 Results and Discussion

The results of this study contain the identification of problems and also the results of each stage of the design process using the Pahl and Beitz method.

Table 1 Operator tasks and complaints

No.	Tasks	Complaints
1	The operator from the Inspection Department must move ± 80 cloth rollers manually in each work shift	Often experience pain in the shoulders and knees because of having to do the manual transfer of the cloth roller
2	The operator brings 1 cloth roller for each cloth roller transfer activity	Operators have to commute frequently to perform roller shifting activities (high cloth roller shifting frequency) and feel fatigued from too much walking and carrying heavy loads
3	The operator must keep the cloth roller in a balanced position during the cloth roller transfer process to prevent the cloth roller from falling	The operator must often bend down every time he pushes the cloth roller moving tool trolley because the size of the tool trolley used today is smaller than the cloth roller dimension
4	The operator must replace the cloth on the inspection table when the cloth roll runs out and at the same time transfer the cloth roll that has been inspected to the folding section	The operator often feels rushed because there are 20 inspection desks in the Inspection Department

3.1 *Problem Identification*

Problem identification is done to explain the problems that exist in existing conditions. This stage is carried out by describing the operator's complaints on the results of the Nordic Body Map questionnaire recap. To obtain further details of the complaint, informal interviews were conducted with 2 operators, the Head of the Inspection Department, and also a leader from the Production Department. The information obtained is in the form of complaints felt while working and hopes for improvements to the process of transporting cloth rollers from the Production Department to the Inspection Department. The results of the interviews that have been conducted is described in Table 1.

3.2 *Planning Phase and Task Explanation*

This stage is used to determine the demand and wish for the design needs. The results of the interview at the previous stage are then called the design requirements. From the design requirements that have been collected, the objectives and sub-objectives are determined, as well as the relationship between the two in the manufacture of fabric roller transfer interventions. The results of these interviews are then used as material for consideration in the preparation of technical specifications according to needs. Technical specifications are prepared in a participatory approach with users related to cloth roller intervention products. The users in question are operators as direct users of the product, the Head of the Inspection Department, and the Head of

Table 2 Design technical specifications

Parameter	Demand (D)	Wishes (W)	Technical specifications
Ergonomics	√		Can reduce the frequency of manual material handling
		√	Can reduce slouching working posture
		√	Dimensions of the tool according to the operator's anthropometry
Energy	√		Using kinetic energy (push/pull)
Operational	√		Easy to operate
	√		Operation is manual
		√	Using 4 wheels
	√		Safe for operators
		√	Safe for the material (cloth roller) that is carried
Manufacturing	√		Simple construction/components
		√	Strong and durable material
	√		Easy to use
	√		Affordable cost
Maintenance		√	Easy to maintain
Capacity	√		Can carry more cloth rollers

the Production Department as the person in charge of production activities in related departments. Design technical specifications can be seen in Table 2.

3.3 *Product Concept Design Stage*

Product concept design is carried out to obtain a product concept that can meet the requirements set out in the technical specifications at the previous stage. At this stage, the product function structure is determined, the solution principle is determined and the product alternative is determined.

From the overall product function (as seen in Fig. 2) that has been created, then the sub-functions of the product are formulated. To determine the sub-functions of the product, it is necessary again to describe the solution concept to meet the design requirements based on the parameters in the technical specifications. The sub-function is a derivative of the main function which will be a component of the tool. To realize the main function of the auxiliary tool as a tool for fabric roller transfer activities and inspection activities, the sub-functions of the auxiliary equipment are determined in Fig. 3.

From these sub-functions, the main components of the tools will be made, that can be seen in Table 3.

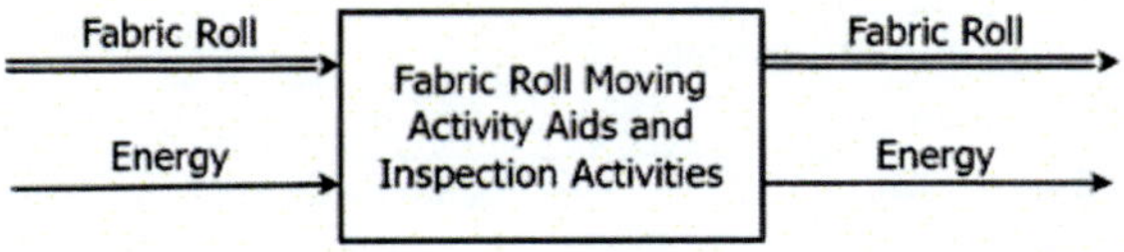

Fig. 2 Overall product function diagram

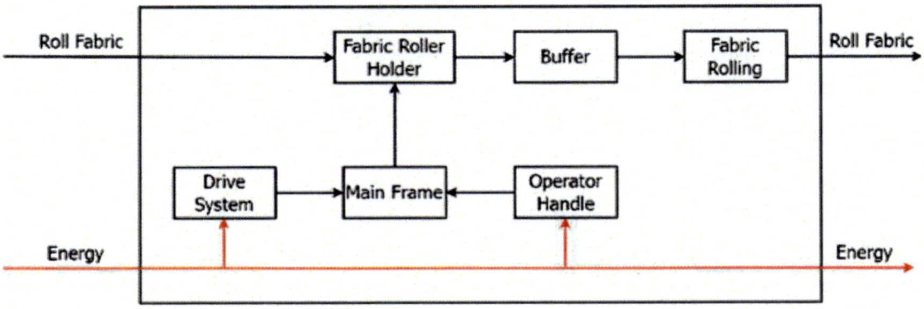

Fig. 3 Product sub-functions diagram

From the existing components, alternative solutions are needed to choose the material to be used. There are candidate materials that match the needs of the tools. Some of these candidate materials will then be combined as a solution principle to solve problems related to materials for the manufacture of tools, that can be see in Fig. 4.

From the principle of the existing solution then a morphological chart is made (as seen in Fig. 5). Table 4 shows the combination of solution principles.

Obtained from the morphological chart. Three alternative concepts were obtained and divided based on three colors, namely blue for design concept 1, green for design concept 2, and red for design concept 3.

The mathematical form of the three concepts is as follows,

a. Alternative 1 (blue): S11 + S22 + S32 + S41
b. Alternative 2 (green): S12 + S21 + S32 + S42
c. Alternative 3 (red): S13 + S21 + S32 + S43

3.4 *Product Design Stage*

At this stage, product design and product design assessment are carried out so that the best design will be produced. There are 3 alternative solutions as follows.

The 1st design alternative (as seen in Fig. 6) is made using H Beam alloy steel. H Beam will be strong enough to support a fairly heavy load. The handle that is used as the operator's grip when carrying out pushing activities on transfers is made of foam coated. This of course increases the comfort of the operator. The wheel material uses

Table 3 The concept for main components

Parameter	Design concept	Sub function description	Component
Ergonomic	Able to carry more cloth rollers	There is a holder for cloth rollers	Holder
	Able to maintain the position of the cloth roller	There is a cloth roller safety	Holder
	Has handling to keep the operator in an upright position	There is an ergonomic handling	Handling
	Dimensions of the tool according to anthropometric measurements	Adjusting the anthropometric size	Frame
Energy	Makes it easy for the operator to move the cloth roller	Light raw material	Frame
Operating	Using strong wheels to support the load	There is space for 4 wheels measuring 6 cm	Wheel housing
	Easy to use on available transportation routes	Size adjusts to conditions	Frame
	Fit to enter the inspection desk area		
	Can be used as a storage rack during the inspection process	It has a rolling pipe and the cloth roller can rotate well	Rolling pipe
	Able to keep the position of the cloth in place	Have safety element	Holder
	Keep the cloth clean	Safe material	Holder and rolling pipe
Manufacturing	Easy to implement	Not complicated	Main frame
	Raw materials are easy to get	Alloy steel	All components
Maintenance	Easy to clean and replace certain components	Connection of components with bolts and nuts	Holder and wheel
	Stainless raw material	Alloy steel	Holder
Capacity	Capable of accommodating multiple fabric rollers	There are 6 holders	Holder
	Strong enough to hold a lot of fabric	Keeping the frame upright	Cantilever

Fig. 4 Solution principle

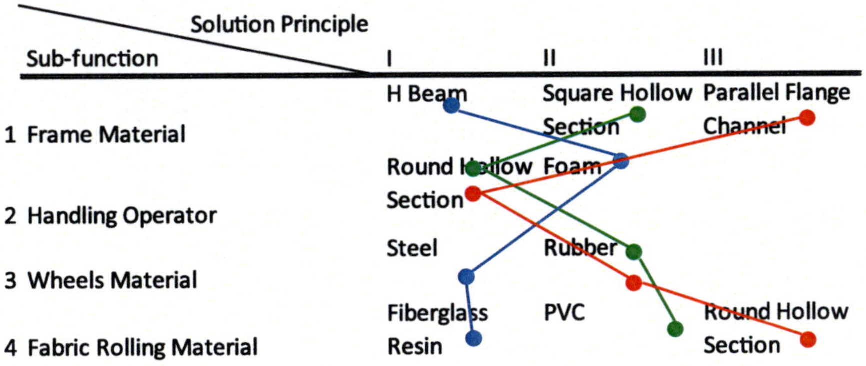

Fig. 5 Solution principles in the morphological chart

Table 4 Combination of solution principles

Solution principle / Sub-function	I	II	III
1	S11	S12	S13
2	S21	S22	
3	S31	S32	
4	S41	S42	S43

steel, which can withstand loads of up to 1000 kg so that the tool will be strong to support the load. The material used for rolling cloth is fiberglass reinforced epoxy resin, where the resin is a fiberglass material that is durable and corrosion resistant, but still safe for workers and the fabric being inspected. Table 5 shows the advantages and disadvantages of the first alternative design.

The second alternative design (as seen in Fig. 7) is made with the main building material using a Square Hollow Section (SHS). The square hollow section used is mixed steel whose structure is hollow so that it can withstand loads but is still light. The handling operator uses a round hollow section following the components that make up the frame. The wheel material uses rubber so that the surface of the factory floor is safe, although repairs must be made periodically to maintain the performance of the tool. Fabric rolling material uses a round hollow section that is covered on the

Fig. 6 Design of alternative 1

Table 5 Advantages and disadvantages of alternative 1

Advantages	Disadvantages
Using H Beam alloy steel so that it can support heavy cloth roller loads	The frame becomes an additional burden when the intervention is used by the operator
Using steel wheels so that it can bear heavy loads	Higher manufacturing and maintenance costs
Using a safe and durable fiberglass resin material	Steel wheels can damage the floor surface in the Inspection Department

outside using a PVC pipe. Table 6 shows the advantages and disadvantages of the second alternative design.

The third alternative design (as seen in Fig. 8) is made using a parallel flange channel or C-Channel. In the high-pressure test when using a double structure (double channel) it is proven to be stronger than SHS steel (square hollow section). It's just that the process of making these tools it takes longer because the preparation is more complicated. Manual tool drive system. Handling operators are made using a round hollow section. The wheel material uses rubber wheels to keep the floor surface safe. While the rolling cloth material uses PVC material. Making it will be quite complicated because you have to pay attention to the constituent parts that are less stable. Table 7 shows the advantages and disadvantages of the third alternative design,

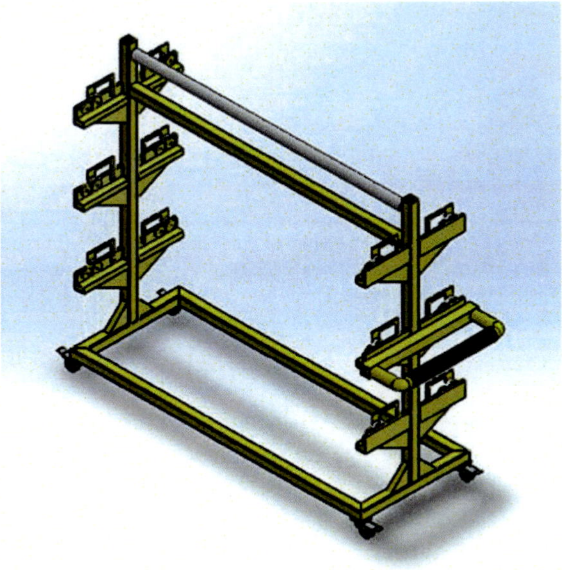

Fig. 7 Design of alternative 2

Table 6 Advantages and disadvantages of alternative 2

Advantages	Disadvantages
Using a square hollow section alloy steel so that it is sufficient to support heavy loads	Needs regular maintenance to ensure the tool's performance remains good
Driven manually so it can save costs	
Using rubber wheels, so it is safer for the floor surface	
Using cheap and safe PVC pipe for the inspected fabric	

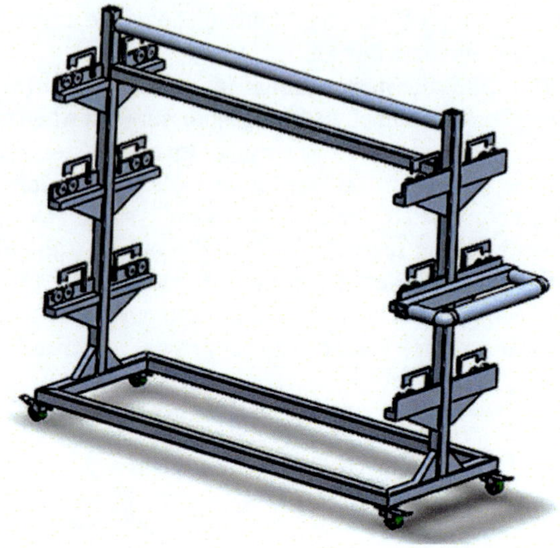

Fig. 8 Design of alternative 3

Table 7 Advantages and disadvantages of alternative 3

Advantages	Disadvantages
Using C-channel alloy steel which is able to support the load better than SHS. steel	The frame becomes an additional burden because the C-channel must be doubled to make it stronger
Lower manufacturing cost	Making tool components is quite complicated and requires special skills
Using rubber wheels so it is safe for floor surfaces	
Using cheap and safe PVC pipe for the inspected fabric	

From the three alternative designs, a product assessment is then carried out. A product design assessment is carried out to get the best design among the three design concepts that have been made previously. The assessment is done by comparing the three product design concepts with several selection criteria using a decision matrix. The weighting of the criteria as well as the assessment of the alternative tools are carried out based on the average weighting value carried out by the Head of the Production Department, the Head of the Inspection Department, and the Head of the Workshop Department. Table 8 shows the result of the design concept assessment table obtained from the average value given.

The total rating is then sorted, starting from the highest rating to the lowest score, that can be seen in Table 9.

From the results of the concept assessment, the best concept is alternative design concept 2. So that product design will use alternative design 2.

Table 8 Alternative solutions' assessment result

No.	Criteria	Weight	Alternative solutions		
			Alt. 1	Alt. 2	Alt. 3
1	Can reduce the frequency of manual material handling	13.67	123	123	123
2	Dimensions of the tool according to the operator's anthropometry	7.33	66	66	66
3	Using kinetic energy (push/pull)	5.67	45.33	45.33	45..33
4	Easy to operate and manual	10.67	96	89.33	96
5	Using 4 wheels	7.67	69	69	69
6	Safe for the operator as well as for the material (cloth roller) being carried	12.00	84	96	96
7	Construction/components are simple and easy to work with	8.67	52	59.33	43.33
8	Strong and durable material	11.67	105	97.67	105
9	Affordable cost	14.00	84	107.3	98
10	Easy to maintain	8.67	60.67	72.67	60.67
Total		100	78.5	82.57	80.2

Table 9 Design concept ranking

Rank	Design concept
1	Alternative 2
2	Alternative 1
3	Alternative 3

3.5 *Product Detail Design Stage*

Product detail design is carried out to obtain product details obtained from anthropometric data and fabric roller specification data to obtain the dimensions of the tool, determine the specifications of the tool and estimate the cost of making the product,

Before making product specifications, first, a bill of material is made to get an idea of the parts that make up the tool. The complete bill of materials can be seen in Fig. 7.

Dimensions or sizes of product designs are made based on fabric roller specifications and anthropometric data. Anthropometric data was carried out to obtain the dimensions of the ergonomic tools. Anthropometric data were obtained from the website database of the Indonesian Ergonomics Association. Anthropometric data used are anthropometric data for the male gender of Javanese ethnicity, with productive age ranging from 25 to 40 years. Anthropometric data retrieval does not use worker anthropometric data because the level of workers who leave work (turnover) of employees, especially production operators at PT XY is quite high. With the

Table 10 Tool dimensions

Component	Size	Terms
Holder	Whole width = 25.55 cm Wheel distance = 20 cm Locker length = 14,465 cm locker height = 10.5 cm	Fabric roller dimensions
Wheel	Minimum diameter = 5 cm Bolt = M10 Dimensions of wheel housing = customize	Dimensions of the frame and able to accommodate the cumulative load of cloth rollers
Frame	Maximum height = 158.5 cm Height of part stand frame = 135 cm 3rd fabric roller height = 143.2 cm 2nd cloth roller height = 97.8 cm Tool width = 55 cm chamfer angle = 45° Pipe Length = 160 cm Diameter = 40 mm	Fabric roller dimensions, anthropometric data dimensions of eye height, shoulder width, standing shoulder height and standing elbow height
Handle	Diameter = 50 cm	Dimensions of hand length, dimensions of the frame
Rolling pipe	Length = 150 cm Inner roller diameter = 30 mm PVC roller diameter = 1.5 in	Frame dimensions
Cantilever	Height = 10 cm Corner = 45°	Frame dimensions

dimensions of the tool being more universal, it is hoped that the tool can be used in the company for a long period.

Table 10 shows the dimensions of the designed products.

Figure 9 shows the final dimensions of the intervention.

Based on Table 11, the estimated cost of making a cloth moving intervention product is Rp.1,855,100.

3.6 Product Design Validation

Product validation is carried out to provide evidence that the design of the tool made can safely support the load of the fabric and the realization of the design can be carried out by making the tool. Product validation is carried out by simulating finite load loading on the main load-bearing components using the Solidworks application. Giving the load will show the value of the strain that occurs in the part of the tool that experiences the most load and will be used as material for calculating the value of the safety factor. Tests were carried out using the Solidworks application on 3

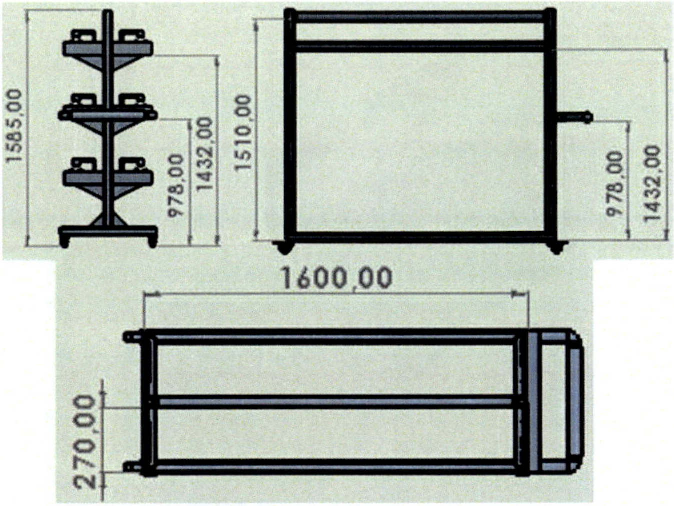

Fig. 9 Overall 3d intervention

Table 11 Cost estimation

Material	Specification	Price	Quantity	Total
Square hollow	4 cm × 4 cm × 0.3 cm × 60 cm	Rp.184.000/rod	3 rods	Rp.552.000
Round hollow	3 cm × 3 cm × 0.15 cm × 60 cm	Rp.92.000/rod	1 rod	Rp.92.000
Mild steel	1 cm × 30 cm × 20 cm	Rp.69.000/sheet	6 sheets	Rp.414.000
PVC	1.5 in × 1 m	Rp.45.000/meter	1.5 m	Rp.67.500
ST 60	Round bar 40 mm	Rp.1.900/pcs	4 pcs	Rp.7.600
Wheel	4 in (2 hrem + 2 die)	Rp.200.000/set	1 set	Rp.200.000
Hex-	M10 X 22	Rp.1.100/pcs	20 pcs	Rp.22.000
Bolt + Hexnut	Gloss Doff finishing	Rp.60.000/tin	1 tin	Rp.60.000
Labor cost				Rp.500.000
Total cost estimate				Rp.1.855.100

main components, namely the holder component, main frame, and wheel housing component.

The test results of testing at the place of placing the cloth roller (holder) with a load of 40 kg (maximum weight of cloth roller) can be seen in Fig. 10.

Based on the test results, the von Mises stress on the support frame is 210 MPa. The value of von Mises is smaller than the tensile strength value of the mixed steel material of 590 MPa so that the support frame can be intervened to be safe. However, the von Mises stress on the roller has a von Mises value that exceeds the tensile strength value, which is 675 MPa. This is not a big problem as it can still be solved with regular replacement of the rollers in the fabric holder.

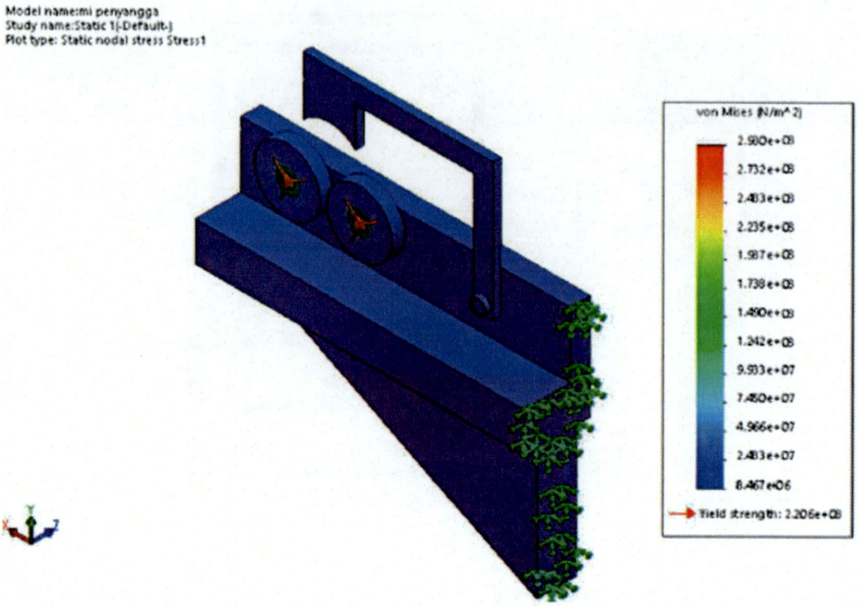

Fig. 10 Von Mises test results for holder components

Tests were carried out on the base frame with a load of 6 × 40 kg (maximum total weight capacity). So the test results can be seen in Fig. 11.

Based on the test results, the von Mises stress on the support frame is 435 MPa. The value of von Mises is smaller than the yield strength value of the iron material of 590 MPa so that the support frame can be intervention to be safe.

The test is carried out on the wheel housing which is useful as a house/place for placing the wheels, with a load of 6 × 40 kg (maximum weight capacity). Because there are 4 similar parts, the load value for each part is 60 kg with the following test. The test results can be seen in Fig. 12.

Based on the test results, the von Mises stress on the supporting wheel housing is 452 MPa. The value of von Mises is smaller than the yield strength value of the iron material of 590 MPa so the wheel housing for the support can be intervention to be safe.

The safety factor value is obtained from the displacement value divided by the von Mises value. The formula for calculating the safety factor is as follows,

$$safety\ factor = \frac{\sigma\ yield}{\sigma\ von\ mises}$$

The construction of a model is intervention to be unsafe if it has a safety factor value of < 1. From the results of testing on the load carried out on the main load

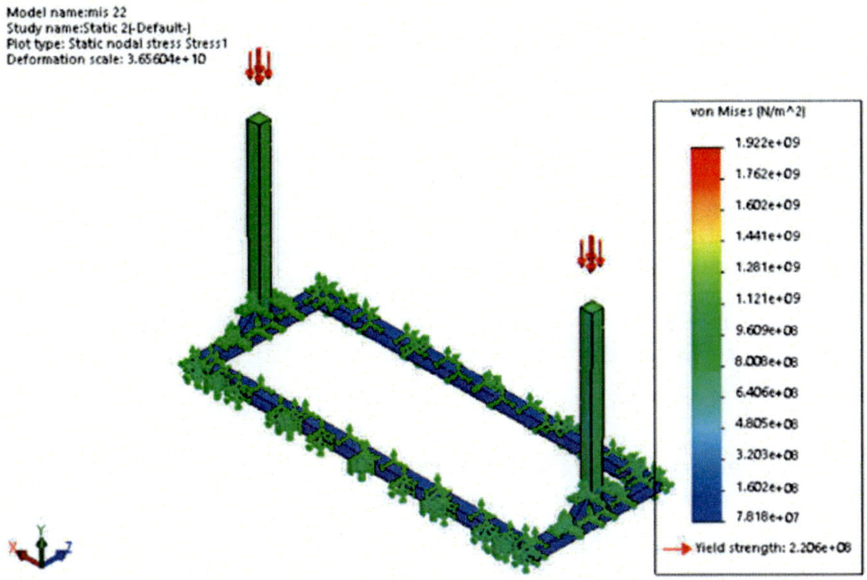

Fig. 11 Von Mises test results of frame components

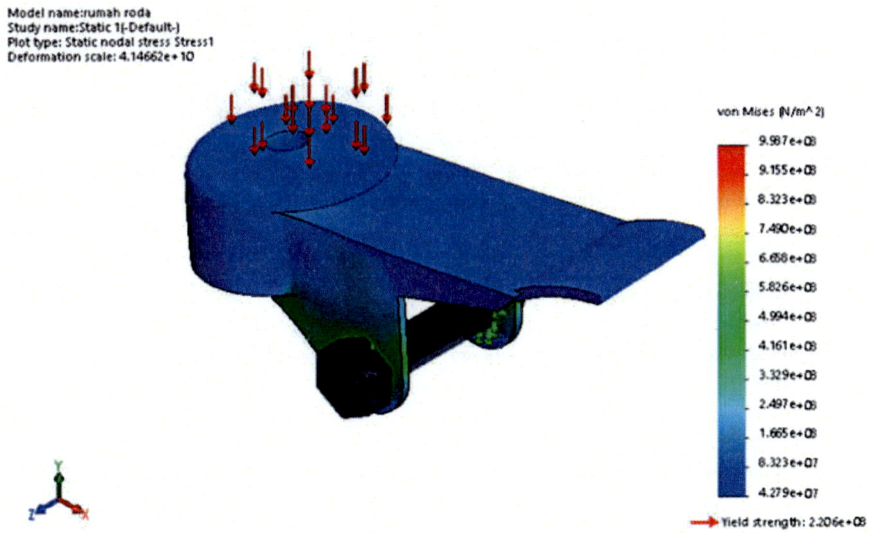

Fig. 12 Von Mises test results of *wheel housing* components

Table 12 REBA assessment result

	Before implementation			After implementation
	Posture 1	Posture 2	Posture 3	
Group A	10	8	6	4
Group B	6	7	8	5
Activity score	2	1	2	2
Final score	13	11	11	7

supporting components, the safety factor value is 8.844; 1377, and 2650 so the cloth roller moving activity intervention is safe to use to carry the cloth roller load.

3.7 Implementation Result Assessment

The selected product design is then created and used in the fabric roll transport process in the Inspection Department. The assessment process of the implementation is carried out using an ergonomic approach using the Rapid Entire Body Assessment (REBA). At this stage, the operator's work posture before the implementation of the tool is assessed and compared with the work posture of the operator after the implementation of the intervention. Table 12 displays the REBA assessment result on conditions before and after implementation.

In the condition before the intervention was implemented, there were 3 working postures observed, namely when the process of loading the fabric rolls into the trolley, transportation of cloth rolls, and inspection. REBA scores for the three processes were 13, 11, and 11. The REBA scores were categorized as very high-risk levels. When the intervention was carried out there was 1 work posture that was assessed, namely loading a roll of cloth into the intervention. This work posture has a REBA score of 7 and is included in the medium risk category. From these results, it is known that there is a decrease in the REBA score from 13 to 7. This result shows an improvement in the operator's work posture in the ergonomics factor.

4 Conclusion

The intervention is made using the Pahl and Beitz method with 4 main stages, namely the task explanation stage, and the product detail design stage. The final result of the design is detailed containing information on the tools that are made based on the alternative concept 2. The tool has 6 main components, namely holder, wheel, frame, handle, rolling pipe, and cantilever. The main materials used in the manufacture of tools are square hollow sections and mild steel. The estimated total cost of making tools consisting of the total cost of materials and labor costs is Rp.1,855,100. From the results of the validation that has been done, the tool has met the safety factor value so it is safe to use in the material transportation process. The tool is implemented and assessed by comparing the operator's work posture using the REBA method. REBA scores before implementation on 3 work postures were 13, 11, and 11. After the implementation of the tool, REBA scores were 7.

References

1. Metev SM, Veiko VP (1998) Laser assisted microtechnology, 2nd edn. Springer, Berlin
2. Breckling J (ed) (1989)The analysis of directional time series: applications to wind speed and direction. Lecture notes in statistics. Springer, Berlin, vol 61
3. Zhang S, Zhu C, Sin JKO, Mok PKT (1999) A novel ultrathin elevated channel low-temperature poly-Si TFT. IEEE Electron Device Lett 20:569–571
4. Wegmuller M, von der Weid JP, Oberson P, Gisin N (2000) High-resolution fiber distributed measurements with coherent OFDR. InProceedings of ECOC'00, paper 11.3.4, p 109

Supply Chain Network Design for e-Groceries Using Clustering and Linear Programming

A. D. P. Prady, G. M. A. Alaudin, T. P. Achmad, and R. T. Cahyono

Abstract This research aims to provide the best strategic supply chain network design decisions for e-groceries to accommodate growing demand at the minimum cost. The decisions include supply chain network configuration, facility location, and facility capacity at each year in the period of analysis. We use combination approach of k-means clustering and linear programming to cater big dataset of historical demand and number of customers. We use a real case from an Indonesian e-grocery provider. A computer-based mathematical model is built using CPLEX Python API. By optimizing the program, this study concludes the recommendation for the company is to operate facilities at certain capacity level for year 2023–2025, along with the service configuration.

Keywords Clustering · Facility location · Linear programming · Supply chain network design

1 Introduction

Recent study stated that companies classified as e-commerce consumer-packaged-goods (CPG) companies has high growth and low profit nature of business. Margins gained from selling CPG are low, but larger business scale will help with this issue. Another strategy being recommended is to lower the supply chain costs of the company. The ratio for shipping-warehousing costs to profit of a product is up to 15%. CPG companies need to evaluate the end-to-end supply chain infrastructure and planning. Therefore, it is important to maintain business' economies of scale while keeping the supply chain costs at minimum [1].

A. D. P. Prady · G. M. A. Alaudin · T. P. Achmad · R. T. Cahyono (✉)
Department of Industrial Engineering, Institut Teknologi Bandung, Jalan Ganesha 10, Bandung 40132, Indonesia
e-mail: r.t.cahyono@itb.ac.id

C. N. Rosyidi et al. (eds.), *Proceedings of the 6th Asia Pacific Conference on Manufacturing Systems and 4th International Manufacturing Engineering Conference*, Lecture Notes in Mechanical Engineering,
https://doi.org/10.1007/978-981-99-1245-2_8

If all demand could be served, the revenue would increase. In order to thrive as a profitable company, their goal is to maximize profit from their received revenue. It is essential for them to minimize their total expenses as companies. This study focuses on the minimization of strategic supply chain costs where their core business processes of production are extremely important. A typical of e-groceries strategic supply chain set up includes the long-term decisions such as investment of facility, both location and capacity, and their service configuration.

In supply chain operations, mathematical modelling can be implemented at any level (strategic, tactical, or operational). The use of mathematical model aims to optimize supply chain. The attention towards its' usage has been increasing, mainly due to their lower cost and greater capability to solve a particular problem [2]. A flexible mathematical SCND model for deterministic, multi-commodity, single-period context has been developed to be used as reference model [3]. The model is a mixed integer linear programming (MILP) problem.

A study proposed a multi-period SCND with multi-products [4]. The study took place in a snacks company in Turkey that aspires to aligning SCND decisions with market growth. The decisions, which spanned over the next few years, are the operational timing of the plants, initial capacity and location of a new plant, the timing and amount of additional capacity to be installed at the new and existing plants, the assignment of demand points to plants and the amount of raw material to be shipped from suppliers to the plants in each period. The objective is to minimize the total cost of the decisions.

Another model is developed to design a sustainable, single product, multi-period, three-echelon supply chain network [5]. This model addresses certain sustainable aspects of perishable product, that is shorter product-age as the value of product deteriorated by the passing of time. The model is formulated using mixed-integer nonlinear programming and solved by non-linear Generalized Reduced Gradient solver in Microsoft Excel.

This research aims to provide the best strategic SCND decisions for generic e-groceries to accommodate growing demand for next few years at minimum cost. The decisions include supply chain network configuration, facility location, and facility capacity at each year in the period of analysis. The motivation of this research is mainly to incorporate the simultaneous approach in seeking best location of new facilities by combining big-datasets of customers' location (demand points) with LP modeling.

2 Modeling

2.1 Linear Programming Models

The first part of the models is the linear programming models used to design the flow of goods among nodes in the supply chain network. The relevant systems is is the SCND with regards to the decision of locating, capacity allocation, and determining service scheme of the MFCs and hubs. Performance measures are determined by the relevance of costs component in supply chain network design decisions. The irrelevant cost components are storage, production, and worker costs. These components are dependent to only the demand. As the demand is excluded in the narrow system of interest, demand acts as a parameter of the problem. Regardless of the decisions, the need for storage, production, and worker costs would remain the same since the demand also is. Thus, these costs are irrelevant to the problem and excluded from the model. The relevant cost components are the cost of facility rent costs and transportation costs between facilities. The first cost components are dependent on the facility location and capacity. The latter is dependent on the service configuration between facilities, as it will impacts on the distance between the nodes. The total relevant cost, which serves as objective function is the sum of facility rent costs and transportation costs. The main decision variables is the flow of goods between nodes.

The models notations are provided in Table 1. The mathematical model developed in this study is a single-objective, multi-product, multi-period, and multi-echelon mixed integer linear programming. The models are as follows. The decision variables are X_{ij}^{ft} which is number of final product f order transported from MFC i to customer k at time-period t, and we have the other transportation flows where j and k represent hubs and customers, respectively. We define HD_{ik} as Haversine distance between MFC i and customer k, where RC_j^t is the yearly rent cost for MFC i at time-period t, and CM is the transportation cost per kilometer for motorcycles.

$$\sum_{t\in T}\left[\sum_{i\in I}\left[Y_i^t RC_i^t + CT\left[\sum_{j\in J}\sum_{f\in F} XT_{ij}^{ft} HD_{ij} + \sum_{k\in K}\sum_{f\in F} X_{ik}^{ft} HD_{ik}\right]\right]\right.$$
$$\left.+\sum_{j\in J}\left[Y_j^t RC_j^t + CM\left[\sum_{k\in K}\sum_{f\in F} X_{jk}^{ft} HD_{jk}\right]\right]\right] \quad (1)$$

$$X_i^{ft} = \sum_{j\in J} X_{ij}^{ft} + \sum_{k\in K} X_{ik}^{ft} \quad \forall i, f, t \quad (2)$$

$$\sum_{i\in I} X_{ij}^{ft} = X_j^{ft} = \sum_{k\in K} X_{jk}^{ft} \quad \forall j, f, t \quad (3)$$

$$D_k^{ft} = \sum_{j\in J} X_{jk}^{ft} + \sum_{i\in I} X_{ik}^{ft} \quad \forall k, f, t \quad (4)$$

Table 1 Mathematical notations

Notation	Description	Unit of measurement
Sets		
I	Set of potential MFC facilities	–
J	Set of potential hub facilities	–
K	Set of customers	–
R	Set of raw materials	–
F	Set of finished products, based on PT X's business services	–
T	Set of planning periods	–
Indices		
i	Potential MFC facility index; $\forall i \in I$	–
j	Potential hub facility index; $\forall j \in J$	–
k	Customer index; $\forall k \in K$	–
r	Raw material index; $\forall r \in R$	–
f	Finished product index; $\forall f \in F$	–
t	Period of analysis; $\forall t \in T$	–
Decision variables		
X_{ij}^{ft}	Number final product f order transported from MFC i to Hub j at time-period t	Pack
X_{ik}^{ft}	Number final product f order transported from MFC i to customer k at time-period t	Pack
X_{jk}^{ft}	Number final product f order transported from Hub j to customer k at time-period t	Pack
Intermediary variables		
Y_i^t	Decision to operate MFC i at time-period t	*Binary*
Y_j^t	Decision to operate Hub j at time-period t	*Binary*
X_i^{ft}	Number of final product f orders allocated on MFC i at time-period t	Pack
X_j^{ft}	Number of final product f orders allocated on Hub j at time-period t	Pack
XA_i^{ft}	Area needed to allocate number of final product f orders MFC i at time-period t	Sqm
XA_j^{ft}	Area needed to allocate number of final product f orders Hub j at time-period t	Sqm
Y_{ij}^{ft}	Decision to connect MFC i to Hub j for transporting final product f at time-period t	*Binary*
Y_{jk}^{ft}	Decision to connect Hub j to customer k for transporting final product f at time-period t	*Binary*

(continued)

Table 1 (continued)

Notation	Description	Unit of measurement
XT_{ij}^{ft}	Number of trips needed from MFC i to Hub j transporting final product f at time-period t	Trip
XT_{ij}^{rft}	Number of trips needed from MFC i to Hub j transporting raw materials r for final product f at time-period t	Trip
Parameters		
D_k^{ft}	Demand of final product f for customer k at time-period t	Order
VND	Volume of final product of Next Day	cm^3/order
VID^r	Volume of raw materials r for final product of Instant Delivery	cm^3/pack
PID^r	Number of packs of raw materials r per final product of Instant Delivery	Pack/Order
DOI^r	Days of inventory for raw material r	Day
NDF	Frequency of ND delivery slot per day	Trip/Day
TCP	Capacity of truck per trip	cm^3/Trip
CT	Transportation cost per kilometre for truck	Rp/(Km.Trip)
CM	Transportation cost per kilometre for motorcycle	Rp/(Km.Order Trip)
HD_{ij}	Haversine distance between MFC i and Hub j	km
HD_{ik}	Haversine distance between MFC i and customer k	km
HD_{jk}	Haversine distance between Hub j and customer k	km
RC_i^t	Yearly rent cost for MFC i at time-period t	Rp
RC_j^t	Yearly rent cost for Hub j at time-period t	Rp
AI^f	Capacity needed to allocate one order of final product f at MFC	sqm/order
AJ^f	Capacity needed to allocate one order of final product f at MFC	sqm/order
CPI_i	Maximum capacity of MFC i	sqm
CPJ_j	Maximum capacity of Hub j	sqm

$$X_{ij}^{ft} - MY_{ij}^{ft} \leq 0 \quad \forall i, j, f, t \tag{5}$$

$$\sum_{i \in I} Y_{ij}^{ft} \leq 1 \quad \forall i, j, f, t \tag{6}$$

$$X_{jk}^{ft} - MY_{jk}^{ft} \leq 0 \quad \forall j, k, f, t \tag{7}$$

$$\sum_{j \in J} Y_{jk}^{ft} \leq 1 \quad \forall j, k, f, t \tag{8}$$

$$X_i^{ft} AI^f = XA_i^{ft} \quad \forall i, f, t \tag{9}$$

$$\sum_{f \in F} XA_i^{ft} \le CPI_i \quad \forall i, t \tag{10}$$

$$\sum_{f \in F} X_i^{ft} - MY_i^t \le 0 \quad \forall i, t \tag{11}$$

$$X_j^{ft} AJ^f = XA_j^{ft} \quad \forall j, f, t \tag{12}$$

$$\sum_{f \in F} XA_j^{ft} \le CPJ_j \quad \forall j, t \tag{13}$$

$$\sum_{f \in F} X_j^{ft} - MY_j^t \le 0 \quad \forall j, t \tag{14}$$

$$XT_{ij}^{1t} \ge 365 \times NDF \times Y_{ij}^{1t} \quad \forall i, j, t \tag{15}$$

The objective function in (1) is to minimize relevant costs regarding multi-period supply chain configuration decisions. The cost components are rent costs of operating facilities and transportation costs between facilities. Constraint (2) shows that the number of orders for final product f stored at MFC i equals to the total amount of outbound final product f to hub j, for ND and ID demand, and to customer k, for B2B demand (X_j^{ft}). Constraint (3) shows that the inbound number of orders for final product f to hub j equals to the amount stored at hub j and number of orders to customer k. Constraint (4) ensures that all the forecasted demand can be supplied by the SC system. The total amount of supply from MFC i and hub j equals to the total demand of final product f order, which calculated from forecasted demand D_k^{ft}.

Based on the SC strategy defined, it is decided that one hub serving one type of final product can only be served by one MFC at one time-period. Constraint (5) ensures that only arc that has flow of material could be connected, where Y_{ij}^{ft} is binary variable connect MFC i to Hub j for transporting final product f at time-period t. Constraint (6) ensures that every flow to hub j could only be coming from one MFC. With the same logic, constraint (7) and (8) ensure that one customer can only be served by one hub at one time-period, where Y_{jk}^{ft} is binary decision to connect Hub j to customer k for transporting final product f at time-period t.

The decisions whether a facility operates are dependent on the decision of allocating orders at the corresponding facility. The number of orders allocated at certain facility is proportional to the facility area needed. The total of allocated area cannot exceed the facility capacity. Then, only facility with allocated order is operated at the corresponding period of analysis. Constraint (9)–(11) ensures the mentioned logics for MFC. Meanwhile, constraint (12)–(14) ensures the same conditions for hubs. The total delivery frequency per year would always be larger than the volume transported divided by the truck capacity. These conditions are represented by constraint (15), where NDF is the frequency of next day (ND) delivery slot per day.

2.2 Customer Clustering

The granularity of the data in e-groceries business is large and is on customer's level, with data points up to 10^5 per year per business service. As the linear programming performs better when handling fewer variables, the granularity of demand data is transformed into clusters of customers. The results will later be used as the values of parameter D_k^{ft} of the mathematical model.

To conduct clustering process, k-means clustering methods is chosen because of our research limitations that cannot yet integrated between optimization of number of facilties and the facilities location. It is a clustering algorithm that divides dataset into a pre-defined number of non-overlapping clusters [6]. Clustering is a machine learning technique that performs grouping of data points. The data points are grouped, or clustered, based on similar properties. K-means' pre-defined number of clusters allows the researcher to balance the data granularity and computing power of computer-based mathematical model. The number of customer index (k) is determined to be 500, in which performs as the pre-defined number of clusters. The cluster's center point will be treated as the demand point for the mathematical model.

3 Numerical Experiments

3.1 Parameters

The experiments were done with a real datasets obtained from the biggest e-groceries platform in Indonesia ranged from January 2021 to January 2022. The values for customer clusters' locations and their number of orders per business services per year (D_k^{ft}) are provided in Table 2. Main Fulfilment Centre (MFC) serves as the supply node of the mid-mile supply chain segment. MFCs produces next day (ND), next day (ID), and business to business (B2B) orders according to their corresponding processes on the mid-mile segment. Varying number of orders are simulated to generate total MFC area needed for the corresponding number of orders. To obtain the parameter values mentioned, the generated total area needed is divided by the number of orders, yielding the amount of area needed to allocate one order. These values are averaged and be used as AJ^f. Data processing and results are shown on Table 3.

Table 2 Average distance of demand point to cluster's centre point

Business service	Average distance (km)		
	2023	2024	2025
ND	0.718	0.720	0.722
ID	0.638	0.642	0.647
B2B	0.768	0.783	0.801

Table 3 MFC area needed per business service order

Business service	Number of order per year	MFC area needed	Area needed per order	Average area needed per order (AI^f)
	#order	sqm	sqm/#order	sqm/#order
ND $f = 1$	1,200,000	1592	0.00133	0.00134
	1,500,000	2142	0.00143	
	1,800,000	2419	0.00134	
	2,000,000	2703	0.00135	
	2,200,000	2737	0.00124	
ID $f = 2$	300,000	800	0.00267	0.00236
	400,000	1055	0.00264	
	500,000	1275	0.00255	
	800,000	1783	0.00223	
	1,000,000	2250	0.00225	
B2B $f = 3$	50,000	1695	0.03390	0.03038
	80,000	2529	0.03161	
	120,000	3630	0.03025	
	180,000	5345	0.02969	
	320,000	9249	0.02890	

3.2 Results

The result of the experiments is presented in Fig. 1. The overall relevant supply chain will be optimal if each facility operates with this capacity and each MFC serves the corresponding hubs according to the order type. After running the model for 20 min and 2 s, a feasible solution is generated. This solution is the optimal solution given the solving model parameters previously mentioned. The total relevant cost of the proposed solution is Rp228,540,911,024.29. To recall, the declared objectives of this study are to determine the optimal location, capacity, and service configuration of the supply chain systems for the next three years.

This study factors out resources that would be spent on facility construction processes. The construction, both opening and closing, of facilities needs resources to be carried out. These resources include time and money. It does not picture the opening and closing processes. When a construction process took place, a facility cannot serve demand just yet. Thus, there are demand within the corresponding period that cannot be fulfilled. Also, there are costs relating to opening and closing facilities, such as construction worker costs and distributing assets costs. They are excluded to simplify the model. The model is proposing the optimal condition of supply chain network design. The proposed solution acts as the ultimate guide of strategic supply chain decisions. Further analysis on resources usage of this optimal solution, such as time and additional costs, needed to be conducted of this study.

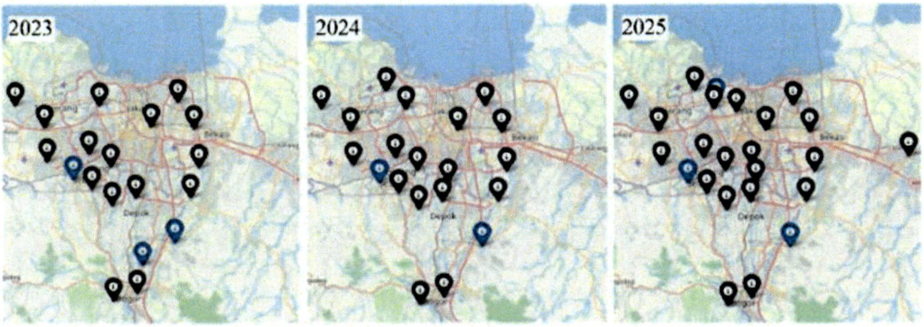

Fig. 1 A figure caption is always placed below the illustration. Short captions are centered, while long ones are justified. The macro button chooses the correct format automatically

The model solving processes in this study is limited by duration. It is only set to run for 1200 s or 20 min. This model parameter is set due to the limitations in computing power and researcher's time in conducting the solving processes. The computing power impacts the computer's speed in the search for the optimal solution. If the computing power is better, the model might get lower total relevant costs with the same duration of solving.

4 Conclusions

We have proposed a model which is based on the multi-product, multi-period, and multi-echelon of e-groceries network and problem with clustering and LP techniques. The models have been tested with large datasets. The mathematical model developed in this research has limitations. These limitations open future research potentials, as follows. Firstly, further development of mathematical model provided in this study could be conducted, considering distribution of goods on first-mile supply chain segment. As previously mentioned, model proposed by this study had excluded sourcing hubs in the mathematical model. Secondly, facility construction time and costs can be taken into account. This study only accommodates these aspects on further analysis. Secondly, A mathematical model and algorithm of vehicle routing problem between MFCs and distribution hubs could be developed.

References

1. Chapple L et al (2021) High growth, low profit: the e-commerce dilemma for CPG companies. McKinsey, New York
2. Lee Y, Golinska-Dawson P, Wu J-Z (2016) Mathematical models for supply chain management. Math Probl Eng 4

3. Bidhandi H, Yusuff R, Ahmad M, Abu Bakar M (2009) Development of a new apporach for deterministic supply chain network design. Eur J Oper Res 121–128
4. Aras N, Bilge U (2018) Robust supply chain network design with multi-products for a company in the food sector. Appl Math Model 526–539
5. Al-Ashhab NO, Afia N (2021) Perishable products supply chain network design with sustainability. Indian J Sci Technol 787–800
6. Dabbura I (2018) K-means clustering: algorithm, applications, evaluation methods, and drawbacks. [Online] Available at: https://towardsdatascience.com/k-means-clustering-algorithm-applications-evaluation-methods-and-drawbacks-aa03e644b48a. Accessed 18 Aug 2022

A Closed-Loop Supply Chain Inventory Model with Remanufacturing and Cleaner Production System

Wakhid Ahmad Jauhari, Anda Iviana Juniani, Wiwik Sulistyowati, and Anik Dwiastuti

Abstract In this paper, a cooperative inventory model for supply chain system under emission reduction and energy usage is formulated. The system consists of a manufacturer who acts as producer or remanufacturer and a buyer who acts as seller of end products. Carbon emissions are released from some activities, i.e. production, waste disposal, transportation, and storage. Electrical energy is consumed by manufacturing system to perform the production and remanufacturing. The model is developed mathematically with the aim of minimizing the total cost. The proposed model is solved by using the combination of analytical approach and an iterative procedure. The results suggest the decision maker to control the review period, shipment frequency and safety factor carefully when the system faces variations in production rate, carbon tax and the demand. Furthermore, we observe that the production rate and carbon tax are the most influential parameters on the level of emissions produced by the system.

Keywords Carbon emissions · Energy consumption · Remanufacturing · Carbon tax

W. A. Jauhari (✉)
Universitas Sebelas Maret, Surakarta, Indonesia
e-mail: wachid_aj@yahoo.com

A. I. Juniani
Shipbuilding Institute of Polytechnic Surabaya, Surabaya, Indonesia

W. Sulistyowati
Universitas Muhammadiyah Sidoarjo, Sidoarjo, Indonesia

A. Dwiastuti
Universitas Negeri Malang, Malang, Indonesia

C. N. Rosyidi et al. (eds.), *Proceedings of the 6th Asia Pacific Conference on Manufacturing Systems and 4th International Manufacturing Engineering Conference*, Lecture Notes in Mechanical Engineering,
https://doi.org/10.1007/978-981-99-1245-2_9

1 Introduction

The most challenging problem facing industries today is how to determine strategic actions to reduce emissions resulting from their operations and to control energy use. In the closed-loop supply chain (CLSC) system, determining the actions is not easy, because it involves many parties with different interests. Transportation, storage, production, waste disposal are the main activities that generate emissions. In addition, production is known as the most energy-consuming activity. Inventory decision making is closely related to these activities, so it has a big impact on emissions reduction and energy consumption.

Inventory management on CLSC system has been widely discussed by some scholars. Christy et al. [1] developed a CLSC inventory model by taking into account price and quality-dependent demand and two recovery activities, i.e. remanufacturing, refurbishing. They formulated the model under centralized and decentralized policies and determined the best policy. Jauhari et al. [2] studied the quality dependent return rate and multiple production and remanufacturing cycles in a three echelon CLSC system. Mawandiya et al. [3] explored the impact of stochastic return on CLSC performance. Ben-Daya et al. [4] considered a consignment stock policy and multi-retailer in a two-stage CLSC. Bazan et al. [5] was the first to incorporate the energy effects and carbon emissions on reverse logistic model. Further, Bazan et al. [6] compared the performance of vendor-managed inventory and classical inventory model under remanufacturing generations restriction. Dwicahyani et al. [7] examined how energy usage and emissions generation influence the CLSC model involved of a depot and distributor. The number of remanufacturing generations is also incorporated in the model and assumed to be finite. Hasanov et al. [8] explored the impact of emissions and energy consumption on the performance of four-stage CLSC model. Zhang et al. [9] proposed a CLSC for deteriorating items under consignment stock policy. The carbon emissions released from supply chain activities is constrained with carbon-cap-and-trade regulation. Asghari et al. [10] utilized a cooperative game theory to analyse the coordination of CLSC under environmental efforts and eco-friendly products.

Although CLSC has been widely investigated, none has developed the model under stochastic demand, cleaner system and remanufacturing and waste disposal policies. In this paper, we propose a CLSC inventory model under situation, in which the remanufacturing is employed to improve the quality of used items and the energy consumption and carbon emissions from some operations are incorporated in the model. The periodic review policy is employed to formulate a two-echelon CLSC system. The objective is to find the optimal production and shipment policies so that the total cost can be minimized. An efficient procedure and numerical example are suggested to show the application of the model.

2 Notations and Assumptions

To formulate the mathematical model, we derive the notations that can be seen in Table 1.

The following assumptions are derived to formulate the supply chain inventory model. (1) The supply chain under investigation comprises of single manufacturer who produces items and single buyer who sells the items to end customers, (2) The buyer faces demand that normally distributed with mean D and standard deviation σ, (3) A periodic review policy is employed to manage the inventories in the buyer side, (4) The manufacturing system can produce final products by converting the raw material or remanufacturing the used items, and (5) Emissions are calculated from main activities, i.e. manufacturing, storage, waste disposal and shipping. (6) The shortages can be fulfilled with backorders.

Table 1 Notations for formulating the model

D	Average demand (units/h)	c_{tr}	Transportation cost ($/truck)
σ	Standard deviation of demand (units/h)	t_{cap}	Truck capacity (unit/truck)
A	Ordering cost ($/order)	c_i	Inspection cost ($/unit)
h_b	Holding cost incurred by buyer ($/unit/h)	ε	Amount of carbon emissions from one gallon of diesel-truck fuel (ton/gallon)
π	Shortage cost ($/unit)	c_e	Energy cost ($/kWh)
L	Lead time (h)	c_m	Raw material cost ($/unit)
W	Carbon emissions generated from storage (kg CO_2/unit/h)	ϑ	Number of gallons per truck per distance travelled (gallon/truck)
E	Carbon tax ($/kg CO_2)	c_p	Cost to collect used items ($/unit)
θ_1	Proportion of remanufacturing	c_r	Recycling cost ($/unit)
θ_2	Proportion of recycling	c_d	Waste disposal cost ($/unit)
θ_3	Proportion of waste disposal	S	Setup cost for the production system ($/setup)
a	Emission parameter for the production system (kg h^2/unit^3)	W_d	Carbon emissions generated from waste disposal activity (kg CO_2/unit)
b	Emission parameter for the production system (kg h/unit^2)	W_{v0}	Power consumed by the production system in idle condition (kW)
c	Emission parameter for the production system (kg/unit)	k_v	Constant for the power used by the production system (kWh/unit)
h_v	Holding cost for the manufacturer ($/unit/h)	T	Review period (h), decision variable
k	Safety factor, decision variable	n	Number of shipments

3 Model Formulation

The buyer employs a periodic policy to control the inventories in the storage system. Thus, a review of inventory will be conducted each T period and the order will be placed with quantity of DT units. The number of backorders is derived by multiplying the number of backorders that occur in each cycle and the number of ordering cycles. Due to the stochastic demand, which is normally distributed, the buyer incurs backorder cost, that is

$$BO = \frac{\pi\sigma\psi(k)}{T}\sqrt{T+L} \tag{1}$$

where $\psi(k) = f_s(k) - k[1 - F_s(k)]$. $f_s(k)$ and $F_s(k)$ denote the probability density function and the cumulative density function of standard normal distribution, respectively. By following Hadley–Within [11], the cost to hold the items in the storage can be derived as follows

$$Hold\,B = h_b\left(\frac{DT}{2} + k\sigma\sqrt{T+L}\right) \tag{2}$$

The carbon cost for the buyer is calculated from two activities, transporting the items and keeping items in the warehouse. The cost is formulated as follows

$$EmB = E\frac{D}{t_{cap}}\vartheta\varepsilon + EW\left(\frac{DT}{2} + k\sigma\sqrt{T+L}\right) \tag{3}$$

Therefore, the total cost for the buyer consisting of ordering cost, transportation cost, backorder cost, holding cost and emission cost, is derived by

$$\begin{aligned} TCB = {} & \frac{A}{T} + c_{tr}\frac{D}{t_{cap}} + \frac{\pi\sigma}{T}\sqrt{T+L}\psi(k) \\ & + (h_b + EW)\left(\frac{DT}{2} + k\sigma\sqrt{T+L}\right) + E\frac{D}{t_{cap}}\vartheta\varepsilon \end{aligned} \tag{4}$$

The manufacturer converts raw material to finished items. The manufacturer system also serves as remanufacturer who converts used items to finished items. The production batch is nDT and the manufacturing time is nT. Thus, the setup cost is $\frac{S}{nT}$. By referring to average manufacturer's inventory level, the cost to hold items in the storage is given by

$$Hold\,M = h_v\frac{DT}{2}\left(n\left[1 - \frac{D}{P}\right] - 1 + \frac{2D}{P}\right) \tag{5}$$

During production run, the system consumed electrical energy which is calculated by

$$En = \left(\frac{W_{v0}}{P} + k_v\right)D \tag{6}$$

The cost related the activity to collect the used items, categorizing the used items, recycling and waste disposal is formulated by

$$Rem = c_p\beta D + c_i D + c_r\theta_2\beta D + c_d\theta_3\beta D \tag{7}$$

The emissions are calculated from the production, keeping items, and waste disposal. The carbon cost is given by

$$EmM = ED\left(aP^2 - bP + c\right) + EW\frac{DT}{2}\left(n\left[1 - \frac{D}{P}\right] - 1 + \frac{2D}{P}\right) + EW_d\theta_3\beta D \tag{8}$$

The total cost for the manufacturer, including the setup cost, holding cost, emission cost, collection cost, inspection cost, raw material cost, recycling cost and waste disposal cost, is formulated as follows

$$\begin{aligned} TCM = {} & \frac{S}{nT} + h_v\frac{DT}{2}\left(n\left[1 - \frac{D}{P}\right] - 1 + \frac{2D}{P}\right) \\ & + EW\frac{DT}{2}\left(n\left[1 - \frac{D}{P}\right] - 1 + \frac{2D}{P}\right) + ED\left(aP^2 - bP + c\right) \\ & + c_m(1 - \theta_1)D + c_e\left(\frac{W_{v0}}{P} + k_v\right)D \\ & + c_i D + c_p\beta D + c_r\theta_2\beta D + c_d\theta_3\beta D + EW_d\theta_3\beta D \end{aligned} \tag{9}$$

Thus, the total cost for the supply chain is given by

$$TCS = TCB + TCM \tag{10}$$

4 Solution Method

The decision variables of the proposed model are the number of shipments (n), review period (T) and safety factor (k) with the objective function is minimizing the supply chain total cost (*TCS*). By obtaining the first partial derivative of total cost with respect to T and k, the optimal values of the decision variables are formulated as follows:

$$T = \sqrt{\frac{A + \pi\sigma\psi(k)\sqrt{T+L} + S/n}{\frac{D}{2}(h_b + EW) + \frac{k\sigma(h_b+EW)}{2\sqrt{T+L}} + \frac{\pi\sigma\psi(k)}{2T\sqrt{T+L}} + (h_v + EW)\frac{D}{2}\left(n\left[1 - \frac{D}{P}\right] - 1 + \frac{2D}{P}\right)}} \tag{11}$$

$$F_s(k) = 1 - \frac{T(h_b + EW)}{\pi} \tag{12}$$

By referring to Jauhari [12] and Glock [13], the algorithm to solve the proposed problem is listed below:

Step 1 Set $n = 1$ and $TCS_{n-1}(T_{n-1}, k_{n-1}, n-1) = \infty$

Step 2 Calculate

$$T = \sqrt{\frac{A + S/n}{\frac{D}{2}(h_b + EW) + (h_v + EW)\frac{D}{2}\left(n\left[1 - \frac{D}{P}\right] - 1 + \frac{2D}{P}\right)}} \tag{13}$$

Step 3 Calculate k using Eq. (13)

Step 4 Calculate T using Eq. (11)

Step 5 Repeat steps 3–5 until no change occurs in the values of T and k.

Step 6 Set $T_n = T$, $k_n = k$. Compute $TCS_n(T_n, k_n, n)$ by utilizing Eq. (10).

Step 7 If $TCS_n(T_n, k_n, n) \leq TCS_{n-1}(T_{n-1}, k_{n-1}, n-1)$ repeat step 2–6 with $n = n + 1$, otherwise go to step 8.

Step 8 Set $TCS_{n-1}(T_{n-1}, k_{n-1}, n-1)$ as the minimum value of joint total cost and T, k, and n are the solutions for the model.

5 Numerical Example and Analysis

Here, we consider the following data: $D = 500$, $\sigma = 20$, $A = 200$, $C_{tr} = 400$, $t_{cap} = 100$, $\vartheta = 375$, $\varepsilon = 0.25$, $L = 0.05$, $\pi = 50$, $P = 1200$, $h_b = 2.5$, $h_v = 1.5$, $K = 800$, $\theta_1 = 0.6$, $\theta_2 = 0.3$, $\theta_3 = 0.1$, $c_p = 0.5$, $c_r = 0.75$, $c_d = 0.35$, $\beta = 0.45$, $a = 0.00012$, $b = 0.0008$, $c = 8.4$, $W_{v0} = 1{,}000{,}000$, $k_v = 1000$, $C_{en} = 0.015$, $W = 10$, $E = 0.0618$, $m = 0.005$, $\theta = 0.8$, $IN_f = 2000$, $\mu = 0.65$, $\theta_t = 0.3$. Here we use Matlab R2018b to run the algorithm. By using the proposed procedure, the following results are obtained. The optimal review period, shipment frequencies, and safety factor, are 0.528; 3; 1.819, respectively. The cost borne by the manufacturer, buyer and supply chain are \$20,875.65, \$2,961.28 and \$23,836.93, respectively.

We further investigate the influence of some key parameters on decision parameters and supply chain performance. Table 2 shows the results of our investigation. As the demand rate is getting higher, the buyer needs more inventories to satisfy the demand from end customer. In addition to the need to increase the ordering lot, the buyer should adjust safety factor to higher level and shorten the review period carefully. Facing this behaviour, the manufacturer must increase production batch and increase delivery frequency to offset the increase in buyer orders caused by the increase in demand. When we investigate the sustainability aspects of the system, we conclude that the increase in demand significantly influences the emissions and energy consumption. As can be seen in Fig. 1, when the demand increases by 80% (500 units/h to 900 units/h), the emissions and electrical energy increase by 77.48 and 80%, respectively.

Our investigation on the production rate reveals that its changes give significant influence on the model's performances. When it is changed gradually, all key parameters and total cost drastically change. When the production goes up by 80% (from 1200 units/h to 2160 units/h), the safety factor and shipment frequencies go down by 5.73 and 33.33%, respectively, while the review period goes up by 26%. The cost borne by the buyer, manufacturer and supply chain go up by 6.59, 43.61 and 39.01%, respectively. Interestingly, the behaviour of carbon emissions and energy consumption caused by changes in the production rate looks different (see Fig. 2). The resulting emissions have a tendency to decrease while energy consumption is the opposite. Further, the investigation on carbon tax shows that it gives negative impact on the safety factor and review period (see Fig. 3). Facing the strengthen green policy

Table 2 Impact of demand, production rate and tax on model's behavior

Parameters	Values	n	T	k	Buyer cost	Vendor cost	Total cost
Demand	100	2	1.464	1.308	1015.78	4338.37	5354.15
	300	2	0.855	1.594	2123.09	12,662.20	14,785.29
	500	3	0.528	1.819	2961.28	20,875.65	23,836.93
	700	4	0.399	1.940	3879.24	28,954.01	32,833.24
	900	6	0.314	2.038	4811.33	36,897.49	41,708.82
	1100	11	0.267	2.101	5751.90	44,652.79	50,404.69
Production rate	720	5	0.440	1.897	2922.72	21,509.73	24,432.45
	1200	3	0.528	1.819	2961.28	20,875.65	23,836.93
	1680	3	0.519	1.827	2961.28	24,204.01	27,165.29
	2160	2	0.666	1.714	3156.69	29,980.42	33,137.10
	2640	2	0.666	1.714	3164.74	37,849.21	41,013.95
	3120	2	0.666	1.714	3170.52	47,630.13	50,800.66
Tax	0.03708	3	0.557	1.832	2912.41	18,590.76	21,503.17
	0.0618	3	0.528	1.819	2961.28	20,875.65	23,836.93
	0.08652	3	0.504	1.806	3008.92	23,157.66	26,166.58
	0.11124	3	0.483	1.793	3055.44	25,437.21	28,492.65
	0.13596	3	0.464	1.781	3100.94	27,714.61	30,815.55
	0.16068	3	0.448	1.770	3145.49	29,990.12	33,135.61

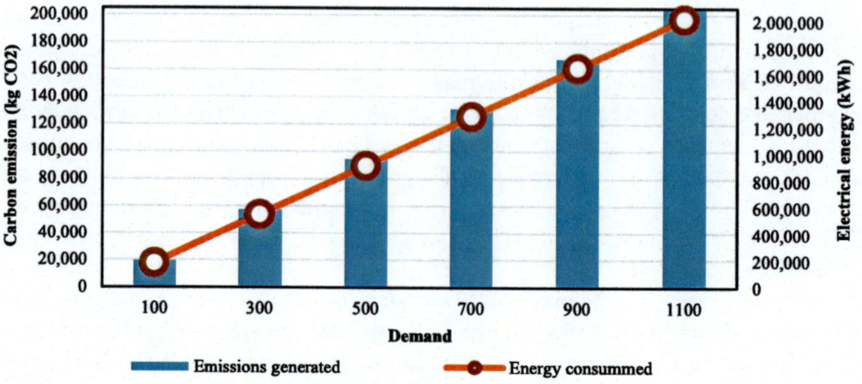

Fig. 1 Impact of demand on emissions and energy

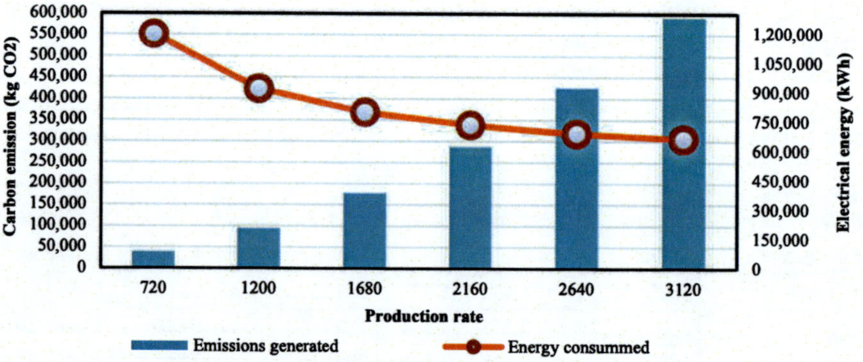

Fig. 2 Impact of production rate on emissions and energy

issued by regulator, the manufacturer is encouraged to lessen the amount of emissions resulted from their operations. We observe that there is a significant decrease on the emissions generated from storage activity in both members of supply chain.

6 Conclusions

The mathematical inventory model under stochastic demand and environmental concerns is formulated. Remanufacturing strategy is adopted to improve the quality of used items taken from the market. Carbon emissions as well as energy consumption are incorporated in the model and investigated in the supply chain context. The objective of the model is to obtain the optimal periodic review, shipment frequencies and

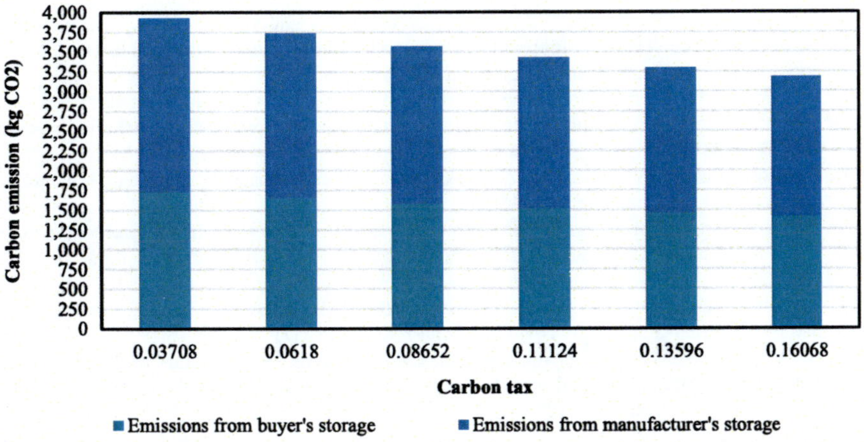

Fig. 3 Impact of tax on carbon emissions

safety factor, such that the total cost borne by the supply chain is minimized. The findings suggest the manager to carefully determine the inventory decisions, including review period, safety factor and shipment frequencies, when there are changes in the demand rate, production rate and tax. Facing an increase in the demand rate, the system should increase the inventories kept in the storage system to increase the capability system in fulfilling the customer's demand. The production rate is the most affecting parameters contributes to the environmental performances of the system. The results also suggest the manager to pay more attention on the changes of production rate to maintain the emissions resulted and energy required to run the manufacturing system. Furthermore, the system should take an appropriate action to lessen the emissions when the carbon tax is getting higher.

References

1. Christy AY, Fauzi BN, Kurdi NA, Jauhari WA, Saputro DRS (2017) A closed-loop supply chain under retail price and quality dependent demand with remanufacturing and refurbishing. J Phys: Conf Ser 855(1):012009
2. Jauhari WA, Dwicahyani AR, Kurdhi NA (2017) Lot sizing decisions in a closed-loop supply chain system with remanufacturing. Int J Procurement Manage 10(3):381–409
3. Mawandiya BK, Jha JK, Thakkar JJ (2020) Optimal production-inventory policy for closed-loop supply chain with remanufacturing under random demand and return. Oper Res Int Journal 20:1623–1664
4. Ben-Daya M, As'ad R, Nabi KA (2019) A single-vendor multi-buyer production remanufacturing inventory system under a centralized consignment arrangement. Comput Ind Eng 135:10–27
5. Bazan E, Jaber MY, El-Saadany AMA (2015) Carbon emissions and energy effects on manufacturing–remanufacturing inventory models. Comput Ind Eng 88:307–316

6. Bazan E, Jaber MY, Zanoni S (2017) Carbon emissions and energy effects on a two-level manufacturer-retailer closed-loop supply chain model with remanufacturing subject to different coordination mechanisms. Int J Prod Econ 183:394–408
7. Dwicahyani AR, Jauhari WA, Rosyidi CN, Laksono PW (2017) Inventory decisions in a two-echelon system with remanufacturing, carbon emission, and energy effects. Cogent Engineering 4(1):1379628
8. Hasanov P, Jaber MY, Tahirov N (2019) Four-level closed loop supply chain with remanufacturing. Appl Math Model 66:141–155
9. Zhang T, Hao Y, Zhu X (2020) Consignment inventory management in a closed-loop supply chain for deteriorating items under a carbon cap-and-trade regulation. Comput Ind Eng 171:108410
10. Asghari T, Taleizadeh AA, Jolai F, Moshtagh MS (2022) Cooperative game for coordination of a green closed-loop supply chain. J Clean Prod 363:132371
11. Hadley G, Within TM (1963) Analysis of inventory systems. Prentice-Hall, Englewood Cliffs
12. Jauhari WA (2012) Integrated inventory model for three-layer supply chains with stochastic demand. Int J Oper Res 13(3):295–317
13. Glock CH (2012) Lead time reduction strategies in a single-vendor single-buyer integrated inventory model with lot size-dependent lead times and stochastic demand. Int J Prod Econ 136(1):37–44

Analysis of Cardboard Sheets Supplier Selection Using Fuzzy AHP and TOPSIS Methods at PT. XYZ

Santika Sari and Dewi Maharani

Abstract Procurement is a process by which companies obtain goods and services to ensure the continuity of the production and logistics processes. The selection of suppliers that have not been optimal has resulted in several problems in the supplier of cardboard sheets PT. XYZ, so the impact on the company's productivity declines. This study aims to determine the best alternative supplier at PT. XYZ uses the F-AHP and TOPSIS methods. Determination of criteria and sub-criteria in supplier selection is carried out using the cut-off point method and obtained five criteria and 16 selected sub-criteria. The Fuzzy AHP method calculates the priority weight of each criterion and sub-criteria. Based on the weighting results with Fuzzy AHP, quality is the main criterion in selecting suppliers of cardboard sheets with a weight of 0.324. In calculating the best alternative based on the preference value of the criteria and sub-criteria with the TOPSIS method, the highest preference value obtained by the PDK supplier is 0.9014.

Keywords Supplier selection · Decision making · Fuzzy AHP · TOPSIS · Cut-off Point

1 Introduction

Procurement is a process by which companies obtain goods and services to support the sustainability of production and logistics systems. The procurement of raw materials is an important stage in supply chain management because it is the beginning of the entire process to be carried out [1]. PT. XYZ is a company engaged in the manufacture of cardboard boxes or cardboard packaging. The production system used in this company is Make To Order (MTO), which means this production system supports customers in ordering products according to the required specifications.

S. Sari (✉) · D. Maharani
Universitas Pembangunan Nasional Veteran Jakarta, Depok, South Jakarta, Indonesia
e-mail: santika.sari@upnvj.ac.id

C. N. Rosyidi et al. (eds.), *Proceedings of the 6th Asia Pacific Conference on Manufacturing Systems and 4th International Manufacturing Engineering Conference*, Lecture Notes in Mechanical Engineering,
https://doi.org/10.1007/978-981-99-1245-2_10

The primary raw materials used by PT. XYZ is a cardboard sheet of various types and sizes.

In 2021, PT. XYZ has used six suppliers who can accept the custom for any type and size of cardboard sheets required. In procuring the primary raw materials, the company is still experiencing problems because, so far, no supplier can meet all the company's needs. The supplier selection process can be very complicated as a provider may only have an outstanding amount of expertise in some areas [2]. So that in determining supplier selection, companies need adequate time and resources to collect data and accurately analyze all good and bad things that affect alternative decisions [3].

One problem that often occurs is delivery delays from a pre-determined date. The results of the 2021 data recapitulation found that 571 of the 758 total orders were delayed for one day to 3 months. The following supplier problem is the number of raw material receipts that are not by the total order because of the large number of rejected products found or the unavailability of the number of products that match the company's needs. In 2021, from 758 total orders to suppliers, 355 or 47% of orders have an incorrect amount of raw material from the total order.

Inappropriate supplier selection will affect the company's productivity decline because raw materials are the main component in the production process and affect the final product [4]—Moreover, the primary raw materials used by PT. XYZ cannot implement raw material inventory control, so when there is an order from a customer, PT. XYZ can only order raw materials from suppliers. Therefore, companies need to be more selective in choosing suppliers. Suppose there is an error in the raw material. In that case, the company must postpone production activities, so it takes longer to complete the finished product, resulting in delays in product delivery or not fulfilling customer requests. The results of the data recapitulation of the delivery of finished goods products to customers in 2021 at PT. XYZ found that 391 of the 659 total orders, or 59% of orders, experienced a delay of 1 to 81 days.

These problems occur because of the process of selecting alternative suppliers at PT. XYZ has not been carried out optimally. Based on the results of interviews with the Procurement Manager of PT. XYZ, it is known that the supplier selection at the company is only based on familial or relationship factors, so there is no clear criteria indicator from the company in supplier selection. In choosing suppliers, criteria are needed to describe the supplier's performance to foster mutually beneficial relationships for both parties and increase the current and future value [5].

The Analytical Hierarchy Process (AHP) method is one of the Multi-Attribute Decision Making (MADM) methods that can solve multi-objective and multi-criteria problems by comparing the preferences of each hierarchical element. However, the AHP method has a subjective assessment and cannot overcome uncertainty. However, the weakness in the AHP method can be overcome by adding fuzzy logic [6]. Developing the AHP method into Fuzzy AHP can calculate the total weight for all specified criteria and reduce subjective assessments of the importance of criteria determined by decision-makers [7].

In this study, the weighting of each criterion and sub-criteria obtained from the Fuzzy AHP calculation will be used for decision-making using TOPSIS. TOPSIS can determine the best alternative based on the closest distance to the positive ideal solution and the furthest from the negative ideal solution [8], so it can help determine the ranking of the best alternative suppliers that match the criteria.

2 Research Methodology

2.1 Data Collection Method

The questionnaires in this study were distributed to 3 company employees who understand the company's conditions and have tasks that are directly related to suppliers, including; The Procurement Manager, PPIC Supervisor, and Warehouse Admin.

The initial data collection stage is by distributing questionnaires for selecting criteria and sub-criteria. The process of determining the criteria and sub-criteria is carried out through a study of journal literature and interviews with experienced companies who understand the criteria considered necessary for the company. The second data collection stage is done by distributing a pairwise comparison questionnaire on a Likert scale of 1–9. The last data collection stage was conducted by distributing a comparison questionnaire between alternatives on a Likert scale of 1–5.

2.2 Data Processing Method

Data processing starts from calculating the weight of the criteria and sub-criteria with a Cut Off Point so that the main criteria and sub-criteria are obtained that will be used for the following research stage. Then Create a comparison matrix, and calculate the value and consistency. In testing the consistency of the hierarchy, if the Consistency Ratio (CR) > 0.1, it cannot perform the subsequent data processing stage, so it is necessary to fill out the questionnaire again. After that, the normalization of vector weights on the criteria and sub-criteria is carried out using a fuzzy AHP to see the rating model, the relationship between criteria and sub-criteria, and the resulting weight value. The final result of this data processing using the TOPSIS method will produce a preference value to determine the best choice in selecting cardboard sheet suppliers at PT. XYZ.

Table 1 Criteria and sub-criteria

Criteria	Code	Sub-criteria	Source
Price (P)	P1	Price offered	[9]
	P2	Discounts	[9]
	P3	Payment method	[9]
Quality (Q)	Q1	Material conformity with specifications	[10]
	Q2	Provision of goods without defects	[11]
	Q3	Quality consistency	[9]
Delivery (D)	D1	On time delivery	[11]
	D2	Material quantity accuracy	[11]
	D5	Type of transportation mode	[12]
Service (S)	S1	Speed in responding to requests	[13]
	S2	Ease of supplier to contact	[13]
	S3	Response in receiving complaints	[13]
	S4	Claim and warranty policy	[14]
Supplier characteristics (SC)	SC1	Reputation	[15]
	SC2	Professionalism	[16]
	SC3	Negotiability	[15]

3 Data Analysis and Discussion

3.1 Determination of Criteria and Sub-criteria with Cut-Off Point

Data were collected based on five criteria. Table 1 below is the result of calculating the natural cut-off point of all criteria and sub-criteria.

3.2 Calculation of the Consistency Ratio Using the Analytic Hierarchy Process (AHP) Method

Table 2 reports the CR value for each respondent.

The Consistency Ratio (CR) for respondents 1, 2, 3 and combined are 0.084, 0.061, 0.054 and 0.025, so it can be declared consistent because the value is less than 0.1 and can be carried out in the following research stage.

Table 2 Consistency ratio for each respondent

Value	Respondent 1	Respondent 2	Respondent 3	Combined
Lambda max	5.377	5.273	5.240	5.113
RI	1.12	1.12	1.12	1.12
CI	0.094	0.068	0.060	0.028
CR	0.084	0.061	0.054	0.025

3.3 Determining the Weight of the Criteria and Sub-criteria with the Fuzzy AHP Method

The global weight is the result of combining the criteria and sub-criteria weights. The merging is done because, hierarchically, the weights of the criteria and sub-criteria have the integrity of their respective values, so they need to be combined into global weights. Table 3 is the results of the weighting recapitulation using the Fuzzy AHP method.

Tabel 3 AHP fuzzy global weight

Criteria	Value	Sub-criteria	Value	Global weight
Price	0.295	P1	0.606	0.179
		P2	0.212	0.063
		P4	0.181	0.053
Quality	0.324	Q1	0.459	0.149
		Q2	0.147	0.048
		Q3	0.394	0.128
Delivery	0.205	D1	0.572	0.117
		D2	0.353	0.072
		D5	0.075	0.015
Service	0.113	S1	0.360	0.041
		S2	0.198	0.022
		S3	0.343	0.039
		S4	0.099	0.011
Supplier characteristics	0.063	SC1	0.201	0.013
		SC2	0.620	0.039
		SC3	0.179	0.011

Table 4 AHP fuzzy supplier characteristics sub-criteria weight

Supplier	Value	Rank
PDK	0.9014	1
CMI	0.7647	2
MI	0.5731	4
KKM	0.6516	3
SMJ	0.2314	6
BMJM	0.3635	5

3.4 Determining the Best Supplier Alternative with TOPSIS Method

The global weight value for each criterion and sub-criteria that has been obtained using the Fuzzy AHP method and the final weight value will be applied to the calculation of the TOPSIS method. Of the 16 sub-criteria, two sub-criteria are categorized as cost, namely sub-criteria P1 and P3, while others are categorized as benefits. There are six alternative suppliers in decision-making. Table 4 is the result of calculating the preference value of each alternative using the TOPSIS method.

4 Conclusion

1. Determination of criteria and sub-criteria in supplier selection is carried out using a cut-off approach to Eliminate less important sub-criteria. Based on data processing results, from the initial 21 sub-criteria designs, 16 sub-criteria were selected.
2. Based on the results of the weighting of the criteria using the fuzzy AHP method,it can be seen that the Quality criteria have the highest weight, which is 0.324. In the Quality sub-criteria, it is known that the quality sub-criteria, according to the specifications, has the highest weight of 0.459. In the Price sub-criteria, it is known that the price sub-criteria offered has the highest weight of 0.606. In the Delivery sub-criteria, it is known that the delivery timeliness sub-criteria has the highest weight of 0.572. In the service sub-criteria, it is known that the speed sub-criteria in responding to requests has the highest weight of 0.360. In the supplier characteristics sub-criteria, it is known that the professionalism sub-criteria has the highest weight of 0.620.
3. In selecting the best alternative using TOPSIS, the highest preference value obtained by PDK suppliers is 0.9014, followed by CMI suppliers at 0.7647, KKM suppliers at 0.6516, MI suppliers at 0.5731, BMJM suppliers at 0.3635, and the last order is SMJ supplier of 0.2314.

References

1. Govindaraju R, Sunulingga JP (2017) Pengambilan keputusan pemilihan pemasok di perusahaan manufaktur dengan metode Fuzzy ANP. J Manajemen Teknologi 16(1):1–16
2. Darmawan H, Setiawan H (2013) Pemilihan Pemasok Bahan Baku Produksi Menggunakan Metode data envelopment analysis. J Teknik Industri 1(2):157–161
3. Ávila P, Mota A, Pires A, Bastos J, Putnik G, Teixeira J (2012) Supplier's selection model based on an empirical study. Procedia Technol 5:625–634
4. Jannah M, Fakhry M, Rakhmawati (2016) Pengambilan Keputusan Untuk Pemilihan Supplier Bahan Baku dengan Pendekatan Analytic Hierarchy Process di PR Pahala Sidoarjo. Agrointek 5(2):88–97
5. Pradipta AY, Diana A (2017) Sistem Penunjang Keputusan Pemilihan Supplier pada Apotek dengan Metode AHP dan SAW (Studi Kasus Apotek XYZ). Sisfotek 1(1):107–114
6. Patil SK, Kant R (2014) A fuzzy AHP-TOPSIS framework for ranking the solutions of Knowledge Management adoption in supply chain to overcome its barriers. Expert Syst Appl 41(2):679–693
7. Chang DY (1996) Applications of the extent analysis method on fuzzy AHP. Eur J Oper Res 95(3):649–655
8. Doaly CO, Moengin P, Chandiawan G (2019) Pemilihan Multi-Kriteria Pemasok Department Store Menggunakan Metode Fuzzy AHP dan TOPSIS. JITI 7(1):70–78
9. Lukmandono L, Basuki M, Hidayat MJ, Setyawan V (2019) Pemilihan Supplier Industri Manufaktur Dengan Pendekatan AHP dan TOPSIS. OPSI 12(2):83–88
10. Baroto T, Utama DM (2020) Integrasi AHP dan SAW untuk Penyelesaian green supplier selection. SENTRA 6:38–44
11. Noviandri MR, Tama IP, Yuniarti R (2015) Analisis Pemilihan Supplier Metallic Box Menggunakan fuzzy analytic hierarchy process (AHP) JRMSI 3(3):453–462
12. Taufik R, Sumantri Y, Farela C, Tantrika M (2014) Penerapan Pemilihan Supplier Bahan Baku Ready Mix Berdasarkan Integrasi Metode AHP Dan TOPSIS (Studi Kasus Pada PT Merak Jaya Beton, Malang). JRMSI 2(5):1067–1076
13. Syamsudin A, Nursanti E, Adriantantri E (2017) Analisis Pemilihan Supplier Yang Tepat Untuk Produk Gigi Palsu (Studi Kasus Di CV. Brother Dent). JTMI 3(2):1–7
14. Rukmin HS, Adianto H, Avianti D (2014) Pemilihan Supplier Bahan Baku Dengan Menggunakan Metode analytical hierarchy process (AHP) (Studi Kasus Di PT Ewindo Bandung). SEMNAS. II Manajemen Rekayasa Kualitas, pp 30–36
15. Lestari S, Fauzi C (2019) Evaluasi Supplier Kemasan Dus Dengan Menerapkan Metode analytical hierarchy process (AHP) (Studi Kasus di PT Innovation). JISS 4(2):60–67
16. Widharto Y, Dewi DS (2018) Penerapan Metode AHP –TOPSIS Dalam Penentuan Supplier Homopolimer XYZ Terbaik (Studi Kasus Pada PT ABC). SNATIF 5(1):96–101

Investigation of Monitoring Electrical Consumption on Load Using IoT

Muhammad Zuhairi Azharudin and Norazlianie Sazali

Abstract Nowadays, usage of electrical energy is overly and uncontrolled used by consumer without any realizing the total usage or actual usage of electrical appliances. Therefore, energy waste is become increase year by year due to increasing of demands such as building, houses, factory and any other usage that are consumes any electrical energy. Besides, electrical energy consumption and demand are increased dramatically. This is because electrical appliances are increasing by times in any building or premises, the energy or load is being overly used without any control or awareness from consumers. The problem that needs to face is the consumption of electrical appliance. This is happened because of growth of technology. The objective for this project is to investigate the concept of energy monitoring system with the fundamental of the concept it could help to reduce the amount of consumption energy. Second, to investigate efficiency of electrical appliances by observing power factor from that the consumer can detect with appliance is use high currents. Next, to create a device that can reduce energy usage by measure and observing how many energies that the appliance will be consumed.

Keywords Monitoring system · Energy efficient · Electrical consumption · Internet of Things (IoT)

1 Introduction

The electricity generation are growing rapidly over the last decade. Energy consumption been overly used by consumers without any observing and realizing the total usage or actual usage of the energy [1]. The energy consumers are being increased by year to year due to the increasing demand includes building, houses, factories, and any other usage that consumes energy. Energy is also used to provide services such as

M. Z. Azharudin · N. Sazali (✉)
College of Engineering, Universiti Malaysia Pahang, 26600 Pekan, Pahang, Malaysia
e-mail: azlianie@ump.edu.my

C. N. Rosyidi et al. (eds.), *Proceedings of the 6th Asia Pacific Conference on Manufacturing Systems and 4th International Manufacturing Engineering Conference*, Lecture Notes in Mechanical Engineering,
https://doi.org/10.1007/978-981-99-1245-2_11

passenger transportation, cooling, and lighting system [2]. For recourse availability is a production that from Malaysia or import from other countries. This happens because of the lack of energy resources. In the growth of demand in commercial and domestic sectors. Malaysia have become stronger based on the demand of electrical consumption and the highest of electrical production in coal. The total electricity production was reported at 122.12TWh in 2012, where gas is still the major fuel source contributing to 52.7% of the total generation fuel mix of electricity followed by coal, 38.9%, hydro, 7.3%, oil, 1% and others, 0.2% [3]. For instance, the consumption of the world's daily of natural gas and oil consumption increased by about 28% and 7%, respectively, from 261 billion cubic feet and 85.4 million barrels in 2009 to around 335 billion cubic feet and 91.2 million barrels in 2013. By the end of 2015, daily oil consumption had increased by about 4.17% over 2013 [4].

Large Household Appliances such as refrigerators and washing machines, IT and Telecom such as PCs, monitors, and laptops, and Consumer Equipment such as televisions and DVD players are the three main categories of e-waste [5]. The energy consumption that does not consider its usage will lead to a big waste of energy. The management of waste, energy, and resources has received special attention from academics and practitioners due to the growing evidence for its effects on the environment [6]. Some of factors energy waste is come from effect of environment and the low efficiency of energy also can be caused the electric bill to boost high. Therefore, energy supplied to the load that has low efficiency will use a high amount of energy to operate compared to the load that has high efficiency. This project aims to examine the efficiency of electrical appliances by observing power factors and electrical bill using IoT-based energy monitoring device.

2 Studies on Power Monitoring System Using IoT

Revolutions of industrial can be divided into four phases. First, the new sources of energy that discover to run the machines and for the second revolution mass production and electricity generation that develop in the industry. The third revolution, the introduction to the technologies, and lastly this revolution. Lack of resources established in the present world is initiating everyone towards energy-efficient technologies. Among all these resources, power is one which needs to be monitored and controlled as per the need since electricity consumption is increasing day by day [7]. The most basic need by everyone in this world is electricity. Therefore, a graph for energy consumption is increasing day by day. The energy sources being uses without any observation and control. In the future, the sources will be decreased.

The policy has been introduced including the addition of power supply, installation system repairs. However, these efforts did not provide the best results unless they were accompanied by efforts to maintain the quality of the electricity itself, because the amount of electricity tends to fluctuate as a result of changes in the load or the power supply itself [8]. This case happens due to the fundamental of saving the energy for the future itself. The consumer is not being introduced to how the electric

Table 1 Classification and benefits

Level	Benefits
Factory level	Process rescheduling
	Supply contract for electricity
Department level	Quantify improvement measures
	Scheduling energy intensive process
	Saving benchmarking, quantify energy saving
Machine level	Energy forecasting in product design
	Energy labelling
	Proactive maintenance
	Power quality data

appliances work and because of that many consumers are just waste electrical energy. Since the late 1970s, there has been an intense discussion on energy efficiency as a result of the 1973 Oil Crisis. The energy prices, environmental regulations, and last the fundamental of energy efficiency for consumers. The energy monitoring system are being classified into three stages. The classified is being divided by categorized and similarity. Therefore, in Table 1 it shows the classification and benefit of this system. And the benefit is considered the calculation of operational cost, energy efficiency, and power quality.

3 Methodology

Energy monitoring systems are considered expensive and uniquely tailored for each building [9]. Microprocessor that fulfills all the requirement which was used in this project is Arduino UNO and module PZEM-004 T. This module provides simplicity of configuration and connection, as well as ease of coding in Arduino UNO. Amount of power from an electrical load or appliance can be monitored from a certain distance. Figure 1 shows the workflow of this project.

Fritzing software was used to design the connection (refer Fig. 2). Fritzing is an Electronic Design Automation tool for people who is not a practicing electrical engineer. It is a free and open-source software toolkit that can be used at several points in prototyping [10]. The circuit have I2C LCD, Node MCU, PZEM-004t module, 5 V battery current transducer and socket 13A with a load of bulb. In the I2C LCD connection, VCC connected to 5 V Node MCU, SDA to D2 Node MCU, SCL to D1 Node MCU and ground. For 5 V battery connects with Vin Node MCU to gives power for the microcontroller to function and turn ON. The output connection PZEM-004T module, TX connects with D5 Node MCU, RX to D6, Vin to 5 V Node MCU and ground for that will connect with the microcontroller. The incoming connection of PZEM-004T, there are 4 terminal incoming which consist current transducer, neutral

Fig. 1 Flowchart of project

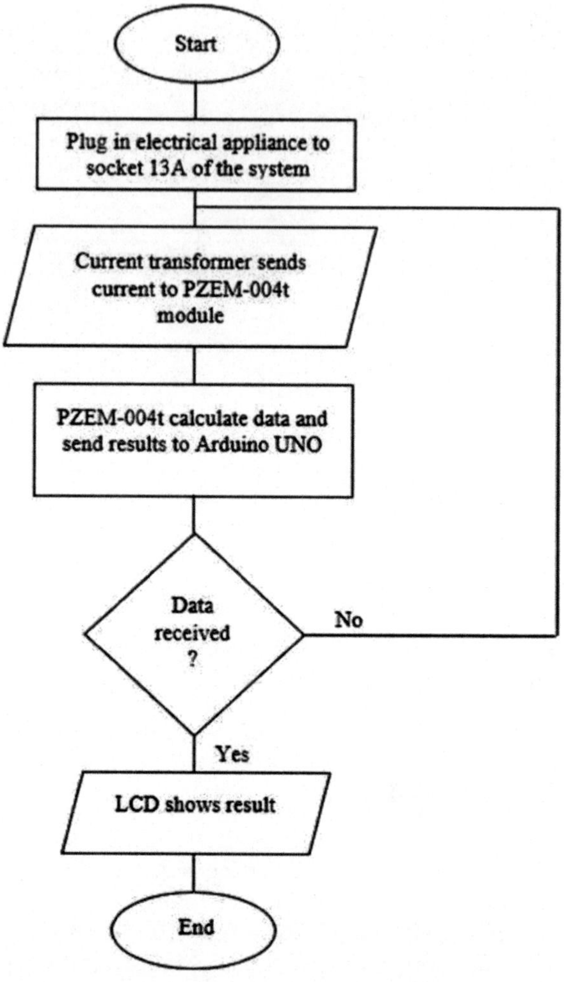

from the socket 13A, live from socket 13A, current transducer and live with neutral to power up the current transducer coil.

As for the results shown in the smartphone which is performed by Blynk apps, the microprocessor sends data to Blynk apps in order to show the results on the smartphone. In Blynk apps, the parameters that shown are power, current, power factor and voltage. Not only that, the Blynk apps can turn out the data gained into graph by real time based on performance of the electrical appliance measured. From that, users can see and monitor the electrical appliance measured and improve electrical usage by changing or replace to a better performance electrical appliance.

The reason using Fritzing software is because of open source and all the tool can be found on the software. The experiment is doing on both side on simulation and actual Arduino. But the result for this experiment is being taken from actual

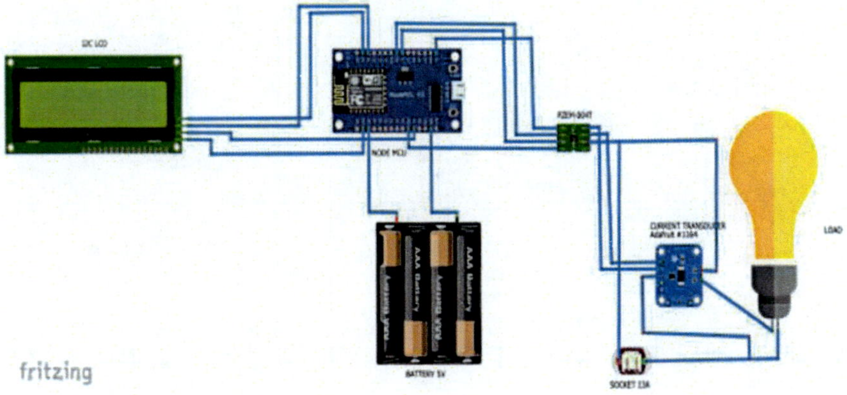

Fig. 2 Connection of whole system

Arduino and being tested on electrical appliance. Both loads are functional when the experiment were done, in which the date and data were being recorded in BLYNK apps and the data also can be converted to graph. For the efficiency calculation of electrical appliance, the power factor that the only parameter that will be considered [4]. When the power factor shows value that is approaching 1, it means the appliance has high efficiency while if the power factor shows values that is not closer to 1, the electrical appliances is not energy efficient. Electrical appliances that were studied in this project were water heater and table lamp.

4 Results and Discussion

Water heater and table lamp were chosen for this project as it can be considered as the most common appliance used in Malaysia, be it in household or even industry. The energy consumed by both electrical appliances gave a massive different in terms of value of energy consumed. Therefore, it can give more benefits for testing out these two appliances. Sample graph generated for water heater tested during the run can be seen in Fig. 3. The data result from Blynk apps which is for parameter current and power factor. Current reading is on left side with an orange colour while power factor is on the right side with a green colour.

Table 2 outlined the value recorded during the test run and power factor calculated. Based on the table above, it was clearly shown that both appliances have different amount of current, power and voltage generated. However, both appliances have moderate efficiency of energy since the power factor for both were calculated to be 0.5. As aforementioned, the efficiency of an electrical appliance is measured by the value of power factor, whether it is closer to 1, or vice versa. In order to avoid higher energy used for electrical appliances and cause the electric bill to go high,

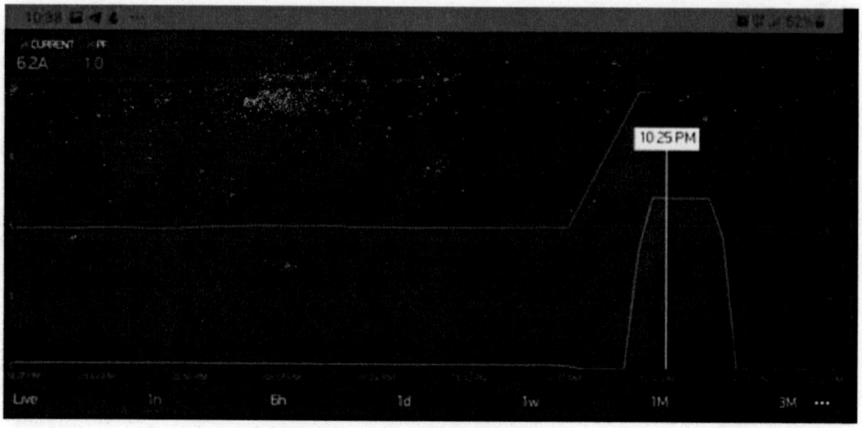

Fig. 3 Graph of current and power factor for water heater

Table 2 Energy used in different load

Parameter	Table lamp	Water heater
Current (A)	0.1	6.2
Power (W)	18.4	1476.8
Voltage (V)	249.2	240.2
Power factor	0.5	0.5

users can monitor and change the electric device to the higher power factor electrical appliances.

5 Conclusion

The device studied in this project is a system that can help consumer to control the electrical usage at any time. This system is very important and useful in order to reduce energy usage in house or building. With observing data on LCD, the consumer can observe their consumption of electrical energy and become more aware about electrical usage. Consequently, energy usage would be reduced and lowered the amount of electricity bill that need to be paid. This system also targets on further controlling of the system which users can turn on or turn off the electrical appliances use at that time.

Acknowledgements Authors would like to thank Ministry of Higher Education Malaysia and Universiti Malaysia Pahang for funding under grant PDU223204.

References

1. Luo X, Wang J, Dooner M, Clarke J (2015) Overview of current development in electrical energy storage technologies and the application potential in power system operation. Appl Energy 137:511–536. https://doi.org/10.1016/j.apenergy.2014.09.081
2. Chong C, Ni W, Ma L, Liu P, Li Z (2015) The use of energy in Malaysia: tracing energy flows from primary source to end use. Energies 8:2828–2866. https://doi.org/10.3390/en8042828
3. Tan CS, Maragatham K, Leong YP (2013) Electricity energy outlook in Malaysia. IOP Conf Ser Earth Environ Sci 16:012126
4. El Sheltawy ST, Al-Sakkari EG, Fouad MMK (2019) Waste-to-energy trends and prospects: a review. In: Waste management and resource efficiency. Springer, Singapore, pp 673–684. https://doi.org/10.1007/978-981-10-7290-1_56.
5. Jaibee S, Abd Rahim AK, Mohamad F, Jamian S, Kiong SC, Seiji Y, Muhd Nor NH (2014) Review on current status of waste electric and electronic product in Malaysia. In: International integrated engineering summit 2014. Trans Tech Publications Ltd., pp 898–907. https://doi.org/10.4028/www.scientific.net/AMM.773-774.898.
6. Ali MH, Zailani S, Iranmanesh M, Foroughi B (2019) Impacts of environmental factors on waste, energy, and resource management and sustainable performance. Sustainability 11:2443. https://doi.org/10.3390/su11082443
7. Sindhuja P, Balamurugan MS (2015) Smart power monitoring and control system through Internet of Things using cloud data storage. Ind J Sci Technol 8. https://doi.org/10.17485/ijst/2015/v8i19/76698.
8. Despa D, Nama G, Muhammad M, Anwar K (2018) The implementation Internet of Things (IoT) technology in real time monitoring of electrical quantities. IOP Conf Ser Mater Sci Eng 335:012063. https://doi.org/10.1088/1757-899X/335/1/012063
9. Sesotya Utami S, Faridah Azizi NA, Kencanawati E, Tanjung MA, Achmad B (2018) Energy monitoring system for existing buildings in Indonesia. E3S Web Conf 42:01003. https://doi.org/10.1051/e3sconf/20184201003.
10. Knörig A, Howell B (2010) Advanced prototyping with fritzing. In: TEI'10—Proceedings of 4th International conference on tangible, embedded, and embodied interaction, pp 341–343. https://doi.org/10.1145/1709886.1709970

Investigation of IoT Devices for Fire Alarm Detection Systems Using Raspberry Pi

Nik Nurjihan Nisrina Nik Zaid and Norazlianie Sazali

Abstract This study investigated the design of a fire alarm detection system based on the Internet of Things (IoT) and the potency of an alarm system with the IoT in reducing time taken for system to alert the user. The design utilizes a system prototype with wireless access in an open-source physical computing platform based on Raspberry Pi technology using a personal computer that provides the communication and user interface for both the control system and the fire alarm systems. This includes wireless connectivity between sensors and apps for monitoring and control. The sensors using to measure the related parameters that is temperature, presence of gas and humidity of the area. The Raspberry Pi is programmed to turn on the buzzer when the temperature and smoke reach a threshold value. This system equipped with WiFi module built in the Raspberry Pi will send the notification to the user via Blynk apps. The trial findings demonstrated the Raspberry Pi's affordability, effectiveness, and responsiveness in reducing the time it takes for the system to alert the user.

Keywords Internet of Things (IoT) · Raspberry Pi · Fire alarm detection · Wireless connectivity

1 Introduction

According to the Malaysia Fire and Rescue Department (BOMBA), 96.7% of reported fires in 2019 were caused by human error. Many products such as smoke detector, fire alarm and automatic sprinkler are designed to prevent the spread of fire. Sensor-based technology has become more popular as technology has improved, and construction costs have decreased. Raspberry Pi, a low cost single board credit card-sized computer, has made it possible to build a range of automated and monitoring devices with low power consumption and higher processing capabilities at a lower

N. N. N. N. Zaid · N. Sazali (✉)
College of Engineering, Universiti Malaysia Pahang, 26600 Pekan, Pahang, Malaysia
e-mail: azlianie@ump.edu.my

C. N. Rosyidi et al. (eds.), *Proceedings of the 6th Asia Pacific Conference on Manufacturing Systems and 4th International Manufacturing Engineering Conference*, Lecture Notes in Mechanical Engineering,
https://doi.org/10.1007/978-981-99-1245-2_12

cost [1]. The proposed fire alarm system in this report combines low-cost equipment, networking, and wireless communication. The early identification of a developing fire incident, as well as the alerting of the house's occupants and fire rescue organizations, is a vital component of fire safety. This is where fire detection and warning devices come in. To begin, the occupant must be warned by a fire alarm fitted with the system as soon as possible in order to detect the presence of fire, smoke or heat in the house. The Blynk app, which was installed to reach users through the Internet, will also alert occupants via their smartphone. Another standard function is the transmitting of an alarm warning signal to the fire department or another emergency response agency. Electrical, air building or special process processes can all be turned off and automatic suppression systems can be activated as a result [2]. The fire detection system comes in a variety of shapes and sizes to help prevent fires from forming. A safe person can be a powerful fire alarm similar to those used in the past that detects flame, smoke and the presence of fire. Humans on the other hand could be less powerful because they are not all present when a fire begins. Automatic detectors are created to mimic one or more of the human senses of touch, smell, and vision. Thermal alarms are similar to our ability to detect high temperatures, and smoke detectors are similar to our ability to detect low temperatures are similar to our sense of smell and flame detectors are electronic eyes. A properly chosen and automatic detectors were placed can serve as a highly accurate fire sensor.

2 Literature Review

Modern fire alarms with automatic activation are used to detect fire as soon as possible. They were designed to work nonstop for 24 h a day, seven days a week. Flame sensors, heat sensors, smoke sensors carbon monoxide sensors, and computer-controlled cameras are all forms of fire detection sensors. Typically, fire detection sensors can activate when a critical value is reached. There are three types of fire detectors: point, linear, and sampling. Point fire detectors collect data in the area where they are placed. The linear one is made up of components that resemble cameras. They are based on laser technology and are mounted in on the opposite walls of the space in front of each other. This form of detector can detect objects in areas up to 100 m long. Sampling fire detectors are the kind of smoke detectors that can be installed in the ventilation system or elsewhere and take air samples at predetermined intervals [3].

2.1 *Arduino Based Fire Detection and Alarm System Using Smoke Sensor*

Previous work from Tun and Htay discussed on the fire alarm system made up of smoke detectors, heat detectors and infrared detectors as well as a control unit and an alarm system [4]. The unit for fire detection is based on temperature and smoke measurements taken at the same time. The alarm fire monitoring system reported fires that were not detected by smoke sensors and were detected for shorter periods of time than when smoke sensors were used alone. Sensor data from smoke, combustion and temperature were used in previous fire detection algorithms. The smoke sensor warns when the analogue output signal reaches or exceeds the threshold value. Analogue smoke, carbon monoxide (CO) and temperature sensors are included in the node. The fire alarm system should notify building occupants of the presence of fire signs such as smoke or high temperatures in a reliable and timely manner.

2.2 *Rebundant for Fire Monitoring System Using Raspberry Pi and Arduino Uno*

A research project involving two LM35 temperature sensors has been used to develop a redundancy principle to enhance the feedback functionality of both temperature sensors to be tested inside the space [5]. At regular intervals, the obtained values are automatically uploaded to the Cloud by Raspberry Pi through the NodeMCU Wi-Fi module. The Thingspeak framework is used for tracking and the user can see the uploaded values in a graphical format. Arduino Uno functioned as interfacing with other Arduino boards, Arduino shields, Raspberry Pi boards and can control relays, LEDs, servos, and motors as an output. To install all of the programs that provide instructions for the proper operation of this computer, this project used Arduino Uno, NodeMCU and Raspberry Pi. Finally, the buzzer sensor's job in this alarm when the temperature rises above the set point. When the temperature sensor detects a reading inside the room, the data is uploaded to the cloud and can be displayed graphically from both sensors on the Thingspeak network. As the temperature rises above 30 °C, the buzzer sounds a warning to alert the user to search for signs of a fire and to take precautions to avoid a fire.

3 Flowchart

Figure 1 depicts the entire project's series. The flow begins and the input flow is read by both sensors (smoke and temperature). The sensor's input will be processed by a microprocessor in the decision stage which will classify the gas concentration and heat from the sensor. The flow will return to input flow if no smoke or fire is observed. If YES, or if the gas concentration and temperature reach the threshold, the process will proceed to the next flow which will activate the alarm/buzzer. When a fire or smoke is detected, a buzzer will sound to warn the building's occupants. In the next step of the procedure, the microprocessor will send a signal to an app on the user's smartphone, alerting the user or occupant to a fire or gas leak in the house. After the whole phase has been completed, the flow will come to an end.

4 Results and Discussion

4.1 *Effectiveness an Alarm System with IoT to Reduce Time Taken to Warn the Users*

Three experiments with varying ranges were set up to measure the time it took for the sensor to detect and send the notification to the user while ringing the buzzer and lighting up the LED. The test approach involves simulating a gas leaking scenario with open air, a sensor capable of picking up the gas in all three tests, and a variety of time responses as in Table 1.

The gas type used in the gas sensor test was butane gas from a pocket lighter. For the first test, the distance between the sensor and the gas source was increased from 1 to 3 to 5 cm. The MQ-2 LED turns on when the sensor detects the gas, indicating that the sensor has detected the gas. The shortest distance resulted in the quickest detection, taking only 1 s to detect, 5 s for the buzzer and LED to beep, and 5.2 s for the user to get the notification. With only 2.3 s for the sensor to detect, 8 s for the buzzer and LED to sound, and 8.4 s for the user to receive the notification, the medium distance gave a different result. The detection time was the slowest at the farthest distance, taking only 4.5, 15 s for the buzzer and LED to beep, and 15.3 s for the user to receive the notification. Because of the open-air test situation, the results of all tests were varied. It tampered with the gas concentration in order for the sensor to notice it.

The threshold for the gas to trigger above 300 ppm was determined by the researchers, and the message "GAS LEAK" was shown. It will print "Safe Area" in addition to the restriction. When the sensor's reading exceeds the threshold value selected, the buzzer and LED on the device will be triggered. The screenshot below shows a smartphone with an alarm issued during the test. If the user has a strong and reliable internet connection on their smartphone, they will be able to get the notification without being limited by range. In this case, butane gas is one sort of gas that is readily available. The MQ-2 sensor can detect six different types of gas, including butane. The test is sufficient to demonstrate the sensor's and system's functionality. With the system's time response documented, this system has improved greatly in comparison to the existing detector on the market, which is expensive.

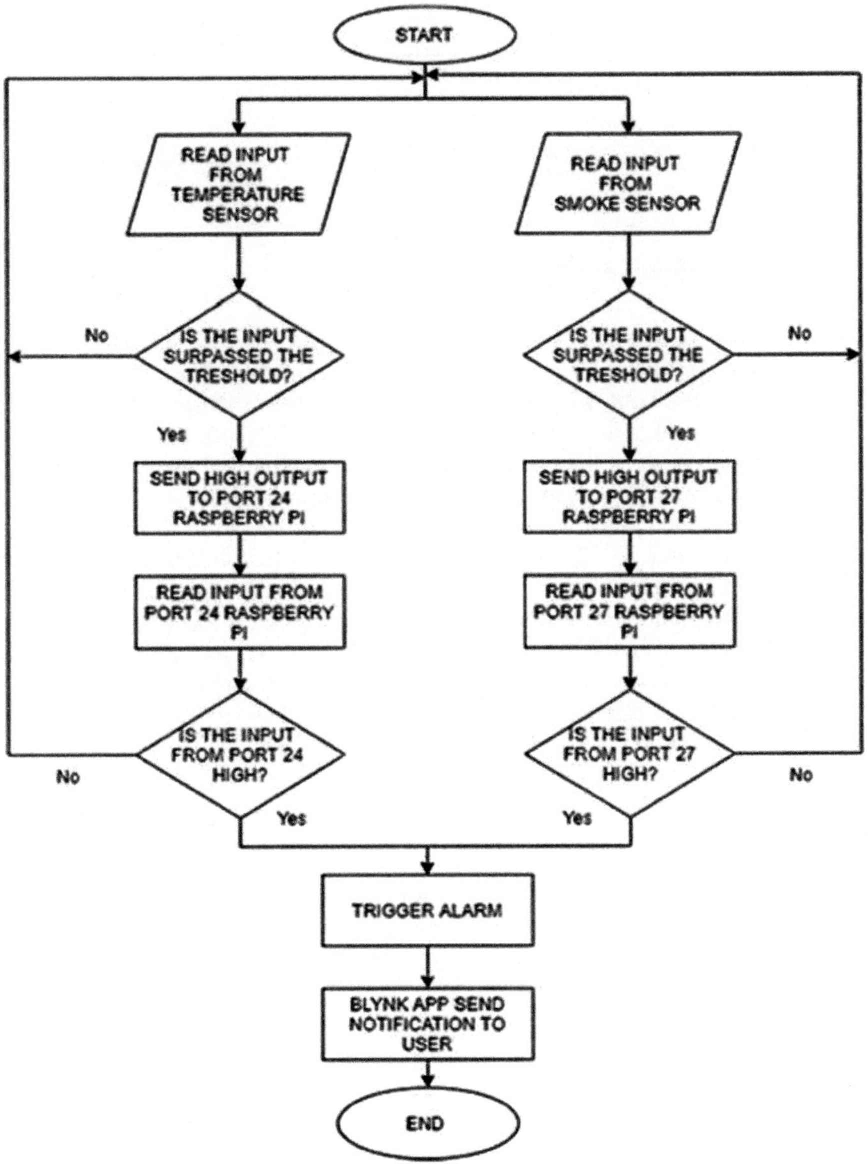

Fig. 1 Flowchart

Table 1 Results of time taken to warn the users

Test	Test 1	Test 2	Test 3
Range fire and sensor (cm)	2	3	5
Time for sensor to detect (s)	1	2.3	4.5
Time to buzzer and LED (s)	5	8	15
Time to send the notification (s)	5.2	8.4	15.3

4.2 *Fire Breakout Detection Test*

The pocket light butane lighter from the previous experiment was utilised as the heat source for the temperature sensor experiment. The research method is to simulate an open-air fire breakout, which allows the sensor to detect temperature rises with distinct time responses in all three trials. The first test distance between detectors and heat sources is 2 cm, then 4 cm, and finally 5 cm. To avoid the sensor melting when it is too close, the beginning point is set at 2 cm.

The maximum range between sensor and firing, as shown in Table 2 is 5 cm, while the minimum range is 2 cm. The longer DHT11 sensor distance takes more time, according to the results. The sensor's range is 2 cm, while the test 1 range is 2 cm. It took Blynk 11 s to convey the message to users. When the distance is raised to 5 cm, test 3 shows that the detecting sensor increases by 7 s from test 1. The increasing distance between the sensor module and the heat source was the condition that altered the discrepancies between all trial results. As a result, the time it takes for Blynk to transmit the message is increasing. Despite the time difference, all three experiments had a much faster response time to notify the tenant. To fix the problem, researchers need to update the sensor to a laser so that it can identify fire outbreaks faster, reducing response time. The researcher set a threshold of 33°, which causes the python shell to print "FIRE WARNING." Both the gas and fire breakout tests were accomplished, and the results were satisfactory, with both sensors and notifications arriving on time. The user can take further action to put out the fire and call the fire service if the residence is unattended after being warned by the buzzer, LED, and message.

Table 2 Result of time taken to warning user based on fire detection

Test	Test 1	Test 2	Test 3
Range fire and sensor (cm)	2	4	5
Time to buzzer and LED (s)	10	15	17
Time to send the notification (s)	11	16	18

5 Conclusion

The major purpose of this project is to solve problems that arise in Malaysia as a result of fires and to reduce the number of people who die as a result of fires. This project is an IoT-based system that can detect a fire in a building and alert the users. Framework for a building fire protection device using Raspberry Pi and IoT for infinite reachability was developed. This microprocessor is a very capable and reliable small computer to compute a security system in the building whether industrial or home automation. It is also very versatile and can be attach with multiple device and sensor attachment. This objective is successfully obtained where the Raspberry Pi able to deliver an instant alert message no matter the condition, with no range restriction applied. The system has been proved can reduce response time taken for user to be notified.

Acknowledgements Authors would like to thank Ministry of Higher Education Malaysia and Universiti Malaysia Pahang for funding under grant PDU223204.

References

1. Nayyar A, Puri V (2015) Raspberry Pi-A small, powerful , cost effective and efficient form factor computer: a review. Int J Adv Res Comput Sci Softw Eng 5
2. Reddy AR, Sai MVR, Vamsi J, Raju CSK (2020) Automatic fire detection and alert system. J Xi'an Univ Archit Technol 12
3. Petrov K (2012) Determining the best location of smoke sensor in office room
4. Tun MZ, Myint H (2020) Arduino based fire detection and alarm system using smoke sensor. Int J Adv Sci Res Eng 6:89–94. https://doi.org/10.31695/IJASRE.2020.33792
5. Sazali L (2019) Redundant for fire monitoring system using Raspberry Pi and Arduino Uno J Comput Technol Creat Content 4

Electric Motorcycle Adoption Research: A Bibliometric Analysis

Yuniaristanto, Wahyudi Sutopo, Muhammad Hisjam, and Hendro Wicaksono

Abstract This paper presents a bibliometric analysis to find the trends and future research opportunities in electric motorcycle (EM) adoption. Bibliometric analysis is performed in three stages: data collection, data visualization, and data analysis. This analysis involves 31 articles published in the Scopus database from 2012 to 2022. The results show that this research area has increased and peaked in 2021. Sustainability (Switzerland) and Transportation Research Part D: Transport and Environment are the two leading journals in this research area. Content analysis reveals future research opportunities and trends related to the integration of choice theory and technology acceptance theory on EM or electric scooter (ES) adoption. In addition, it is interesting to study the sharing system on EM or ES in countries with high motorcycle density levels.

Keywords Bibliometric analysis · Electric motorcycle adoption · Choice theory · Technology acceptance theory · Sharing system

1 Introduction

In many Asian countries, motorcycles are the fundamental transportation mode because they are cheap, have low operating costs, and can adapt well to existing road conditions [1, 2]. Asia has the world's largest motorcycle population and contributes nearly 85% of global motorcycle sales. The motorcycle population in East and South Asian countries is growing at around 8% every year, where in 2017, it reached 554.7 million units [3]. The number of motorcycles certainly impacts increasing air pollution and carbon emissions which are relatively high compared to cars. So the

Yuniaristanto (✉) · W. Sutopo · M. Hisjam
Universitas Sebelas Maret, Ir. Sutami St. No. 36A, Surakarta 57126, Indonesia
e-mail: yuniaristanto@ft.uns.ac.id

H. Wicaksono
Jacobs University Bremen gGmbH, Campus Ring 1, 28759 Bremen, Germany

C. N. Rosyidi et al. (eds.), *Proceedings of the 6th Asia Pacific Conference on Manufacturing Systems and 4th International Manufacturing Engineering Conference*, Lecture Notes in Mechanical Engineering,
https://doi.org/10.1007/978-981-99-1245-2_13

electrification of vehicles, especially motorcycles, is becoming more urgent to be implemented in the Asian motor vehicle market.

Studies on adopting electric motorcycles (EM) are still rare [3]. This paper focuses on the study of the adoption of EM. EM differs from electric bicycles (e-bikes), where EM does not need a pedal drive. EM and e-bike are subgroups of the powered two-wheelers (PTW) category. Based on the previous description, the research questions can be written as follows:

Q1: What is the trend of scientific publications regarding EM adoption?

Q2: What information can be found from this research trend?

Q3: What is the direction of future research in this area?

This bibliometric analysis has the following objectives:

- Providing bibliometric information on scientific research extracted from Scopus;
- Obtaining quantitative data from selected articles;
- Using variables such as number of publications, number of citations per article, number of documents per author, author network map, and keywords to identify leading journals and authors in this research area;
- Use content analysis to identify future research directions and opportunities in this research area.

Finally, this paper is arranged with the following structure. Section 2 describes the methodology. Section 3 presents the results of the bibliometric analysis as well as discussions that discuss opportunities and future research directions. Section 4 concludes the paper and implications for future research.

2 Methodology

Bibliometrics is an analytical method used to identify scientific trends and systematize research [4]. The documents in this study were collected from the Scopus database, an ideal database for bibliometric analysis, including information published in indexed journals. The stages of bibliometric analysis refer to the research of Wangsa et al. [5] and Barbosa et al. [6]. Bibliometric analysis was performed in three stages: data collection, data visualization, and data analysis. At the data collection stage, there are two processes: initial search and refinement. In the initial search, two rules related to EM adoption and the publication year limit were used as follows:

- Rule 1: TITLE-ABS-KEY ("electric motorcycle" OR "electric scooter" OR "electric powered two-wheeler" OR "e-motorcycle" OR "e-scooter").
- Rule 2: TITLE-ABS-KEY("adoption" OR "intention" OR "willingness to").

Then, articles were limited in the year of publication: 2012 – 2022, and found 84 articles. Next, the first refinement process was performed by limiting it to document type: article; source type: journal; language: English and found 56 articles. The second refinement process was carried out by reading the abstract and full text of 56 articles, resulting in 31 articles.

In the second stage, data visualization is analyzed on the following elements: annual publication, top journal list, and most articles cited using MS-excel. In the third stage, data analysis is performed to identify opportunities and directions for future EM adoption research by conducting content analysis. The content analysis includes five categories: models or theories, factors, research objects, research methods, and countries. Then, opportunities and directions for future EM adoption research are sought based on content analysis.

3 Results and Discussion

The bibliometric analysis begins with the visualization of the main bibliometric data. This visualization includes the following elements: (1) annual scientific publications, (2) journal sources, and (3) the number of citations per article. Then proceed with bibliometric data analysis through content analysis, including models or theories, factors considered, research objects, research methods, and country of study.

3.1 Data Visualization

Article Developments and Trends Bibliometric data was downloaded from the Scopus database in June 2022. Based on this data, the growth in annual publications with the EM adoption research from 2012 to 2022 shows a polynomial function, $y = 0.0133 \times 3 - 0.085 \times 2 + 0.237 \times$ with an R-value of $2 = 0.68$ and can be illustrated in Fig. 1a. The number of published articles peaked in 2021 at 13 articles. By referring to the polynomial function at the end of 2022, it is estimated that the total publications will only reach ten articles, which is still lower than in 2021. Figure 1b shows the number of annual publications and cumulative annual publications, which show an increasing trend from 2012 to 2022. However, 2012, 2014, and 2016 found no articles regarding EM adoption. In addition, between 2012 and 2019, the number of publications in this research area is still rare, with only one to two publications. This result shows that research interest in this period is still low.

Meanwhile, since 2020, the number of publications has increased significantly, reaching its peak in 2021, with 13 publications. This fact may be influenced by the emergence of government policies in Asian countries such as Indonesia, India, Thailand, and Vietnam to migrate from fossil fuel vehicles to electric vehicles (EV). In addition, Taiwan, the earliest country in the migration to electric vehicles, especially EM, has begun to develop an e-scooter sharing system (ESS) to accelerate the migration process. This condition is caused because Taiwan is a country that has a high density of motorcycles [7]. This fact causes much research on EM adoption or ESS

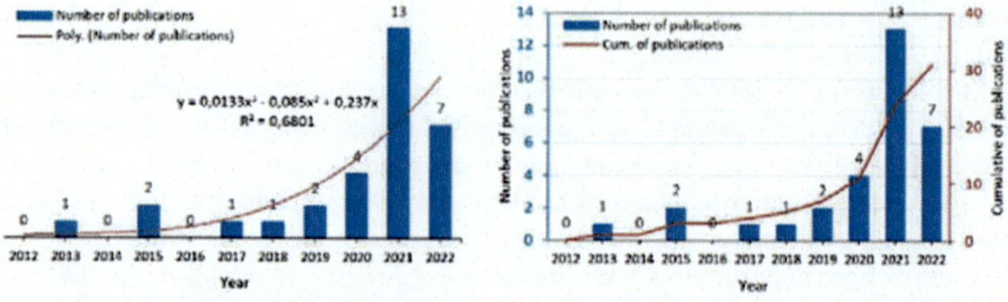

(a) Trends in the number of annual publications (b) Cumulative publications on EM adoption

Fig. 1 Annual publication growth of EM adoption area

adoption from Taiwan. The number of publications in 2022 is still lower than in 2021 because the 2022 data is still being updated. The facts show that research in this area will begin to grow in 2020 and is predicted to increase in the coming years.

Journal Productivity Furthermore, the sources of journals that have published the most papers from 2012 to 2022 can be identified. The top five leading journals identified are Sustainability (Switzerland) (6), Transportation Research Part D: Transport and Environment (6), Transportation Research Part A: Policy and Practice (3), Energies (2), and Journal of Business Research (2). The Journal of Transportation Research Part D: Transport and Environment is the journal that publishes the most papers in 2021, reaching four articles.

Highest Citation Table 1 shows the five most cited articles from 2012 to 2022. The most cited articles are Wu et al. [8] who studied purchase intention on EM in Taiwan by adopting the technology acceptance model (TAM) with 65 citations in seven years or 9.29 citations/year (C/Y). The second most cited article is Eccarius and Lu [7], which investigates the intention to use e-scooter sharing services for university students in Taiwan by adopting the theory of planned behavior (TPB). This article was cited 52 times in two years or 26 citations/year and is the article with the highest citations/year ratio. Jones et al. [9] is the third most cited article examining the effects of economic incentives and technological improvements on EM adoption in the household in Vietnam, with 46 citations in nine years or 5.11 citations/year. Another interesting fact is that five of the ten main articles took up cases in Taiwan, equivalent to 50%.

3.2 Data Analysis

Types of Theory The types of theories or models that researchers most often adopt can be classified into choice, theory of planned behavior (TPB), technology acceptance model (TAM), and unified theory of acceptance and use of technology

Table 1 The top five citations of the article from 2012 to 2022

Rank	Author(s)	Article Title	TC	C/Y
1	Wu et al. [8]	Green purchase intentions: An exploratory study of the Taiwanese electric motorcycle market	65	9.29
2	Eccarius and Lu [7]	Adoption intentions for micro-mobility—Insights from electric scooter sharing in Taiwan	52	26.00
3	Jones et al. [9]	The effect of incentives and technology on the adoption of electric motorcycles: A stated choice experiment in Vietnam	46	5.11
4	Guerra [10]	Electric vehicles, air pollution, and the motorcycle city: A stated preference survey of consumers' willingness to adopt electric motorcycles in Solo, Indonesia	40	13.33
5	Bieliński and Ważna [11]	Electric scooter sharing and bike sharing user behaviour and characteristics	38	19.00

(UTAUT). In addition, None means that researchers develop their conceptual models without adopting existing theories. While, Other means that several other theories are applied, each represented by one article. The number of articles that do not adopt the existing theory is six articles (19.4%), the Choice theory is six articles (19.4%), TPB, TAM, and UTAUT are three articles each (9.7%), and other theories a total of 10 articles (32.3%). Other theories applied include transaction cost economics (TCE), innovation resistance, domain-specific innovativeness, diffusion of innovation (DOI), customer lifetime value, consumer innovativeness, behavioral reasoning theory (BRT), and behavioral change theory.

Meanwhile, only one article integrates two theories, namely the integration of the technology readiness and acceptance model (TRAM) and the norm activation model (NAM) [12]. The studies referring to TPB, TAM, UTAUT, Other, and None focused on investigating the factors that significantly influence the adoption of EM or electric scooters (ES), dominating 25 articles. Meanwhile, six articles on choice theory focus on estimating consumer choices for several alternative vehicles. So there are still opportunities to conduct future research by integrating choice theory and technology acceptance theory such as TPB, TAM, UTAUT, and DOI on EM and ES objects. Another research opportunity is to develop a conceptual model or integrate several existing theories to determine the factors that significantly influence the adoption of EM or ES.

Categories of Factor The following classification considers the categories of factors considered in the research. This classification includes demographic, psychological, technical, environmental concerns, monetary, and contextual. The number of studies that consider demographic factors is 31 articles; in other words, all researchers consider them in their research. The number of studies that consider psychological factors is 18 articles from 31 articles (58.1%), technical factors are 16 articles from 31 articles (51.6%), environmental concerns and monetary factors are 12 articles from 31 articles each (38.7%), as well as the context factor of 11 articles out of 31 articles (35%). One study considers these six factors by Öztaş Karlı et al. [13]. However, this study examines the factors that influence the interest in using ESS and have not considered consumer choices for several existing EM alternatives. So there are still research opportunities that examine consumer choices for several EM alternatives by considering all these factors.

Types of Research Method Research methods can be classified into six groups: structural equation model (SEM), choice model, linear regression, logistic regression, analysis of variance (ANOVA), and Other. Other means that several other methods are applied, each represented by only one article. The number of articles using the SEM method is 17 articles (47.2%), the choice model is six articles (16.7%), logistic regression is five articles (13.9%), linear regression is four articles (11.1%), ANOVA is two articles (5.6%), and Other is two articles. Other research methods are descriptive statistics and fuzzy set qualitative comparative analysis (fsQCA). SEM, logistic regression, linear regression, and ANOVA methods are used to investigate factors influencing EM or ES adoption intention. Meanwhile, the choice model is used to predict consumer choices for several alternative vehicles, such as choices for e-scooters and e-bikes [11, 14–16], choices for EM or CM [9, 10], and the choice of e-cars or e-scooters [17]. In addition, it can also be used to select EM attributes that are important to consumers when adopting EM [2]. From the previous explanation, it can be found that there are research opportunities by combining or integrating two research methods, such as SEM and the discrete choice model.

Types of Research Object Research objects can be classified into six types of objects: e-scooter sharing (ESS), e-motorcycle (EM), e-scooter (ES), e-bike sharing (EBS), hydrogen e-motorcycle (HEM), conventional motorcycle (CM), and others. Other objects include bike-sharing (BS), e-bike (EB), and e-car (EC) sharing. Although this bibliometric analysis focuses on EM, several articles discuss EM or ES in conjunction with other vehicles. The number of articles that examine the object of ESS is 17 articles (37.5%), EM is eight articles (20%), ES is six articles (15%), EBS is four articles (10%), HEM or CM is two articles (5%), and Other is three articles (7.5%). In addition, several articles studied several types of vehicles together, such as ES and EB [11, 14–16]; EM and CM [9, 10], and EC and ES [17]. Studies on ESS appear to be more dominant than other types of vehicles. So that future research trends will lead to sharing systems including EC, EM, ES, or EB. This research trend is certainly in line with government programs in the world to reduce carbon emissions to prevent climate change and reduce air pollution to improve the quality of public health.

Countries Several countries have become study cases for studies on EM adoption. Based on the country where the study is conducted, it can be classified into seven countries: Taiwan, the United States (US), Sweden, Indonesia, Austria, Macau, and others. The number of articles conducted a study in Taiwan is eight articles (25.8%), the US is three articles (9.7%), Sweden, Indonesia, Macau, or Austria is two articles (6.5%), and other 12 articles (10%). Other means that there are 12 other countries studied by one article each. Taiwan is the country that most often becomes the study case for EM adoption because it has already started studies on EM adoption compared to other countries and has a high density of motorcycles [3]. So that this research opportunity opens for other countries with high CM density, such as India, Indonesia, Vietnam, and Thailand. The EM adoption research in these countries can offer a solution for governments and manufacturers to accelerate EM diffusion and adoption.

4 Conclusion and Future Research

The bibliometric analysis in this paper consists of three parts: data collection, data visualization, and data analysis. Data visualization shows that the leading journals are Sustainability (Switzerland) and Transportation Research Part D: Transport and Environment, with six publications each. Publications in these two journals are growing and reaching a peak in 2021. The article with the highest total citations is Wu et al. [8], with a total of 65 citations, while Eccarius and Lu [7] have the highest total citation ratio per year of 26. Eccarius is the most prolific writer with two articles.

Based on data analysis, the following opportunities and directions for future research can be identified: (1) integrating choice theory and technology acceptance theory such as TPB, TAM, UTAUT, DOI, or others on EM objects, (2) developing conceptual models or integrating several existing theories to investigate the factors that affect the adoption of EM, (3) conducting a study of consumer choice of several EM alternatives by considering all factors including demographic, psychological, environmental concern, technical, monetary, and contextual, (4) integrating two research methods such as SEM and the discrete choice model, (5) applying the vehicle sharing system on EC, EM, ES, or EB, and (6) case studies could be conducted in countries with high CM density levels such as India, Indonesia, Vietnam, and Thailand. This research limitation is attributed to excluding other document sources such as PubMed or Web of Science.

Acknowledgements This research was funded by the Institution of Research and Community Services of Universitas Sebelas Maret through Applied Research Grant number: 254/UN27.22/PT.01.03/2022

References

1. Dorocki S, Wantuch-Matla D (2021) Power two-wheelers as an element of sustainable urban mobility in Europe. L 10:618. https://doi.org/10.3390/LAND10060618

2. Zhu L, Song Q, Sheng N, Zhou X (2019) Exploring the determinants of consumers' WTB and WTP for electric motorcycles using CVM method in Macau. Energy Policy 127:64–72. https://doi.org/10.1016/j.enpol.2018.12.004
3. Eccarius T, Lu CC (2020) Powered two-wheelers for sustainable mobility: a review of consumer adoption of electric motorcycles. Int J Sustain Transp 14:215–231. https://doi.org/10.1080/15568318.2018.1540735
4. Zhao X, Wang S, Wang X (2018) Characteristics and trends of research on new energy vehicle reliability based on the web of science. Sustain 10:3560. https://doi.org/10.3390/SU10103560
5. Wangsa ID, Vanany I, Siswanto N (2022) Issues in sustainable supply chain's futuristic technologies: a bibliometric and research trend analysis. Environ Sci Pollut Res 29:22885–22912. https://doi.org/10.1007/S11356-021-17805-8/TABLES/8
6. Barbosa W, Prado T, Batista C, et al (2022) Electric vehicles: bibliometric analysis of the current state of the art and perspectives. Energies 15:395. https://doi.org/10.3390/EN15020395
7. Eccarius T, Lu CC (2020) Adoption intentions for micro-mobility—Insights from electric scooter sharing in Taiwan. Transp Res Part D Transp Environ 84:102327. https://doi.org/10.1016/j.trd.2020.102327
8. Wu J-H, Wu C-W, Lee C-T, Lee H-J (2015) Green purchase intentions: an exploratory study of the Taiwanese electric motorcycle market. J Bus Res 68:829–833. https://doi.org/10.1016/j.jbusres.2014.11.036
9. Jones LR, Cherry CR, Vu TA, Nguyen QN (2013) The effect of incentives and technology on the adoption of electric motorcycles: a stated choice experiment in Vietnam. Transp Res Part A Policy Pract 57:1–11. https://doi.org/10.1016/j.tra.2013.09.003
10. Guerra E (2019) Electric vehicles, air pollution, and the motorcycle city: a stated preference survey of consumers' willingness to adopt electric motorcycles in Solo, Indonesia. Transp Res Part D Transp Environ 68:52–64. https://doi.org/10.1016/j.trd.2017.07.027
11. Bieliński T, Ważna A (2020) Electric scooter sharing and bike sharing user behaviour and characteristics. Sustain. https://doi.org/10.3390/su12229640
12. Ho C-W, Wu C-C (2021) Exploring intention toward using an electric scooter: integrating the technology readiness and acceptance into norm activation model (tra-nam). Energies. https://doi.org/10.3390/en14216895
13. Öztaş Karlı RG, Karlı H, Çelikyay HS (2022) Investigating the acceptance of shared e-scooters: empirical evidence from Turkey. Case Stud Transp Policy. https://doi.org/10.1016/j.cstp.2022.03.018
14. Flores PJ, Jansson J (2022) Being innovative, fun, and green? Hedonic and environmental motivations in the use of green innovations. J Mark Manag. https://doi.org/10.1080/0267257X.2022.2062426
15. Flores PJ, Jansson J (2021) The role of consumer innovativeness and green perceptions on green innovation use: the case of shared e-bikes and e-scooters. J Consum Behav. https://doi.org/10.1002/cb.1957
16. Reck DJ, Axhausen KW (2021) Who uses shared micro-mobility services? Empirical evidence from Zurich, Switzerland. Transp Res Part D Transp Environ. https://doi.org/10.1016/j.trd.2021.102803
17. Brezovec P, Hampl N (2021) Electric vehicles ready for breakthrough in MaaS? Consumer adoption of E-car sharing and E-scooter sharing as a part of mobility-as-a-service (MaaS). Energies 14:1088. https://doi.org/10.3390/en14041088

Design of Halal Traceability System Model for Edge Sewing Rectangular Veil Products at PT. YKI and PT. KPS Bandung

Gina Rahayu Wardiani, Mohammad Mi'radj Isnaini, and Dradjad Irianto

Abstract Based on the mandate of Law Number 33 of 2014 concerning guarantees for halal products, that all products circulating in the market including goods used or utilized by the public must be halal, one of which is textile and clothing products. Currently, not many textile and clothing products are halal-certified. Whereas textile and clothing products have great potential to develop their business in supporting the strengthening of the value chain in the halal industry. Halal traceability is one of the halal requirements that must be met in the manufacture of halal products, this is stated in the Halal Product Guarantee (JPH) HAS 23000. Halal traceability is also an important factor in maintaining halal integrity. There are various methods used in developing a halal traceability system. One of them is the IDEF0 method. The IDEF0 method can be used in determining plans to capture traceability information needed by internals. In addition, IDEF0's simple notation construction consisting of input, control, output, mechanism (ICOM) becomes the main strength as a communication tool that is easily understood by everyone. Therefore, this study develops a model of a halal traceability system based on critical activities that can change the halal status of a product with the IDEF0 method in the Small Medium Enterprises (SMEs) of Muslim women's clothing. Based on the results of the study, there are six critical activities that affect the halal status of products during the production process, namely, material purchasing activities, material receiving and storage activities, supply to production activities, production activities, finished product storage activities, and outgoing transportation activities.

Keywords Halal traceability · Small and medium-sized Muslim clothing industry · Model IDEF0

G. R. Wardiani (✉) · M. M. Isnaini · D. Irianto
Faculty of Technology Industry, Industrial Engineering and Management, Institut Teknologi Bandung, West Java, Indonesia
e-mail: ginarwardiani@gmail.com

C. N. Rosyidi et al. (eds.), *Proceedings of the 6th Asia Pacific Conference on Manufacturing Systems and 4th International Manufacturing Engineering Conference*, Lecture Notes in Mechanical Engineering,
https://doi.org/10.1007/978-981-99-1245-2_14

1 Introduction

Indonesia is a country that is considered in the Muslim clothing industry. Based on data from the Ministry of National Development Planning [1] in the Indonesian Islamic Economic Masterplan 2019–2024, 30% of SMEs in Indonesia are controlled by the Muslim clothing industry. In addition, Indonesia ranks second globally in the production of Muslim clothing and ranks third as the country with the highest spending on Muslim clothing in 2018 [1]. Therefore, the Muslim clothing sector produced by SMEs has great potential to develop its business in supporting the strengthening of the value chain of the halal industry.

In Islam, clothing is not only used in daily activities but is also used for worship. One of the requirements for clothing used in worship that it is clean from unclean and does not pose a danger to its users [2]. Based on the Decree of the Minister of Religion (KMA) Number 464 of 2020, textiles and clothing are products that must be certified halal because they are included in the category of used goods. According to [3], halal requirements for the textile and clothing industry consist of two, namely technical requirements and management requirements. In technical requirements, halal critical points include raw materials, processed materials, additives, auxiliary materials, production processes, and packaging. Meanwhile, management requirements include the duties and functions of halal supervisors [3].

Halal is a way of life in Islam which includes behavior, speech, clothing, etiquette and food, while haram is something that is forbidden in Islam, but its meaning can change depending on certain contexts because haram can also represent purity [4]. Halal and haram usually refer to food, but nowadays halal is not only about food, halal has covered other areas, such as logistics and supply chain [4]. In line with [5], that halal is in the realm of business and trade, as well as being a global symbol for quality assurance and lifestyle choices.

Halal traceability is a system that allows for the identification of the halal status of products and product ingredients both backward (tracing) and forward (tracking) along the production chain [6]. According to [7], the foundation of halal supply chain management is determined by three factors, namely direct contact with haram, risk of contamination, and perceptions of Muslim consumers. According to [4, 8], with the traceability system, the critical point of halal control can be fully monitored and if the product is suspected to be contaminated by non-halal elements, detailed information can be retrieved so that the point of contamination can be identified and action can be taken to deal with it. Thus, traceability plays an important role in the success of halal supply chain management, besides halal traceability effectively identifies products, processes, logistics, and supply chain actors from upstream to downstream and this information is recorded and stored in a database [9].

According to [10], the development of traceability systems involves various phases from a practical context, including planning, design, analysis and others, and this process involves coordination and communication within and between industries. Furthermore [10] explained that IDEF0 modeling was used to determine plans to capture traceability information needed by internal parties. Therefore, the IDEF0 model is often created as one of the first tasks of system development efforts [11]. In addition, the IDEF0 method can identify the functions that are performed and what is needed to run the system or activity [12]. IDEF0 also focuses on functional relationships to represent what things are done in the process with its ICOM-based modeling diagrams [13]. This is supported by [12], that IDEF0 supports process modeling by decomposing high-level ICOM into a more detailed model that describes the decomposition of the activity hierarchy. Therefore, in developing a halal traceability system, the IDEF0 method was chosen because this method is communicative and can capture the information needed as a company's internal halal traceability tool.

Research on halal traceability, especially for the clothing industry in Indonesia, has not been widely carried out, so it needs a general model that can be used by the industry with reference to HAS 23000. In developing the halal traceability system model, a case study approach is carried out on two SMEs namely PT. KPS and PT. YKI with the object of research is the hood with the edge sewing motif. PT. KPS and PT. YKI have not yet carried out halal certification for hoods with sewing motif on the edges. Therefore, the question of this research is how to develop a model of halal traceability in small and medium enterprises (SMEs) of Muslim women's clothing, especially the edge sewing motif veil product using the IDEF0 method. This research was conducted to help companies fulfill one of the halal requirements in HAS 23000 with the IDEF0 method. Based on an explanation of the research background, the definition of halal traceability in this study is the ability to identify the origin of the product and follow the product trail based on critical halal activities along the supply chain based on company information records. Thus, there is hoped that SMEs of Muslim clothing will be able to produce Muslim clothing that meets the applicable requirements and regulations regarding halal products, as well as being of added value for the company to compete with other competitors.

2 Research Model Development

2.1 Research Reference Model

In developing the halal traceability model, this study refers to the research of [7, 14], and [15]. The study of the three researches became a reference in the development of the halal traceability model, especially in the development of halal traceability information items. To facilitate understanding, the following is an image of the research model of [7, 14, 15] can be seen in Figs. 1, 2 and 3.

HAS 23000 critical activities include material purchasing activities, material and product receiving and storage activities, production activities, material and product transportation activities. The three studies describe the halal integrity factors that companies must pay attention to. The next step is to map each halal integrity factor

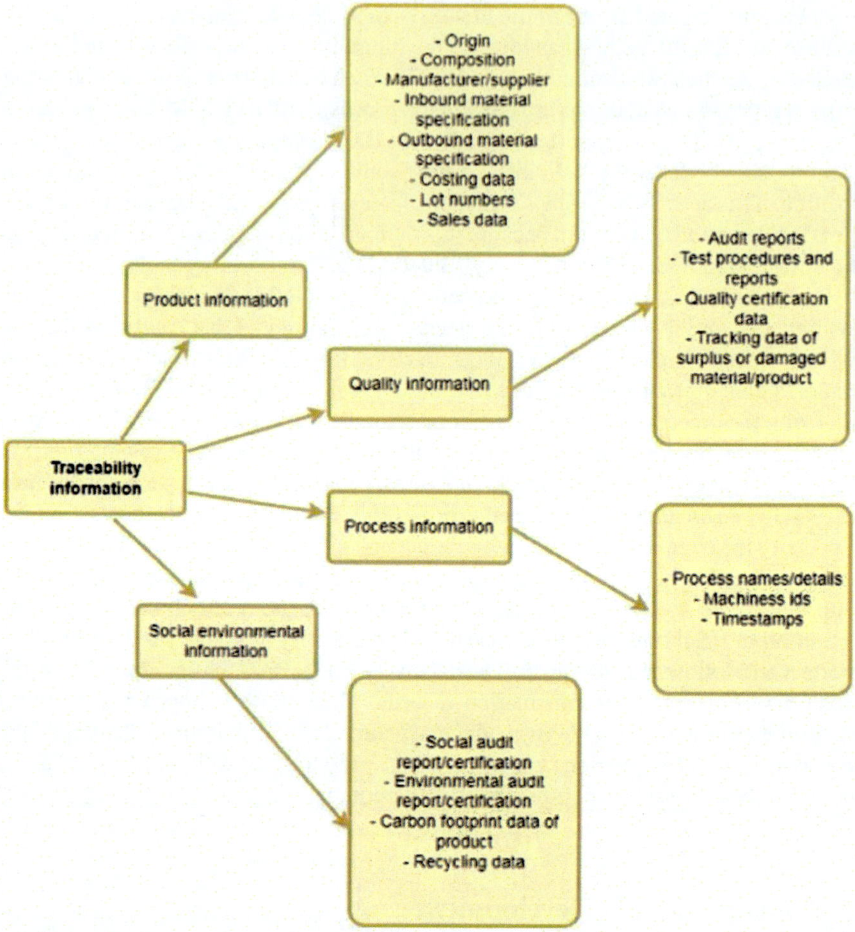

Fig. 1 Traceability information set [15]

related to the critical activity of HAS 23000. The results of mapping information items in developing the halal traceability system model can be seen in Table 1.

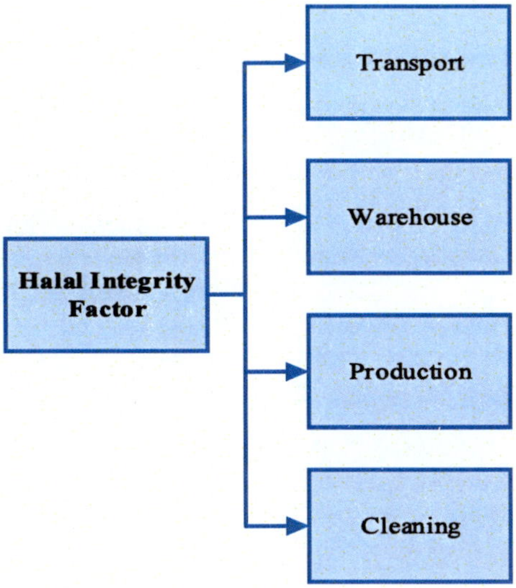

Fig. 2 Halal integrity factor [7]

2.2 Halal Traceability Model Development

As explained in the introduction that in developing the halal traceability model, the IDEF0 approach is carried out. The stages of developing the halal traceability model consist of:

1. Identification of halal requirements related to halal product materials and processes based on Law Number 33 of 2014 concerning halal product guarantees (JPH) related to critical activities of HAS 23000.
2. Mapping of existing business processes for hoods with a sewing pattern in both SMEs.
3. Identification of the availability of information based on existing documents related to critical activities, namely material purchasing activities, incoming transportation activities, material receiving and storage activities, production activities, finished product receiving and storage activities, and outgoing transportation.
4. Based on the identification of the availability of information from existing documents, then develop a proposed model of halal traceability using the IDEF0 method.
5. To test the sensitivity of the proposed model.
6. Perform verification based on the analysis of differences in actual conditions with the proposed model.
7. Conduct an assessment to determine the suitability of SMEs to the proposed model.
8. Integration of points 1, 2, 3, 4, 5, 6, and 7 to ensure whether the proposed model is verified or not.

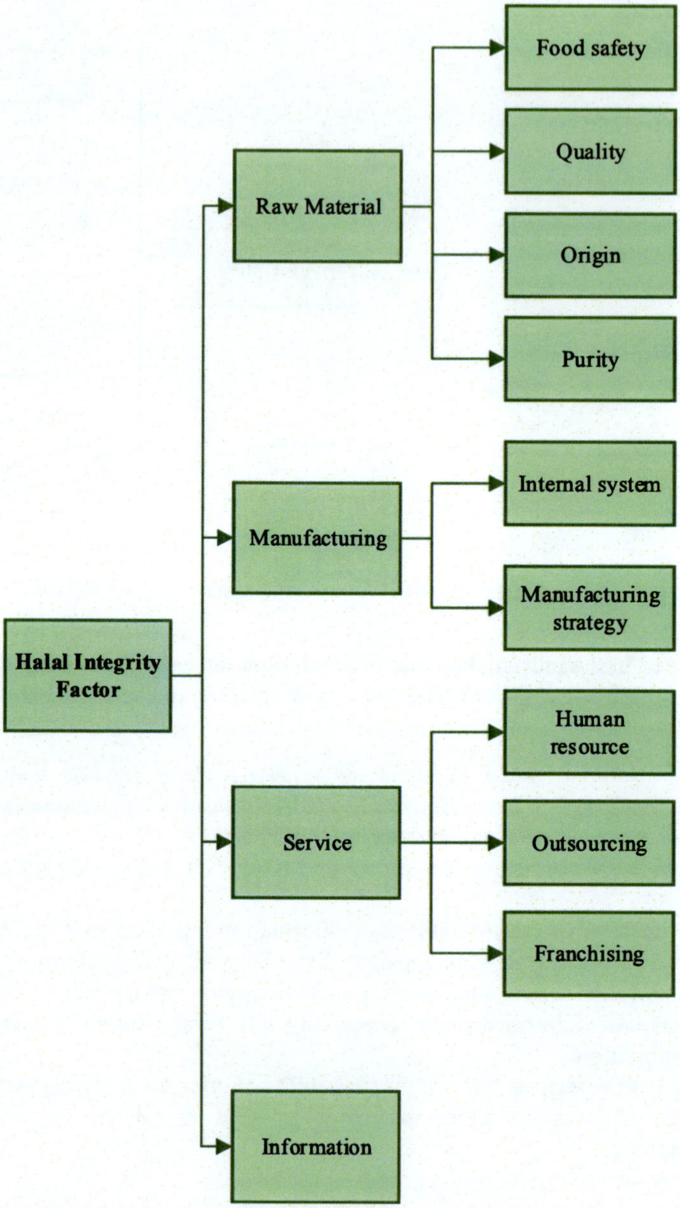

Fig. 3 Halal integrity factor [14]

Table 1 Halal traceability information item mapping based on halal critical activities HAS 23000

Material purchase	Material transportation, material inspection, and material storage	Process production	Product storage and product transportation
Critical activity of HAS 23000			
Supplier [15]	• Transportation and warehouse [7] • Composition, origin, inbound [15] • Origin, purity, safety, and quality [14]	• Quality certification data, tracking data of surplus or damaged material product, process names/details, lot numbers, machiness ids, time stamps [15] • Internal system, manufacturing strategy, and outsourcing [14] • Production [7]	• Transportation and warehouse [7] • Outbound [15]

9. Validate the proposed model based on the fulfillment of halal requirements related to halal product materials and processes (PPH) Law no. 33 of 2014 and the critical activity of HAS 23000 for each SMEs.
10. Integration of points 4, 5, 6, 7, and 8 to ensure whether the proposed model is valid or not.

Based on the explanation of the critical activity information carried out on the two SMEs, then this information becomes the basis for developing a halal traceability system in this study by adding information that is not yet available, especially information related to halal based on previous research. Figure 4 is an overall depiction of the halal traceability system for the edge-sewing motif veil product. The middle box shows the function of the system. In the whole halal traceability system, recorded data relating to material purchasing activities, material receipt and inspection data, production data, material and product storage data, as well as material and product transportation data are depicted in the form of arrows on the left of the box. While the output arrows depicted on the right of the box are the results issued from the halal traceability process. Then the arrow from the top of the box is the control arrow, which means things that must be controlled in the halal traceability process. Finally, there is an arrow from the bottom of the box that defines the person associated with the halal traceability process. Because all processes in halal traceability are related and directly carried out by employees, the arrow shows employees.

3 Result

Based on the results obtained from the development of the halal traceability system, a model of the halal traceability system has been created for the edge sewing motif veil product to meet one of the halal requirements of HAS 23000, namely halal traceability

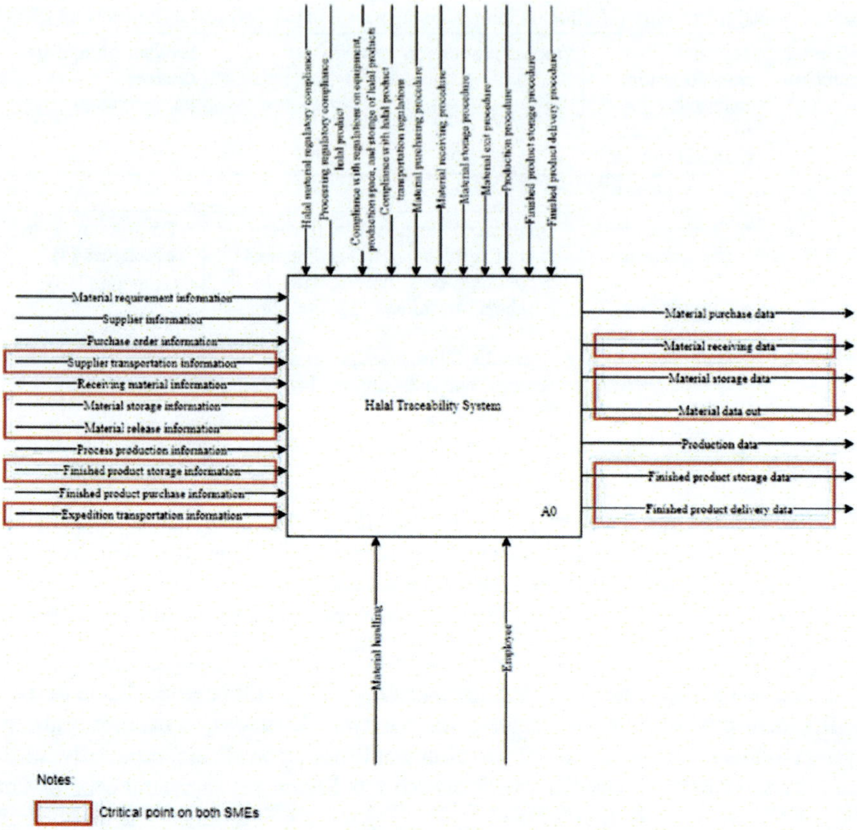

Fig. 4 IDEF A0 halal traceability system

using the IDEF0 method. In developing the model of the halal traceability system, it has been identified that there are six critical activities in producing the hood of the edge sewing motif, namely material purchasing activities, material receiving and storage activities, supply-to-production activities, production activities, product storage activities, and product delivery activities.

Based on the assessment results, both SMEs have low scores on supply to production activities, transportation, and storage which is indicated by red box in Fig. 4. Some points such as information on material requirements, information on suppliers, and information on the production process are not critical points for the two SMEs, because in both SMEs the information can be provided quite completely by the company. So that the halal search process can be carried out properly on this information.

Supply to production activity is part of internal transportation which in good field conditions at PT. YKI and PT. KPS had an error in sending materials from the material warehouse to production. Based on this, it is an input for both SMEs that they must have special records for supply to production activities. Because in these activities there is a material transfer process where the process is also prone to contamination either from the equipment used or from the transfer process. In supply to production activities, it is necessary to record information regarding materials released from the warehouse and their quantities, along with who carries out the expenditure process. Because in the process of moving, even though it is carried out without assistance, the risk of contamination can occur from the hands of the workers who carry out the transfer.

In addition to supply to production activities, even in transportation activities, both incoming transportation (sending of materials from suppliers) and outgoing transportation (delivery of finished products to expeditions), information that must be recorded relating to halal provisions such as transportation conditions that are free from unclean and haram has not been carried out by the two SMEs. Both SMEs have not paid attention to the use of transportation for materials and products. According to [7], transportation activities are activities that are prone to contamination. Therefore, guarantee activities must be carried out on this transportation activity, such as the company ensuring that the transportation used is clean from unclean and haram. If at the time of arrival of the material or during the delivery of the product there is a smell or dirt of unknown origin on the vehicle being used, it must be suspected of being unclean and haram. Therefore, it is necessary to clean it first according to the provisions of the Sharia to avoid contamination that can change the halal status of the material or product.

Then in the activity of receiving materials, the two SMEs do not yet have special records in verifying that the materials used are clearly halal. Although at PT. KPS information regarding the composition and origin can be found from the shipping documents, but both of PT. KPS and PT. YKI has not verified the halal status of the ingredients. Therefore, in the second acceptance activity, SMEs must make special records regarding the completeness of documents related to halal, be it the composition of the ingredients contained on the packaging label, or halal certification data that can be obtained at the time of delivery.

Meanwhile, in storage activities for both materials and finished products, the two SMEs do not yet have special records for the storage of materials and products. Currently, storage separation is carried out based on the type of material and type of product. However, there is no special numbering on the shelves used to store materials and products. In addition, based on the actual conditions in the two SMEs, especially at PT. KPS, some materials such as paper for the printing process are still placed on the floor. To avoid contamination, the company must make improvements in the form of providing special racks for paper materials, as well as numbering the storage racks. This is necessary to facilitate the tracing process if contamination occurs if it is indicated that it comes from storage. In addition, based on observations in the field both at PT. KPS and PT. YKI, the storage of finished products is not all stored on the shelves. This allows for unclean contamination from floor dirt or from animals such as rats or lizards that are susceptible to being in storage. Because the two SMEs do not yet have specific regulations for periodic maintenance of storage areas, preventive measures must be taken such as regular cleaning and ensuring all materials and finished products are stored on shelves to avoid contamination. This is an input for both companies that storage for both materials and finished products is one of the critical points in the halal product process. Storage must be a concern of the company because the storage is very likely to occur contamination.

In addition to the above explanation regarding the results of the assessment of the suitability of existing conditions with model components, there are main things that the two SMEs do not yet have, namely standard procedures for each critical activity. Standard procedures must be owned by the company as a guarantee that all production activities have complied with halal provisions that are free from the risk of unclean and haram contamination. Each standard procedure must include a number as a reference in carrying out activities. Thus, all activities will be traceable to standard procedures and employee awareness to know, understand, and comply with standard procedures in every activity carried out will increase.

Based on the development of the halal traceability model, there are new findings that in carrying out halal tracing, not only critical activity requirements must be met, but also every detail of activities that allow contamination to occur must be paid attention to by the company. As happened in the two SMEs, namely the process of moving materials from the warehouse to the production floor and also the process of moving products from the production floor to the warehouse. This activity is part of a critical activity of transportation. Each activity must be recorded in detail, because the transfer process is done manually. The manual process does not rule out the possibility of contamination. Therefore, in this study, the process of moving materials from the warehouse to the production floor is used as a separate activity for which information must be recorded.

4 Conclusion

Based on the results of the developed model, to realize halal traceability along the supply chain, stakeholders must focus on internal traceability and external traceability. The halal traceability system model developed has met one of the criteria for the HAS 23000, namely the company's traceability in producing halal products. In the process of designing halal traceability system, the captured data can be done by modeling IDEF0. All activities in the supply chain must be recorded and stored in the database management system. The implementation of IDEF0 can be used in the development of a halal traceability system because it provides the ability to simplify processes, document, integrate, and facilitate the exchange of information between stakeholders. Next, for future research is to develop a technical model that is ready for testing, validation, and implementation to finally produce a traceability system that can be implemented if the company is going to implement a digitalization system in its internal company.

References

1. Deputy of Economy Indonesian Ministry of National Development Planning (ed) (2018) Indonesia islamic economic masterplan 2019-2024. PT Zahir Syariah, Indonesia
2. Muflihin M (2018) Konstruksi Indikator Halal Dalam Perkembangan Industri Halal Fashion. Program Magister, Universitas Islam Negeri Sunan Ampel, Surabaya
3. Mastuki (2021) Perspektif Halal Dalam Tekstil dan Fashion. Textalk Daring, 21 Juni 2021. Balai Besar Tekstil Bandung
4. Ab Talib MS, Hamid AB, dan Zulfakar M (2015) Halal supply chain critical success factors: a literature review. J Islamic Market 6
5. Lada S, Tanakinjal GH, dan Amin H (2009) Predicting intention to choose halal products using theory of reasoned action. Int J Islam Middle East Financ Manag 2(1):66–76
6. Zainuddin N, Saifudin A, Deraman N, dan Osman A (2020) The effect of halal traceability system on halal supply chain performance. Int J Supply Chain Manage 9:490–498
7. Tieman M, Vorst J, dan Ghazali M (2012) Principles in halal supply chain management. J Islamic Market 3:217–243
8. Zulfakar M, Anuar M, Ab Talib M (2012) Conceptual framework on halal food supply chain integrity enhancement
9. Khan S, Haleem A, Khan M, Abidi M, Al-Ahmari A (2018) Implementing traceability systems in specific supply chain management through critical success factors
10. Kumar V, Hallqvist C, Ekwall D (2017) Developing a framework for traceability implementation in the textile supply chain
11. Kim S-H, dan Jang K-J (2002) Designing performance analysis and IDEF for enterprise modelling in BPR. I J Prod Econ
12. Giaglis G (2001) A taxonomy of business process modeling and information systems modeling techniques. In J Flexible Manuf Syst
13. Lin F-R, Yang M-C, Pai Y-H (2002) A generic structure for business process modeling. Bus Process Manag J 8(1):19–41
14. Ali M, Tan K, dan Ismail M (2017) A supply chain integrity framework for halal food. Br Food J 119:20–38
15. Agrawal TK, dan Pal R (2019) Traceability in textile and clothing supply chains-classifying implementation factors and information sets via Delphi study

The Design of Cultural Tourism Potential Indicators at Rembang Regency Using SECI Method

Mia Amelia, Lutfia Septiningrum, and Augustina Asih Rumanti

Abstract This research aims to design indicators and measurement instruments to determine the potential of cultural tourism in Rembang Regency. The Rembang Regency Government has not developed cultural tourism potential according to standards. A common practice for determining tourism potential is to design indicators and measurement tools for cultural tourism potential as a valuable tool to assess the situation of destinations to identify and evaluate problems in cultural tourism. The design of cultural tourism potential indicators guided by tourism planning consists of 4A, namely (Attraction, Amenities, Ancillary Services, and Accessibility). SECI is knowledge created through the interaction between tacit knowledge and explicit knowledge. The interaction is the four phases of knowledge conversion: socialization, externalization, combination, and internalization. The SECI method is used to design indicators because the data obtained is in the form of tacit knowledge, so it is necessary to convert knowledge into explicit knowledge. The benefits of this research are that they can find indicators and measuring instruments for cultural tourism potential in Rembang Regency and make it easier for stakeholders to set indicators and measurement tools for cultural tourism potential in Rembang Regency.

Keywords Indicators · 4A · SECI · Cultural tourism · Knowledge conversion

1 Introduction

According to the World Tourism Organization (UNWTO) [1], tourism is one of the most relevant sectors in the world with sustainable global economic growth, which demonstrates its enormous potential to generate development opportunities but also presents significant challenges in terms of sustainability. Tourism development is essential because it can become one of the mainstay sectors expected to contribute to

M. Amelia (✉) · L. Septiningrum · A. A. Rumanti
Telkom University, Bandung 40257, Indonesia
e-mail: miaameliaa@student.telkomuniversity.ac.id

C. N. Rosyidi et al. (eds.), *Proceedings of the 6th Asia Pacific Conference on Manufacturing Systems and 4th International Manufacturing Engineering Conference*, Lecture Notes in Mechanical Engineering,
https://doi.org/10.1007/978-981-99-1245-2_15

Regional Original Income. According to the Central Bureau of Statistics for Rembang Regency, the total area of Rembang Regency is 1036.70 km^2, a reasonably large regency area compared to other regencies or cities in Central Java Province. The following is the number of tourists from Rembang Regency from 2016 to 2020. Data on the number of tourists from Rembang Regency in 2016 was 7137 foreign and 1,222,448 domestic tourists.

In 2017 was a decreasing number of foreign tourists, as many as 23 people and domestic tourists, as many as 987,193 people. One of the causes of the lack of interest and the decline in the number of tourists in Rembang Regency in 2017 was the lack of media that presented tourist information needs [2]. In 2019 there was a decline in foreign tourists to 251 people, and domestic tourists to 915,780 people. In 2020 the number of tourists experienced a drastic decline, with the number of foreign tourists visiting only ten people and domestic tourists 351,731 people. The decline in the number of tourists in 2019 and 2020 in the Rembang Regency was due to the COVID-19 pandemic [3].

Tourist visitors to Rembang Regency experienced fluctuations from 2016 to 2020. However, they tended to decrease, which was not significant, so there was a need for development and additional tourism potential to boost public interest in traveling. A common practice for determining tourism potential is to design indicators and measurement instruments of cultural tourism potential as a valuable tool to identify and evaluate problems in cultural tourism so that appropriate decision-making can be made [4]. An indicator is a regularly selected collection of information necessary for tourism development and management. Much work has been done on clarifying critical issues in tourism sustainability and how indicators can support better decisions and actions. The development and use of indicators are increasingly seen as a fundamental part of overall destination planning and management. They are integral in efforts to promote sustainable development for the tourism sector at all scales [1].

Efforts to design cultural tourism potential indicators require several components of tourism development related to the tourism planning approach consisting of 4A, namely Attractions, Amenities, Ancillary Services, and Accessibility which will be used as the basis for this research [5]. So that it can be managed clearly and accurately measured and tourism potential can be said to be feasible, it must first design indicators so that tourism managers can find out what must be developed for cultural tourism potential. So later, it is expected to become a tourist destination that can help the economy of the Regency Rembang. This research uses the SECI method (Socialization, Externalization, Combination, and Internalization). Based on the introduction that has been submitted, the formulation of the problem for the research is "How is the design of indicators and measuring tools to be able to determine the potential of cultural tourism in Rembang Regency?" and intend to produce a design indicator and measuring instrument to determine the potential for cultural tourism in Rembang Regency.

2 Literature Review

2.1 Knowledge Management

According to Becerra-Fernandez and Sabherwal [6] state that *Knowledge Management* can be defined as an action on what is needed to get the most out of knowledge sources. According to Dalkir [7] there are two main types of knowledge: Tacit knowledge is difficult to say clearly and challenging to enter in words, text, or images. Usually, this is like an idea or idea, and explicit knowledge fills in the picture captured in tangible forms such as words, voice recordings, or pictures, and Explicit Knowledge is knowledge that can be transferred in formal and systematic language, such as through documentation, reports, and databases.

2.2 Knowledge Sharing

Knowledge Sharing is a systematic process of sending, distributing, and disseminating knowledge and multidimensional contexts from a person or organization to other people or organizations who need it through various methods and media [8].

2.3 Knowledge Management Implementation Strategy

Tiwana [9] explains the strategy for implementing KM by providing two alternative strategies that focus on using information technology. The two strategies are:

1. Codification Strategy

The Knowledge Management implementation strategy emphasizes using information technology as a supporting tool because the premise is that all knowledge is codified as a source of knowledge.

2. Personalization strategy

A strategy that emphasizes the role of humans as a source of knowledge. The acquisition, sharing, utilization, and storage of knowledge here rely heavily on humans.

3 Methodology

According to Nonaka and Takeuchi [10] knowledge is created through the interaction between tacit and explicit knowledge. These interactions are expressed in four phases of knowledge conversion (see Fig. 1).

SECI is knowledge created through the interaction between tacit knowledge and explicit knowledge. The interaction is the four phases of knowledge conversion, namely Socialization as a process of transfer of knowledge or experience between individuals that produces tacit knowledge without using language in which there are activities in collecting and sharing knowledge, Externalization is the process of tacit knowledge being articulated into explicit knowledge, Combination is the process of increasing and transferring explicit knowledge by comparing and combining it with the explicit knowledge of others, and Internalization changes explicit knowledge into tacit knowledge and will document knowledge in the form of documents [11].

This study uses a qualitative research method with the SECI Model approach. In this study, the design of indicators of cultural tourism potential in Rembang Regency was carried out. The data needed as input from this research are secondary data, namely the results of literacy studies, and primary data, namely interview data with the Rembang Regency Culture and Tourism Office and interview data with tourism managers at the R.A Kartini Museum. The design of cultural tourism potential indicators uses the SECI model. This knowledge conversion method can generate new knowledge in the form of potential tourism indicators from the tacit knowledge of the tourism office in Rembang Regency.

There are stages in the systematics of indicator design consisting of the preliminary stage in the form of searching for literacy studies, the stage of determining the formulation of problems and research objectives, the stage of data collection and

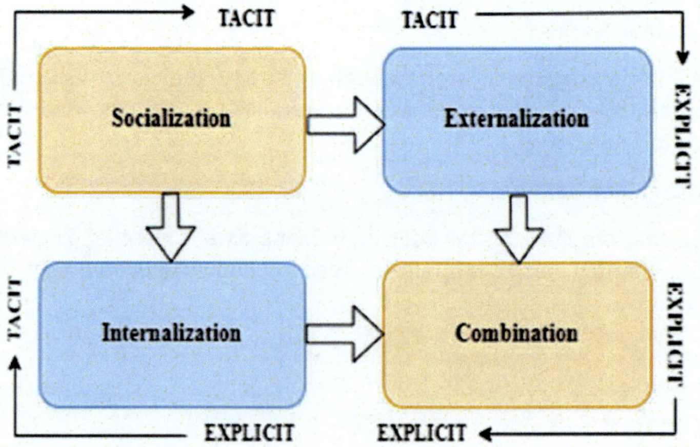

Fig. 1 SECI model [10]

Table 1 Interview results

Narrator	Interview results
Tour manager at the R.A. Kartini Museum	Observation and analysis of tourist attractions based on the 4A component (Attraction, Amenities, Ancillary Service, and Accessibility)
Rembang Regency Culture and Tourism Office	The staff of the Culture and Tourism Office of Rembang Regency in the Field of Tourism Destinations, namely the Sub-Coordinator of Tourism Empowerment and Development, analyzes and explains the indicators and measuring tools considered in the development of cultural tourism potential, for example, in the Attraction parameter, states an opinion on indicators about the uniqueness of cultural tourism with a measuring instrument, namely there are historical relics in every tourist area

processing, the stage of design using the SECI model, the stage of conclusions and suggestions.

4 Research Result

The design standards in this study refer to 4A, namely Attractions, Facilities, Additional Services, and Accessibility. The results of the design of indicators and measuring instruments are carried out using the SECI method with four stages: Socialization, Externalization, Combination, and Internalization.

4.1 Socialization

The socialization stage was carried out by conducting literacy studies, interviewing resource persons, and observing one of the cultural tours. The first resource person is the tour manager at the R.A Kartini Museum in Rembang Regency, which aims to obtain information about the museum's condition in the field so that interviews and observations can be carried out. The second resource person from the Culture and Tourism Office of Rembang Regency aims to obtain information on the indicators needed to build cultural tourism. Table 1 shows the interview results.

4.2 Externalization

In the externalization stage, knowledge changes from tacit knowledge to explicit knowledge [7]. The results of literacy studies and interviews from each resource person at the socialization stage will be documented into explicit knowledge to make

Table 2 Attraction

Indicators	Measuring instruments
The uniqueness of cultural tourism [12]	There are historical relics in the tourist area
	There are unique culinary delights in the tourist area
	There are museums and art galleries
Panorama	There is natural scenery in the tourist area
Cleanliness [13]	There are no doodles in the tourist area building
	Cleanliness is maintained, and no trash is scattered in the tourist area
	Air pollution in tourist areas is 0–50 (good) or 51–100 (moderate)
Convenience [14]	Noise in tourist areas with a sound intensity of 85 dB (decibels)
	Room temperature 18–30 °C in all tourist areas

it easier to learn because when carrying out this stage, the interview results are documented using language that can be understood without losing the meaning of the interviews that have been carried out. The following are the results of literacy studies from component 4A, namely Attractions, Amenities, Ancillary Service and, Accessibility as shown in Tables 2, 3, 4 and 5. Attraction is something that can attract tourists to visit a tourist area. Attractions are considered an essential fundamental component of tourism because every tourist can enjoy tourist attractions.

Amenities are supporting facilities needed in tourism activities and aim to make visiting tourists feel comfortable.

Ancillary Service is Additional services are supporting services that the government or tourism destination managers must provide to organize tourism activities.

Accessibility is access is the ability of tourists to visit tourist destinations through services such as vehicle rental and local transportation from their place of origin.

4.3 Combination

The combination stage is an advanced stage from the Externalization stage, where there is a process of converting explicit knowledge into new explicit knowledge by applying it to explicit knowledge and information. In the Combination stage, explicit knowledge is combined from the results of the literacy study, the results of interviews and observations with each documented resource person, and then summarized in the externalization stage in the form of indicators and measuring tools adjusted. After getting the indicators and measuring tools, then an assessment of the weight of each parameter is carried out by tourism experts based on their level of importance; there are four parameters, namely Attractions (16%), Amenities (22%), Ancillary Service (13%), and Accessibility (49%).

Table 3 Amenities

Indicators	Measuring instruments
Facilities [15]	There is public transportation in the tourist area
	Food and beverage services (restaurants, restaurants, or rice stalls) exist in the tourist area
	There is a travel agent in the tourist area
	There is a tour guide in the tourist area
	There are storage facilities and luggage storage in the tourist area
	There are souvenir stalls selling souvenirs or mementoes of a distinctive culture or local wisdom of a tourist destination in a tourist area
	There is an art or performance stage with the minimum seating width required for one person being 1.5 or about 55 cm in the tourist area
	There is a light fire extinguisher in the tourist area
Infrastructure [15, 16]	There are separate female and male toilets and providing toilets for persons with disabilities
	There is a theatre in the tourist area
	There is an electrical installation or generator in the tourist area
	There is a guide path or marker as a circulation path for people with disabilities
	There is an internet and telephone access network in the tourist area
	There is drainage or sewerage in the tourist area
	There is an ATM in the tourist area
	There are warning signs about cleanliness in the tourist area
	There are closed trash cans (organic and non-organic trash cans) in the tourist area
	There is a health center and pharmacy in the tourist area
	There is a road to the tourist area that the mode of transportation can pass

4.4 Internalization

At the internalization stage, knowledge changes from explicit in the form of designing indicators and measuring tools for cultural tourism potential in Rembang Regency to tacit knowledge in the form of knowledge for cultural tourism managers and the Regional Tourism and Culture Office of Rembang Regency. At this stage, the indicators and measuring instruments are subjectively validated by the tourism office by discussing and confirming four parameters: Attractions with four indicators and nine measuring instruments, Amenities with two indicators and 19 measuring instruments, Ancillary Service with three indicators and eight measuring instruments, and Accessibility with two indicators and ten measuring instruments. With a sequence of parameters by the level of importance, i.e., Accessibility, Amenities, Attraction, and Ancillary Service.

Table 4 Ancillary service

Indicators	Measuring instruments
Information centre	There is a tourism information service and a warehouse storage space for brochures and other goods in the tourist area
Security post [17]	There is a security post in the tourist area
	The tourist area has a CCTV security system and other security equipment
Guide sign [15, 18]	There are signposts in the tourist area at the crossroads position. The horizontal distance between the end of the road and the end of the signpost is 1200 mm, but if there is limited land, the distance can be 450 mm in urban areas and 600 mm in rural areas
	There are warning signs in the tourist area as warnings of possible dangers
	There are command signs in the tourist area that road users must obey
	There are prohibited signs in the tourist area to state actions that are prohibited by road users, for example, no-stop signs, no parking
	There are embossed signs in the tourist area for the disabled

Table 5 Accessibility

Indicators	Measuring instruments
Transportation	The tourist area can be passed by two-wheeled transportation mode
	The tourist area can be passed by four-wheeled transportation mode
	The tourist area can be passed by bus/truck transportation mode
	There is an airport in the tourist area
	There is a train station in the tourist area
	There is a terminal in the tourist area
	There is a seaport in the tourist area
Street	There is a road that has been cast/on the asphalt
	Tourists can easily access roads from one cultural tourism site to another
	There is a pedestrian circulation network system in the tourist area

5 Conclusion

From this research, it can be concluded that knowledge is fundamental and needs to be appropriately managed to support the spread of information and knowledge for various purposes. In particular, storing various experiences, problems, and their solutions and reused to solve cases and form new knowledge. In the design aspect, indicators and measuring instruments were successfully designed and identified using the SECI method based on four parameters, namely Accessibility, Amenities, Attraction, and Ancillary Service.

The research resulted in the design of 4 parameters, 11 indicators, and 46 measuring instruments to build cultural tourism potential in Rembang Regency. According to the tourism office in Rembang Regency, Rembang Regency tourism does not yet have indicators and measuring tools to determine cultural tourism potential. So with the design of cultural tourism potential indicators, it is hoped that the Culture and Tourism Office can establish and use indicators and tools to determine the potential for cultural tourism in the area of Rembang Regency. In building and developing tourism potential, stakeholders can review parameters based on the level of importance of each parameter, starting from Accessibility, Amenities, Attraction, and Ancillary Services.

References

1. WTO (2004) Indicators of sustainable development for tourism destinations a guidebook. The World Tourism Organization, Madrid
2. Prasetyo SD (2020) Media Informasi Berbasis Website Bagi Wisatawan Pada Objek Wisata Jembatan Merah Hutan Mangrove di Desa Pasarbanggi Kabupaten Rembang. UNNES J 2
3. Mukharomah W, Raharja CP (2021) Segmentation, targeting, and positioning Destinasi Wisata Kabupaten Rembang Yang Terkena Dampak covid-19. University Research Colloquium (URECOL), p 132
4. Lee TH, Jan F-H, Liu J-T (2021) Developing an indicator framework for assessing sustainable tourism: evidence from a Taiwan Ecological Resort. Elsevier J 2
5. Cooper J, Fletcher D, Gilbert, Wanhill S (1995) Tourism: principles and practice. Longman, London
6. Becerra-Fernandez I, Sabherwal R (2014) Knowledge management systems and process. Routledge, USA
7. Dalkir K (2011) Knowledge management in theory. Elsevier Butterworth-Heinemann, Burlington
8. Lumbantobing P (2011) Manajemen knowledge sharing Berbasis Komunitas. Knowledge Management Society Indonesia, Bandung
9. Tiwana A (1999) The knowledge management toolkit. Prentice Hall PTR, New Jersey
10. Nonaka I, Takeuchi H (1995) The knowledge-creating company: how Japanese companies create the dynamics of innovation. Oxford University, New York
11. Nakamori Y (2019) Knowledge construction methodology: fusing systems thinking and knowledge management. Springer, Germany
12. Sari UAIPM (2017) Analisis Potensi Objek Wisata di Kota Jambi. e-J UNP 11–12
13. Kementerian Lingkungan Hidup dan Kehutanan (2020) Peraturan Menteri Lingkungan Hidup dan Kehutanan Republik Indonesia Nomor P.14/Menlhk/Setjen/KUM.1/7/2020 Tentang Indeks Standar Pencemar Udara. Jakarta
14. Kementerian Tenaga Kerja Republik Indonesia (1999) Keputusan menteri tenaga kerja Republik Indonesia Nomor : KEP.51/MEN/1999 Tentang Nilai Ambang Batas Faktor Fisika di Tempat Kerja. Kementrian tenaga kerja, Jakarta
15. Kementerian Pariwisata dan Ekonomi Kreatif (2022) Peraturan Menteri Pariwisata dan Ekonomi Kreatif Republik Indonesia Nomor 3 Tahun 2022 Tentang Petunjuk Operasional Pengelolaan Dana Alokasi Khusus Fisik Bidang Pariwisata Tahun Anggaran 2022. Jakarta
16. Suwena IK, Widyatmaja IGN (2017) Pengetahuan Dasar Ilmu Pariwisata. Pustaka Larasan, Denpasar
17. Sumantri D (2018) Strategi pengembangan desa wisata di Kelurahan Jelekong, Kabupaten Bandung. Jurnal Geografi Lingkungan Tropik 30–31
18. Hayati R, Achmadi NS, Adelia S (2021) Implementasi Konsep 6A di Wisata Alam Rammang-Rammang Kabupaten Maros. Hospital Gastron Res J 158

Packaging Design for Kelincidunia Cassava Chips Using Ergonomic Function Deployment (EFD) in UD. Wahyu Pendawa Lestari

Niken Utami Tyastuti, Bambang Suhardi, Rahmaniyah Dwi Astuti, and Annisa Syahliantina

Abstract The visual aspect of packaging is essential to attract customers' attention. The packaging design is the primary identity that distinguishes a product from similar products. One product that pays attention to packaging in maintaining its products is cassava chips. UD. Wahyu Pendawa Lestari is an SME that produces Kelincidunia cassava chips in Rambeanak Village, Magelang. The current packaging uses transparent polypropylene (PP) plastic, and the information is written by screen printing, so it is easily lost. This research uses the Ergonomic Function Deployment (EFD) method in packaging design through House of Quality (HOQ) because the SMEs want to expand market segmentation and meet customer demands. This method can help to identify the customer need, compare with competitors, provide feedback on design results, and strengthen ergonomic aspects from an environmental point of view when choosing materials. The cassava chips packaging design needed was a standing pouch made of paper foil consisting of aluminum foil as an inner layer and art paper as an outer layer. The packaging dimension is 18 cm × 24 cm, accommodating 100 g of chips. Cover packaging using a zip lock and tear notch. The packaging colors used are red–orange gradations and yellow, while the font colors are black and white so that the color contrasts with the background color. The font used is a sans serif type with a size of more than 6 pt. The information displayed on the packaging is in according with the processed food label guidelines from BPOM of 2020.

Keywords Packaging design · Cassava chips · Ergonomic function deployment · House of quality

---

N. U. Tyastuti · B. Suhardi (✉) · R. D. Astuti · A. Syahliantina
Department of Industrial Engineering, Universitas Sebelas Maret, Jl. Ir. Sutami 36A, Surakarta 57126, Indonesia
e-mail: bambangsuhardi@staff.uns.ac.id

C. N. Rosyidi et al. (eds.), *Proceedings of the 6th Asia Pacific Conference on Manufacturing Systems and 4th International Manufacturing Engineering Conference*, Lecture Notes in Mechanical Engineering,
https://doi.org/10.1007/978-981-99-1245-2_16

1 Introduction

Hermawan Kartajaya said that packaging sells what it protects, and packaging protects what it sells [1]. Packaging can inform the product's contents and maintain the quality of the product in it. In the packaging, the visual aspect becomes a significant attraction in designing to attract customers' attention [2]. According to the Undang-Undang RI Number 18 of 2012 concerning food, food packaging is a material used to contain and or wrap food, whether in direct contact with food or not [3]. Almost all products have good packaging for food products, cosmetics, toys, and others. One product that pays attention to packaging in maintaining its products is cassava chips. Cassava chips are snack products made from cassava tubers (Manihot sp) sliced/chopped, fried with or without adding other food ingredients, and permitted food additives [4]. The packaging method strongly affects the physical resistance of the chips, so they are not crushed and overcome the rancidity of cassava chips [5]. The packaging method of cassava chips has been regulated in SNI 01-4305-1996, namely packing the product in a tightly closed container, not being influenced and affecting the contents, and safe during storage and transportation. Packaging that is suitable for chips is plastic packaging and aluminum foil because the packaging has a strong adhesive power to protect chips from changes in texture [6].

UD. Wahyu Pendawa Lestari is one of the SMEs that produces cassava chips in Rambeanak Village, Mungkid District, Magelang Regency. The owner has the desire to be able to expand his target market so that chips can be enjoyed as snacks by all groups and marketed in stalls and marketplaces. Based on the data, the volume of purchases in the last three years has consistently decreased. Therefore, a strategy is needed to increase sales, such as through packaging.

The packaging used is transparent polypropylene (PP) plastic and is closed using a sealing machine, so the packaging is difficult to open and close again, and the chips are not crispy. The product's identity and information about the product are written in red screen printing. Therefore the packaging information is easily erased. Screen printing plastic packaging is also commonly used to wrap snacks so that it does not have any differences from other products. Meanwhile, according to Muhitha and Ulumuddin, packaging must be different from other packaging or have specific characteristics that represent the product's identity so that it is easier to recognize [7]. Besides that, processed food labels contain information at least regarding the name of the product, list of materials used, net weight, manufacturer's name and address, halal for those required, date and code of production, expiration date, and distribution permit number [8]. Meanwhile, the packaging of Kelincidunia cassava chips used now does not include the net weight and the latest halal label.

Three aspects are needed in designing the packaging: functionality, wearability, and emotion [9]. These aspects can be reviewed using Kansei Engineering (KE) and Quality Function Deployment (QFD). According to Nagamachi Sukmaningsih and Asih (2020), Kansei Engineering (KE) is considered superior in translating emotional customer needs into product attribute parameters. The advantages of using QFD include prioritizing consumers, planning and shaping products according to the client's requirements, getting to know other companies' advantages and weaknesses, shortening the product development cycle duration, lower production costs, and a structured, systematic approach [10]. Therefore, in this study, the Ergonomic Function Deployment (EFD) method was chosen which is the development of Quality Function Deployment (QFD) by adding the relationship between consumer desires and the ergonomic aspects of the product [11].

Research on the design of cassava chips packaging has been done by Riki Dwi Setiawan (2015), who uses the define, design, develop method in the form of Research and Development (R&D) research which describes issues regarding the design of cassava chips packaging at UD. Aji Jaya Makmur. However, the study only conducted a visual analysis and comparison to its competitors as an idea for making packaging designs without considering the customer wishes [12]. The book entitled "The Packaging That Sells" states that poor product packaging will also create a bad impression on a good product and without saving any sales. However, a good packaging can be useful and increase customer attractiveness [13].

2 Literature Review

2.1 Quality Function Deployment (QFD)

The idea of QFD was first created by Yoji Akao in Japan in 1966 through a work published in 1972 entitled Hinshitsu Tenkai System or popularly known as Quality Deployment. In addition, it was first applied to a Mitsubishi company in 1978 (Cohen 1995). QFD stands for Quality Function Deployment. The origin of the term is based on Japanese which has the following meaning: Hinshitsu means "quality", "attributes", "features", or "qualities", Kino which means "function" or "mechanization", and Tenkai which means "deployment", "development", "diffusion", or "evolution". Therefore in English, the word is interpreted as Quality Function Deployment (QFD) (Utami 2018). QFD can be regarded as the practice of building a process in response to customer needs and wants [14].

The stages of Quality Function Deployment (QFD) are divided into 4 phases (Fig. 1), including:

1. Product Planning
   This product planning matrix is known as the House of Quality (HOQ), which compares customer needs and product characteristics.
2. Design Planning
   The second house matrix is a matrix that serves to identify technical factors that have a sharp influence on product development.
3. Process Planning
   The third house matrix is a matrix used to analyze the development process of creating a product.

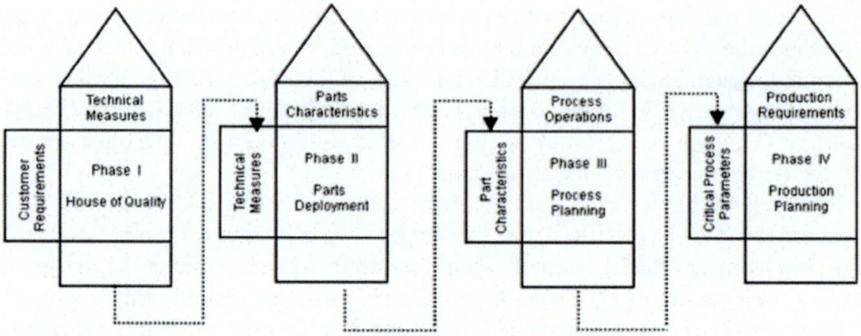

Fig. 1 Four phases of quality *function deployment* [15]

4. Production Planning
   The fourth house matrix is a matrix that describes the actions that must be taken when making improvements to the production process of a product.

2.2 Ergonomic Function Deployment (EFD)

Ergonomic Function Deployment (EFD) is a method that aims to ease the design process, decision making is recorded in the form of matrices so that it can be rechecked and modified in the future. Then usually to find out ergonomic or not the result of a design. EFD is a development method of QFD that not only pays attention to the needs and desires of consumers in product design but also pays attention to the ergonomic aspects of the product when used. EFD is the development of the concept of QFD (Quality Function Deployment) by including the relationship between consumer needs and the ergonomics of the product [11].

2.3 Ergonomics

Ergonomics consists of 5 kinds of studies [16]:

1. Physical ergonomics is related to the physical environment, anthropometry, and biomechanics.
2. Cognitive ergonomics examines the relationship between display and control. Relevant points in cognitive ergonomics are workload, decision making, and work stress.
3. Organizational ergonomics is related to communication in the environment, an organization that provides a comfortable atmosphere when working.
4. Environmental ergonomics relates to lighting, air, and temperature.

5. Applied psychology results from experimental psychology studies in the design are needed to determine the needs of the dimensions of the human body (eg human behavior and habits when sitting, standing, and moving).

2.4 *House of Quality*

House of Quality (HOQ) is the initial phase of the 4 phases of making EFD known as the product planning phase. HOQ can be interpreted as a product planning matrix that outlines customer needs and wants, company targets, and product comparisons to competing products [17]. The main components of the HOQ are shown in the Fig. 2.

3 Methods

This research started with observation and interview with the owner of Cassava Chips UD. Wahyu Pendawa Lestari in Rambeanak Village, Magelang. Then read the literature study related to the problems taken. Data collection was done using interviews and distributing questionnaires.

There are three types of questionnaires distributed, including:

1. Questionnaire related to current packaging complaints and consumer needs for packaging. This questionnaire is in the form of an open questionnaire which allows respondents to write their personal opinion and presents a checklist option. These attributes are grouped into several categories, including protection, ergonomics, information, and price [18]. The category of protection includes

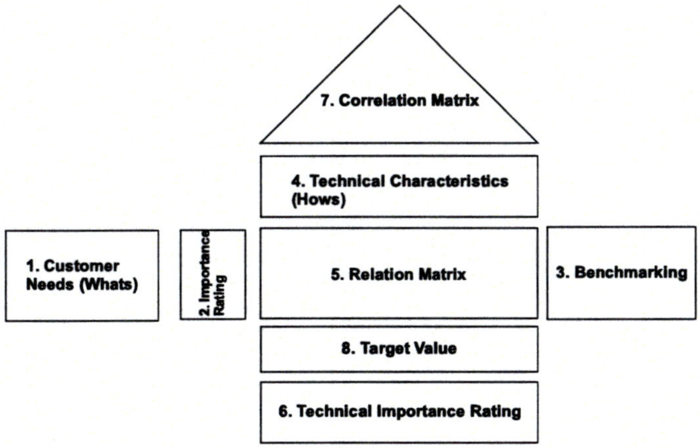

Fig. 2 House of quality

packaging attributes related to its ability to protect and maintain product content. The ergonomic category includes attributes related to the ease and convenience of using packaging. Attributes of design, color, and shape are also included in the ergonomic category because they are the application of cognitive ergonomics. Information category is a group of attributes related to the existence of the information listed on the packaging. While the price category includes attributes related to packaging prices.

2. Questionnaires related to the level of importance of each attribute of voice of customer results from the first questionnaire, product comparisons with competitors, and consumer perceptions of packaging design. This questionnaire is a closed questionnaire using a Likert scale of 1–5.
3. Questionnaire related to packaging color selection preferences. The colors used as options are a combination of red-yellow, green-yellow, and orange-yellow. The color was chosen because it is a color that can attract the attention of customers according to Lv, 2018 [19].

3.1 Determination of Sample

In this study, people or consumers who buy cassava chips products are not known with certainty, so to calculate the minimum sample, we can use the Lemeshow formula [20].

$$\mathrm{n} = \frac{Z_{1-\alpha/2}{}^{2} \times P(1-P)}{d^{2}}$$

Description:

n = Number of sample

Z = Z score

p = Maximum estimate

d = Alpha (0.10) or *sampling error.*

The target market for the packaging that will be designed is all groups ranging from children, teenagers, and adults. However, the children were not involved because in viewing the packaging, they only observed the color, shape, and size [21]. Therefore, the respondents in this study were classified into the categories of early adolescents aged 12–16 years, late adolescents 17–25 years, early adults 26–35 years, late adults 36–45 years, and elderly over 45 years [22].

The sample in this study used a z score at 95% confidence (Z = 1.96), a maximum estimate of 0.5 (p = 0.5), and an alpha of 10%. So that the sample calculation uses equation as follows:

$$n = \frac{1.96^2 \times 0.5(1 - 0.5)}{0.1^2}$$
$$n = \frac{0,9604}{0,01}$$
$$n \approx 96 \text{ respondent.}$$

3.2 Arrangement QFD Matrix Phase 1 (House of Quality)

1. Identification of customer needs (what)
   Identification of consumer needs begins with collecting data through interviews and questionnaires. Interviews were conducted with owners UD. Wahyu Pendawa Lestari and questionnaires were distributed to consumers of cassava chips.
2. Determining the level of importance of packaging attributes
   The level of consumer interest is the consumer's perception of the attributes of a product based on the importance of these attributes for design. The importance of each attribute was assessed using a second-stage questionnaire obtained from the average of each attribute.
3. Benchmarking
   At this stage, an assessment of competing product packaging is carried out using a questionnaire as an evaluation and reference in designing Kelincidunia packaging.
4. Determining technical characteristics (Hows)
   This section contains a list of technical responses that can be given to respond to the voice of customers. This section can be called the "hows" section, which contains how to respond to customer needs obtained with the relevant technical characteristics. The translator tool for customer needs used is a tree diagram.
5. Determination of the Relationship between Consumer Needs and Technical Characteristics (Relation Matrix)
   This section describes the relationship between customer voice and technical response. The following symbols denote the relationship as shown in Fig. 3.
   There are three types of relationships: strong relation has a value of nine, moderate relation has a value of three, and weak relation has a value of one.
6. Determination of the Level of Importance of Technical Characteristics

Fig. 3 Relationship symbol

Relationships	
Strong	●
Moderate	○
Weak	∇

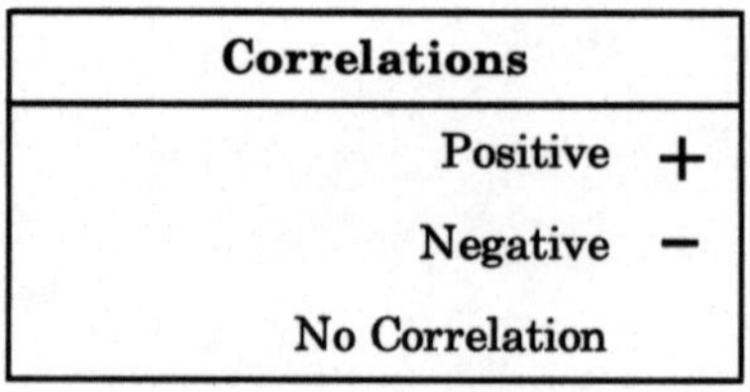

Fig. 4 Correlation matrix symbol

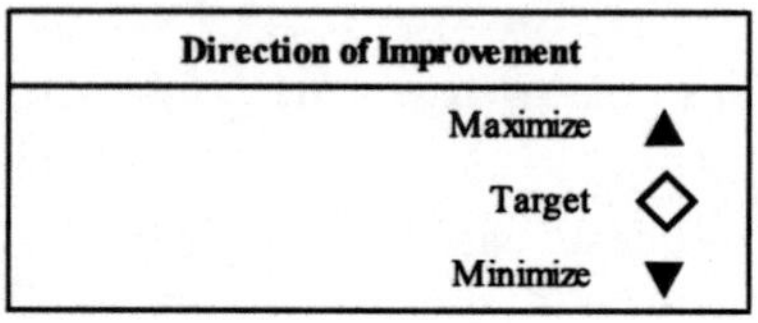

Fig. 5 Direction of improvement symbol

The level of importance uses the following formula:

$$W_j = \sum_{i=1}^{n} d_i.r_{ij}$$

$$R = \frac{W_j}{\sum W_j}.100\%$$

7. Determination of Relationship Between Technical Characteristics (Correlation matrix) and Direction of Improvement
   This section is a relation or relationship between each technical characteristic. In this section, relationships are depicted in the symbols in Fig. 4.
   Positive symbols indicate a relationship between technical characteristics. The negative symbol indicates an opposite relationship between the technical characteristics. In contrast, the empty symbol indicates the absence of a relationship.
   Then also determine the direction of improvement on each attribute of the technical characteristics using the symbols in Fig. 5.
8. Target values
   This section defines the target of each technical characteristic attribute. It can act as a basis for comparison. Targets are specific and measurable. In addition, targets are determined based on the company's technical resources.

3.3 Visualization of Packaging Design and Customer Perception

Visualization of packaging design is made to clarify the packaging of cassava chips according to the needs and desires of the customer. The elements or variables that a designer can use to design a packaging are shape, size, color, text, graphic, and material [23]. Visualization of this design proposal using CorelDRAW software for 2D design and Photoshop software for 3D design.

After designing the packaging design, a questionnaire was distributed to assess consumer perceptions of the design results. The assessment results are used as evaluation material on whether the packaging can be used or still needs to be improved. The rating scale used is 1–5.

4 Results and Discussion

4.1 *Voice of Customer*

The results of interviews and questionnaires were recapitulated and used as a reference in detailing the respondent's needs for cassava chips packaging. VOC attributes will later be used as the basis for determining the level of importance and assessment of competitors. Table 1 shows the important rating of customer needs that can be used as input to the QFD Phase I.

The calculation shows that each VOC attribute has an importance value exceeding four, which means it is essential for each cassava chip package.

Table 1 Importance rating customer needs

No.	Customer needs	Importance rating
1	Maintain crispness	4.78
2	Not easily damaged	4.72
3	Easy to hold and carry	4.50
4	Can be opened easily and closed again	4.64
5	Attractive and clear colored fonts	4.35
6	The shape and color of packaging is attractive	4.42
7	The packaging background color contrasts with the font	4.16
8	Product identity is clear (product name and logo)	4.63
9	The packaging displays flavor variant, composition, net weight, manufacturer's address, expired date, nutrition facts, and production code	4.78
10	The packaging displays halal label, distribution permit number, and 2D barcode clearly	4.78
11	The information not easily vanished	4.57
12	Affordable prices	4.50

4.2 Result of Validity and Reliability

The validity and the reliability of the data obtained from the questionnaire were tested. The result of the validity test shows that all variables are declared valid. After the variable is declared valid, the next step is to perform a reliability test with SPSS 24 software. The variable is declared reliable if the Cronbach's Alpha value is higher than 0.6. Cronbach's Alpha value is 0.828 > 0.6, so it can be said that the questionnaire is reliable.

4.3 Benchmarking

Table 2 shows the results of consumer assessments through the second stage of the questionnaire. The packaging of competing products used is the packaging of cassava chips brand AB and XY.

Table 2 Benchmarking

No.	Customer needs	Competitive assessment	
		AB	XY
1	Maintain crispness	3.84	3.90
2	Not easily damaged	3.90	3.95
3	Easy to hold and carry	3.96	3.94
4	Can be opened easily and closed again	2.97	3.50
5	Attractive and clear colored fonts	4.35	3.75
6	The shape and color of packaging is attractive	4.14	3.69
7	The packaging background color contrasts with the font	4.13	3.89
8	Product identity is clear (product name and logo)	4.58	4.23
9	The packaging displays flavor variant, composition, net weight, manufacturer's address, expired date, nutrition facts, and production code	4.56	4.25
10	The packaging displays halal label, distribution permit number, and 2D barcode clearly	4.43	4.38
11	The information not easily vanished	4.53	4.36
12	Affordable prices	4.39	3.66

4.4 Interpretation of Respondent's Needs in Technical Characteristics

This section contains a list of technical responses that can be given to respond to the voice of customers. The translator tool for customer needs used is a tree diagram as shown in Fig. 6

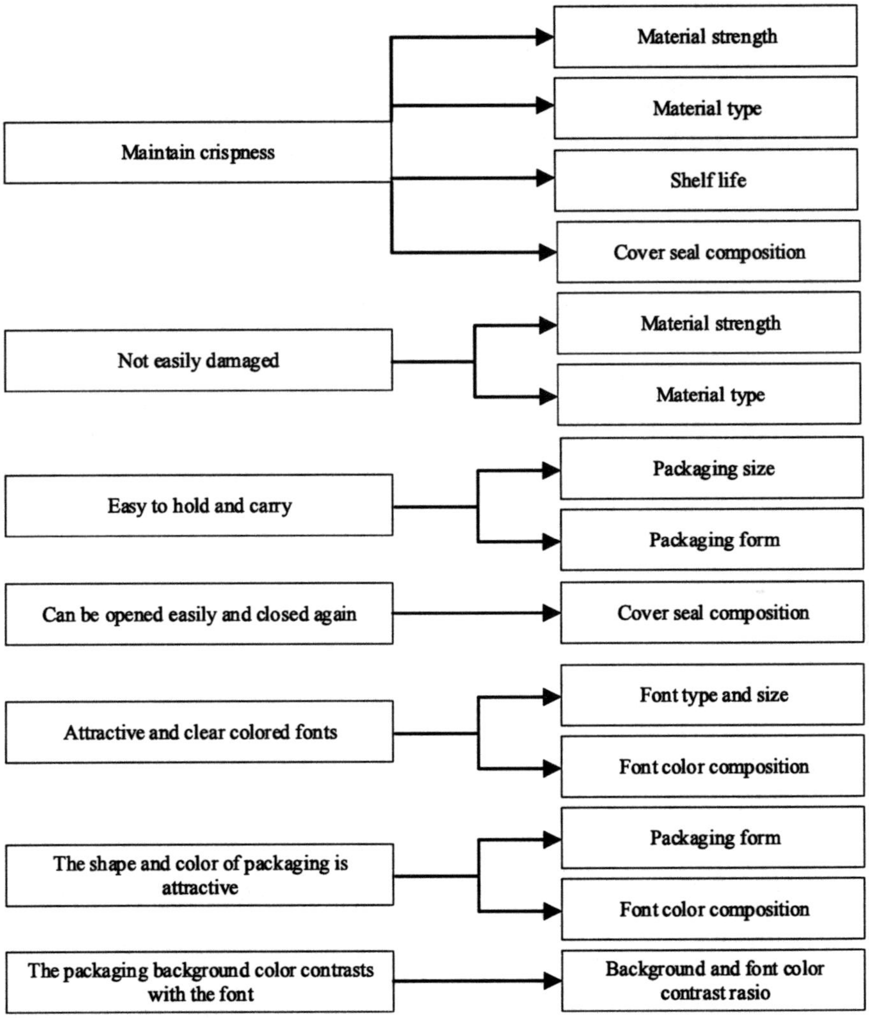

Fig. 6 Tree diagram technical characteristics

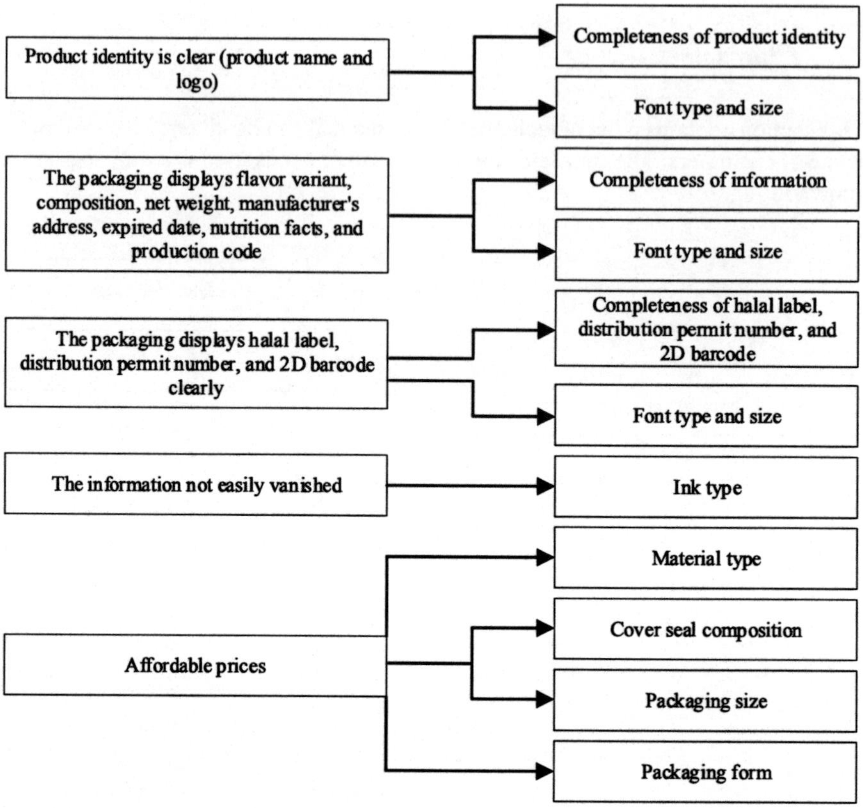

Fig. 6 (continued)

4.5 Relation Matrix

The relationship between consumer needs and technical requirements is shown in the Fig. 7.

4.6 The Level of Importance of Technical Characteristics

The following results from calculating the importance of technical characteristics and relative weights after translating the relationship between requirements and technical characteristics. An example of calculating the level of importance of technical characteristics is as follows.

$$\text{The importance of material strength} = (9 \times 4.78) + (9 \times 4.72) + (1 \times 4.50) = 90$$

Customer Needs \ Technical Characteristics	Material Strength	Shelf Life	Material Type	Packaging Size	Cover Seal Composition	Font Type and Size	Font Color Composition	Packaging Form	Packaging Color Compoisition	Background and Font Color Contrast Ratio	Completeness of product identity	Completeness of information	Completeness of halal label, distribution permit number, and 2D barcode.	Ink Type
Maintain crispness.	●	●	○		●									
Not easily damaged	●	○	●											
Easy to hold and carry.				●				●						
Can be opened easily and closed again.					●									
Attractive and clear colored fonts.						●	●			○	▽	▽	▽	○
The shape and color of packaging is attractive.				▽				●	●	○				○
The packaging background color contrasts with the font.							○		○	●				▽
Product identity is clear (product name and logo)						○	▽			▽	●			▽
The packaging displays flavor variant, composition, net weight, manufacturer's address, expired date, nutrition facts, and production code.						○	▽			▽		●		▽
The packaging displays halal label, distribution permit number, and 2D barcode clearly.						○	▽			▽			●	▽
The information not easily vanished.											▽	▽	▽	●
Affordable prices.	▽		●	○	○			●						▽

Fig. 7 Relationshhip between customer needs and technical requirement

$$\text{Relative weight} = \frac{90}{(90 + 97.32 + 58.42 + 98.28 + 81.72 + 65.82 + 120.8 + 52.26 + 77.94 + 50.59 + 51.94 + 47.59 + 90.29)} \times 100\% = 9\%$$

Based on these calculations, the highest level of importance is in the technical requirements of the packaging form.

4.7 *Correlation Matrix*

Figure 8 shows the relationship between technical characteristics with each other. In addition, the direction of improvement for each technical is also determined.

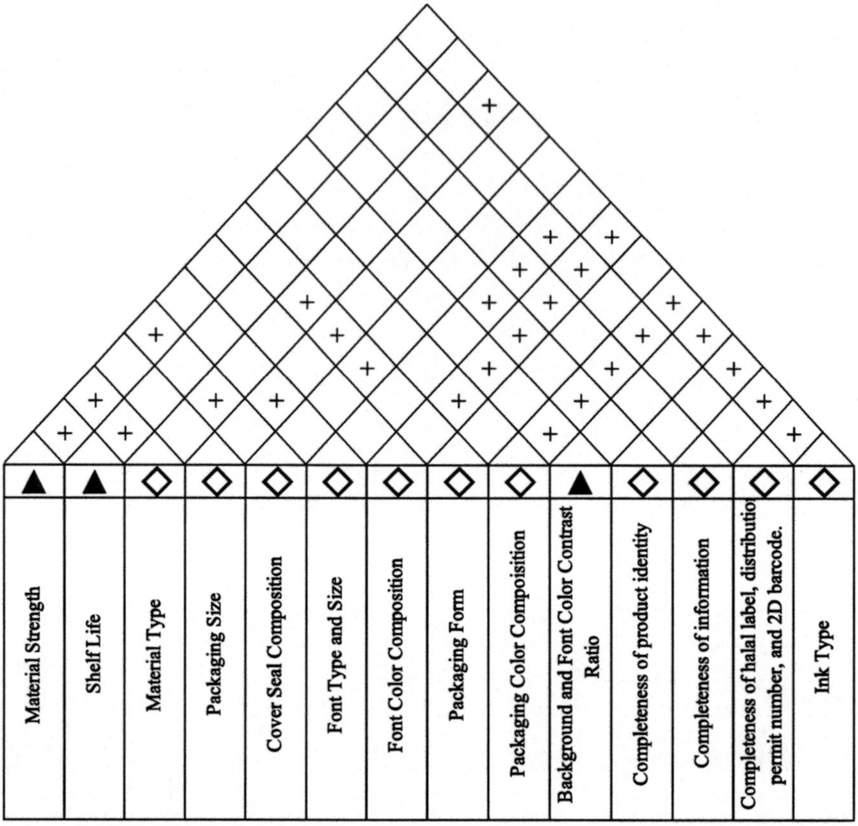

Fig. 8 Correlation matrix

4.8 *Target Values*

The target of each technical requirement is determined based on literature studies, VOCs, and the results of discussions with visual communication design experts, as shown in the Table 3.

4.9 *Visualization of Packaging Design and Customer Perception*

Table 4 shows the design based on the respondents' needs, literature studies, and targets.

Table 3 Target values

Technical requirement	Target values
Material strength	Moisture, oil, and water resistant
Shelf life	2 Months
Material type	Paper foil (food grade) and thick >0.0375 mm
Packaging size	18 cm × 24 cm
Cover seal composition	Ziplock and Tear Notch
Font type and size	Sans Serif type and dimensions >6 pt
Font color composition	Bold color
Packaging form	Standing Pouch
Packaging color composition	Bright color and attractive
Background and font color contrast ratio	4.5: 1
Completeness of product identity	There is a product name and logo
Completeness of information	There are variants of taste, composition, net weight, manufacturer's address, nutritional facts, expired date, and production code
Completeness of halal label, distribution permit number, and 2D barcode	There is halal label, distribution permit number, and 2D barcode
Ink type	UV ink

Table 4 Design specification

Form	Standing pouch
Material	Paper foil (food grade) and thickness 0.05 mm
Cover seal	Ziplock and tearnotch
Size	18 cm × 24 cm
Background color	Combination of yellow and red–orange gradations
Font color	Black and white
Ratio	>4.5
Font size and type	Sans Serif type and dimensions >6 pt
Information display	The front of the packaging displays the product name, logo, product image, PIRT number, MUI label, flavor variant, manufacturer's address, and net weight information. In comparison, the back of the packaging displays the composition, nutritional value information, expiration date, manufacturer's social media, and the date and production code
Packaging price	Rp 3200

1. Form
   The form of packaging chosen is a standing pouch according to consumer needs so that it is easy to hold and carry, as well as being opened and closed again so that it can maintain the crispness of the chips.
2. Material Selection
   The material used is food grade so it is not harmful to food. The type of material is paperfoil which consists of a layer of aluminum foil and art paper as the outer layer according to the target and needs of respondents who want a material that is not easily damaged and can maintain crispness.

   Aluminum foil packaging is difficult to degrade, so that it will pollute the soil and waters for a long time. The waste will decompose naturally for up to 180 years [24]. Meanwhile, if allowed to continue, it will lead to the accumulation of hazardous waste. Therefore, the use of aluminum waste can be done by recycling. However, the packaging needs to be separated from the paper layer to process it. The nature of art paper is also easy to recycle and decompose within 2-5 months [25].

   Aluminum foil packaging waste can be recycled into waste processing aids, namely alum which helps the coagulation process. This has a good impact on the environment, namely reducing the amount of solid waste and reducing the use and purchase of commercial alum for WWTP processing so that waste management costs can be reduced. In addition, it can also be applied as a substitute for wood particle board. At the same time, the paper recycling process requires several other chemical processes to remove the ink on the paper (deinking process).

   Most packaging materials on the market use metalized plastic materials. Although this material is cheaper than paper foil and thinner, Metalized Plastic Waste (MPW) is unsuitable for reuse or recycling and can harm the environment. However, it can be used for sustainable construction materials [26].
3. Cover Seal
   The packaging cover uses a ziplock model as an adhesive so that it is easy to open and close again and there is a tear notch so that the packaging is tight and keeps the chips crisp.
4. Size
   The packaging will be used to accommodate circular chips with a diameter of ±4 cm, a thickness of 1.0–1.5 mm, and a weight of 100 grams. In addition, air is added to the empty space to protect the contents inside so that it measures 18 cm long and 24 cm high.
5. Background Color
   The colors used are bright, a combination of red-orange gradations and yellow. According to research, the combination of red and orange can attract attention and increase consumer tastes [19]. Yellow is the most easily noticed color. Yellow signifies happiness, optimism, and warmth but also caution [27]. The main advantage of yellow is its eye-catching features, so the combination of red-orange and yellow is perfect.

6. Font Color, Size, and Type
   The colors used for the font are black and white because the colors are bold so that the information is clearly legible. On a orange background, the font color is white. While the background color is yellow, the font color is black.
   The font used for writing is Malgun Gothic, Trebuchet MS, and Gill Sans MT. All types of fonts used belong to the Sans Serif category. Font size exceeds 6 pt for easy reading. Writing the title is made manually using the pen tool.
7. Ratio
   Color ratio checks are performed using WebAIM: Contrast Checker (2022). The results indicate that the combination of orange background color and white font produces a ratio of 4.7: 1. At the AA level, the font size of 18 pt (large text) and 14 pt (standard text) has exceeded the minimum limit. While the AAA level font size 18 pt (large text) has passed, the font 14 pt (standard text) has not.
   Meanwhile the combination of yellow background color and black font produces a ratio of 13.52: 1. At the AA level, the font size of 18 pt (large text) and 14 pt (standard text) has exceeded the minimum limit. Meanwhile, for the AAA level, the font size of 18 pt (large text) and 14 pt (standard text) has also passed the minimum limit.
8. Price
   Chips are currently sold at a price of Rp. 12,000,—with a net weight of 200 g. Meanwhile, the packaged chips are sold at a price of IDR 11,000 with a net weight of 100 g. This is due to the difference in packaging prices. The plastic packaging used now costs Rp. 2,000, while the stand-up pouch is Rp. 3200 with a minimum order of 500 pcs.

Figure 9 shows a design visualization in 3D made by Photoshop.
The elements in the packaging include:

1. UD. Wahyu Pendawa Lestari Logo
2. Product name
3. Halal label and number
4. Marketing authorization number
5. Manufacturer's address
6. Variety of flavors
7. Net weight
8. Production date and code
9. Expiration Date
10. The symbol of throwing garbage in place
11. Recycle symbol
12. 2D barcodes
13. Social media contacts
14. Nutritional value information
15. Composition.

Next is customer perception, which assesses Kelincidunia packaging design from each packaging attribute resulting from VOC. The assessment was carried out using

Fig. 9 Packaging in 3D

a questionnaire. Table 5 shows the results of customer perception of the packaging design:

Table 5 Customer perception

No.	Customer needs	Kelincidunia
1	Maintain crispness	4.67
2	Not easily damaged	4.55
3	Easy to hold and carry	4.61
4	Can be opened easily and closed again	4.83
5	Attractive and clear colored fonts	4.36
6	The shape and color of packaging is attractive	4.36
7	The packaging background color contrasts with the font	4.36
8	Product identity is clear (product name and logo)	4.46
9	The packaging displays flavor variant, composition, net weight, manufacturer's address, expired date, nutrition facts, and production code	4.55
10	The packaging displays halal label, distribution permit number, and 2D barcode clearly	4.54
11	The information not easily vanished	4.56
12	Affordable prices	4.24

The customer perception shows that all attributes are highly valued in the packaging design. Then the value results are entered into the benchmarking section to compare their value with competing products.

4.10 *Ergonomic Function Deployment*

The results of the ergonomic function deployment matrix stage 1 or house of quality are as Fig. 10.

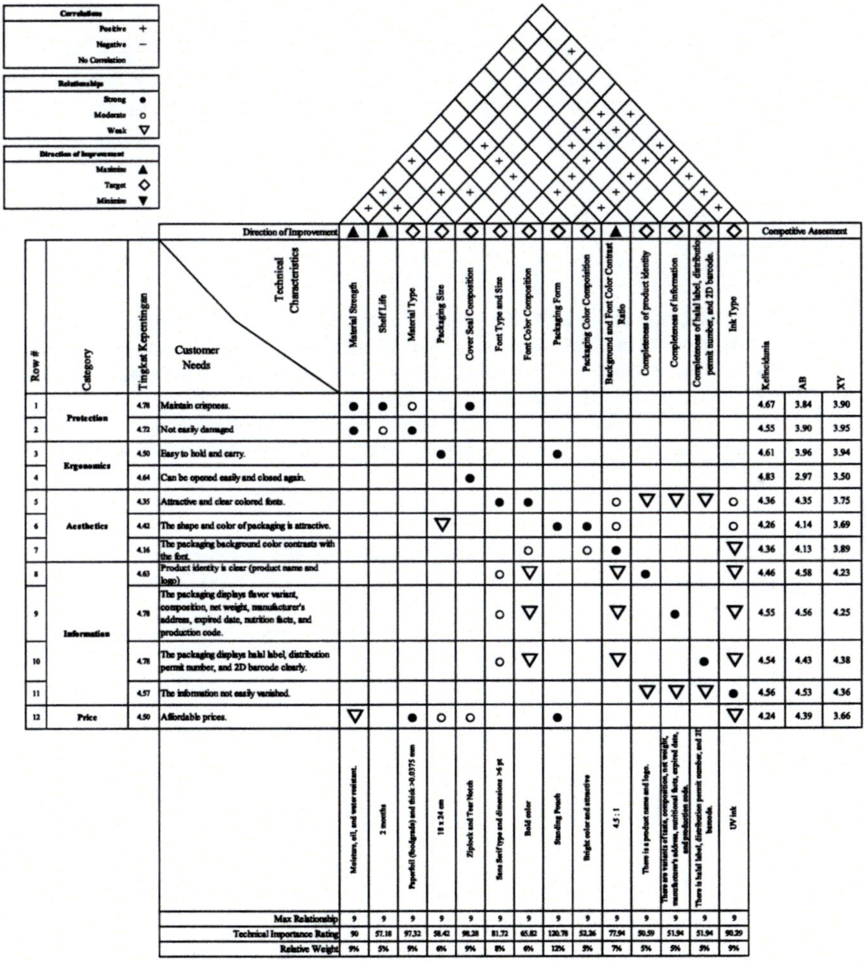

Fig. 10 Ergonomic function deployment

5 Conclusion

Based on the questionnaire results from 97 respondents, there were 12 attributes of the respondent's needs for Kelincidunia cassava chips packaging. These attributes are grouped into protection, ergonomics, information, and price categories. The packaging is 18 cm × 24 cm standing pouch with paper foil material consisting of an inner layer of aluminum foil and an outer layer of art paper and glossy laminates. Packaging cover using ziplock and tear notch. The packaging colors are gradations of red and orange, and yellow. The font is a sans serif type with a size of more than 6 pt. The information displayed is in accordance with BPOM Article 10 No. 31 2018.

This study chose aluminum foil which is relatively expensive and difficult to decompose so it is expected to be considered for further research with cheaper and environmentally friendly materials. Further research is also recommended to test the performance of the packaging so that the reliability of the product for use can be more guaranteed quality. In addition, it is recommended to consider the material with the lowest environmental impact.

References

1. Suherlan Y, Muttaqin H, Anam C, Widiyanti E (2016) Model Balai Pengembangan Kemasan Ramah Lingkungan Untuk Meningkatkan Daya Saing Produk Lokal UMKM Pangan Olahan Menghadapi Pasar Global. In: Proceeding seminar Nasional Peningkatan Kapabilitas UMKM Dalam Mewujudkan UMKM Naik Kelas, pp 290–301
2. Sari IGAW, Setiawan PY (2017) Pengaruh Green Marketing dan Packaging Terhadap Brand Image dan Loyalitas Pelanggan Pada Konsumen Starbucks Coffee. E-Jurnal Manajemen Unud 6(7):3820–3849
3. Undang-Undang RI Pasal 1 (2012) Undang-Undang Republik Indonesia Nomor 18 Tahun 2012 Tentang Pangan
4. Badan Standardisasi Nasional (1996) Badan Standardisasi Nasional Keripik Singkong
5. Deglas W, Fransiska (2019) Pengaruh Jenis-Jenis Kemasan Terhadap Umur Simpan Produk Keripik Ubika. Jurnal Pertanian Dan Pangan, 1(1):23–27
6. Asih CER, Restuari L (2019) Pengaruh Penyimpanan Keripik Buah Nanas Pada Berbagai Kemasan Dengan Metode Accelerated Shelf-Life Testing (ASLT)
7. Muhitha IF, Ulumuddin DII (2020) Studi Visual Kemasan VS Cita Rasa Indonesia Pada Kemasan Produk Keripik Singkong Rasalokal. Jurnal Citrakara 2(2):112–126
8. Badan Pengawas Obat dan Makanan RI (2020) Pedoman Label Pangan Olahan 2020. Badan Pengawas Obat dan Makanan RI
9. Aurelli B (2013) Evaluasi dan Perancangan Ulang Produk Striped Clipboard Berdasarkan Aspek Usability dan Emotion Skripsi
10. Wolniak R (2018) The use of QFD method advantages and limitation. Prod Eng Arch 18(2):14–17
11. Surya RZ, Badruddin R, Gasali M (2014) Aplikasi Ergonomic Function Deployment (EFD) Pada Redesign Alat Parut Kelapa Untuk Ibu Rumah Tangga. Jurnal Optimasi Sistem Industri 13(2):771–779
12. Setiawan RD, Oemar EAB (2015) Pengembangan Desain Kemasan Keripik Singkong UD. Aji Jaya Makmur Di Kecamatan Cerme Di Kabupaten Gresik. Jurnal Pendidikan Seni Rupa 3(3):183–189

13. Wirya I (1999) Kemasan Yang Menjual. Gramedia Pustaka Utama, Jakarta
14. Tutuhatunewa A (2010) Aplikasi Metode Quality Function Deployment Dalam Pengembangan Produk Air Minum Kemasan. Arika 4(1):1–13
15. Chan LK, Wu ML (2002) Quality function deployment: a comprehensive review of its concepts and methods. Qual Eng 15(1):23–35
16. Safenla G, Agustin F, Novelin W (2017) Perancangan Mebel Kantor Kezia Karin Surabaya Berdasarkan Ilmu Ergonomi. Jurnal Desain Interior 2(2):63–72
17. Devani V, Kartikasari DD (2012) Usulan Perbaikan Kualitas Pelayanan Administrasi Mahasiswa Menggunakan Metode Quality Function Deployment (QFD). Jurnal Ilmiah Teknik Industri UIN Sultan Syarif Kasim 11(2):185–197
18. Harahap MFB, Mubarak A, Suzianti A (2020) Designing a green food delivery packaging with QFD for environment (QFDE) and TRIZ. In: IOP conference series: earth and environmental science, vol 464, no 1
19. Lv C (2018) Research on color application in children's food packaging design. In: International conference on education, management, and information technology, vol 4, pp 42–46
20. Lwanga SK, Lemeshow S (1991) Sample size determination in health studies. World Health Organization
21. Akbari Z (2014) The dilemma of flavor, shape, and color in the choice of packaging by children. Int J Acad Res Bus Soc Sci 4(1)
22. Al Amin M, Juniati D (2017) Klasifikasi Kelompok Umur Manusia Berdasarkan Analisis Dimensi Fraktal Box Counting Dari Citra Wajah Dengan Deteksi Tepi Canny. Jurnal Ilmiah Matematika 2(6)
23. Pramono RG, Iqbal M, Wulandari S (2015) Design Improvement of packaging of Kerupuk Strawberry Kencana Mas using quality function deployment method. E-Proc Eng 2(1):956–965
24. Sukanto H, Triyono, Raharjo WW (2015) Aplikasi Low Pressured Sitering Untuk Pengolahan Limbah Kemasan Aluminium Foil Menjadi Papan. In: Proceeding Seminar Nasional Tahunan Teknik Mesin, vol 14
25. Terekhova N, Burenina V (2019) Philosophical foundations of the concept of green economy. In: 2nd International conference on contemporary education, social sciences and ecological studies
26. Bhogayata AC, Arora NK (2019) Utilization of metalized plastic waste of food packaging articles in geopolymer concrete. J Mater Cycles Waste Manage 21(4):1014–1026
27. Deliya MM, Parmar BJ (2012) Role of packaging on consumer buying behavior-Patan district. Glob J Manage Bus Res 12(10)

Six Sigma Application on Software Improvement Process

Tatinia A. R. Amalia, Mohammad Mi'radj Isnaini, and Dradjad Irianto

Abstract This study proposes a quality development method based on six sigma to solve software quality problems of the software. The six sigma DMADV framework is used to analyze the root causes that have a major influence on user complaints against the software. Based on the results of joint elaboration with the company stakeholders, proposed improvements will be designed that can reduce user complaints. The suggestions for improvement that will be designed are the design of a junior programmer training program and the design of technical documents. The program will be designed by adapting the Windows training framework and developing the SKKNI 615-2012 programmer curriculum. Meanwhile, technical documents will be designed based on the IEEE 1063-2001 standard for software user documentation. The results of the proposed design are expected to increase the effectiveness of the script writing of the software and minimize technical repair activities.

Keywords Six sigma · DMADV · Software

1 Introduction

Software is one of the complex technologies that have a high probability of error consequently [1]. Like many manufacturing products, software must be concerned about its quality during its development. According to ISO 25010, the quality of the software is the ability to attain the required objectives based on consumers, users, and stakeholders. The quality level of a product, in most cases, is often considered from the number of issues that might come across, such as defective or malfunctioned product. There are three types of problems in software systems, namely error, defect, and failure [2]. This paper highlights the error and defect issues only, since the

T. A. R. Amalia (✉) · M. M. Isnaini · D. Irianto
Engineering Management Study Program, Bandung Institute of Technology, Bandung 40132, Indonesia
e-mail: tatiniaarda@gmail.com

C. N. Rosyidi et al. (eds.), *Proceedings of the 6th Asia Pacific Conference on Manufacturing Systems and 4th International Manufacturing Engineering Conference*, Lecture Notes in Mechanical Engineering,
https://doi.org/10.1007/978-981-99-1245-2_17

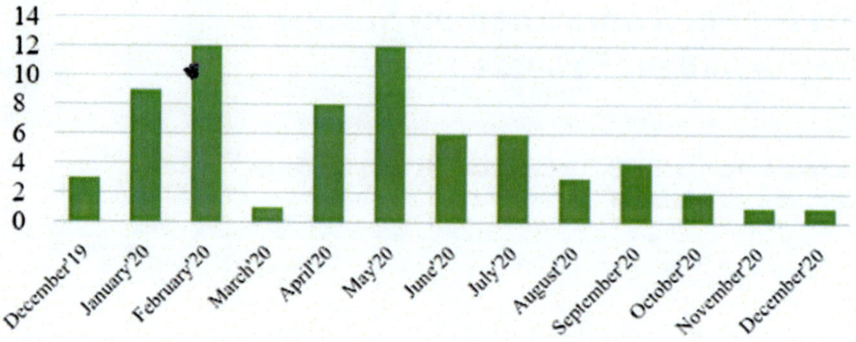

Fig. 1 Number of compliments of Software Y during December 2019–December 2020

failure is related to the inability to run the fundamental functions or, in other words, the software is defaced.

From December 2019 until December 2020, 68 complaints have risen regarding the Y software (see Fig. 1), while the company's target is 12 complaints annually. In other words, this software is far from satisfactory. Patel and Zu [3] successfully proved the application of the six-sigma method in e-government application, combined with development life cycle (SDLC) stage and define, measure, analyze, design, verify (DMADV) tools [3]. Based on that discovery, this study will utilize six sigma DMADV to determine the proposed improvement of the software development process that the company can use to reduce the number of complaints against Y software.

2 Literature Review

In ISO 24765 (ISO 17a), software is all or part of the programs, procedures, rules, and documentation regarding an information system. The software development phase is a cyclic process often called the software development life cycle (SDLC), which consists of six independent and isolated stages, requirement, analysis, design, code, test, and maintenance [4]. Another software development method is six sigma, which is a quality enhancement method focusing on minimizing the main characteristic variability of product quality to reach the least failure probability [5]. This method has been used by Motorola Company since 1986 to measure and calculate the defect intensity and strengthen their product quality [6]. Practically, the method comprises two distinct methods, namely DMAIC (Define, Measure, Analyze, Improve, Control) and DMADV (Define, Measure, Analyze, Design, Verify) or design for six sigma (DFSS). The first method is a structured problem solver and is often used in quality improvement and process, while the second method is a customer need-oriented method and used for the latest or radically redesigned services, product quality improvement, and process. DMADV or DFSS is based on the Research Methodology software quality concept where the reliable software must be able to fulfill the user's demand.

The define stage aims to identify and validate improvement opportunities that represent all potential problems that can be solved. At this stage the project charter design, CTQ determination and SIPOC diagrams are made. The next stage in DMADV is a measure, at this stage a measurement of the performance of existing conditions is carried out. There are two things that must be considered in the measurement stage, namely determining the characteristics or processes that can be measured and calculating the related characteristic or process data. In software development, software quality can be measured through two methods, namely based on the size of the software and the time between failures that occur. In this study, software quality is measured based on the size of the software or commonly called the defect rate or defect per unit. In this study, the size of the software used is the number of lines of non-commented source code (LOC). LOC is a source of instructions that includes executable lines and data definitions, but does not include comments and program prologues [7].

The third stage carried out in DMADV is the analyze stage. In the analysis stage, a search for potential causes of defects, quality problems, consumer complaints, cycle times, and inefficiency will be carried out. The focus at this stage is on analyzing and benchmarking against core product or process performance metrics [8]. The analysis stage begins with identifying the potential causes of the problem using Fishbone diagrams and 5 whys. Fishbone diagrams are generally used in evaluating process performance, especially in identifying the causes of deviations from ideal conditions. The 5 whys technique is a technique developed by Sakichi Toyoda to help solve problems by reflecting on what is the cause of the problem through 5 whys questions [9]. The root causes of the problems that have been found are used to analyze possible improvements that can be made using FMEA. Failure mode and effect analysis (FMEA) is one of the six sigma tools used to identify things that can be improved and developed.

The design stage is concerned with designing new products or processes that can help improve the quality of the main product. To determine the repair process in detail, in this study the 5W1H method was used. 5W1H (what, when, why, who, where, how) is a method that uses several questions to further analyze the improvements that will be made to the product or process [10]. There are several tools that can be used in the improvement stage, namely flowchart, value stream mapping, design of experiment (DOE), measurement system evaluation (MSE), and lean. However, DOE, MSE, and lean tend to be less applicable in the software development process than in the manufacturing process. Flowchart or process mapping is a chronological sequence that describes the steps of a process or workflow. The last stage of Six Sigma DMADV is the verify stage. The verify stage aims to ensure that the improvement plan can be implemented and managed properly by the company. In short, this stage begins with the preparation of a solution implementation plan. After that, at this stage a solution management plan is also designed so that it can continue to run and have a good impact on product quality.

The six sigma method is widely used as an approach in manufacturing product development, especially in terms of quality. However, in fact six sigma can not only be used in my manufacturing industry, but can also be applied to software companies. According to Tayntor, there are two misunderstandings in the application of six sigma in software companies [11]. Firstly, six sigma is a statistical analysis so it is not suitable for use in software. This misunderstanding can be resolved by ensuring the "three rights" are implemented correctly. Secondly, six sigma must be applied to all parts of the company. In practice, the implementation of six sigma is indeed easier for software companies that adopt the six sigma philosophy in running their business. However, the implementation of six sigma can also be useful in software development even in companies that do not adopt six sigma as a whole. The benefits in question are low defects and increased customer satisfaction.

3 Research Methodology

3.1 Define

During this stage, project charter design, the business process analysis, and determining critical to quality (CTQ) were done. There are things to be done within three objectives. Designing the project charter requires project description, goals, probability, scope, implementation plan, and implementation team. Business process analysis is done by interviewing one of the clients. After determining the company business process and problem grouping based on SDLC, CTQ of the software is then identified by prior voice of customer (VOC) identification. VOC in this study is the user's complaint of the software performance which is then grouped and used to define the number and types of CTQ. Based on 4 groups of VOC 3 CTQs are identified, which are requirement error, design error, and coding error.

3.2 Measure

Furthermore, the measure step is done by examining the line of code (LOC) of the software. Secondary data from the company was used, containing the LOC quantity from the user's complaint from December 2019 until December 2020 and logical LOC of the software. Both data types were used to analyze the defect rate.

$$Defect\ per\ unit = \frac{Defect\ of\ LOC}{Total\ LOC} \tag{1}$$

$$DPMO = \frac{Defect\ per\ unit}{Opportunities\ for\ error\ per\ unit} \tag{2}$$

$$Sigma = ABS\left(NORMSINV\left(\frac{\frac{DPMO}{2}}{10^6}\right)\right) \quad (3)$$

In this study, the LOC comprises 30,000 lines excluding comments that generate 11 modules in the software. Furthermore, based on the user's complaint data the defect of LOC was generated, which is the LOC that could cause defect as complaints. There are 157 LOC that could cause 68 complaints. Later, the LOC amount and LOC defect were utilized to produce the defect rate and sigma level using the equation. Thus, the level sigma is 3.47 which is relatively low in comparison with ideal condition which is 6.

3.3 Analyze

This step is designed to examine the major causes of complaints. Firstly, these causes could be identified using two tools, which are fishbone diagrams and 5 whys (see Fig. 2). Later on, failure mode and effect analysis (FMEA) were used to obtain the risk priority number (RPN) of every major cause. These values were then applied to the Pareto chart to obtain the priority level among the major causes. By considering the latest group discussion with the company's representative, two major causes opted into focused solution design, which are the absence of a training program among junior programmers and the absence of a technical document. The data of these causes were then collected by using interviews, discussion, and brainstorming.

3.4 Design

The improvement solution design was done based on the potential failure mode that has been prioritized in the prior steps. 5W1H was used to accomplish this, as shown in Table 1.

Complaints that are also given by many users are the discrepancy of the data displayed on the Y software when compared to the data that should be. In some Y software modules, such as the accounting module, a calculation formula is needed to process the data entered by the user. In some cases, calculation formulas that are translated into programming languages sometimes cause coding errors. These errors can be caused by the programmer's lack of experience in designing coding results. Generally, these inexperienced programmers are junior programmers who have only been given responsibility for designing one part of the Y software module. This error can be avoided if the senior programmer provides structured training to senior programmers before being given the big responsibility. However, in fact, currently there is no training program specifically for junior programmers in the ERP division. The transfer of knowledge that has been carried out so far has only relied on the

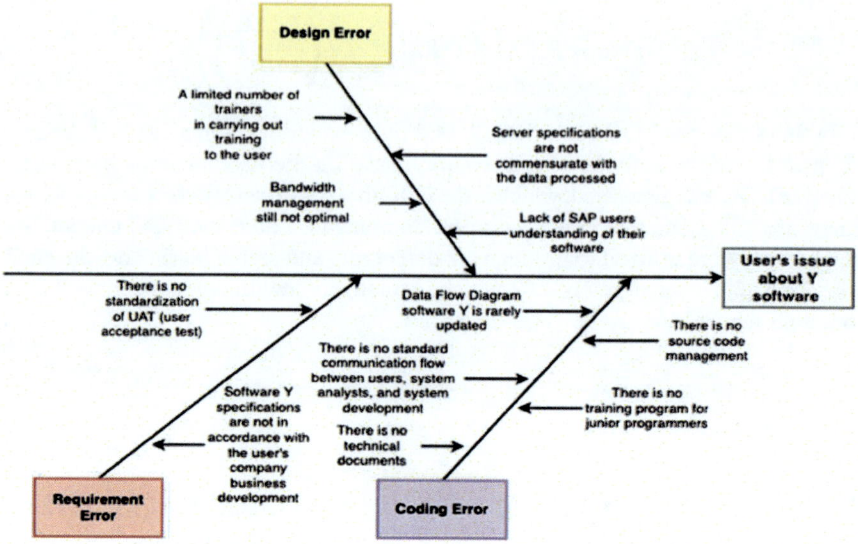

Fig. 2 Fishbone diagram of the user's issue about Y software

initiative of junior programmers to ask senior programmers about user complaints that they do not understand.

The difficulty of junior programmers in understanding the flow of making each Y software module can also cause errors in the script structure of Y software. Currently, junior programmers learn the structure of Y software through working on each error case and creating new features. When junior programmers have problems when working on the case, they will take the initiative to ask senior programmers. Ideally, a junior programmer should learn the structure of the software first before trying to write code in a software script. For this reason, it is necessary to design a document containing the specifications and structure of the Y software module. This document is generally referred to as a technical document.

The junior programmer will apply a framework from the already established junior training program by Windows namely project based learning. Three distinct stages will construct the program; language phase, application phase, and project phase. Beside the framework training, some competences are also fashioned specifically as a standard junior training program of the X corporation. This standardization is taken from *Standar Kompetensi Kerja Nasional Indonesia* (SKKNI) no. 615 in 2012, which is elaborately determined in the workflow in Fig. 3. The second solution is technical document absorbed based on IEEE standard 1063-2001 about software user documentation (see Fig. 4).

Table 1 5W1H of Y software quality improvement design

No.	Potential failure mode	What (solution)	Why	Where	Who	When	How
1	There is no training program for junior programmers	Management training design for junior programmers	There is no IT development governance	System development department	Senior programmer	2022	Determination of competence, design of training implementation flow, curriculum design, and training evaluation
2	There is no technical document	Technical document design	There is no interest in making a technical document from the developer team	System development department	Programmer team	2022	Overall document structure and content design

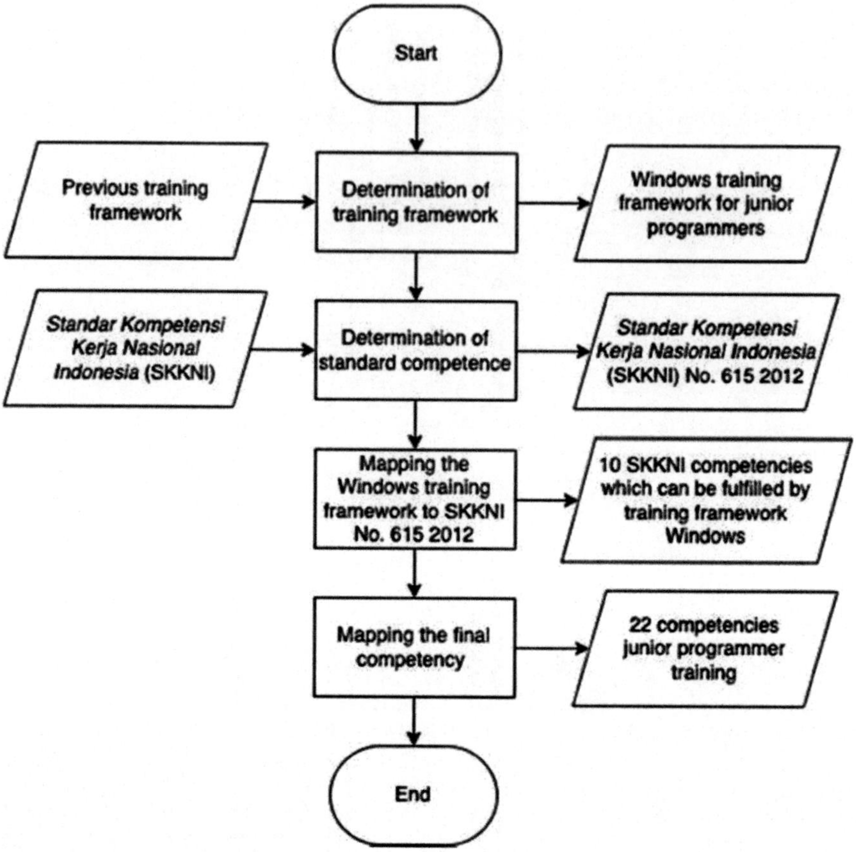

Fig. 3 Junior programmer training design process

3.5 Verify

Within this step, the implementation plan and design was done by documenting the application procedure and maintenance of the solution by taking into account the company's condition. This study suggests the evaluation of program training for junior programmers and nurturing the suitability between technical documents and the software by generating the standard operating procedure (SOP).

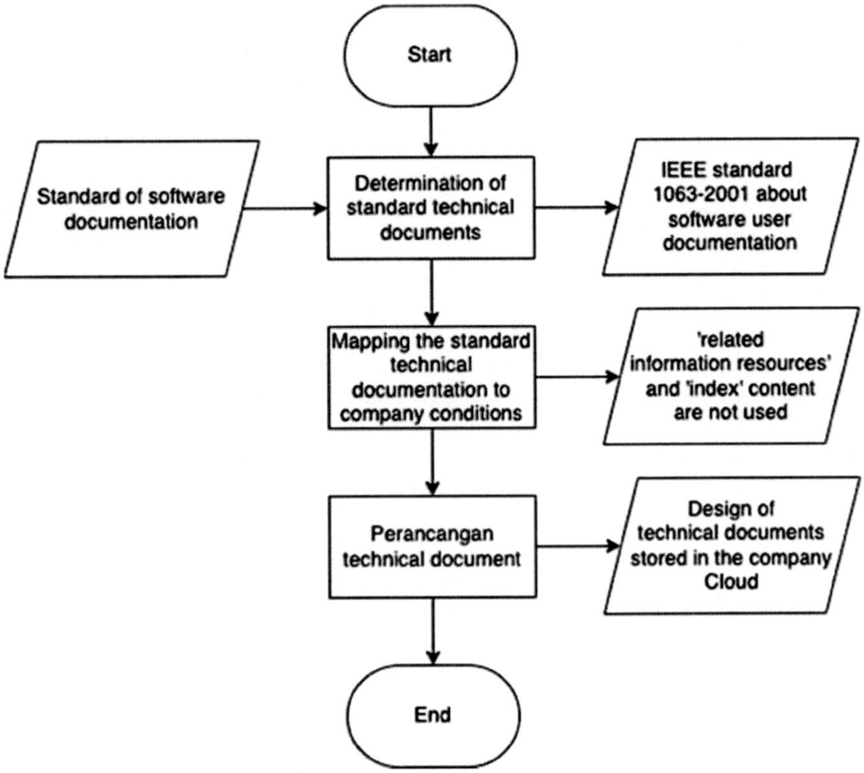

Fig. 4 Technical document design process

4 Results, Analysis, and Discussion

After the application of the six sigma method in the development of Y software, it can be concluded that this method has several advantages. The first advantage is that it can help identify the causes of defects (user complaints) as a whole. In one of the six sigma stages, namely the analyze stage, it supports a comprehensive search for the cause of the problem because it uses several tools, such as fishbone diagrams and 5 whys which encourage the identification process to be carried out in depth. In addition, six sigma also helps analyze the root causes of problems that have the greatest impact on the quality of Y software. This is because FMEA is carried out so that the analysis of the root causes of problems becomes more in-depth and detailed. Not only that, the six sigma method also supports continuous improvement of software Y. This is in accordance with software that is agile, so it must continue to be improved and developed. However, the implementation of a six sigma project takes a long time. For example in this study, to implement the six sigma method takes 4 (four) months or more. This is contrary to the software that continues to change along with user needs, so it is feared that the solution design is no longer relevant. Therefore, the six sigma stage must be carried out more effectively in the future.

The design of the junior programmer training program aims to reduce user complaints caused by code writing errors made by junior programmers. With the training, it is hoped that junior programmers will have competencies that match the needs and standards of the company. For this reason, it is necessary to first determine a training method or framework that can be applied to junior programmers. In this study, the junior programmer training program will adapt the training framework that has been used by Windows for their junior programmers, namely project based learning. There are three (3) stages of training in project based learning, namely the language phase, application phase, and project phase. In addition to the training framework, the competence of junior programmers was also designed as a standard set by the company. In this study, the competence of junior programmers was adapted from the Indonesian National Work Competency Standard (SKKNI) number 615 of 2012. The development of competencies in SKKNI 615-2012 is carried out by identifying competencies based on three competency groups, namely knowledge, skills, and attitudes. In SKKNI 615-2012, the knowledge competency group related to the company's business processes and software that will be operated by junior programmers in the future was not discussed. In addition, the attitude competency group was not discussed in detail by SKKNI 615-2012. Therefore, it is necessary to identify other competencies that are included in knowledge and attitude and obtain three competencies that are classified as knowledge and two competencies that are classified as attitude.

The second improvement solution design is the technical document. A technical document is a document that contains information related to the company's business processes, display structure and technical software, as well as errors that often occur in software and their solutions. This document is useful as a knowledge transfer tool between members of the programmer. One of its uses is in the junior programmer training program, more precisely at the application phase, this document is used as material for supporting junior programmers to understand Y software technically. Technical documents can also be used as a communication tool between the programmer team and the system analyst when discussing the development of the current software design. In this research, the technical document is designed according to the IEEE standard 1063-2001 regarding software user documentation. However, there are some components that are optional, such as related information resources and indexes. Thus, these two components were not included in the draft technical document because they adapted to the company's conditions, where the programmer team was more interested in reading documents that were more concise and clear. The technical document will be designed as an online document and stored in the Cloud owned by the ERP division. This is done to facilitate the revision of technical documents and minimize the cost of procurement and storage of technical documents if they are paper-based. The technical document is also a technical description of the software so that when there is a change in the software, this document must also be revised according to the change. To ensure that the revision of the technical document is carried out, in this study, an SOP was designed that can regulate the schedule and method of revising the technical document so as to minimize errors and omissions in data input. In addition, revision activities will also be used as the key performance index (KPI) of the PIC appointed by the system development section head. As a result, if the revision does not work properly, the PIC will be held accountable and this can result in an increase in salary or bonuses.

Solution design implementation could affect the company's performance, such as the increased responsibilities of the system development team to produce technical documents and implement continuous evaluation. However, there is no additional manpower needed as these responsibilities could practically be done by the programmer who modified the software. Meanwhile, the training program could increase the effectiveness of learning time up to 40 days of work and performance for young programmers rather than self-learning. In addition, both solutions could reduce the intensity of software reparation activities, or even remove it for a period of time. To ensure the purpose of the study could be established, a company should deploy the six sigma DMADV regularly, in order to encounter the actual factors of complaint and prevent the further factors of complaints impacting the software. Continuous improvement of the software could enhance its performance and sustain the user's fulfillment.

5 Conclusions

To ameliorate the Y software quality, this study utilizes the six-sigma method with DMADV as the analysis framework. This framework is widely employed when the development activity has some objectives to design or modify the process design or product so that the probability of defect is extremely low. In software development, DMADV is more convenient to be applied than DMAIC due to its limitation to repair the software technically. As a consequence, the reparation could only be done by re-designing the product or process that could affect the software quality. Initially, preparation activity in X company was implemented independently in each division. Fortunately, the reparation could be done holistically by dealing with all of the divisions in the software development. After applying the six-sigma method to Y software development, it could be concluded that this methodology could have several benefits. Firstly, this method could strengthen the defect identification (user's issues) thoroughly. Secondly, this method could advance the analysis of the major causes that have the biggest impact on software quality.

In addition, this study has several limitations, namely the software problems studied are only those related to errors and defects, the data used in this study is user Y software complaint data in the range of December 2019 to December 2020, the design stage carried out is only the design of proposed improvements and the results of the proposed improvements that are selected to be implemented are the result of brainstorming between the author and the company and do not change the structure or form of the software. For further research, it is expected to use the six sigma DMADV method on other research objects, such as company business processes. In addition, it is also expected to conduct a feasibility test of the proposed solution design by considering the condition of the company.

References

1. Mahanti R, Antony J (2005) Confluence of six sigma, simulation and software development. Manag Audit J 20(7):739–762
2. Laporte C, April A (2018) Software quality assurance, 1st edn. Wiley, Hoboken
3. Patel S, Zu X (2009) E-government application development using the Six Sigma approach. Electron Government Int J 6(3):295–306
4. Guthrie R (2003) Program design, coding, and testing. In: Encyclopedia of information system. Elsevier, pp 529–543
5. Montgomery D (2013) Introduction to statistical quality control, 7th edn. Wiley, Hoboken
6. Al-Qutaish R, Al-Sarayreh K (2008) Applying six sigma concepts to the software engineering: myth and facts. In: 7th WSEAS international conference on software engineering, parallel, and distributed system, pp 178–183
7. Hong G, Goh T (2003) Six sigma in software quality. TQM Mag 15(6):364–373
8. Park SH (2003) Six sigma for quality and productivity promotion. The Asian Productivity Organization, Tokyo
9. Serrat O (2009) The five whys technique. Asian Development Bank, Metro Manila
10. Knop K, Mielczarek K (2018) Using 5W1H and 4M methods to analyze and solve the problem with the visual inspection process. In: MATEC web of conference 183. EDP Sciences
11. Khan AR, Zhang L (2008) Applying six sigma in software companies for process improvement. Blekinge Institute of Technology, Karlskrona

Multi-skilled Operator Assignment Model for Long Cycle Time and Low-Volume Assembly Line: A Case Study of Aircraft Assembly Line

Anas Ma'ruf and Shinta Uli Agnes Monica Gultom

Abstract This research aims to determine the number of operators and the optimal assignment of operators in a typical center fuselage assembly line. The unique characteristics of an aircraft assembly line generally include long cycle time, low production volume, and operations involving one or two types of skills employed simultaneously based on assembly operation needs. The model proposed in this study is an analytical model with the objective of minimizing total idle time to ensure the utilization of workers. The output of the proposed model is the number of operators of each skill type needed to complete the whole assembly line. A case study was undertaken in an aircraft company. The computation result verifies the feasible implementation of the proposed model by considering practical constraints, including the availability of skilled operators, the number of assembly operations, assembly precedence constraints, and targeted tack time.

Keywords Operator assignment · Multi-skilled operator · Assembly line

1 Introduction

This research is a case study of aircraft assembly process in which the assembly duration is lengthy and the production volume is low. Conventionally, the assembly is performed using a fixed layout and operators carry out the assembly according to precedence constraints. The company in which the research was carried out applied a fixed layout concept to assemble aircraft but the production target could not be achieved. Its inability to meet production targets was caused by delays in the assembly of center fuselage. The manufacture of one unit of center fuselage should have been completed in a period of two months, however in reality the process took three months. Currently, the assembly of the center fuselage is carried out using the barrel

A. Ma'ruf (✉) · S. U. A. M. Gultom
Bandung Institute of Technology, Bandung 40132, Indonesia
e-mail: maruf@itb.ac.id

C. N. Rosyidi et al. (eds.), *Proceedings of the 6th Asia Pacific Conference on Manufacturing Systems and 4th International Manufacturing Engineering Conference*, Lecture Notes in Mechanical Engineering,
https://doi.org/10.1007/978-981-99-1245-2_18

method, in which all assembly operations are carried out in one large workstation (jig).

The application of the barrel method results in an assembly process with a lengthy cycle time due to the fact that many components are focused on merely one workstation. Various available alternatives include changing the plant layout, material handling process, manufacturing process, and resource allocation [1]. Changes in the evaluation of work methods and layout facilities have been initiated through the application of panelization method. The panelization method aims to break down the center fuselage into smaller subcomponents and divide them into several workstations (jigs). Sub-components are arranged in such a way that the division of work elements between stations is balanced. This approach is also proposed by Eberlein and Freitag [2] and Pilati et al. [3]. The panelization method was conducted by experts from related departments of the company. The center fuselage assembly operation would be divided into five jigs that have independent relationships with one another. In addition, there were other practical constraints that needed to be considered. A typical center fuselage assembly operation requires five operator skills, namely level A, level B, sealer, painter, and quality inspector skills. Nearly all assembly operations require two operators in the process. Operations in the center fuselage assembly process may involve one or two types of skills simultaneously according to the operation's needs.

Based on the aforementioned background, the objective of this research is to determine the number of operators and their assignments in the five center fuselage jigs by minimizing total operator idle time. In the following section, studies related to the subject of research will be described. Section 3 describes the modification of the model that is appropriate for the case at hand, while Sect. 4 discusses the results of the implementation of the proposed model and the last section concludes the study with the results.

2 Related Works

Aircraft assembly is a complex process since it requires optimization measures both in process technology and operation management [4]. Optimization of the process technology aims to fulfill the technical specifications of the aircraft [5–7]. This research examines the optimization of operation management, including the assignment of tasks to workstations, staffing issues for each workstation, and the assignment of tasks to operators at each workstation.

Wang proposed a mobile operation of assembly personnel in an aircraft assembly line [8]. Fang also proposed a flexible worker allocation by applying the cyber-physical production system to collect real-time disturbances [9]. Arkhipov propose an operator assignment by taking into account economic and ergonomics constraints [10]. Neither Wang, Fang, nor Arkhipov skill type of workers in their model, which in this case study is a practical constraint to be considered.

There are two studies closely related to this case study, namely Giglio's model [11] and Martignago's model [1]. Giglio proposed a model to allow workers in each multi-manned workstation to perform different assembly tasks of similar products simultaneously. It does not consider that one job can be performed by more than one operator with different skills. Martignago's model enhances Giglio's model and complies with the case study described in the research background with several discrepancies as follows. First, the objective of the reference model is to minimize costs, while the objective of the real system is to minimize idle time to ensure proper utilization of workers assigned to operations across the center fuselage panel assembly. Rios considers workload smoothing appropriate in addressing the final assembly line due to specific features of an aircraft [12]. Second, the model divides workers into single-skilled and cross-trained groups, while in the real system, each operation in the assembly line panelization of center fuselage consists of only single-skilled operators. Third, the skill set in the model consists of only three skills, while the real system requires five types of skills. Lastly, each operation in the reference model can only be performed by one skill group, while operations in the real system can be performed by two different skill groups simultaneously. Based on the discrepancies, this study modified Martignago's model based on the aforementioned four aspects.

3 Proposed Model

The notation and model are described as follow.

Notation	
I	Set of operation i
A	Set of operation i requiring skill Level A
B	Set of operation i requiring skill Level B
C	Set of operation i requiring skill sealer
D	Set of operation i requiring skill painter
E	Set of operation i requiring skill quality inspector
P	Set of precedence constraint operation i prior to operation j
K	Set of available station (1, …, Station)
Station	Number of available stations
Takt time	Cycle time between two units of product (hours)
Total time	Time spent for a product in the assembly line (hours)
p_i	Operation time of task i (hours)
$n_{i,l}$	Number of workers requiring skill l to conduct task i
N_A	Number of workers having skill Level A
N_B	Number of workers having skill Level B
N_C	Number of workers having skill sealer

(continued)

(continued)

Notation	
N_D	Number of workers having skill painter
N_E	Number of workers having skill quality inspector
N_W	Total number of workers available
M	Big number
S_i	Start time of operation *i* (hours)
$x_{i,m}$	1 if operation *i* is assigned to operator *m*, else 0
$u_{i,j}$	1 if operation *i* and operation *j* is conducted in sequence and assigned to the same operator, else 0
$z_{i,j}$	1 if operation *i* is conducted prior to operation *j*, else 0
$y_{i,s}$	1 if operation *i* is conducted at station *s*, else 0
$j_{m,s}$	1 if operator m is assigned at station *s*, else 0

The objective of the model is to minimize the total time inconveniencing operators.

$$\min \sum_{s \in K} \left(\sum_{m=1}^{N_w} j_{m,s} x \left(Takt\ time - \sum_{i \in I} x_{i,m} x p_i \right) \right) \tag{1}$$

Subject to:

$$\sum_{m \leq N_A} x_{i,m} = n_{i,l} \quad \forall i \in A, \quad l = A \tag{2}$$

$$\sum_{m > N_A \wedge m \leq (N_A + N_B)} x_{i,m} = n_{i,l} \quad \forall i \in B, \quad l = B \tag{3}$$

$$\sum_{m > N_A + N_B \wedge m \leq (N_A + N_B + N_C)} x_{i,m} = n_{i,l} \quad \forall i \in C, \quad l = C \tag{4}$$

$$\sum_{m > N_A + N_B + N_C \wedge m \leq (N_A + N_B + N_C + N_D)} x_{i,m} = n_{i,l}, \forall i \in D, \quad l = D \tag{5}$$

$$\sum_{m > N_A + N_B + N_C + N_D \wedge m \leq N_W} x_{i,m} = n_{i,l} \quad \forall i \in E, \quad l = E \tag{6}$$

$$S_{Last\ Task} + p_{Last\ Task} \leq Total\ Time \tag{7}$$

$$S_j - S_i \geq p_i \quad \forall (i, j) \in P \tag{8}$$

$$u_{i,j} - u_{j,i} \leq 1 \quad \forall (i, j) \in I^2, j > i \tag{9}$$

$$u_{i,i} = 0 \quad \forall i \in I \tag{10}$$

$$u_{i,j} + u_{j,i} \geq x_{i,m} + x_{J,m} - 1 \quad \forall m = 1, \ldots, N_w, \quad \forall i, j \in I^2, i > j \tag{11}$$

$$S_j - S_i \geq p_i - M\left(1 - u_{i,j}\right) \quad \forall (i, j) \in I^2 \tag{12}$$

$$\sum_{s \in K} y_{i,s} = 1 \quad \forall i \in I \tag{13}$$

$$\sum_{s \in K} s.\left(y_{i,s} - y_{j,s}\right) \leq 0 \quad \forall (i, j) \in P \tag{14}$$

$$\sum_{s \in K} j_{m,s} \leq 1 \quad \forall m \in 1, \ldots, N_w \tag{15}$$

$$\sum_{m=1}^{N_w} x_{i,m} = n_i . y_{i,s} \quad \forall i \in I, \quad \forall s \in K \tag{16}$$

$$x_{i,m} \leq \sum_{s \in K} j_{m,s} \quad \forall \in I, \quad \forall m \in 1, \ldots, N_w \tag{17}$$

$$x_{i,m} + y_{i,s} \leq 1 + j_{m,s} \quad \forall i \in I, \forall m \in 1, \ldots, N_w, \forall s \in K \tag{18}$$

$$\sum_{s \in K} (j_{m,s} - \left(j_{m+1,s}\right) \geq 0 \quad \forall m = 1, \ldots, (N_A - 1) \tag{19}$$

$$\sum_{s \in K} (j_{m,s} - \left(j_{m+1,s}\right) \geq 0 \quad \forall m = (N_A + 1), \ldots, (N_A + N_B - 1) \tag{20}$$

$$\sum_{s \in K} (j_{m,s} - \left(j_{m+1,s}\right) \geq 0 \quad \forall m = (N_A + N_B + 1), \ldots, (N_A + N_B + N_C - 1) \tag{21}$$

$$\sum_{s \in K} (j_{m,s} - \left(j_{m+1,s}\right) \geq 0 \quad \forall m = (N_W - N_E - N_D + 1), \ldots, (N_W - N_E - 1) \tag{22}$$

$$\sum_{s \in K} (j_{m,s} - \left(j_{m+1,s}\right) \geq 0 \quad \forall m = (N_W - N_E + 1), \ldots, (N_W - 1) \tag{23}$$

$$S(i) \geq \left(-1 + \sum_{s \in K} (s.y_{i,s})\right) . Takt\,Time \quad \forall s \in K \tag{24}$$

$$S(i) + p(i) \leq \sum_{s \in K} (s.y_{i,s}).Takt\ Time \quad \forall s \in K \tag{25}$$

$$\sum_{i \in A} x_{i,m} \leq 0 + M.\sum_{s \in K} j_{m,s} \quad \forall m = 1, \ldots, N_W \tag{26}$$

Equations (2)–(6) guarantee that every operation that requires certain skills is carried out by one or more workers with these skills in accordance with the required number of operators. Equation (7) guarantees that the completion time of the last operation in the assembly line is not more than the total time. Equation (8) guarantees that the start time of each operation does not violate precedence constraints. Equation (9) guarantees that there is only one execution sequence for every two operations using one worker. Equation (10) guarantees that the assignment of two operations to one worker is not performed on similar two operations. Equation (11) guarantees that if operations i and j are assigned to one worker, then one of the variables $u_{i,j}$ and $u_{j,i}$ must be one. Equation (12) guarantees that two operations (i, j) performed by one worker are not performed at the same time. Equation (13) guarantees that each operation is assigned to only one workstation. Equation (14) guarantees that the precedence constraint relationship between stations is met. Equation (15) guarantees that one worker can only be assigned to a maximum of one workstation. Equation (16) guarantees that every operation placed at a station must be performed by the number of workers required by the operation. Equation (17) ensures that operations are assigned to active workers. Equation (18) ensures that the operation and the assigned workers are at the appropriate workstation. Equations (19)–(23) aim to ensure that each operation that requires skills A, B, C, D, and E is assigned to the worker with the smallest number. This limitation is useful in reducing the computation time of the model. Equation (24) guarantees that the start time of an operation must be greater than or equal to the start time of the station where the operation is located. Equation (25) guarantees that the completion time of an operation must be less than or equal to the completion time of the station where the operation is located. Equation (26) guarantees that a worker assigned to a station can be assigned to several operations as long as the worker is active.

4 Results and Discussion

This research is undertaken at an aircraft industry in Indonesia. Prime data of the assembly process in the center fuselage assembly line consists of 123 operations using five different jigs in which each jig is already assigned certain operations to be performed as shown in Table 1. Each operation requires at least one type of skill to carry out the assembly operation process. The available of operators for each skill type is presented in Table 2. The precedence diagram is presented in Fig. 1, while the operation time and the required skill type of each operation is omitted in this paper due to the bulk of data.

Table 1 Panelization jigs used in the center fuselage assembly line

Jig number	Jig name
1	Barrel assembly
2	Left panel center fuselage
3	Lower panel center fuselage
4	Right panel center fuselage
5	Upper panel center fuselage

Table 2 Availability of operators

Skill type	Availability
Level A	4 operators
Level B	6 operators
Sealer	4 operators
Painter	6 operators
Quality Inspector	2 operators

The computation of the proposed solution was carried out using the help of the CPLEX Optimization Studio solver. The computational results show that it requires 12 operators across the center fuselage assembly line with total assembly time of 281.5 h. The details of the computational results are presented in Table 3. There is a requirement of three operators each for Level A and Level B. Both Level A and Level B operators are required to be active in the assembly line for 87% of the required total time to complete the center fuselage. Sealers, painters, and quality inspectors have a low utilization due to the limitation of technical assignment in the assembly operation. It has also been found that there are excess of operators having skill type Level A—1 operator, Level B—3 operators, Sealer—2 operators, and Painter—4 operators. This information is important to the management since it enables shifting multi-assignment of operators using similar skill types in other aircraft assembly lines.

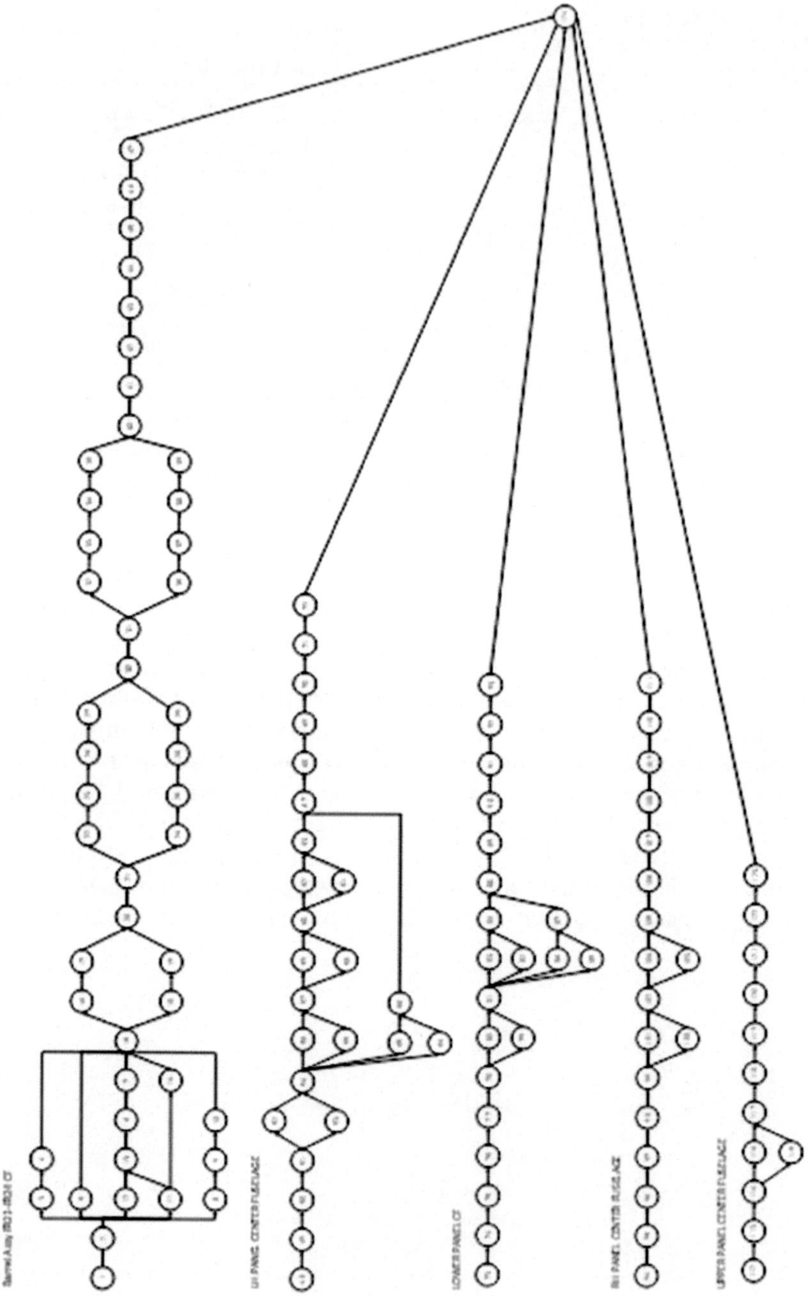

Fig. 1 Precedence diagram of the center fuselage

Table 3 Computational results

Operator	Skill type	Sub total	Utilization (%)
1	Level A	3	91
2			79
3			72
4	Level B	3	87
5			71
6			75
7	Sealer	2	15
8			15
9	Painter	2	14
10			14
11	Quality Inspector	2	5
12			11

5 Conclusion

This study proposes a design method for determining the number of operators and assigning operators in accordance with the assembly line characteristics of the aircraft's center fuselage. The proposed model is a mixed integer linear programming mathematical model with the objective of minimizing idle time. Practical constraints considered in the model include the availability of skilled operators, the number of assembly operations, assembly precedence constraints, and takt time to assemble the aircraft. The case study undertaken in this research result the need of five different skill type and a total of 12 operators. The computational results verify the feasible solution. The output of the model provides useful information for multi-assignment of operators for aircraft assembly line. The current research considers one aircraft model, while it is commonly found multi-model aircraft are manufactured in a company. Future research will be conducted for multi-model aircraft assembly line. Due to the non-polynomial nature of the assembly line and problem size, direction of a meta-heuristic algorithm is expected.

References

1. Martignago M, Battaïa O, Battini D (2017) Workforce management in manual assembly lines of large products: a case study. IFAC-PapersOnLine 50(1):6906–6911
2. Eberlein S, Freitag M (2022) A pull principle for the material supply of low-volume mixed-model assembly lines. Procedia CIRP 107:1385–1390
3. Pilati F, Faccio M, Cohen Y (2021) Absenteeism and turnover as motivation factors for segmenting assembly lines. IFAC-PapersOnLine 54(1):613–616

4. Wan XQ, Yan H, Sen, Wang Z (2015) Scheduling and self-reconfiguration of an aircraft engine assembly line in knowledgeable manufacturing. Zidonghua Xuebao/Acta Automatica Sinica, 41(1)
5. Li T, Lockett H (2017) An Investigation into the interrelationship between aircraft systems and final assembly process design. Procedia CIRP 60:62–67
6. Formentini G, Cuiller C, Dereux PE, Bouissiere F, Favi C (2021) Impact assessment of design guidelines in the conceptual development of aircraft product architectures. Procedia CIRP 100:223–228
7. Zhao D, Wang G, Guo W, Zhang Q (2019) An integrated framework for aircraft digital assembly coordination design and simulation. Procedia CIRP 83:330–334
8. Zhang J, Wang P, Xin B (2016) Study on aircraft assembly line balancing problem based on mobile workers. In: Proceedings of the ASME design engineering technical conference, vol 4
9. Fang P, Yang J, Liao Q, Zhong RY, Jiang Y (2021) Flexible worker allocation in aircraft final assembly line using multiobjective evolutionary algorithms. IEEE Trans Ind Inf 17(11)
10. Arkhipov D, Battaïa O, Cegarra J, Lazarev A (2018) Operator assignment problem in aircraft assembly lines: a new planning approach taking into account economic and ergonomic constraints. Procedia CIRP 76
11. Giglio D, Paolucci M, Roshani A, Tonelli F (2017) Multi-manned assembly line balancing problem with skilled workers: a new mathematical formulation. IFAC-PapersOnLine 50(1):1211–1216
12. Ríos J, Mas F, Menéndez JL (2012) Aircraft final assembly line balancing and workload smoothing: a methodological analysis. Key Eng Mater 502

Analysis of Employment Increasing for Potential Tourism Industry in Indonesia: An Exploration Method Approach

Lutfia Septiningrum, Fandi Achmad, Augustina Asih Rumanti, and Iwan Inrawan Wiratmadja

Abstract This research aims to increase employment in the industry tourism potential through variable form Support Systems and Facilities (SSF) model. Based on Statistics 2021, the Rembang district is one of the regions in Central Java that has great tourism potential and high unemployment rates. This is in line with the government's policy of reducing unemployment in the Rembang district, namely increasing employment opportunities for people around tourist attractions. The indicators of readiness to expand the business field with categorical data types can be done using the multinomial logistic regression method. The advantages of the analysis using the multinomial logistic regression method are knowing variables that are significant and affect the increase in the potential for employment in the tourism industry area and can be used to predict the pattern of success in increasing employment by improving the significant variable components. The results of the test model show that the model can be used to measure the evaluation of expanding employment. This means that the government and tourism managers are quite ready to open new jobs for creative communities around tourist attractions.

Keywords SSF · Multinomial regression · Employment

L. Septiningrum
University of Yogyakarta, DI Yogyakarta, Indonesia
e-mail: lutfiaseptiningrum@uny.ac.id

F. Achmad (✉) · A. A. Rumanti
Telkom University, Bandung, Indonesia

I. I. Wiratmadja
Bandung Institute of Technology, Bandung, Indonesia

C. N. Rosyidi et al. (eds.), *Proceedings of the 6th Asia Pacific Conference on Manufacturing Systems and 4th International Manufacturing Engineering Conference*, Lecture Notes in Mechanical Engineering,
https://doi.org/10.1007/978-981-99-1245-2_19

1 Introduction

The development of a region is very dependent on the success of the region in managing the potential of natural resources and available human resources. One indicator to assess the development of a region can be seen from the increase in production activities that have economic value and the ability of the region to create job opportunities. Rembang district is one of the districts that have a high level of job opportunities (TKK), but the population listed as job applicants is relatively low [1]. TKK shows how much of the working-age population is working or temporarily not working. At the same time, registered job seekers are the workforce who are unemployed and looking for work, as well as those who are already working but want to move. Central Java is one of the provinces that absorbs the most labor. The number of the workforce in Central Java is 256,706 people, while the available job opportunities are 223,377 [1]. In January 2021, the number of unemployed in Central Java was 1.12 million. This is a problem that must be handled by the government, where job vacancies must be added as much as possible to accommodate the number of unemployed in Central Java which is increasing year by year. One of the districts in Central Java that has the lowest registered job vacancies is Rembang Regency. Meanwhile, Rembang Regency is one of the regencies in Central Java that has few job opportunities in the tourism sector. The Rembang Regency Tourism Office, in partnership with the tourism village community by forming a Village Owned Enterprise or hereinafter abbreviated as BumDes. To develop the industry tourism potential, some of the facilities around Rembang regency tourism objects are adequate with Support System and Facilities (SSF) [2]. Support System Facilities include several indicators of success, there is a 4G signal, there are places of worship, toilet facilities, lodging, gift shops, food stalls, surfboard rentals, emergency health areas, and parking lots. Support System and Facilities is a service that is developed either directly or indirectly to support and encourage tourism potential [2, 3]. As tourism develops, it will be balanced by an adequate level of facilities because the needs of tourists for carrying capacity and facilities reflect the development of tourism in the area [3, 4]. But it is unfortunate that the tourism development potential of Rembang regency, which is very promising, cannot be optimized to create jobs. To see the readiness for expansion and improvement of business fields in the tourism sector, this research will look at factors that affect readiness and suggest policy improvements.

2 Methodology

Support System and Facilities (SSF) is a service that is developed either directly or indirectly to support and encourage tourism potential [2]. As tourism develops, it will be balanced by an adequate level of facilities because the needs of tourists for carrying capacity and facilities reflect the development of tourism in the area [3]. The variables used in this study refer to the SSF research model [2]. This type of research is descriptive explorative research with quantitative and qualitative research paradigms. This study uses primary data. Data can be in form of text, picture, voice, and video. Social media is one factor of the data increase as everybody expresses, gives opinion, and even complains in social media [5]. where in this research, the data obtained are obtained from interviews and observations with agencies or parties that have direct links and authority [6]. The objects of the interview are the tourism office, tourist site managers, and workers in tourist areas. The survey method was carried out by means of structured observation of as many as 32 tourist attractions as a sample in April–July 2022. Sampling was carried out in this study using the accidental sampling method. The advantage of this method is that it gives freedom to determine the sample those deemed to be the most qualified in this case, the criteria in question are that the respondent is currently visiting tourist attractions in Rembang, has visited more than two times to tourist attractions in Rembang, and can answer questions about tourism.

Evaluating the variables that have been determined in Tables 1 and 2 is solved by multinomial logistic regression. Multinomial Logistics Regression (MLR) is a model used to predict the probability of a different possible outcome of the dependent variable being categorically distributed when a set of independent variables is given, and to analyze relationships between non-metric dependent variable and metric or dichotomous independent variables [7]. MLR can be applied in different cases. This will be implemented to the variables which will be depending on the dichotomous (For example yes/no) and can prolonging to accept a dependent variable with other categories [8]. Multinomial logistic models are suitable for full factorial models or researcher-defined models. Parameter estimation is carried out through a repeated maximum likelihood algorithm [9–11]. To explain in more detail the multinomial logistic regression model formed, odd ratios analysis was used, then the results of the descriptive explorative carried out would produce a policy suggestion for the Rembang district government. This explanation can be seen in Fig. 1.

Table 1 Variable dependent

Notation	Variable	Scale data
Y	Readiness to expand business field	1. Need improvements and policies in opening business land
		2. Tourist attractions have the potential to open business land
		3. No potential to open a business area

Table 2 Variable independent

Notation	Variable	Scale data
X1	Telecommunication	Average internet speed
X2	Power source	Total power source
X3	Transportation	Distance of tourist attractions from city center
X4	Waste management	Number of trash cans
X5	Location	Area of tourism
X6	Clean water source	Number of clean water sources at tourist sites
X7	Supporting industry	Number of business areas in tourist area
X8	Spatial	Number of supporting tours around tourist attractions
X9	Hospitality	The number of public facilities in the tourist area
X10	Safety and security	Total availability of security posts and medical equipment such as first aid kit
X11	Stakeholder	Number of tourist attractions

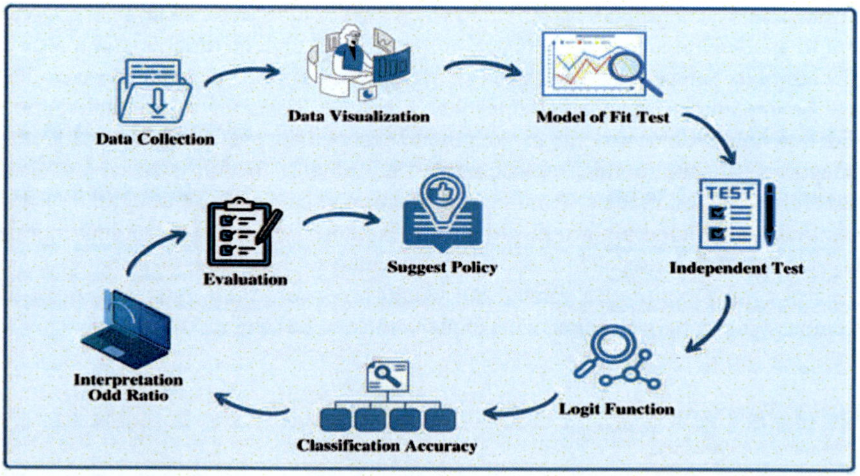

Fig. 1 Research flowchart

3 Result and Discussion

Rembang Regency is the Easternmost Regency in Central Java Province and is located on the northern coast of Central Java. In the north, the Rembang area is in direct contact with the Rembang bay (Java Sea), which makes the Rembang district have high tourism potential. The impact of this is that the Rembang district can open wide employment opportunities, especially for people around tourist attractions. One form of evaluating the potential for job expansion can use the statistical regression method because the dependent variable of this study is a variable with the categorical data type. The appropriate statistical test uses multinomial logistic regression. The result of model testing shows in Table 3.

Based on Table 3, to ascertain what predictor variables have a significant effect, it is necessary to do a partial test. The formation of a multinomial logistic regression model using the Maximum Likelihood Estimation (MLE) method. MLE method is a method by converting the probability density function into a likelihood function. The results of the analysis of the formed multinomial logistic model are shown in Table 4.

Table 4 shows there is 3 variables that are not significant, which are Spatial, Waste Management, and Safety and Security. To evaluate further, it is necessary to have a mathematical model that supports the analysis by estimating the parameters of the multinomial logistic model. The value of the model estimation results can be seen in Table 5.

The model that is formed from all decision models for increasing employment in the tourism sector, it is known that the equation results from the effect of the variable support system facilities on expand business field (Y) in Rembang Regency as follows:

$$\begin{aligned} g_1(x) = 4.622 + 1.074X1 + 4.426X2 + 2.402X3 + 5.681X5 \\ + 2.786X6 + 3.605X7 - 5.960X9 + 3.080X11 \end{aligned} \quad (1)$$

Table 3 Result of goodness of fit and simultaneous test

Test results	*P*-value	Criteria	Decision
Goodness-of-fit	0.987	>0.50	Model fit
Simultant test	0.001	<0.05	There is at least 1 independent variable that affects the increase in the tourism sector workforce

Table 4 Result of partial test

Notation	Variable	*P*-value	Decision
X1	Telecommunication	<0.001	Significant influence
X2	Power source	0.036	Significant influence
X3	Transportation	0.038	Significant influence
X4	Waste management	0.095	No significant effect
X5	Location	<0.001	Significant influence
X6	Clean water source	<0.001	Significant influence
X7	Supporting industry	0.005	Significant influence
X8	Spatial	0.263	No significant effect
X9	Hospitality	<0.001	Significant influence
X10	Safety and security	0.061	No significant effect
X11	Stakeholder	<0.001	Significant influence

Table 5 Result value of the model estimation

Logit	Parameter	Coefficient	*P*-value	Odds ratio
1	Constant	4.622	0.021	
	X1	1.074	0.042	21.757
	X2	4.426	0.043	0.192
	X3	2.402	0.063	0.000
	X5	5.681	0.028	0.857
	X6	2.786	0.004	0.001
	X7	3.605	0.002	0.000
	X9	−5.960	0.093	0.000
	X11	3.080	0.047	0.006
2	Constant	4.743	0.153	
	X1	1.399	0.050	0.228
	X2	−1.702	0.137	0.015
	X3	−1.986	0.076	0.008
	X5	2.900	0.034	1.001
	X6	2.055	0.001	0.007
	X7	1.630	0.000	0.008
	X9	2.361	0.324	0.003
	X11	1.869	0.000	0.011

$$g_2(x) = 4.743 + 1.399X1 + 2.900X5 + 2.055X6 + 1.630X7 + 1.869X11 \quad (2)$$

The multinomial logistic regression approach not only finds out the significant variables that affect the increase in employment potential in the tourism industry, but the results of this modeling can be used to predict the pattern of success in increasing employment by improving the significant variable components. As an example, the Rembang Regency Government will increase employment in tourism object areas with category 1, namely the need for improvement of tourist attractions in opening business fields, it is necessary to improve and procure related to Telecommunication, Power Source, Transportation, Location, Clean Water variables Source, Supporting Industry, Hospitality, and Stakeholders. Where every increase in tourism sector employment must be supported by improvements in telecommunication facilities, 4.426% addition of power sources, 2.4% expansion of tourist areas, 5.7% addition of clean water source areas, and 2.8% addition of supporting industry areas. This can increase the selling area and add stakeholders as business managers/ employees. While in category 2, which has the potential to create employment opportunities, local governments only need to improve or prepare telecommunications in tourist attractions, increase tourist land by 2.9%, increase the amount of clean water

Table 6 Results indicated by the accuracy of classification

Observation	Predicted			Percentage correct (%)
	1	2	3	
1	6	1	2	28.125
2	2	8	1	34.375
3	1	4	7	37.5
Percentage	28.125%	40.625%	31.25%	65.625

available at tourist sites by 2.1%, increase business land by 1.6% in tourist attractions or around tourist areas and increasing the management of tourist attractions by 1.8%.

Table 6 shows the results indicated by the accuracy of classification to evaluate the readiness to expand employment for the community around the tourist attractions of Rembang Regency by applying the multinomial logit model of 65.625%. This value shows that the absorption of labor in the potential tourism sector in Rembang Regency in the period of 2022 is quite ready because the value is more than 50%. So that the Rembang district government is expected to be able to assist the management of tourist attractions to prepare for increasing employment opportunities for the community around Rembang district tourist attractions. With the readiness of tourist attractions and local governments, it is necessary to make policies. The policy suggestion if the Rembang district government will implement the model results in this research is that the Manager can implement the CHSE-based health protocol strictly and end-to-end discipline as an improvement in Stakeholder indicators, and the Manager can determine the Crisis point as an effort to improve Hospitality in tourist areas.

4 Conclusion

Based on the results of the study, information was obtained to increase employment in the potential tourism industry sector. It is necessary to have modeling, one of the variables used to support this is the support system facilities. The assumption is that if a tourist site has good support system facilities, the local government can assist BumDes in expanding employment for the community around tourist attractions. This can be supported by several policies to regulate and manage business fields in tourist attractions. From the multinomial logistic regression model, several indicators that affect category 2 are obtained tourist attractions have the potential to open business fields. In this study, the method used is still multinomial logistic regression, which in future studies will be tested with a comparison of test methods to create more accurate results.

Acknowledgements The researchers thank the tourism office and the manager of tourist attractions in Rembang Regency, Indonesia, for their availability in preparing data and being respondents in this study.

References

1. Statistik BP (2021) Keadaan Angkatan Kerja di Indonesia Agustus 2021. https://www.bps.go.id/publication/2021/12/07. Last accessed 01 Aug 2022
2. Septiningrum L, Rizana AF, Soesanto RP, Rumanti AA (2022) Success factors for elevating Indonesian tourism area potential. In: IEEE international conference on industrial engineering and engineering management
3. Lickorish LJ, Jenkins CL (2007) Introduction to tourism. Routledge Taylor & Francis Group, London and New York
4. Lohmann G, Netto AP (2016) Tourism theory: concepts, models and systems. Cabi
5. Ramadhan WP (2017) Sentiment analysis using multinomial logistic regression
6. Ghozali I (2018) Aplikasi Analisis multivariate Dengan program IBM SPSS 25, 9th edn. Badan Penerbit Universitas Diponegoro
7. Pramesti W, Damayanti I, Asfani DA (2016) Stator fault identification analysis in induction motor using MLR
8. Rekha SN, et al (2019) Multinomial logistic regression for fault type detection in benchmark fault model of wind energy conversion system
9. Xiao MX, Lu CH, Ta N, Wei HC, Yang CC, Wu HT (2022) Toe PPG sample extension for supervised machine learning approaches to simultaneously predict type 2 diabetes and peripheral neuropathy. Biomed Signal Process Control 71:103236. https://doi.org/10.1016/j.bspc.2021.103236
10. Agresti A (1990) Categorical data analysis. Wiley, New York
11. Hosmer DW Jr, Lemeshow S, Sturdivant RX (2013) Applied logistic regression, vol 398. Wiley

Development of an Optimization Model for Home Health Care Routing and Scheduling Problem

Annisa Puja Ramadhandyna, Pringgo Widyo Laksono, and Cucuk Nur Rosyidi

Abstract This paper formulated a Mixed Integer Non-Linear Programming (MINLP) for HHC daily planning. The main issue of HHC planning problems is to assign multiple caregivers to serve patients during working hours. This problem is modelled to minimize the total operational costs, consisting of vehicle fuel costs, patient preference violation penalties, and carbon tax costs. The model is tested with real data at RSUD Dr Moewardi to provide an overview of the model that can be used to solve an existing problem. Ten patients must be served in their homes by two caregivers, where each patient has a time window and caregiver's gender preference. The model was solved using LINGO 18.0 and obtained the optimal result of the order of visits by female caregivers is 1-10-4-11-9-7-1, while for male caregivers is 1-5-2-3-6-8-1. The objective function is 123,393 IDR with a total fuel cost of 52,980 IDR, total carbon emission tax of 412.89 IDR, and total the patient preference violation penalty of 70,000 IDR. This results can be used as a decision-making tool by the HHC coordinator.

Keywords Home health care · Routing and scheduling · Daily planning

1 Introduction

Home health care (HHC) is a health service the hospital team provides to patients at home [1]. HHC is usually cheaper, more convenient, and as effective as services in a hospital [2] so in the last decades, the number of organizations providing HHC services has increased. In many countries, HHC planning is usually done manually, which takes time and effort, leading to inefficient solutions [3]. HHC operations planning is a complex problem, making it interesting for planning by operation research techniques [4]. Routing and scheduling HHC can be solved by the exact

A. P. Ramadhandyna (✉) · P. W. Laksono · C. N. Rosyidi
Department of Industrial Engineering, Universitas Sebelas Maret, Surakarta, Indonesia
e-mail: annisapuja06@student.uns.ac.id

C. N. Rosyidi et al. (eds.), *Proceedings of the 6th Asia Pacific Conference on Manufacturing Systems and 4th International Manufacturing Engineering Conference*, Lecture Notes in Mechanical Engineering,
https://doi.org/10.1007/978-981-99-1245-2_20

methods, such as branch and price algorithm [5], a column generation-based gossip algorithm [6], SP-timed logic-based benders decomposition (ST-LBBD) approach [7] and others.

Developing models on routing and scheduling problems in HHC has been done before. For example, a linear integer scheduling model was developed to support human resource short-term planning in HHC [8]. A mixed linear program to minimize the total travelled distance in French homecare and the model is solved by LINGO [9]. Daily planning in HHC focuses on patient satisfaction, such as service time and gender [10]. Bi-objective optimization model HHC routing problem to minimize travel distance and carbon emission to reduce environmental pollution [11]. A mixed integer programming model for medium-term HHC planning problems minimizes the total operational cost [12]. And based on the literature review by Mascolo et al. [13] the future research topics on HHC planning issues are patient preferences, sustainability aspects, multi-objective approaches, and others.

The HHC team carries out transportation activities to the patient's home using the vehicles available at the hospital, such as ambulance, motorbike and others. This activity requires fuel oil that emits compounds such as CO (carbon monoxide), THC (Total HydroCarbon), TSP (dust), Nox (Nitrogen Oxide), Sox (Sulphur Oxide), and CO2 (Carbon Dioxide). These emissions are the main source of pollution in big cities in Indonesia [14]. So it is necessary to control the pollution which can increase the risk of disease and greenhouse gases (GHG) emissions due to transportation activities in HHC [11].

Sustainable transportation refers to the minimization of gas emissions produced and environmentally friendly means of transport. Environmental issues require consideration of environmental factors, including HHC activities that involve transportation to the patient's home. One of the policies that can address carbon emissions is setting a tax on each unit of carbon emissions [15]. The carbon tax policy uses free carbon emissions (no limit) but will be taxed for each unit of carbon produced. So, in this study, applying a carbon emission tax was considered to develop a model for HHC routing and scheduling.

To improve the efficiency of HHC planning, this paper formulated a Mixed Integer Non-Linear Programming (MINLP) for HHC daily planning that can be used by HHC coordinators as a decision-making tool. This study uses the MINLP model and solves it with LINGO software, because it can determine the schedule and route of HHC caregivers with state global optimal. The model proposed in this paper is a combination of models in research conducted by [10–12], which can minimize the total operational costs, consisting of vehicle fuel costs [12], patient preference violation penalties [10], and carbon tax costs generated by the caregiver's vehicle [11].

2 Model Development

2.1 Problem Description

The main issue of HHC planning problems is to assign multiple caregivers to serve multiple patients during working hours. The problem considered in this paper is the daily planning of the caregivers that can minimize the total operational costs, consisting of vehicle fuel costs, patient preference violation penalties, and carbon tax costs generated by the caregiver's vehicle. The model proposed can determine the optimal order of patient visits and caregiver schedules.

Several assumptions in this model are:

1. Each patient is served by one caregiver
2. Each caregiver uses a different type of vehicle
3. The duration of care for each patient is deterministic
4. google maps obtain the travel distance between two locations and the travel time.

2.2 Mathematical Model

The notations used in the mathematical model are listed as follows:

Sets

P locations, $\{g, j, l \in P\}$.
K caregivers, $\{k \in K\}$.

Parameters

tP_j	Service duration of patient *j*.
tJ_{gj}	Travel time from location *g* to location *j*.
D_{gj}	Travel distance from location *g* to location j.
P_Gen_j	The wish of caregiver's gender for the patient at the location *j* (0 for female, 1 for male and -1 for not specified gender).
K_Gen^k	Gender of caregiver *k* (0 for female and 1 for male).
$tStart$	Starting time point of caregiver at the hospital.
tF	Maximum ending time point of caregiver at the hospital.
dM	Maximum working duration per day for caregiver.
$\{PL_j, PU_j\}$	Lower/Upper bound of a patient *j*'s time window.
C^k	Fuel cost per unit distance for caregiver's transport system *k*.
CG_j	Penalty cost if the wish of caregiver's gender for the patient at the location *j* is not respected.
$Ccarbon$	Carbon tax per carbon unit.

E^k The CO_2 emission rate per unit distance for caregiver's transport system k.

Decision Variables

x_{gj}^k Binary variable (equals 1 if location j is visited after location g by caregivers k, 0 otherwise).
$Care_j^k$ Binary variable (equals 1 if location j is visited by caregivers k, 0 otherwise).
NR_j Binary variable (equals 1 if the wish of caregiver's gender for the patient at the location j is not respected, 0 otherwise).
ts_j The time point when the caregiver at location j.
tE_j The time point when the caregiver finishes at location j.
tB^k The time point of the caregiver k at the last location (hospital).
$tEnd^k$ Ending time point of the caregiver k to work.
TA^k Total time services by caregiver k.

The mathematical model is formulated as a multi-objective mixed integer non-linear programming (MOMINLP) problem as follows:

Objective Function

$$Minimize(S + TC + CE) \tag{1}$$

$$S = \sum_{g \in P} \sum_{j \in P} \sum_{k \in K} C^k \times x_{gj}^k \times D_{gj} \tag{2}$$

$$TC = \sum_{j \in P} CG_j \times NR_j \tag{3}$$

$$CE = Ccarbon \times \sum_{g \in P} \sum_{j \in P} \sum_{k \in K} E^k \times D_{gj} \times x_{gj}^k \tag{4}$$

Subject To

$$\sum_{j \in P} x_{gj}^k = 1, \quad \forall g \in P, g = 1; \quad \forall k \in K \tag{5}$$

$$\sum_{g \in P} x_{gj}^k = 1, \quad \forall j \in P, j = 1; \quad \forall k \in K \tag{6}$$

$$\sum_{k \in K} x_{gj}^k = 1, \quad \forall g, j \in P, \quad j \neq 1, \tag{7}$$

$$\sum_{k \in K} x_{gj}^k = 1, \quad \forall g, j \in P, \quad g \neq 1 \tag{8}$$

$$\sum_{g\in P, l\neq g} x_{gl}^{k} - \sum_{j\in P, l\neq j} x_{lj}^{k} = 0, \quad \forall l \in P; \quad \forall k \in K \tag{9}$$

$$x_{gj}^{k} = 0, \forall g, \quad j \in P, g = j; \quad \forall k \in K \tag{10}$$

$$TA^{k} = \sum_{g,j\in P} x_{gj}^{k}\left(tJ_{gj} + tP_{j}\right), \quad \forall k \in K \tag{11}$$

$$TA^{k} \leq dM, \quad \forall k \in K \tag{12}$$

$$tEnd(k) = tStart + TA(k), \quad \forall k \in K \tag{13}$$

$$tEnd(k) \leq tF, \quad \forall k \in K \tag{14}$$

$$tE_{g} = tStart + tP_{g}, \quad \forall g \in P, \quad g = 1 \tag{15}$$

$$ts_{j} \geq \left(tE_{g} + tJ_{gj}\right) \times x_{gj}^{k}, \quad \forall g, j \in P, \quad g \neq j, j > 1; \quad \forall k \in K \tag{16}$$

$$tE_{j} = ts_{j} + tP_{j}, \quad \forall g, j \in P \quad j > 1 \tag{17}$$

$$tB^{k} \geq \sum_{g,j\in P} \left(tE_{g} + tJ_{gj}\right) \times x_{gj}^{k} \quad \forall g, j \in P, \quad g \neq j, \quad j = 1 \tag{18}$$

$$tB^{k} \leq tEnd^{k}, \quad \forall k \in K \tag{19}$$

$$PL_{j} \leq ts_{j}, \quad \forall g, j \in P, \quad g \geq 1, \quad j \geq 1 \tag{20}$$

$$ts_{j} + tP_{j} \leq PU_{j}, \quad \forall g, j \in P, \quad g \geq 1, \quad j \geq 1 \tag{21}$$

$$ts_{j} \geq 0, \quad \forall g, j \in P, \quad g \neq 1, \quad j \neq 1 \tag{22}$$

$$PL_{j} \leq ts_{j} \leq PU_{j}, \quad \forall j \in P, vj \neq 1; \quad \forall g \in P, \quad g \neq j \tag{23}$$

$$\sum_{g\in P, g\neq j} x_{jg}^{k} = \sum_{j\in P}\sum_{k\in K} Care_{j}^{k} \quad \forall g \in P, \quad \forall k \in K \tag{24}$$

$$\sum_{k\in K} Care_{j}^{k} = 1 \quad \forall j \in P, \quad j \neq 1 \tag{25}$$

$$\sum_{k \in K} Care_j^k \times K\_Gen^k = NR_j \forall j \in P, P_{Gen} = 0 \quad (26)$$

$$\sum_{k \in K} Care_j^k \times (1 - K\_Gen^k) = NR_j \quad \forall j \in P, \quad P_{Gen} = 1 \quad (27)$$

$$tB^k, tE_j, ts_j, u_j \geq 0 \text{ and integer} \quad (28)$$

$$x_{gj}^k, Care_j^k, NR_j \in \{0, 1\} \quad (29)$$

The objective function is integrated as Eq. (1). The first objective function shown in Eq. (2) explains the fuel costs, the second objective function shown in Eq. (3) explains the penalty cost due to the wish of the caregiver's gender for the patient is not respected, the third objective function shown in Eq. (4) explains the carbon emissions tax.

There are also a set of constraints used in the model. Constraints (5) (6) ensure that each caregiver must start and end their work at the hospital. Constraint (7) (8) reveals that each patient is visited once only. Constraint (9) means the caregiver visits the patient and then leaves them. Constraint (10) ensures no route from the patient to the same patient. Constraint (11) defines the total time HHC services. Constraint (12) guarantees that the working time of the caregivers on each day must be less than the maximum daily working duration. Constraints (13) and (14) indicate the end of the caregiver's working time. Constraints (15)–(17) indicate the time point when the caregiver starts and finishes serving at location j. Constraints (18) and (19) indicate the time point of the caregiver at the last location (hospital). Constraints (20)–(23) ensure the feasibility of each visit according to the patient's time window. Constraints (24) and (25) define the location j s visited by caregivers k. Constraints (26) and (27) ensure that if the gender preference of the patient at the location j not respected, the decision variable corresponding to the non-satisfaction of the wish NR_j is equal to one. Constraints (28) and (29) enforce the non-negativity restrictions and binary variables.

3 Numerical Examples

The model is tested with real data at RSUD Dr Moewardi to provide an overview of the use of the model to solve a real problem. Ten patients must be served in their home, where each patient has a time window and caregiver's gender preference. We assume that there are two available caregivers, i.e. male and female. Male caregivers use a motorcycle and female caregivers use an ambulance. The operational time of home health care is from 07.30 AM to 03.30 PM. So, the caregiver's maximum working duration per day is 8 h. Patients can request to be served by caregivers of a certain gender. If the patient is not served by caregiver according to their gender preference, the hospital must pay a penalty fee. We assume that the caregiver's gender preference penalty is 50,000 IDR for a female caregiver and 20,000 IDR for a male caregiver.

Based on study by Nurdjanah [14] the ambulance fuel consumption is 0.1160 liter per km, while motorcycle fuel consumption is 0.0266 liter per km. Each vehicle is assumed to use pertalite fuel and the cost of pertalite fuel in Indonesia is 10,000 IDR per liter. And based on study by Sihotang and Assomadi [16] the carbon emission factors for passenger vehicle and motorcycle with gasoline are 259,786 g per liter respectively.

The calculation of fuel costs and the carbon emission factors for each vehicle used by the caregivers as follows:

Ambulance fuel cost

$$C^1 = 0.1160\frac{\text{L}}{\text{km}} \times \frac{10{,}000 IDR}{\text{L}} = \frac{1160\, IDR}{\text{km}} \tag{30}$$

Motorcycle fuel cost

$$C^2 = 0.0266\frac{\text{L}}{\text{km}} \times \frac{10{,}000\, IDR}{\text{L}} = \frac{266\, IDR}{\text{km}} \tag{31}$$

Ambulance carbon emission factors

$$E^k = 0.1160\frac{\text{L}}{\text{km}} \times 259{,}786\frac{\text{gram}}{\text{L}} = 30{,}135\frac{\text{gram}}{\text{km}} \tag{32}$$

Motorcycle carbon emission factors

$$E^k = 0.0266\frac{\text{L}}{\text{km}} \times 259{,}786\frac{\text{gram}}{\text{L}} = 6910\frac{\text{gram}}{\text{km}} \tag{33}$$

The carbon tax cost per unit of carbon in Indonesia is 30 IDR per kilogram. The other parameters used in the numerical example are given in Table 1.

We used LINGO 18.0 to solve the model and determine the optimal schedule and the order of patient visits. The optimal result of the objective function is 123,393 IDR with a total fuel cost of 52,980 IDR, total carbon emission tax of 412.89 IDR, and total the patient preference violation penalty of 70,000 IDR because patient 2 and patient 4 are not served by caregiver according to their gender preference.

To find the global optimum solution, it takes 7 s. The result of determining the order of visits by female caregivers is 1-2-10-9-5-11-7-1. That is, starts from location 1 in RSUD Dr. Moewardi to patient 2 at Jl. Veteran No.15, then to patient 10 at Jl. Imam Bonjol No.56, then to patient 9 at Jl. Veteran, then to patient 5 at Kepatihan Kulon, then to patient 11 at Jl. Siwalan No.12, then to patient 7 at Jebres, Surakarta, and then back to RSUD Dr. Moewardi. While for male caregivers is 1-4-3-8-6-1. That is, starts from location 1 in RSUD Dr. Moewardi to patient 4 at Fajar Indah, then to patient 3 at Bayan, Kadipiro, then to patient 8 at Gedangan, Grogol, then to patient 6 at Jl. Manggis VI, Jajar, and then back to RSUD Dr. Moewardi. Optimum

Table 1 Numerical example parameter

Indeks	Locations	P_Gen	TP	LB	UB
J1	RSUD Dr Moewardi, Jl. Kolonel Sutarto Jebres	−1	0		
J2	Jl. Veteran No.15, Tipes, Serengan, Surakarta	1	15	07.30	08.40
J3	Bayan, Kadipiro, Kec. Banjarsari, Surakarta	1	20	08.30	09.30
J4	Fajar Indah, Baturan, Colomadu, Karanganyar	0	15	08.00	09.00
J5	Kepatihan Kulon, Jebres, Surakarta	0	20	07.30	12.00
J6	Jl. Manggis VI, Jajar, Laweyan, Surakarta	1	20	07.30	15.30
J7	Jebres, Jebres, Surakarta	0	15	07.30	15.00
J8	Gedangan, Grogol, Sukoharjo	1	25	07.30	15.00
J9	Jl. Veteran, Tipes, Serengan, Surakarta	0	30	08.00	15.30
J10	Jl. Imam Bonjol No.56, Ps Kliwon, Surakarta	0	20	07.50	15.30
J11	Jl. Siwalan No.12, Kerten, Laweyan, Surakarta	0	15	07.30	15.30

schedule for female caregivers is from 07.30 to 11.13, while for male caregivers is from 07.30 to 11.35.

4 Conclusions

This research proposes a mathematical model for the HHC daily planning. The objective function of the model is to minimize the total operational costs, consisting of vehicle fuel costs, patient preference violation penalties, and carbon tax costs. The proposed model can increase the efficiency of HHC planning and be used as a decision-making tool by the HHC coordinator. The model is tested with real data at RSUD Dr Moewardi. The optimal result of the objective function is 123,393 IDR with a total fuel cost of 52,980 IDR, total carbon emission tax of 412.89 IDR, and total the patient preference violation penalty of 70,000 IDR. And the result of determining the order of visits by female caregivers is 1-2-10-9-5-11-7-1, while for male caregivers is 1-4-3-8-6-1. Optimum schedule for female caregivers is from 07.30 to 11.13, while for male caregivers is from 07.30 to 11.35.

LINGO 18.0 software is effectively used to solve scheduling and routing for HHC with a number of patients less than 20 people. While scheduling with 20 patients, LINGO 18.0 software takes 4 h 24 min to determine the optimal solution. As a direction for future research, applying metaheuristic algorithms to solve larger-sized cases more efficiently. In the real case, the data have high variability due to uncertainty, which may cause the original schedule to be invalid. Then the problem must change into a stochastic problem to overcome the uncertain constraint.

References

1. Pratiwi UD, Andayanie E, Rizqiani Rusydi A (2021) Evaluasi Pelaksanaan Pelayanan Home Care di Puskesmas Antang Perumnas Kota Makassar Tahun 2020. Wind Publ Heal J 2(1):835–843. https://doi.org/10.33096/woph.v2i1.117
2. Ferdowsi A, Taghanaki RB, Abraham A (2021) Novel routing-scheduling problem for home health care network, vol 1372 AISC, no. April. Springer
3. Alves F, Costa L, Rocha AMAC, Pereira AI, Leitão P (2019) A multi-objective approach to the optimization of home care visits scheduling. In: ICORES 2019—Proceedings of 8th international conference on operations research and enterprise system, no. 2016, pp 435–442. https://doi.org/10.5220/0007565704350442.
4. Nickel S, Schröder M, Steeg J (2012) Mid-term and short-term planning support for home health care services. Eur J Oper Res 219(3):574–587. https://doi.org/10.1016/j.ejor.2011.10.042
5. Liu R, Yuan B, Jiang Z (2017) Mathematical model and exact algorithm for the home care worker scheduling and routing problem with lunch break requirements. Int J Prod Res 55(2):558–575. https://doi.org/10.1080/00207543.2016.1213917
6. Riazi S, Wigstrom O, Bengtsson K, Lennartson B (2018) A column generation-based gossip algorithm for home healthcare routing and scheduling problems. IEEE Trans Autom Sci Eng 1–11. https://doi.org/10.1109/TASE.2018.2874392
7. Naderi B, Begen MA, Zaric GS, Roshanaei V (2021) A novel and efficient exact technique for integrated staffing, assignment, routing, and scheduling of home care services under uncertainty. SSRN Electron J. https://doi.org/10.2139/ssrn.3836827
8. Borsani V, Matta A, Beschi G, Sommaruga F (2006) A home care scheduling model for human resources. In: Proceedins—ICSSSM'06 2006 international conference on service systems and service management, vol 1, pp 449–454. https://doi.org/10.1109/ICSSSM.2006.320504.
9. Bachouch B, Guinet A, Hajri-Gabouj S (2009) A model for scheduling drug deliveries in a French homecare. Available https://hal.archives-ouvertes.fr/hal-00385516
10. Di Mascolo M, Espinouse ML, Haddadene SRA (2018) Taking patients' wishes into account for daily planning in the Home Health Care contex. In: IFAC-PapersOnLine conference paper archive, vol 51, no 11, pp 1010–1015. https://doi.org/10.1016/j.ifacol.2018.08.475
11. Fathollahi-Fard AM, Hajiaghaei-Keshteli M, Tavakkoli-Moghaddam R (2018) A bi-objective green home health care routing problem. J Clean Prod 200:423–443. https://doi.org/10.1016/j.jclepro.2018.07.258
12. Liu W, Dridi M, El Hassani AH, Fei H (2019) A mathematical model for medium-term home health care planning problem. In: Proceedins of 2019 International conference on industrial engineering and systems management, IESM 2019, pp 1–6. https://doi.org/10.1109/IESM45758.2019.8948184.
13. Di Mascolo M, Martinez C, and M. L. Espinouse, "Routing and scheduling in Home Health Care: A literature survey and bibliometric analysis," *Comput. Ind. Eng.*, vol. 158, 2021, doi: https://doi.org/10.1016/j.cie.2021.107255.
14. Nurdjanah N (2015) Emisi CO_2 Akibat Kendaraan Bermotor di Kota Denpasar. J. Penelit. Transp. Darat 17(1):1–14
15. Lamba K, Singh SP, Mishra N (2019) Integrated decisions for supplier selection and lot-sizing considering different carbon emission regulations in big data environment. Comput Ind Eng 128:1052–1062. https://doi.org/10.1016/j.cie.2018.04.028
16. Sihotang SR, Assomadi AF (2015) Pemetaan Distribusi Konsentrasi Karbon Dioksida (CO_2) Dari Kontribusi Kendaraan Bermotor Di Kampus ITS Surabaya. Available http://digilib.its.ac.id/ITS-Undergraduate-3100011041874/14134.

Development of a Sustainable Supplier Selection Model and Order Allocation Considering Transportation Alternatives and Quantity Discounts

Rosifa Sahara Prasiwi, Cucuk Nur Rosyidi, and Pringgo Widyo Laksono

Abstract To increase their competitive advantage, many organizations compete through efficient supply chain management. One of the company's efforts to improve supply chain management is by choosing the right supplier to provide the right product at the right price at the right time. Not enough with that, other needs arise due to environmental and social issues that require companies to consider sustainable aspects or Triple Bottom Line (economic, social, and environmental). This paper aims to develop an optimization model for sustainable supplier selection and order allocation that can help decision-makers. This study focuses on sustainable supplier selection and order allocation with multi-product, multi-supplier, multi-period on a 4-level supply chain network (supplier-warehouse-3PL-wholesaler) by considering transportation alternatives and quantity discounts. A Multi Objective Mixed Integer Linear Programming (MOMILP) is developed to represent the model and an optimal results is obtained by using the Weighted Sum Method with the help of LINGO 18.0 software. The results obtained for the objective function of minimizing the total cost of 17,892,950,000 IDR; minimization of carbon emissions as much as 7,124,513 kgCO_2; and the maximum value of social responsibility is 4.275441 with a multi-objective function value of 0.0915218.

Keywords Supply chain management · Sustainable supplier selection · Order allocation

1 Introduction

Companies must constantly enhance their performance, and competitive advantage plans to compete in the global market in light of business globalization [1]. According to [2], effective supply chain management will boost a company's competitive edge.

R. S. Prasiwi (✉) · C. N. Rosyidi · P. W. Laksono
Industrial Engineering Department, Sebelas Maret University, Surakarta, Indonesia
e-mail: rosifasp@student.uns.ac.id

C. N. Rosyidi et al. (eds.), *Proceedings of the 6th Asia Pacific Conference on Manufacturing Systems and 4th International Manufacturing Engineering Conference*, Lecture Notes in Mechanical Engineering,
https://doi.org/10.1007/978-981-99-1245-2_21

Table 1 Summary of research gap

	Tirkolaee et al. (2020)	This article
Sustainable factors		
Economic	✔	✔
Environment	–	✔
Social	–	✔
Alternative transportation	–	✔
Quantity discount	–	✔

One of them involves choosing the appropriate supplier. Choosing the right supplier will help a company to deliver the right product, at the right price, in the right quantity, and at the right time [3]. Concerns about environmental degradation, the growing global population, and the decreasing availability of resources, according to [4], cause businesses to revise their perspectives by considering sustainability aspects (economic, environmental, and social). It is impossible to separate logistics activities like distribution or transportation from supply chain management. Choosing the best transportation alternatives may save costs while simultaneously reducing carbon emissions. In the study of [5], the Supplier Selection and Order Allocation (SSOA) model was developed by considering the choice of transportation alternatives with several constraints. Apart from that, other considerations involving quantity discount schemes are also worth considering. Research by [5, 6] provides discounts with a quantity discount scheme or when buyers make purchases in larger quantities than usual.

This study refers to the research of [7] developed an SSOA model on a three-level supply chain network (suppliers, warehouses, and wholesalers) with the stated objectives of minimizing total costs. Table 1 shows the gap between this paper and Tirkolaee's study. Referring to [5] where the SSOA model they developed considers the selection of transportation alternatives and quantity discounts to minimize costs. Additionally, Sustainable SSOA was developed based on a study by [6] with the objectives of minimizing costs, minimizing the total carbon emissions, and maximizing the value of social responsibility. This research aims to create a 4-level supply chain network (supplier-warehouse-3PL-wholesaler) based sustainable SSOA optimization model that takes alternative transportation modes and quantity discounts into account.

2 Assumption and Notation

The following are the assumptions used in this paper:

1. At the beginning and the end of the period, each product's total inventory is equal to zero.

2. Each product's request volume is determined.
3. A lost sales cost will be charged if the demand for a set period cannot be fulfilled.
4. A 3PL can handle the shipping of all items (there is no limit to the number of trucks available).

The notation used in this paper is as follows:

Decision Variables.

X_{ijmt} Quantity of product j ordered from supplier i in period t (units)
XW_{kjpt} Quantity of product j shipped to wholesaler k using type p trucks in period t (units)
OX_t Number of temporary warehouses built in period t (warehouses)
Inv_{jt} Total inventory of product j in period t (units)
LOS_{jt} Number of lost sales of product j in period t at supplier level (units)
LOW_{jt} Number of lost sales of product j in period t at warehouse level (units)

Parameter

UR Maximum number of temporary warehouses (warehouses)
C_{ijt} Capacity of supplier i to supply product j in period t (units)
SC_{jt} Maximum capacity of central warehouse for product j in period t (units)
SO_j Maximum capacity of temporary warehouse for product j (units)
PO Cost of established a temporary warehouse (Rp)
P_{imj} Purchase price per unit of product j at price level m from supplier i (Rp/unit)
st_{ij} Setup time for product j by supplier i (hours)
LT_{jt} Delivery lead time for product j (hours)
O_{ij} Ordering cost of supplier i for product j (Rp/order)
TC_{kp} Transportation costs to wholesaler per unit of product j using truck type p (Rp/truck)
DK_{kjt} The demand of wholesaler k for product j in period t (units)
H_j Holding cost per unit for product j (Rp/unit)
LS_j Cost of lost sales per unit of product j at supplier level (Rp/unit)
LW_j Cost of lost sale per unit of product j at wholesaler level (Rp/unit)
Up Delivery capacity of truck type p (unit)
Qa_i Average defect rate from supplier i (%)
Qaa_j Maximum acceptable defect rate for product j (%)
$Lmax$ Maximum rate of lost sales (%)
Et_{kp} CO_2 emissions produced by truck type p from transportation activities to wholesaler k (kg CO_2/unit)
Eh CO_2 emissions produced by facilities (warehouses) from storage activities (kg CO_2 /unit)
Fjg Number of job opportunities created for fixed jobs (e.g. managers, etc.)
Flg Number of lost days of work due to work damages by fixed jobs (e.g. managers, etc.)
$Varjg$ Number of job opportunities created for variable jobs
$Varlg$ Number of lost days of work due to work work damages by variable jobs
Bp_{ijmt} Breakpoint of product j at price level m from supplier i in period t (units)

BM Big Number

3 Model Development

The Multi-Objective Mixed Integer Linear Programming (MOMILP) model formulation for the Sustainable SSOA issue is as follows:

$$\begin{aligned} Minimize f_1 = & \sum_{j=1}^{J}\sum_{i=1}^{I}\sum_{m=1}^{M}\sum_{t=1}^{T}(P_{imj} \cdot X_{ijmt}) + \sum_{j=1}^{J}\sum_{t=1}^{T}(H_j \cdot Inv_{jt}) \\ & + \sum_{i=1}^{I}\sum_{j=1}^{J}\sum_{t=1}^{T}(O_{ij} \cdot Y_{ijt}) + \sum_{j=1}^{J}\sum_{t=1}^{T}(LS_j \cdot LOS_{jt}) + \sum_{j=1}^{J}\sum_{t=1}^{T}(LW_j \cdot LOW_{jt}) \\ & + \sum_{k=i}^{K}\sum_{j=1}^{J}\sum_{p=1}^{P}\sum_{t=1}^{T} TC_{kp} \cdot Z_{kjpt} + \sum_{t=1}^{T}(PO \cdot OX_t) \end{aligned} \quad (1)$$

$$Minimize f_2 = \sum_{k=1}^{K}\sum_{j=1}^{J}\sum_{p=1}^{P}\sum_{t=1}^{T} Et_{kp} \cdot Z_{kjpt} + \sum_{j=1}^{J}\sum_{t=1}^{T} Eh \cdot Inv_{jt} \quad (2)$$

$$\begin{aligned} Maximize f_3 = & \zeta_1 \left\{ fjg \cdot OX_t + varjg \cdot \left[\left(\frac{X_{ijt}}{SC} \right) \right] \right\} \\ & - \zeta_2 \left\{ f\lg \cdot OX_t + varlg \cdot \left[\left(\frac{X_{ijt}}{SC} \right) \right] \right\} \end{aligned} \quad (3)$$

Subject to:

$$\sum_{i=1}^{I}\sum_{m=1}^{M} X_{ijmt} + Inv_{njt-1} + LOS_{jt} - \sum_{k=1}^{K}\sum_{p=1}^{P} XK_{kjpt} = Inv_{jt} \quad (4)$$

$$D_{jt} = \sum_{k=1}^{KR} DK_{kjt} \cdot B_{kjt} \quad (5)$$

$$LOS_{jt} \le BM(1 - B_{kjt}) \quad (6)$$

$$\sum_{j=1}^{J}\sum_{t=1}^{T} LOS_{jt} \le L_{\max} \cdot \sum_{j=1}^{J}\sum_{t=1}^{T} D_{jt} \quad (7)$$

$$LOW_{jt} \le \sum_{k=1}^{KR} DK_{kjt} \quad (8)$$

$$\sum_{p=1}^{P} XK_{kjpt} + LOW_{jt} = DK_{kjt} \tag{9}$$

$$\sum_{i=1}^{I}\sum_{m=1}^{M}(X_{ijmt} \cdot Qa_i) \le Qaa_j \cdot D_{jt} \tag{10}$$

$$\sum_{i=1}^{M}(st_{ij} \cdot Y_{ijt}) \le LT_{jt} \tag{11}$$

$$\sum_{i=1}^{I}\sum_{m=1}^{M} X_{ijmt} + Inv_{jt-1} \le SC_{jt} + SO_j \cdot OX_t \tag{12}$$

$$OX_t \le UR \tag{13}$$

$$\sum_{i=1}^{I} X_{ijt} + Inv_{j,t-1} \ge \sum_{k=1}^{K} XK_{kjpt} \tag{14}$$

$$XK_{kjpt} \le U_p \cdot Z_{kjpt} \tag{15}$$

$$\sum_{p=1}^{TR} Z_{kjpt} \le 1 \tag{16}$$

$$Bbp_{ij(m-1)t} \cdot S_{ijmt} \le X_{ijmt} \le Bbp_{ijmt} \cdot S_{ijmt} \tag{17}$$

$$X_{ijmt} \le C_{ijt} \cdot S_{ijmt} \tag{18}$$

$$\sum_{m=1}^{M} S_{ijmt} = Y_{ijt} \tag{19}$$

$$OX_t \in Z^+, X_{ijmt}, XK_{kjpt}, Inv_{jt}, LOS_{jt}, LOW_{jt}, D_{jt} \ge 0 \; dan \; integer \tag{20}$$

$$Y_{ijt}, B_{kjt}, S_{ijmt}, Z_{kjpt} \in \{0, 1\} \tag{21}$$

The objective function is shown by Eq. (1) and is the minimization of total costs, which consist of purchase costs, holding costs, supplier lost sales costs, wholesaler lost sales costs, transportation costs, and temporary warehousing costs. Minimizing carbon emissions and maximizing social responsibility value are the following objective functions (2) and (3). The constraint of inventory balance is given by Eq. (4). Calculating the total demand is constrained by Eqs. (5). Equation (6)–(7) the constraint of lost sales at the supplier level. Equation (8)–(9) the lost sales constraint at the wholesaler level. Equation (10) quality constraint. The delivery time is represented by Eq. (11). The capacity of the central and temporary warehouses is defined by Eq. (12). Equation (13) limits the maximum number of temporary warehouses. Equation (14) constrains the number of shipments to wholesalers. Equation (15) is the constraint of the delivery truck capacity. Equation (16) determines only one type of truck that can be selected for one product delivery to the wholesaler. Equations (17)–(19) is a constraint to choose the price level. Equations (20)–(21) shows the type of decision variables. It is necessary to transform the multi-objective model into a single objective function to find the optimal solution.

4 Numerical Example

The numerical example in this paper uses a lighting warehouse that provides LED lamps, fluorescent lamps, and incandescent lamps to demonstrate sustainable SSOA. Seven supplier candidates, each with their weight and characteristics, will receive orders from the warehouse. In addition, the warehouse will distribute products to wholesalers using trucks provided by Third Party Logistics (3PL), each of which has certain qualities that must be considered. There are 16 wholesalers, and four different truck types (van, small box, CDE, CDD) are provided. Data from the study of [7] were utilized in this paper, along with additional assumptions made by the researcher.

It is necessary to change the objective function with the objective of transforming from a multiple objective function to a single objective function in order to get the optimum value from the multi-objective model's form. The idea of multi-objective optimization (MOO) is employed in this article, and the weighted sum technique is the approach used to get the results. The objective function must be normalized to transform as the next step; this paper uses Eq. (22) to normalize the objective function.

$$\min f = 0.333\frac{F_1(x) - F_1^o}{F_1^{\max} - F_1^o} + 0.333\frac{F_2(x) - F_2^o}{F_2^{\max} - F_2^o} + 0.333\frac{F_3^o - F_3(x)}{F_3^o - F_3^{\min}} \tag{22}$$

Using the equation above, the first step to finding the optimal value of the model is to find the maximum and minimum values of each objective function shown in Table 2.

The multi-objective function's value can be derived by identifying the maximum and minimum values of each objective function. The optimum global value was calculated using the LINGO 18.0 software application during the 548th iteration, which took eight seconds to complete, as shown in Table 3.

Under optimal model conditions, the minimum cost is 17,892,950,000 IDR with carbon emissions produced as much as 7,124.513 $kgCO_2$, and the value of warehouse social responsibility is 4,275,441. Table 4 illustrate the value of the order allocation to selected suppliers based on the chosen price level.

Table 2 The maximum and minimum value of objective function

Value objective function		
f_1	Maximum	89,909,760,000 IDR
	Minimum	16,743,280,000 IDR
f_2	Maximum	11,122 $kgCO_2$
	Minimum	572,635 $kgCO_2$
f_3	Maximum	42,754
	Minimum	37,171

Table 3 Optimum value of each objective function

Optimum value of each objective function	
f	0.0915218
f_1	17,892,950,000 IDR
f_2	7,124,513 $kgCO_2$
f_3	4,275,441

Table 4 Order allocation for selected suppliers

X_{ijmt} (units)						
Supplier	Product	Price level	Period			
			1	2	3	4
1	1	3	–	130.000	–	60.000
	2	3	150.000	188.700	147.300	–
	3	3	–	–	148.800	–
4	3	3	213.000	–	–	–
5	3	3	–	133.300	170.000	–
7	1	3	186.400	112.700	173.600	144.900
	2	3	135.000	–	170.000	–
	3	3	–	100.000	160.000	–

As said in the table, suppliers 1 and 7 were chosen to supply product types 1 and 2, and suppliers 1, 4, 5, and 7 supplied product type 3. Table 5 shows the value of the decision variable from XK_{kjpt} for allocating product shipments to wholesalers using the selected truck type. Table 6 shows the number of temporary warehouses required to deal with fluctuations in demand from the wholesaler.

Based on the above results, only one temporary warehouse is required for the 1st, 2nd, and 4th periods. While the 3rd period requires two temporary warehouses, it causes demand in the 3rd period is higher than in periods 1, 2, and 4. Furthermore, the result of this model's application indicates that are no lost sales at the supplier level, there were 11,000 units of the third product type at the warehouse level in the first period which is product demands from wholesaler 14 that cannot be fulfilled.

Table 5 Allocation of Shipments to wholesaler

XK_{kjpt} (units)						
Wholesaler	Product	Truck type	Period			
			1	2	3	4
1	1	2	0	9.000	0	0
	2	2	0	0	0	9.000
		3	0	0	17.000	0
	3	4	0	0	33.000	0
16	1	2	0	9.000	0	0
	2	2	0	0	0	9.000
		3	0	0	17.000	0
	3	4	0	0	33.000	0

Table 6 Number of temporary warehouse

Period	OX_t
1	1
2	1
3	2
4	1

Table 7 shows the total inventory of every product for each period In order to prevent an excess of products in the central warehouse, there must be no inventory in the final or fourth period.

Based on the results, the model could be run and provides reasonable results, allowing it to be applied as an alternate solution to the problems of supplier selection and order allocation while taking the sustainability issue into account.

Table 7 Total inventory

I Inv_{jt} (units)				
Product	Period			
	1	2	3	4
1	0	0	0	0
2	68,000	130,900	230,000	0
3	0	0	197,800	0

5 Conclusions

The MOMILP model was developed in this paper to solve the problem of sustainable SSOA by considering transportation alternatives and quantity discounts. The objective of this model is to minimize the total cost, minimize carbon emissions, and maximize social responsibility's value. By using LINGO 18.0 software, the optimum value is achieved in the 548th iteration with the optimum value of each objective function being 17,892,950,000 IDR; 7,124.513 $kgCO_2$; and 4.27544. The selected suppliers are suppliers 1, 4, 5, and 7. The number of temporary warehouses required in each period is 1, 1, 2, and 1. There are recommendations for future studies that use fuzzy approaches to consider the uncertainty closer to the condition in the real system. It may also take late penalties or backorder conditions into account.

References

1. Setiawan A (2020) Pengaruh Intellectual Capital Terhadap Kinerja Keuangan Dengan Process Capital Sebagai Variabel Mediasi Pada Perusahaan Manufaktur Yang Terdaftar Di Bursa Efek Indonesia. Universitas Airlangga
2. Suharto R, Devie (2013) Analisa Pengaruh Supply Chain Management terhadap Keunggulan Bersaing dan Kinerja Perusahaan. Bus Acc Rev 1(2):161–171. http://eprints2.binus.ac.id/id/eprint/24110
3. Cengiz AE, Aytekin O, Ozdemir I, Kusan H, Cabuk A (2017) A multi-criteria decision model for construction material supplier selection. Proc Eng 196(June):294–301. https://doi.org/10.1016/j.proeng.2017.07.202
4. Cheraghalipour A, Farsad S (2018) A Bi-Objective sustainable supplier selection and order allocation considering quantity discounts under disruption risks: a case study in plastic industry. Comput Ind Eng 118(February):237–250. https://doi.org/10.1016/j.cie.2018.02.041
5. Choudhary D, Shankar R (2014) A goal programming model for joint decision making of inventory lot-size, supplier selection and carrier selection. Comput Ind Eng 71(1):1–9. https://doi.org/10.1016/j.cie.2014.02.003
6. Moheb-Alizadeh H, Handfield R (2019) Sustainable supplier selection and order allocation: a novel multi-objective programming model with a hybrid solution approach. Comput Ind Eng 129(January):192–209. https://doi.org/10.1016/j.cie.2019.01.011
7. Tirkolaee EB, Mardani A, Dashtian Z, Soltani M, Weber GW (2020) A novel hybrid method using fuzzy decision making and multi-objective programming for sustainable-reliable supplier selection in two-echelon supply chain design. J Clean Prod 250:119517. https://doi.org/10.1016/j.jclepro.2019.119517

Optimization Model for Lot Sizing and Scheduling Problem Considering Multiple Suppliers, Quantity Discounts, Carriers, and Production Time Constraints

Hani Aninda Intan Permatasari, Cucuk Nur Rosyidi, and Pringgo Widyo Laksono

Abstract This research developed a MILP model to integrate lot sizing and scheduling problems by considering multiple suppliers, quantity discounts, carriers, and production time constraints. The objective function of this model is to maximize total sales revenue minus total costs, which consist of the cost of purchasing raw materials by considering discounts, ordering costs, transportation costs, production costs for all lines, holding costs for raw materials and products, product defect penalties, and setup costs. A numerical example was conducted to evaluate the model and to illustrate how the problem was solved. Lingo 18.0 software is used to complete the search for the best solution. The results show that the optimal decision was determined by obtaining optimal demand, determining the optimal amount of each raw material that must be purchased from the optimal supplier, determining the optimal amount of each product that must be produced, and obtaining the optimal amount of inventory from product and raw material. Furthermore, the production beginning and completion times are obtained, providing more detailed scheduling information.

Keywords Supplier · Quantity discount · Carrier · Lot sizing · Scheduling

1 Introduction

Lot sizing can be defined as a production quantity decision of a particular product produced by a specific machine and in a single production process [1]. Lot sizing is often associated with efficient production planning for a product which can be

H. A. I. Permatasari (✉) · C. N. Rosyidi · P. W. Laksono
Department of Industrial Engineering, Faculty of Engineering, Universitas Sebelas Maret, Jl. Ir Sutami 36 A, Surakarta 57126, Indonesia
e-mail: permatasarihani301@student.uns.ac.id

C. N. Rosyidi et al. (eds.), *Proceedings of the 6th Asia Pacific Conference on Manufacturing Systems and 4th International Manufacturing Engineering Conference*, Lecture Notes in Mechanical Engineering,
https://doi.org/10.1007/978-981-99-1245-2_22

achieved by solving lot allocation problems based on demand to be met and stock availability [2]. The lot sizing purpose is to minimize production costs by determining the optimal production quantity [3]. According to Pinedo [4], scheduling is expressed as planning to carry out several activities within a particular time, where these activities are related to the distribution of resources such as machines, workers, and materials. The purpose of scheduling is to minimize production time by determining the optimal work sequence [5].

Lot sizing and scheduling problem are one of the most critical problems in the field of production planning. Lot sizing and scheduling are categorized into two levels of decision-making (planning). According to Almeder et al. [6], lot sizing is included in medium-term planning (tactical planning), while scheduling is included in short-term planning (operational planning). Decision-making for these two levels is usually carried out hierarchically so that the results of the lot sizing problem are applied to the scheduling model [7]. Due to the relationship between these levels, planning decisions must be taken simultaneously to obtain a globally optimal solution. In addition, when planning production, it is necessary to calculate the amount of raw materials used. Therefore, raw material purchasing decisions must also be integrated with production planning to achieve an optimal global solution [8].

In purchasing raw materials, multiple suppliers allow manufacturers to buy the same material from different sources. Suppliers can provide quantity discounts. Quantity discounts are a common and effective practice for suppliers to promote their products, and manufacturers can purchase raw materials at lower unit prices when the order quantity exceeds a certain amount [9]. Source and lot size decisions are also influenced by economies of scale in transportation costs, which depend on the geographical distance between the manufacturer and supplier and the carrier capacity [10]. In most previous studies, transportation costs are usually only considered implicitly as setup costs with the assumption that they do not depend on the shipment size [11].

In scheduling production, it is important to know the beginning and completion times to make the schedule on time. Generally, the scheduling is carried out only at the period stage; for example, the amount of products produced by product A in period two is 5000 pieces, while product B in period two is 6000 pieces. So it cannot be known in detail which product was produced first. In addition, the lack of time-scheduling information can also provide opportunities for production delays. This is due to ignorance or lack of awareness of the beginning time of the next production. With the constraints of production time, it can provide more detailed scheduling information where not only information on how many quantities are produced but also can know when the beginning and completion time of production and the sequence.

Mohammadi et al. [12] developed a model for determining lot size and multi-level scheduling with a complex setup and considering the selection of suppliers with quantity discounts and various modes of transportation. Georgiadis et al. [13] researched optimal production scheduling in the dairy industry. In this model, there is a production time constraint that can result in the decision variables of processing time and the completion of the product family. Sutrisno et al. [14] developed a multi-objective optimization model to solve production planning and raw material supplier selection problems simultaneously. The model has considered the supplier quality factor in the procurement of raw materials, described by the presentation of the level of raw material defects.

This study refers to the research model that has been developed by Mohammadi et al. [12], Georgiadis et al. [13], and Sutrisno et al. [14]. Based on these three studies, this research will develop a model for lot-sizing and scheduling problem by considering multiple suppliers, quantity discounts, carriers, and production time constraints. The developed model provides multi-level and multi-period lot sizing for multiple products and production lines. The form of the model developed in this research is Mixed Integer Linear Programming (MILP) with the search for the optimum solution using LINGO 18.0 software.

2 Assumption and Notation

The assumptions used in this paper are as follows:

1. Limited supplier capacity.
2. Limited product and raw material warehouse capacity.
3. There is no raw material or product inventory at the beginning of the period.
4. Production line setup must be completed in the same period as it was started.

The following are the notations used in this paper:

Indices

t	periods
f	raw materials
s	suppliers
k	products
d	discounts level
I, j	families
l	lines
m	carriers

Parameter

Ld_{kt}	lower bound of demand (Unit)
Ud_{kt}	upper bound of demand (Unit)
O_k^K	occupied space of product (kg/Unit)
O_f^F	occupied space of raw material (Palette/Unit)
df_{fs}	defect rate (%)
dfc_{fs}	penalty cost for defect rate of raw material (Rupiah/unit)
Cap_{ft}^F,	warehouse capacity of raw material (kg)
Cap_{kt}^K	warehouse capacity of product (Palette)
Cap_{lt}	available capacity (Minute)

DIS_{sd} discount rate (%)
u_{sd} upper bound of discount (Unit)
h_f^F holding cost of raw material (Rupiah/Unit)
h_k^K holding cost of product (Rupiah/Unit)
oc_s order cost (Rupiah)
bc_{fs} purchase cost (Rupiah/Unit)
trc_{fsm} transport cost (Rupiah/Unit)
r_{kt} sale revenue (Rupiah/Unit)
e_{fkl}^F consumption of raw material (Unit/Unit)
cp_{kl} production cost (Rupiah/Unit)
p_{kl} production time (Minute)
scp_{kl} setup cost of product (Rupiah)
stp_{kl} setup time of product (Minute)
scb_{ijl} setup cost of family (Rupiah)
stb_{ijl} setup time of family (Minute)
LBL_{lk} minimum production (Unit)
M^{big} a big number

Decision Variable

D_{kt} accepted demand for product (Unit)
Q_{fsdtm}^F quantity of raw material (Unit)
Q_{klt} quantity of product produced (Unit)
I_{ft}^F, I_{kt}^K inventory of raw material and product (Unit)
ST_{ilt}, CL_{ilt} beginning and completion time of family (Minute)
V_{jlt} auxiliary variables for scheduling sub-cycle elimination
X_{ijlt} 1, if on production line l in period t a switchover from family i to family j takes place (0, otherwise)
A_{jlt} 1, if, at the start of period t, production line l is set up for family j (0, otherwise)
P'_{klt} 1, if product k is produced on line l in time period t (0, otherwise)
Z_{sdt} 1, if a supplier acquisition occurs in period t at discount level d (0, otherwise)
Y_{st} 1, if any raw material is bought from supplier s during period t (0, otherwise)

3 Model Development

Companies engaged in manufacturing meet their production needs by procuring raw materials. These raw materials can be purchased from several suppliers. Multiple suppliers allow companies to buy the same material from different sources at a lower price. Each supplier provides different levels of discounts based on the amount of raw materials purchased. Not all raw materials purchased from suppliers are perfect, so there is a percentage of the level of raw material defects when purchasing raw materials from suppliers. To send raw materials from suppliers to companies, there are several choices of carriers with different transportation costs depending on the type of carrier used and the location between the supplier and the manufacturer. Orders of raw materials that have arrived at the manufacturer are stored in the raw material warehouse. Furthermore, these raw materials are used to produce various products.

There are many products produced by manufacturing companies. These products are grouped into several families based on similarities in design characteristics or manufacturing characteristics. Design characteristics include size, shape, and function. In contrast, the manufacturing or process characteristics are based on the type and sequence of operations required. When doing production, there is a sequence-independent setup for each product. Sequence-independent setup is when the order of the products is different, and then the setup time will remain the same. The changeover from one family to another also requires setup. The changeover setup between families in this model is a sequence-dependent setup. Sequence-dependent setup is a setup when the order of the family is different, and then the setup time is also different. The finished product is stored in the product warehouse. Figure 1 is an overview of the description of this research system in general.

This research is developing a model for lot sizing and scheduling problem by considering multiple suppliers, quantity discounts, carriers, and production time constraints. In each period, the planner must decide the amount of purchases of each raw material, suppliers and carriers, inventories of raw materials and products, production quantities of each product, and the order of lots that need to be produced. In addition, the beginning and completion times of production and their sequence are also determined, so detailed scheduling information is obtained.

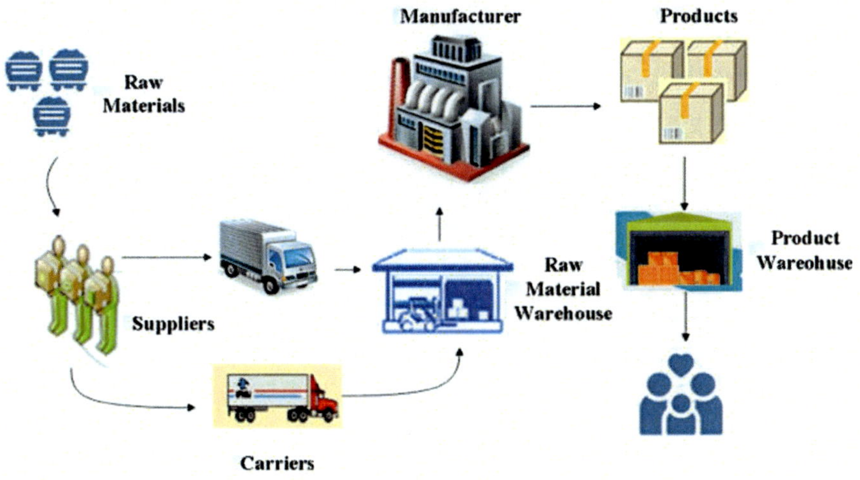

Fig. 1 Model description

The objective function of this model (is) is to maximize the total sales revenue minus the total cost, which consists of (i) the cost of purchasing raw materials where the discount is considered; (ii) the cost of the order, which is a fixed shipping cost regardless of the size of the order; (iii) transportation costs; (iv) production costs of all lines; (v) the cost of storing raw materials and products; (vi) penalty fees for raw material defects; (vii) setup costs that represent the readiness of the production line to produce various products (product setup costs and setup costs for changeover between product families).

$$\begin{aligned}
\text{Max} \sum_{k}\sum_{t} r_{kt}D_{kt} \\
&- \left(\sum_{s}\sum_{d}\sum_{t}\left((1-\text{DISsd})\sum_{f}\sum_{m}(bc_{fs}\,Q^{F}_{fsdtm})\right)\right. \\
&+ \sum_{s}\sum_{t}\sum_{m}(oc_{s}\,Y_{st}) + \sum_{f}\sum_{s}\sum_{d}\sum_{t}\sum_{m} trc_{fsm}\,Q^{F}_{fsdtm} \\
&+ \sum_{k}\sum_{l}\sum_{t} cp_{kl}\,Q_{klt} + \sum_{f}\sum_{t} h^{F}_{f} I^{F}_{ft} + \sum_{k}\sum_{t} h^{K}_{k} I^{K}_{kt} \\
&+ \sum_{f}\sum_{s}\sum_{d}\sum_{t}\sum_{m} dfc_{fs}\,df_{fs}\,Q^{F}_{fs\,dtm} + \sum_{k}\sum_{l}\sum_{t} scp_{kl}\,P'_{klt} \\
&\left. + \sum_{i}\sum_{j}\sum_{l}\sum_{t} scb_{ijl}\,X_{ijlt}\right)
\end{aligned} \tag{1}$$

Subject to:

$$u_{s,d-1}Z_{sdt} < \sum_{f}\sum_{m} Q^{F}_{fsdtm} \le u_{s,d}Z_{sdt} \quad \forall s, d, t \tag{2}$$

$$\sum_{d} Z_{sdt} \le Y_{st} \ \forall s, t \tag{3}$$

$$\sum_{F}\sum_{d}\sum_{m} Q^{F}_{fsdtm} \le M^{big}Y_{st} \ \forall s, t, \tag{4}$$

$$I^{F}_{f.t-1} + \sum_{s}\sum_{d}\sum_{m}(1-df_{fs})Q^{F}_{f,s,d,t,m} = \sum_{k\in K^{F}(f)}\sum_{l} e^{F}_{fkl}Q_{klt} + I^{F}_{ft} \ \forall s, t \tag{5}$$

$$I^{K}_{kt} = I^{K}_{k,t-1} + \sum_{l} Q_{klt} - D_{kt} \ \forall k, t \tag{6}$$

$$Ld_{kt} \le D_{kt} \le Ud_{kt} \ \forall k, t \tag{7}$$

$$\sum_{k \in K_j} P'_{klt} \le |K_j| \left(A_{jlt} + \sum_i X_{ijlt} \right) \quad \forall j, l, t \tag{8}$$

$$Q_{klt} \le \frac{Cap_{lt}}{p_{kl}} P'_{klt} \quad \forall k, l, t \tag{9}$$

$$\sum_j X_{ijlt} + A_{i,l,t+1} = \sum_j X_{jilt} + A_{ilt} \quad \forall i, l, t \tag{10}$$

$$\sum_j X_{ijlt} \le \sum_j X_{jilt} + A_{ilt} \quad \forall i, l, t \tag{11}$$

$$\sum_j A_{jlt} = 1 \quad \forall l, t \tag{12}$$

$$V_{ilt} + NX_{ijlt} - (N - 1) \le V_{jlt} \quad \forall i, j, l, t \tag{13}$$

$$\sum_{k \in K_j} (p_{kl} Q_{klt} + stp_{kl} P'_{klt}) + \sum_i \sum_{j \ne i} stb_{ijl} X_{ijlt} \le Cap_{lt} \quad \forall l, t \tag{14}$$

$$Q_{klt} \ge LBL_{lk} P'_{klt} \quad \forall k, l, t \tag{15}$$

$$ST_{ilt} = CL_{ilt} - \sum_{k \in K_i} (p_{kl} Q_{klt} + stp_{kl} P'_{klt}) \quad \forall i, l, t \tag{16}$$

$$CL_{ilt} + stb_{ijl} X_{ijlt} \le ST_{jlt} + Cap_{lt}(1 - X_{ijlt}) \quad \forall i, j, l, t \tag{17}$$

$$ST_{ilt} \ge \sum_{j \ne i} stb_{ijl} X_{ijlt} \quad \forall i, l, t \tag{18}$$

$$CL_{ilt} \le Cap_{lt}(1 - X_{ijlt}) \quad \forall i, j, l, t \tag{19}$$

$$O^F_f I^F_{ft} \le Cap^F_{ft} \forall f, t \tag{20}$$

$$O^K_k I^K_{kt} \le Cap^K_{kt} \quad \forall k, t \tag{21}$$

$$D_{kt}, Q_{klt}, I^F_{ft}, I^K_{kt}, Q^F_{fsdtm}, V_{jlt}, ST_{ilt}, CL_{ilt} \ge 0 \quad \forall f, k, t, l, m, s, d, i, j \tag{22}$$

$$X_{ijlt}, A_{jlt}, P'_{klt}, Z_{sdt}, Y_{st} \in \{0, 1\} \quad \forall k, t, l, m, s, d, i, j \tag{23}$$

Constraints (2) are to set the discount level d offered by supplier s based on the amount of raw materials purchased from supplier s. Constraints (3) indicate that in each period, there is at most one discount level d that supplier s can activate. Constraints (4) indicate whether there are orders for raw materials f to supplier s in each period. Constraints (5) and (6) show the inventory balance of raw materials and products, respectively. Constraints (7) indicate the upper and lower bound of product demand. Constraints (8) show that the production of k products in family j cannot start in period t unless a previous setup has been made for that family at the beginning of this period or there is a changeover for that family during this period. Constraints (9) indicate that the amount of product production must be less than equal to the production capacity. Constraints (10) aim to ensure the balance of the setup flows as a connected network, i.e., the input flow must be equal to the output flow of each production line. Constraints (11) indicate that the changeover from family i to other families can occur in production line l in period t when the transition from other families to family i has occurred, or production line l is arranged for product family i at the beginning of period t. Constraints (12) ensure that there is only one setup for one family at the start of each period. Constraints (13) are to eliminate scheduling sub-cycles. Constraints (14) aim to check the total production time and arrangements not to exceed the production line capacity in each period. Constraints (15) indicate the minimum quantity of a product when it is produced on a production line. Constraints (16) show that the beginning time of family i is equal to the completion time of family i minus the processing time of family i. Constraints (17) show that the beginning time of the following family must be greater than equal to the completion time of the previous family plus the setup time for changing to that family. Constraints (18) indicate the lower bounds of the family beginning time. Constraints (19) show the upper bounds of the family completion time. Constraints (20) and (21) check the warehouse capacity of raw materials and products, respectively. Constraints (22) and (23) show the various types of variables used in the proposed model.

4 Numerical Example

We provide a numerical example with four planning periods to apply the developed model. One planning period consists of six working days, with a working time of 8 h in one day, so the total available time capacity in one planning period is 2880 min. There are five types of products produced and grouped into three product families (F1 = MKAP, MKHC; F2 = MIFGH, MISE; F3 = MISCC). All products can be made in two production lines, namely, line A and line B. There are three raw materials used, namely Wheat Flour (TT), Tapioca Flour (TP), and Salt (G). These raw materials can be purchased from two suppliers: Cahaya Kimia (CK) and Mega Kimia Jaya (MKJ). Each supplier provides two discount rates, namely d1 and d2. To send raw materials from suppliers to manufacturers, there are two choices of carriers, namely Tronton Box (TA) and Tronton Wing Box (TB). The parameters are shown in Tables 1, 2, 3, 4, and 5.

Table 1 contains data on the upper bound of demand (Ud), lower bound of demand (Ld), product space (OK), product warehouse capacity (CapK), sale price (r), and

Table 1 Product demand bounds, product space, product warehouse capacity, sale price, and holding cost

Product	Ud	Ld	OK	CapK	r	HK
MKAP	9375	3884	0.0089	95	42,000	40
MKHC	8250	3884	0.0089	95	42,000	40
MIFGH	6000	2436	0.01	80	66,300	64
MISE	3750	1585	0.0182	80	74,600	72
MISCC	19,125	4969	0.0069	122	17,500	17

Table 2 Production cost, production time, setup cost, setup time, production lower bound, and raw material consumption

Product	cp; p; scp; stp; LBL		Ef (TT, TP, G)
	Line A	Line B	
MKAP	829;0.13;3193;10;3884	882;0.14;3513;11;3651	0.02644;0.0044;0.00026
MKHC	829;0.13;3193;10;3884	882;0.14;3513;11;3651	0.02644;0.0044;0.00026
MIFGH	1172;0.21;3832; 2; 2290	1101;0.2;3513;11;2436	0.16728;0.00112;0.00084
MISE	1801;0.33;4151;13;490	1693;0.31;3832;12;585	0.15052;0.00126;0.00098
MISCC	509;0.11;5110;16;4671	478; 0.1;4790;15;4969	0.03896;0.00086;0.0002

Table 3 Raw material space, raw material warehouse capacity, raw material holding cost, purchase cost, defect penalty cost, and defect rate

Raw material	OF	CapF	HF	bc; drc; dr	
				Supplier CK	Supplier MKJ
TT	25	200,000	239	256,000;25,600;4	242,000;24,200;6
TP	50	40,000	269	268,000;26,800;5	293,000;29,300;3
G	50	20,000	99	110,000;1100;6	97,000;9700;8

Table 4 Supplier order cost, discount levels, and transportation cost

Supplier	oc	DIS; U		Trc	
		d1	d2	TA (TT;TP;G)	TB (TT;TP;G)
CK	7000	0;1100	12;2200	2990;5980;5980	3115;6230;6230
MKJ	6000	0;1000	8;2000	2743;5485;5485	2286;4571;4571

Table 5 Family setup time and cost

Family	stb; scb		
	F1	F2	F3
F1	–	85;40,717	80;38,321
F2	90;43,112	–	75;35,926
F3	85;40,717	80;38,321	–

product holding cost (HK). Based on the table, it can be seen that MKAP product has an upper bound of demand of 9375 units, a lower bound of demand of 3884 units, product space of 0.0089 palettes, product warehouse capacity of 95 palettes, the sale price of Rp. 42,000/unit, and the product holding cost of Rp. 40/unit/period.

Table 2 contains data on production cost (cp), production time (p), setup cost (stc), setup time (stp), production lower bound (LBL), and raw material consumption (Ef). Based on the table, it can be seen that to produce MKAP product on line A requires a production cost of Rp. 829/unit, the production time of 0.13 min/unit, setup cost of Rp. 3.183/unit, setup time is Rp. 10, and production lower bound (LBL) of 3884 units. The raw materials needed for MKAP products in lines A and B are the same, namely for TT of 0.02644 units/unit, TP of 0.0044 units/unit, and G of 0.00026 units/unit.

Table 3 contains data on raw material space (OF), raw material warehouse capacity (CapF), raw material holding cost (HF), purchase cost (bc), defect penalty cost (drc), and defect rate (dr). Based on the table, it can be seen that TT raw material at CK supplier has a purchase price of Rp. 256,000/unit, a defect penalty cost of Rp. 25,600/unit, and a defect rate of 4%. Meanwhile, TT raw materials for both CK and MKJ suppliers have a raw material space of 25 kg/unit, a raw material warehouse capacity of 20,000 kg, and a raw material holding cost of Rp. 239/unit/period.

Table 4 contains data on supplier order cost (oc), discount levels (DIS), the upper bound of discount (U), and transportation cost (trc). Based on the table, it can be seen that for CK supplier, when the order size of raw materials exceeds 1100 units up to 2200 units, the manufacturer will get a quantity discount for all units of 12%. While for MKJ supplier, when the order size of raw materials exceeds 1000 units up to 2000 units, the manufacturer will get a quantity discount for all units of 8%. The order size of raw materials in each period can not exceed 2200 units for the CK supplier and 2000 units for the MKJ supplier because the supplier's capacity to provide raw materials is limited. Ordering costs for CK and MKJ suppliers are Rp. 7000 and Rp. 6000. For transportation costs, each carrier has different costs for each type of raw material; for example, the cost of transporting TA to CK supplier for TT raw material is Rp. 2990/unit, TP raw material is Rp. 5980/unit, and TG raw material is Rp. 5980/unit.

Table 5 contains data on family setup time (stb) and family setup cost (scb). There is no setup for switching between types of the same family, such as F1 to F1. Based on the table, it can be seen that the required family setup time from F1 to F2 is 85 min with a family setup cost of Rp. 40,717.

The optimal solution of the numerical example presented is searched using the Lingo 18.0 software. The maximum profit is Rp. 3,512, 278,000. Optimal Solutions from accepted product demand, produced product quantity, product inventory, raw material inventory, and purchased raw material quantity are shown in Tables 6, 7, 8, 9, and 10. The production beginning and completion times are summarized in the Gantt chart shown in Fig. 2.

Table 6 shows that the amount of accepted product demand for MKAP, MKHC, MIFGH, and MISE products in the four periods is always the same. As for MISCC products, the amount of accepted product demand had different values in the four

Table 6 Optimal solution of accepted product demand

Product	D (unit)			
	Period			
	1	2	3	4
MKAP	9375	9375	9375	9375
MKHC	8250	8250	8250	8250
MIFGH	2436	2436	2436	2436
MISE	3750	3750	3750	3750
MISCC	10,711	16,105	19,125	17,754

Table 7 Optimal solution of produced product quantity

Product	Qk (unit)			
	Period			
	1	2	3	4
MKAP	13,750	5271	9173	9306
MKHC	8250	8250	8252	8248
MIFGH	2667	4787	0	2290
MISE	3750	3750	3750	3750
MISCC	10,711	16,105	20,774	16,105

Table 8 Optimal solution of product inventory

Product	Ik (unit)			
	Period			
	1	2	3	4
MKAP	4375	271	69	0
MKHC	0	0	2	0
MIFGH	231	2582	146	0
MISE	0	0	0	0
MISCC	0	0	1649	0

Table 9 Optimal solution of raw material inventory

Raw material	If (unit)			
	Period			
	1	2	3	4
TT	584	114	159	0
TP	281	198	98	0
G	40	26	13	0

Table 10 Optimal solution of purchased raw material quantity

Raw material	Qf (unit)	
	Supplier	
	CK	MKJ
TT	3189	6000
TP	4157	
G	57	

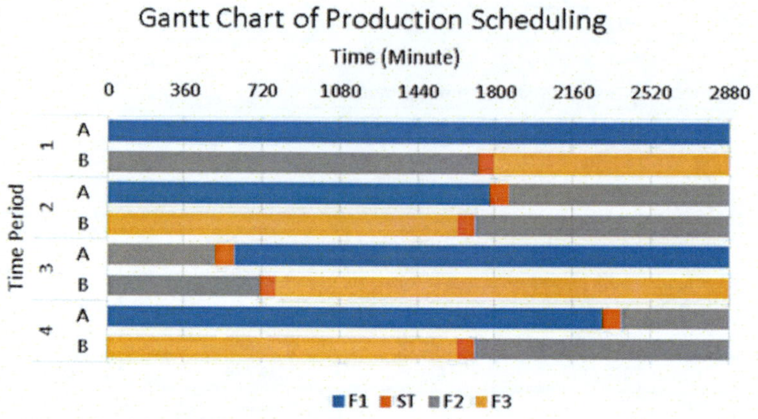

Fig. 2 Gantt chart of production scheduling

periods, with the highest demand in period three of 19,125 units and the lowest in period one of 10,711 units. Table 7 shows that the quantity of products produced for MKAP, MKHC, MIFGH, and MISCC products in the four periods is different. As for MISE products, the quantity of products it produces has the same value in the four periods, which is 3750 units. In period three, MIFGH product was not produced. Based on Table 8 it can be seen that MKAP and MIFGH products have product inventories from period one to period three. MKHC and MISCC products only have product inventory in period three. Meanwhile, MISE products do not have product inventory in the four periods.

Table 9 shows that there are inventories for all raw materials with different values from periods one to three. Meanwhile, in period four, there were no raw materials in stock. Table 10 shows that the total raw materials during the four planning periods that manufacturers can purchase from CK suppliers for TT raw materials are 3189 units, TP raw materials are 4157 units, and G raw materials are 57 units. Meanwhile, the optimal solution for MKJ suppliers shows that manufacturers can only buy one type of raw material, TT, with a total of 6000 units for four planning periods.

Based on Fig. 2 families F1, F2, and F3 have different beginning and completion times in lines A and B in periods one to four. There is a changeover setup between families, symbolized by the letter ST. The production sequence in periods two and four on lines A and B is the same, from F1 to F2 on line A and F3 to F2 on line B. The production order in periods one and three on line B is also the same, from F2 to F3. The production sequence in period three on line A, from F2 to F1. Meanwhile, in period one on line A, there was no change in production, so there was only one family, F1.

5 Conclusion

This research developed a MILP model to integrate lot sizing and scheduling problems by considering multiple suppliers, quantity discounts, carriers, and production time constraints. A numerical example was conducted to evaluate the model and to illustrate how the problem was solved. The results show that the optimal decision was determined by obtaining optimal demand, determining the optimal amount of each raw material that must be purchased from the optimal supplier, determining the optimal amount of each product that must be produced, and obtaining the optimal amount of inventory from product and raw material. Furthermore, the production beginning and completion times are obtained, providing more detailed scheduling information. This model can be further developed for further research by developing production time constraints at the product level because this model is still at the product family level. Another suggestion is to consider other types of discounts, carrier capacity, and lead time from suppliers.

References

1. Badri HM, Khamis HM, and Ghazali MJ (2020) Integration of lot sizing and scheduling models to minimize production cost and time in the automotive industry. Int J Ind Optim 1(1):1–14
2. Almada-Lobo B, Klabjan D, Carravilla MA, and Oliveira JF (2010) Multiple machine continuous setup lotsizing with sequence-dependent setups. Comput Optim Appl 47(3):529–552. https://doi.org/10.1007/s10589-009-9235-8
3. Ramezanian R, Saidi-Mehrabad M, and Teimoury E (2013) A mathematical model for integrating lot-sizing and scheduling problem in capacitated flow shop environments. Int J Adv Manuf Technol 66:1–4:347–361. https://doi.org/10.1007/s00170-012-4329-3.
4. Pinedo ML (2008) Scheduling: theory, algorithms, and systems
5. Liu M, Wang S, and Chu C (2013) Scheduling deteriorating jobs with past-sequence-dependent delivery times. Int J Prod Econ 144(2):418–421. https://doi.org/10.1016/j.ijpe.2013.03.009
6. Almeder C, Klabjan D, Traxler R, and Almada-Lobo B (2015) Lead time considerations for the multi-level capacitated lot-sizing problem. Eur J Oper Res 241(3):727–738. https://doi.org/10.1016/j.ejor.2014.09.030
7. Quadt D (2004) Lot-sizing and scheduling for flexible flow lines, 1st edn. Springer, Berlin, Heidelberg
8. Cunha L, Santos MO, Morabito R, and Barbosa-Póvoa A (2018) An integrated approach for production lot sizing and raw material purchasing. Eur J Oper Res 269(3):923–938. https://doi.org/10.1016/j.ejor.2018.02.042

9. Lee HI, Kang HY, Lai CM, and Hong WY (2013) An integrated model for lot sizing with supplier selection and quantity discounts. Appl Math Model 37(7):4733–4746. https://doi.org/10.1016/j.apm.2012.09.056
10. Choudhary D, and Shankar R (2013) Joint decision of procurement lot-size, supplier selection, and carrier selection. J Purch Supply Manag 19(1):16–26. https://doi.org/10.1016/j.pursup.2012.08.002
11. Ertogral K, Darwish M, and Ben-Daya M (2007) Production and shipment lot sizing in a vendor-buyer supply chain with transportation cost. Eur J Oper Res 176(3):1592–1606. https://doi.org/10.1016/j.ejor.2005.10.036
12. Mohammadi M, Esmaelian M, and Atighehchian A (2020) Design of mathematical models for the integration of purchase and production lot-sizing and scheduling problems under demand uncertainty. Appl Math Model 84:1–18. https://doi.org/10.1016/j.apm.2020.03.021
13. Georgiadis GP, Kopanos GM, Karkaris A, Ksafopoulos H, and Georgiadis MC (2019) Optimal production scheduling in the dairy industries. Ind Eng Chem Res 58(16):6537–6550. https://doi.org/10.1021/acs.iecr.8b05710
14. Sutrisno S, and Wicaksono PA (2020) Multi-objective optimisation approach as decision-making support for product mixing and raw material allocation planning. Int J Procure Manag 13(2):180–198. https://doi.org/10.1504/IJPM.2020.106511

An EOQ Model for Perishable Product Considering Carbon Emissions Resulting from Refrigerated Truck and Cold Storage

Hanin Tsimar Andriyani, Cucuk Nur Rosyidi, and Pringgo Widyo Laksono

Abstract This study discusses the development of an Economic Order Quantity (EOQ) model for perishable products by considering carbon emissions produced by refrigerated trucks and cold storage. The carbon emissions of refrigerated trucks depend on the fuel needed during product delivery activities from the distribution area to the retailer, while the carbon emissions of cold storage depend on the retailer's cold storage energy consumption during product storage activities. The products sent are multi-product perishable products with varying optimal storage temperatures. This model aims to minimize the costs involved in shipping and storing perishable products, including ordering costs, storage, penalties, operations, and carbon taxes. A penalty fee is imposed if the product is allocated in a place that is not in accordance with the optimal storage temperature. This study assumes that lead time is negligible and costs related to carbon emissions are imposed on the retailer. This study is expected to assist retailers in making decisions to determine the number of economic orders, optimal temperature, and allocation of perishable products on refrigerated trucks and cold storage by minimizing the costs involved in shipping and storing products. The optimization results show that this model produces a high utility (98%), the higher the utility the greater the benefits and profits obtained by the retailer with a total cost of $619.82 and carbon emission worth 0.1271 ton.

Keywords Perishable product · Cold chain logistic · EOQ · Multi product

1 Introduction

Storage activities, shipping transportation, logistics, and warehousing operations are the leading supply chain activities that contribute to various industries' greenhouse gas emissions [1]. According to Tiwari et al. [2], transportation emissions depend

H. T. Andriyani (✉) · C. N. Rosyidi · P. W. Laksono
Industrial Engineering Department, Sebelas Maret University, Surakarta, Indonesia
e-mail: hanintsimar@student.uns.ac.id

C. N. Rosyidi et al. (eds.), *Proceedings of the 6th Asia Pacific Conference on Manufacturing Systems and 4th International Manufacturing Engineering Conference*, Lecture Notes in Mechanical Engineering,
https://doi.org/10.1007/978-981-99-1245-2_23

on fuel consumption and vehicle mileage, warehouse emissions depend on total inventory and warehouse energy consumption per unit, and goods storage emissions depend on products that experience deterioration related to their disposal. One of the categories of deterioration products is perishable products. Perishables are products that decompose quickly or have a short life span, such as fruit, vegetables, and meat [3]. The product's perishable nature, which is easily decomposed, reduces consumer interest in buying as the product ages until, at the end of its useful life the item is completely damaged and has no use for consumers [4]. Ordering and storing perishable products can also produce carbon emissions [5].

Perishable products require a controlled temperature environment during the production, storage, transportation, and sale to ensure food quality and reduce food waste [6]. According to Aung and Chang [7], facilities such as refrigerated containers, cold storage, and refrigerated display cabinets are needed in the cold chain to store perishable products properly. However, product allocation to suitable refrigerated facilities is complicated due to product variations and different temperature requirements.

One solution that can be offered for the problem of determining the optimal order quantity in refilling activities is to use the EOQ method. EOQ is an inventory control method that can optimize the company's inventory costs [8]. Also, the solution to the problem of cold product logistics activities is to control the temperature of refrigerated trucks and cold storage in transportation and storage activities. According to Aung and Chang [7], controlling temperature in the food supply chain is the most crucial factor in extending product shelf life. Temperature control can be done by allocating cold products on refrigerated trucks to match the optimal storage temperature.

The development of EOQ on perishable products was carried out by Li and Teng [9] which discusses pricing decisions and lot sizes for perishable products when demand depends on the selling price, reference price, product freshness, and the number of products displayed. Dobson et al. [4] developed an EOQ model on perishable products by considering the demand that depends on the lifetime of the product. The results of his research show that the increasing demand for the product during its lifetime can increase the company's profits by reducing storage costs. Tiwari et al. [2] conducted a study on inventory management on deteriorating products by considering aspects of carbon emissions. The results of his research state that by reducing the frequency of shipments and increasing the quantity of shipments, carbon emissions from transportation activities can be reduced.

Amaruchkul et al. [10] conducted a study on the problem of refilling perishable products which discussed the allocation of multi-products in single-cooled trucks and multiple-cooled trucks with due regard to penalty costs. The purpose of this research is to determine the allocation of the product in the temperature zone that corresponds to the optimal storage temperature of the product so as to extend the shelf life of the product by minimizing the total cost involved. This research model has not considered the carbon emission aspect. Shamayleh et al. [11] conducted a study related to an economic model that pays attention to environmental factors in cold products with demand at different times. Environmental factors that are considered are aspects of carbon emissions resulting from transportation and storage activities with the implementation of carbon tax policies. The main objective is to determine the optimal order quantity, number of trucks, and cold storage so as to minimize the company's operational costs and the resulting carbon emissions. The research model has not considered the allocation of multiproducts on refrigerated trucks and cold storage.

This study is a development of the previous model by Amaruchkul et al. [10] and Shamayleh et al. [11]. The contribution of this research is the development of the EOQ model for perishable products by considering the implementation of carbon tax policies on carbon emissions produced by refrigerated trucks and cold storage which are expected to assist retailers in making decisions to determine the number of economical orders, optimal temperature, and allocation of perishable products on refrigerated trucks and cold storage so as to minimize the costs involved in shipping and storing products. On the other hand, if the number of economical orders, temperature, and allocation of perishable products on refrigerated trucks and cold storage is not optimally regulated, it will cause the product to deteriorate and increase the cost of ordering and storing products.

2 Assumption and Notation

This paper has the assumptions, which are :

1. The lead time is negligible.
2. The carbon emissions-related costs imposed on the retailer.
   The following notations are used in this paper:
   Decision variables

m_i	Integer number of T intervals
T	Family cycle (day)
q_i	Order quantity of item i (unit)
x_{ij}	Allocation quantity of item i on truck j(unit)
x_{iz}	Allocation quantity of item i on cold storage z (unit)
y_j	Temperature point of truck j (°C)
y_z	Temperature point of cold storage z (°C)
u_j	Binary variable (=1), if truck j is used
u_z	Binary variable (=1), if cold storage z is used
T_i	Cycle time of item i to fulfill q_i (day)

Parameters

k_i	Minor setup of item i each delivery ($)
K_j	FTL cost truck j each delivery ($)
K_o	Other fixed major setup cost each delivery ($)
t_i^U	Upper bound of item i's optimal temperature (°C)
t_i^L	Lower bound of item i's optimal temperature (°C)

p_i^U	Penalty cost from exceeding t_i^U (\$/°C/unit)
p_i^L	Penalty cost from falling below t_i^L (\$/°C/unit)
v_i	Volume of item i (m^3)
c_j	Capacity of truck j (m^3)
c_z	Capacity of cold storage z (m^3)
y_j^L	Minimum temperature point on truck j (°C)
y_j^U	Maximum temperature point on truck j (°C)
tx	Carbon tax (\$/ton)
pg	Fuel price per gallon (\$/gallon)
EP	Electricity price per kWh (\$/kWh)
GPM_j	Gallon per mile for a full truckload of truck j (gallon/mile)
GPM_j^0	Gallon per mile for an empty truckload of truck j (gallon/mile)
$TECS_j$	Total energy consumption of cold storage z (kWh/day)
$CEGF$	Carbon emission level per gallon of fuel (ton/gallon)
$TCFE$	Total carbon footprint of one kWh energy (ton/kWh)
T	Base cycle (day)

3 Model Development

In this mathematical model, we develop the lot sizing and allocation problem of perishable products considering carbon emissions resulting from transportation and storage activities. The system we discussed is shipping and storage activities of multi-product perishable products with varying optimal storage temperatures. The perishable product is shipped from the distribution center to the retailer's storage facility via full-truckload (FTL) trucking, which is refrigerated trucks with controlled temperature. In this system, the penalty fee for refrigerated trucks and cold storage is also one of the considerations in minimizing the total costs involved. A penalty fee is imposed if the product is allocated to a refrigerated truck and cold storage with a temperature that exceeds the upper limit or less than the lower limit of the optimal storage temperature for the product. Therefore, the allocation of product placement adjusted to the temperature of the refrigerated truck or cold storage is very influential so that the product does not deteriorate during transportation and storage activities. Another cost component also considered in this system is the carbon tax rate on carbon emissions generated by refrigerated trucks and cold storage in product delivery and storage activities. Figure 1 shows system description of this model.

Generally, the cost components considered in minimizing the total costs involved consist of ordering and holding costs. The ordering costs include major setup costs, minor setup costs, refrigerated truck carbon tax rates, refrigerated truck penalty costs, and refrigerated truck fuel costs. Meanwhile, storage costs include refrigerated truck holding costs, cold storage holding costs, cold storage carbon tax rates, cold storage penalty costs, and cold storage operational costs. This study has several constraint functions: the filling cycle constraint, the order quantity constraint, the refrigerated truck volume constraint and cold storage, the refrigerated truck temperature

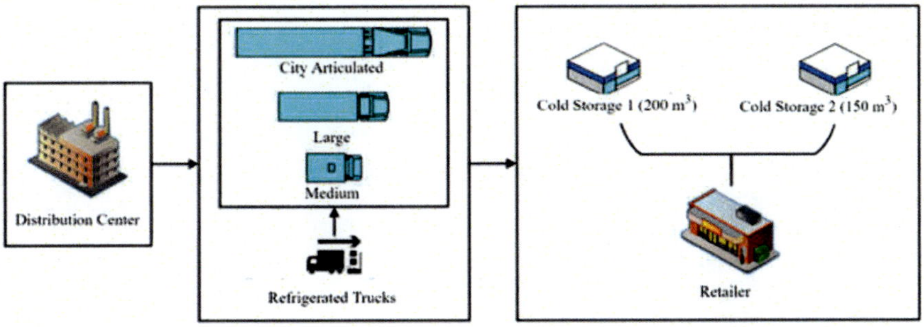

Fig. 1 System description

constraint, and the refrigerated truck and cold storage penalty fee constraint. The aim is to determine order quantity, allocation plan, and optimal temperature on refrigerated trucks and cold storage in one period with Mix Integer Linear Programming (MINLP) by minimizing total cost.

$$\text{Minimize } \frac{1}{T}\left(K_O + \sum_{j=1}^{m} u_j K_j + \sum_{i=1}^{n}\left(\frac{k_i}{T_i} + h_i q_i\right)\right. \tag{1}$$

$$+\sum_{i=1}^{n}\left(\frac{1}{T_i}\sum_{j=1}^{m}\left(p_i^U S_{ij}^{U+} + p_i^L S_{ij}^{L-}\right)x_{ij}\right) \tag{2}$$

$$+\sum_{i=1}^{n}\left(\frac{1}{T_i}\sum_{z=1}^{w}(p_i^U S_{iz}^{U+} + p_i^L S_{iz}^{L-})x_{iz}\right) \tag{3}$$

$$+\sum_{j=1}^{m}[(pgL(GPM_j - GPM_j^0)u_j) \tag{4}$$

$$+(pgLGPM_j^0 u_j)] + \sum_{Z=1}^{w} EPTECS_Z u_Z$$

$$+ tx\left[\sum_{j=1}^{m}((CEGFL(GPM_j - GPM_j^0)u_j)\right. \tag{5}$$

$$+(CEGFLGPM_j^0 u_j))]$$

$$\left.+tx\left[\sum_{Z=1}^{w} TCFETECS_Z u_z\right]\right) \tag{6}$$

Subject to :

$$T_i = m_i T \quad \forall_i \in n \tag{7}$$

$$q_i = d_i T_i \quad \forall_i \in n \tag{8}$$

$$y_j - S_{ij}^{U+} + S_{ij}^{U-} = t_i^U \quad \forall_i \in n, \forall_j \in m \tag{9}$$

$$y_j - S_{ij}^{L+} + S_{ij}^{L-} = t_i^L \quad \forall_i \in n, \forall_j \in m \tag{10}$$

$$y_z - S_{iz}^{U+} + S_{iz}^{U-} = t_i^U \quad \forall_i \in n, \forall_z \in w \tag{11}$$

$$y_z - S_{iz}^{L+} + S_{iz}^{L-} = t_i^L \quad \forall_i \in n, \forall_z \in w \tag{12}$$

$$y_j^L \le y_j \le y_j^U \quad \forall_j \in m \tag{13}$$

$$\sum_{i=1}^{n} v_i x_{ij} \le z_j \quad \forall \quad \forall_j \in m \tag{14}$$

$$\sum_{j=1}^{m} z_j \le c_j u_j \quad \forall_j \in m \tag{15}$$

$$\sum_{i=1}^{n} v_i x_{iz} \le z_z \quad \forall_z \in w \tag{16}$$

$$\sum_{z=1}^{w} z_z \le c_z u_z \quad \forall_z \in w \tag{17}$$

$$\sum_{j=1}^{m} x_{ij} = q_i \quad \forall_i \in n \tag{18}$$

$$\sum_{z=1}^{w} x_{iz} = q_i \quad \forall_i \in n \tag{19}$$

$$u_j \in \{0, 1\} \quad \forall_j \in m, \forall_z \in w \tag{20}$$

$$u_z \in \{0, 1\} \quad \forall_j \in m, \forall_z \in w \tag{21}$$

$$m_i, x_{ij}, x_{iz} \ge 0\&\ integer \quad \forall_i \in n, \forall_j \in m, \forall_z \in w \tag{22}$$

$$x_{ij}, S_{ij}^{U+}, S_{ij}^{U-}, S_{ij}^{L+}, S_{ij}^{L-} \geq 0 \quad \forall_i \in n, \forall_j \in m \tag{23}$$

$$x_{iz}, S_{iz}^{U+}, S_{iz}^{U-}, S_{iz}^{L+}, S_{iz}^{L-} \geq 0 \quad \forall_i \in n, \forall_z \in w \tag{24}$$

In the objective function (1), the first term is the average of major setup cost, second term is the average of minor setup cost and holding cost, (2) is the penalty cost from violating the temperature range of item *i* on all trucks, (3) is the penalty cost from violating the temperature range of item *i* on all types of cold storage, (4) is the sum of the refrigerated truck's diesel consumption to deliver item *i* from distribution center to retailer, (5) is the sum of the cold storage's energy consumption to keep item *i*, and (6) is the sum of the carbon tax resulting from carbon emission of refrigerated trucks and cold storages.

In the constraints (7) states about the cycle time of item *i*, where m_i is the integer number of T intervals that the replenishment quantity q_i will last. Constraints (8) states that the order quantity q_i must be equal to the daily demand d_i in the cycle time. Constraints (9)–(12) together with the non-negativity constraint are used to define the deviational variables. Note that S_{ij}^{L+}, S_{ij}^{U-}, S_{iz}^{L+}, S_{iz}^{U-} does not appear in the objective function, since we do not get penalized from being over the lower bound t_i^L and under the upper bound t_i^U. Constraint (13) states that the truck temperature must be around minimum temperature point y_j^L and y_j^U. Constraint (14) and (16) states that the total volume across all items in truck *j* and cold storage *z* is at most its planned volume z_{jk} and z_z. The truck and cold storage capacity constraint is given in (15) and (17). Constraint (18) and (19) states that the sum of units of item *i* in all trucks and all cold storage must be equal to the replenishment quantity q_i. In (20), the binary u_j must be 1 if truck *j* is used and in (21), the binary u_z must be 1 if cold storage *z* is used. Since item *i* will be included in every m_ith replenishment of the family, m_i, x_{ij}, and x_{iz} needs to be a natural number as in (22).

4 Numerical Example

We present an example with one-period delivery, and we consider a lot sizing and allocation problem of $n = 12$ items of perishable product sold in the retailer. This retailer is replenished from the distribution center. Numerical data used in this research is secondary data obtained from the reference literature. The input parameter for all items is given in Table 1 including item description, minimum temperature, maximum temperature, daily demand, unit volume, and unit holding cost of item *i*. These data were obtained from research data by Aung and Chang [7] and Amaruchkul et al. [10].

The company has three types of refrigerated trucks, one medium truck, one large truck, and one city articulated truck for delivery from the distribution center to the retailer's cold storage facility. The retailer provides two cold storage which has different volume. Table 2 shows truck information, cold storage information, and

Table 1 Product parameters

Item	Description	Min temp	Max temp	Daily demand	Unit volume	Unit holding cost
I		t_i^L	t_i^U	d_i	v_i	h_i
		°C	°C	Unit/day	m^3/unit	$/unit/day
1	Chocolate	18	21	168	0.002044	0.0028
2	Chili paste	5	25	198	0.003288	0.0028
3	Dessert	5	15	268	0.002327	0.0084
4	Dry goods	15	25	267	0.23501	0.0028
5	Egg	7	20	231	0.046059	0.0028
6	Frozen food	−20	0	257	0.032325	0.014
7	Candy	5	13	357	0.044995	0.0084
8	Guavas	7	10	238	0.056634	0.0084
9	Lemons	11	13	256	0.028317	0.0084
10	Lychees	4	7	148	0.084950	0.0028
11	Mango	10	13	287	0.056634	0.014
12	Melon	10	13	269	0.084950	0.014

other parameters. These data were obtained from research data by Amaruchkul et al. [10], Shamayleh et al. [11], Asad et al. [12], and Tassou et al. [13].

Table 2 shows the optimal solution for lot sizing and allocation problem with T = 1. The allocation plan includes optimal quantity order, cycle time item *i*, temperature of truck *j*, and cold storage *z*. We put * to indicate the temperature violation.

Table 3 shows the result of the optimization model, which describes the volume utility and the total costs involved in product shipping and storage activities. * indicates that there is a penalty cost for the item because it is not in the optimal temperature range.

Table 2 Other parameters

Truck information				Other parameters	
Parameters	Value			Parameters	Value
	Medium	Large	City artic	k_i ($/delivery)	2.9
C_j (m^3)	33.42	61.15	78.79	p_i^U ($/°C/unit)	0.015
K_j ($/delivery)	24.66	29.74	34.82	p_i^L ($/°C/unit)	0.015
y_j^L (°C)	−5	−5	−5	T (day)	1
y_j^U (°C)	20	20	20	K_o ($/delivery)	2.9
$GPMO_j$ (gallon/mile)	0.064	0.093	0.081	n (item)	12
GPM_j (gallon/mile)	0.081	0.118	0.102	m (unit)	3

(continued)

Table 2 (continued)

Other parameters (advanced)					
Truck information				Other parameters	
Parameters	Value			Parameters	Value
	Medium	Large	City artic		
$CEGF_j$ (tons/gallon)	0.002787	0.004025	0.003526	pg ($/gallon)	10.849
Cold Storage Information				L (miles)	100
	Large	Medium		tx ($/ton)	25
C_z (m^{3})	100	80		EP ($/kWh)	0.17
$TECS$ (kWh/day)	15.78	12.624		$TCFE$ (ton/kWh)	0.000744

Table 3 Allocation plan

Item	q_i	m_i	T_i	Carbon tax policy	Refrigerated truck			Cold storage	
					Med	Lar	Arc	Lar	Med
i	Unit		Day	y_{jk} (°C)	11	18	10	10	15
1	168	1	1	x_{ijk} (unit)	–	168	–	*16	*152
2	198	1	1		6	96	96	127	71
3	268	1	1		4	–	264	251	17
4	267	1	1		–	227	*40	–	267
5	231	1	1		–	–	231	228	3
6	257	1	1		*46	*20	*191	*255	*2
7	357	1	1		1	*7	349	293	*64
8	238	1	1		*16	*100	122	215	*23
9	256	1	1		66	–	*190	*256	–
10	148	1	1		–	–	*148	*147	*1
11	287	1	1		128	–	159	95	*192
12	269	1	1		257	*5	7	269	–

Based on Table 4 the optimization results show that this model produces a high utility (98%), the higher the utility the greater the benefits and profits obtained by the retailer with a total cost of $619.82 and carbon emission worth 0.1271 ton.

Table 4 Result summary of the optimization model

Description	Value
Cycle time (day)	1
Volume utilization	98%
Carbon footprint resulting from truck (ton)	0.1060
Carbon footprint resulting by cold storage (ton)	0.0211
Total cost	$ 619.82

5 Conclusion

This research developed an optimization model in the form of a mixed-integer non-linear programming model to resolve lot sizing and allocation problem. The objective is to minimize the total cost consisting of major setup cost, minor setup cost, refrigerated truck carbon tax rates, refrigerated truck penalty cost, refrigerated truck fuel costs, refrigerated truck holding costs, cold storage holding costs, cold storage carbon tax rates, cold storage penalty fees, and cold storage operational costs. The model considered carbon emissions resulting from transportation and storage activities. The aim is to determine order quantity, allocation plan, and optimal temperature on refrigerated trucks and cold storage. Suggestions that can be considered for further research are to consider the implementation of other carbon policies, such as carbon cap, cap-and-trade and cap-and-offset and to consider other parties involved in the system such as carrier transportation providers or 3PL (Third Party Logistics).

References

1. Schaefer B, Konur D (2015) Economic and environmental considerations in a continuous review inventory control system with integrated transportation decisions. Transp Res Part E 80:142–165
2. Tiwari S, Daryanto Y, Wee HM (2018) Sustainable inventory management with deteriorating and imperfect quality items considering carbon emission. J Clean Prod 192:281–292
3. Widyadana IGA (2013) Penentuan Ukuran Pemesanan Material dengan Memperhatikan Decay Inventory dan Quantity Unit Discount. 1(1):57–62
4. Dobson G, Pinker EJ, Yildiz O (2017) An EOQ model for perishable goods with age-dependent demand rate. Eur J Oper Res 257(1):84–88
5. Yu C, Qu Z, Archibald TW, Luan Z (2020) An inventory model of a deteriorating product considering carbon emissions. Comput Ind Eng 148:106694
6. Ma G (2009) The application research of cold-chain logistics delivery schedule based on JIT 368–370
7. Aung MM, Chang YS (2014) Temperature management for the quality assurance of a perishable food supply chain. Food Control 40(1):198–207
8. Lahu EP, Sumarauw JSB, Ekonomi F, Manajemen J, Sam U, Manado R, Belakang L (2017) Analisis Pengendalian Persediaan Bahan Baku Guna Meminimalkan Biaya Persediaan Pada Dunkin Donuts Manado. Jurnal EMBA: Jurnal Riset Ekonomi, Manajemen, Bisnis Dan Akuntansi 5(3):4175–4184
9. Li R, Teng JT (2018) Pricing and lot-sizing decisions for perishable goods when demand depends on selling price, reference price, product freshness, and displayed stocks. Eur J Oper Res 270(3):1099–1108
10. Amaruchkul K, Pongsathornwiwat A, Bantadtiang P (2022) Constrained joint replenishment problem with. 26(1):75–91
11. Shamayleh A, Hariga M, Asad R, Diabat A (2019) Economic and environmental models for cold products with time varying demand. J Cleaner Prod 212:847–863
12. Asad R, Hariga M, Shamayleh A (2020) Sustainable dynamic lot sizing models for cold products under carbon cap policy. Comput Ind Eng 149(July 2020):106800
13. Tassou SA, De-Lille G, Ge YT (2009) Food transport refrigeration—approaches to reduce energy consumption and environmental impacts of road transport. Appl Therm Eng 29(8–9):1467–1477

Web Design: Develop the Decision Support System for Judgement Additional Order (JAO) Using Waterfall Method

Yoshan Ardhi Pratama, Pringgo Widyo Laksono, and Eko Liquiddanu

Abstract Judgement additional orders (JAO) are a problem faced by a company that occurs because the process of working on them takes a long time. Therefore, the company needs a web-based decision support system to measure the feasibility of additional orders to be received. The considerations of the decision-making process are the processing time duration and the availability of materials. The decision-making model is described in a mathematical model and implemented into the system using the waterfall method. There are five stages of the waterfall method, including requirement, design system, implementation, verification, operation and maintenance. The results of the system design are also tested using black-box testing. Based on the trials, the system design is able to run by the model and can solve the JAO problem.

Keywords Judgement additional order (JAO) · Decision support system · Waterfall method · Black-box testing

1 Introduction

The development of the manufacturing industry leads to a manufacturing system that produces various products in slightly smaller quantities for each product. This condition requires a company to operate with a short lead time to complete customer demand [1]. Customer satisfaction is the company's primary goal, followed by a sufficient profit level. PT. XYZ is a manufacturing company that implements the MTO system and prioritizes customer satisfaction. The company still encounters difficulties in carrying out its business process, especially in order acceptance management. The acceptance management process becomes a problem because of limited resources.

Y. A. Pratama (✉) · P. W. Laksono · E. Liquiddanu
Industrial Engineering Department, Universitas Sebelas Maret, 57126 Surakarta, Indonesia
e-mail: yoshanardhi1@student.uns.ac.id

C. N. Rosyidi et al. (eds.), *Proceedings of the 6th Asia Pacific Conference on Manufacturing Systems and 4th International Manufacturing Engineering Conference*, Lecture Notes in Mechanical Engineering,
https://doi.org/10.1007/978-981-99-1245-2_24

There are two types of orders received by the company; the first one is regular order, and the other is an additional order. The focus of this research is on solving additional order problems. Currently, the company manages additional orders manually using excel. The drawback of the current system occurs in time management. Decision maker take a long time to collect data that is used to judge the feasibility of accepting additional orders. Based on this problem, the company wants to make an integrated system to manage the acceptance of additional orders.

The decision support system development based on a website can solve the company's problem of managing additional orders [2]. A website designed using the client–server model to complete the user's wishes. The client and the user, can communicate directly with the server when using the program [3]. Therefore the client–server model can minimize the management time duration because of the excellent information sharing. The client acts as a user or requester, and the server gives services.

Based on the background, this research focuses on designing a web based-decision support system for judgement additional orders using the waterfall method. The decision model of this research considers processing time duration and material availability. Several assumptions are used in this study including ignoring setup time and overtime, Producing each part number in one machine minimum, and each process after the first process is considered one day.

2 Literature Review

Judgement Additional Order (JAO) is the process of assessing the feasibility of accepting additional orders requested by customers. The company uses two considerations to make a decision. The first consideration is the duration of processing time, and the second one is material availability. To make the JAO process easier, the company built a web based-decision support system for managing additional orders.

2.1 Decision Support System

A decision support system (DSS) is an interactive system created to support a decision maker in making the decision. The decision making process is carried out by integrating data and processing it into a model. Based on characteristics, DSS has four components: Data management, model management, communication, and knowledge management [4]. Data management is one of the sub system that organizes all data in every condition. Model management is a part of sub system is able to provide analytical capabilities to the system. Communication is a part of sub system that can connect between the user and the system, this sub system produces two-way communication in a user interface design. At the same time, knowledge management is an optional sub systems and acts as a support for another sub system.

2.2 *Waterfall Method*

A waterfall method is a method used to develop software. This method works with a systematic and structured approach. The approach of this method is often referred to as a linear sequence model [5]. The stages of the waterfall method are requirement, design system, implementation, verifications, operation and maintenance [6]. The first stage, the requirement, is an initial step in planning software development. The activities in this stage are discussed with the decision maker or requester to collect needs for software development. The second stage is the design system, activities in this stage are building a design of data structure and software architecture. The third stage of the waterfall method is implementation. In this stage, the developer divides into small units to create the software Each unit develops the system due to its function. The fourth stage is verification, in this stage, the system is tested to measure the results of the system design. The last stage is operation and maintenance. The activities in this stage are mass system usage and periodic maintenance.

2.3 *Mathematical Model*

A model represents real conditions, and someone can solve the existing problem with a model. At the same time, a mathematics model is a mathematical formulation used to solve complex problems [7]. Three components that compose a mathematics model are variable, parameter, and function. The function represents the relation between variable and parameter. There are three classifications of mathematical models, including the statistical model, deterministic model, and probabilistic model. The statistical model usually functions with one or more variables. The deterministic model is a model with a high degree of certainty, this model has a definite initial value and limit value. While the probabilistic model or often referred to as the stochastic model has a probabilistic value on the variable or in the parameter, this value is uncertain.

3 Methodology

The design of the decision support system in the judgement additional order process is carried out using the waterfall method. This method goes through several stages, namely requirements, system design, implementation, verification, operation and maintenance (Fig. 1).

1. Requirement

The requirement is an initial stage in system design. The activities here are collecting and analyzing system needs. Based on discussions conducted with a decision maker,

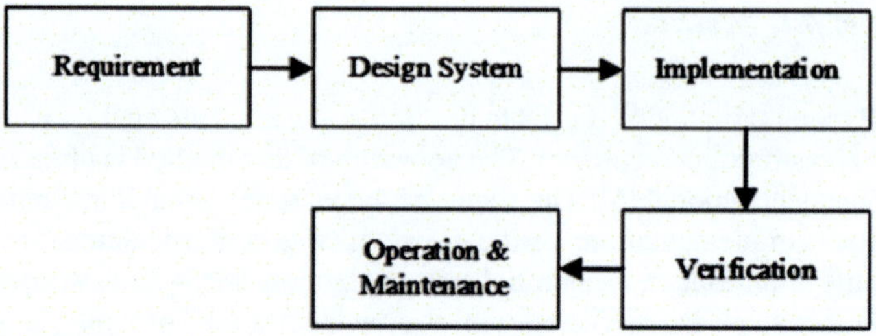

Fig. 1 Research Stages

some data was obtained that was needed in the processing of judgement additional orders. The data includes additional order data, stock part number, arrival of imported parts, the cycle time of each part number, flow process, number of running machines, weight consumed of each part number, stock raw material, additional running machines, and working days.

2. Design System

The design system is a stage of designing the architectural design of the system. The process design using a UML (Unified Modeling Language) Model. The UML Model is applied to specify and visualize the system's needs to be designed. In this study, the UML model used a use case diagram and activity diagram.

3. Implementation

The implementation is the stage of creating a database and coding. Database management is done using phpMyAdmin software. Meanwhile, the decision support system is designed using the CodeIgniter framework. The CodeIgniter framework is one of the PHP frameworks that developers often use to build a web-based application with the PHP programming language [8]. The CodeIgniter framework works by using the MVC (Model-View-Controller) Architecture (Fig. 2).

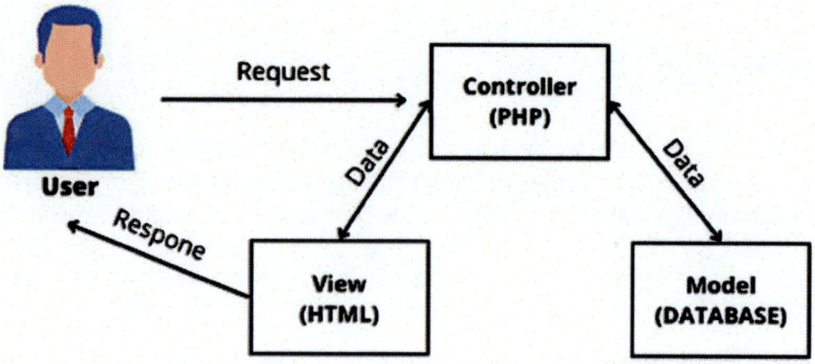

Fig. 2 MVC Architecture

4. Verification

The verification explains the trial of the JAO system design. The trial was conducted using black-box testing to assess the success of the program. Trials using black-box are carried out by designing a table containing test scenarios, test cases, expected results, test results, and conclusions.

5. Operation and Maintenance

This stage is the last stage in the waterfall model, in this section the design of the decision support system is used directly by decision makers to overcome JAO problems. Besides that, DSS maintenance is also carried out to maintain that the system design runs optimally.

4 Result and Discussion

The results obtained from the research that has been carried out are divided into four parts. The first part describes the data flow diagram, the second part describes the results of the mathematical model, the third part describes the architecture system and the last describes the results of the user interface design of the JAO System.

4.1 Data Flow Diagram

Data flow diagram (DFD) is a part of the requirements stage in the waterfall method. Based on the identifications, some data is used in decision-making process of additional order acceptance. The following is a figure of data flow diagram from the judgement additional order process has been made.

Figure 3 explains about seven parties involved in the judgement additional order process. The seven parties include customer, customer service, engineering, warehouse, production, procurement and decision makers. They are involved in the judgement additional order process as a data provider. The data will be considered by decision maker on additional order request. The customers has the authority to provide the request order data and the products delivery schedule data. Moreover, the customers also gets information related to the results of the request order that has been sent to the company. Customer service is a liaison between decision maker and customers. All informations regarding the request order will be forwarded to the decision maker to be assessed for the feasibility of each orders received. The engineering department is authorized to determine the specifications of each products requested by the customer. The specification data used include process flow data and the cycle time of each product. The warehouse department controls the stock of raw materials, both current and future stock condition. The production department is directly related to the production process so they are able to provide stock data for each products. The

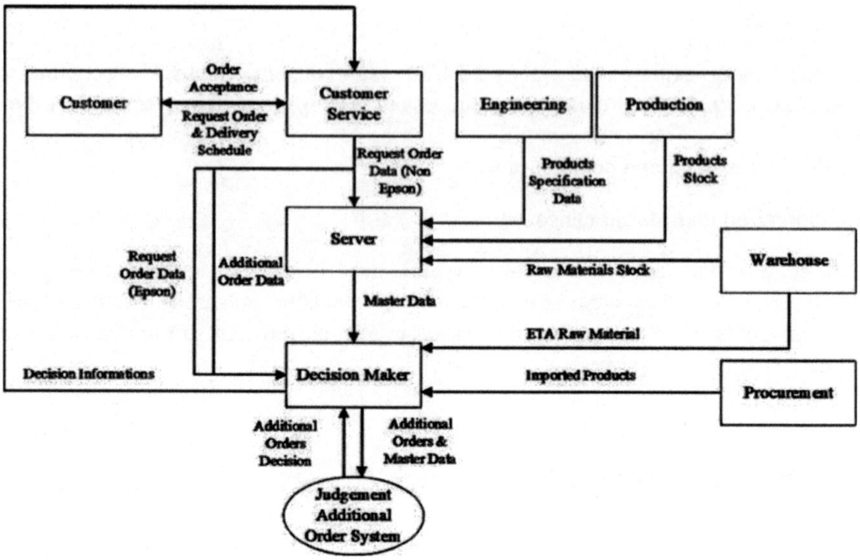

Fig. 3 Data flow diagram

procurement department is authorized to purchase materials, one of the involvements of the procurement department in the JAO process is to provide data on products imported from abroad. The last party is the decision maker whose operate the JAO system and determine decisions related to additional orders submitted by customers.

4.2 Mathematical Model

This section describes the mathematical models that have been created for the decision making process. There are eight calculations used to represent the relationships between the data that has been collected.

1. Safety Stock

$$SS_{it} = 0.3 \times FO_{it} \tag{1}$$

2. Balance

$$BLC_{it} = \sum_{t=1}^{T} SD_{it} - PO_{it} \tag{2}$$

3. Daily Production Output

$$OP_i = \frac{3600}{CT_i} \times OH \tag{3}$$

4. Production Targets

$$TP_{it} = BLC_{it} - AO_{it} - SS_i + STC_{it} + IMP_{it} \tag{4}$$

5. Processing Time Duration

$$PT_i = \left[\left(\left|\frac{TP_{it}}{OP_i \times (MC_{it} + AMC_{it})}\right|\right) + (FP_i - 1)\right] \tag{5}$$

6. Total Material Requirements

$$TRM_{jt} = \sum_{i=1}^{I} \left(WC_{ij} \times \left(BLC_{ijt} - SS_{ijt} + STC_{ijt} + IMP_{ijt}\right)\right) \tag{6}$$

7. Material Availability

$$\Delta RM_{jt} = SRM_{jt} - (TRM_{jt} \times -1) \tag{7}$$

8. Conversion of Additional Orders to Raw Material

$$AORM_{ijt} = WC_{ij} \times AO_{it} \tag{8}$$

Based on the eight models above, the authors determine the cut off value used as the basis for decision making. The cut off value can be seen as follows.

- The Decision of Processing Time Duration
  - Orders accepted due to overstock

$$TP_{it} \geq 0\text{, or} \tag{9}$$

  - Orders accepted, if the production processing time is less than the remaining working days

$$PT \leq SH \tag{10}$$

- The Decision of Material Availability
  - Orders are accepted if the result of the conversion of part number to raw material is less than the remaining raw material (Table 1).

$$AORM_{ijt} \leq \Delta RM_{jt} \tag{11}$$

Table 1 Notations used in developing model

Notations			
i	Part number type	SD_{it}	Delivery quantity (unit)
j	Raw material type	PO_{it}	Purchase order quantity (unit)
t	Time of decision making	BLC_{it}	Balance part number (unit)
SS_{it}	Safety stock part number (unit)	OP_i	Production output (unit)
FO_{it}	Forecast order (unit)	CT_i	Cycle time (sekon)
OH	Working hour (hour)	SH	Remaining working day effective (day)
AO_{it}	Additional order quantity (unit)	STC_{it}	Part number stock (unit)
IMP_{it}	Imported part number (unit)	MC_{it}	Number of running machine (unit)
TP_{it}	Production target (Unit)	AMC_{it}	Additional machine (unit)
FP_i	Number of flow process	PT_i	Processing time duration (day)
WC_{ij}	Weight consume (kg)	TRM_{jt}	Total raw material requirement (kg)
SRM_{jt}	Stock raw material (kg)	$AORM_{ijt}$	Raw material for additional order (kg)
ΔRM_{jt}	Remaining raw material (kg)		

4.3 Architecture System

This section explains the design system of JAO. The architecture system of JAO described through UML (*Unified Modelling Language*). Three diagrams are used to explain the system architecture of the JAO system namely use case diagram and activity diagram.

Use case diagram describes the functions contained in the JAO system. Figure 4 explains that the JAO system has five main menus, namely master data, additional PO, Judgment raw material, loading machine and history. The master data is used to view or update data in JAO Process. The additional PO is a menu used to input the additional order requests. The judgement raw material menu contains details of the calculation seen from the availability of materials. Otherwise, the loading machine menu contains details of the calculation seen from the processing time duration. The history menu contains decision regarding acceptance or rejection of additional order requests.

Activity Diagram describes the flow of activities that can be carried out in the system from start to finish. The following is an activity diagram created in the JAO system design (Fig. 5).

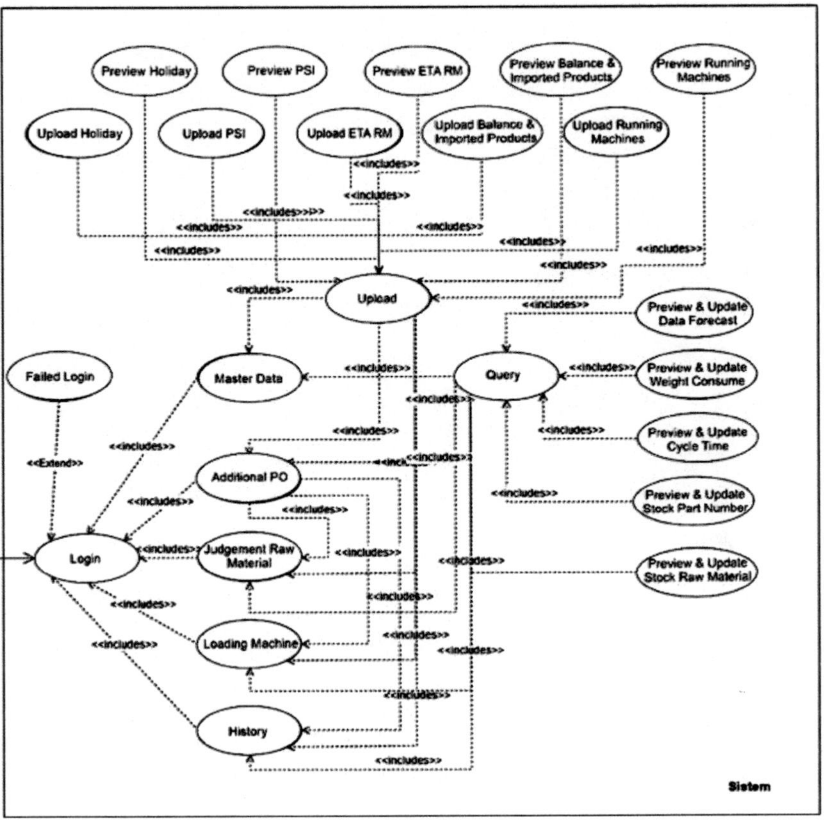

Fig. 4 Use Case Diagram JAO

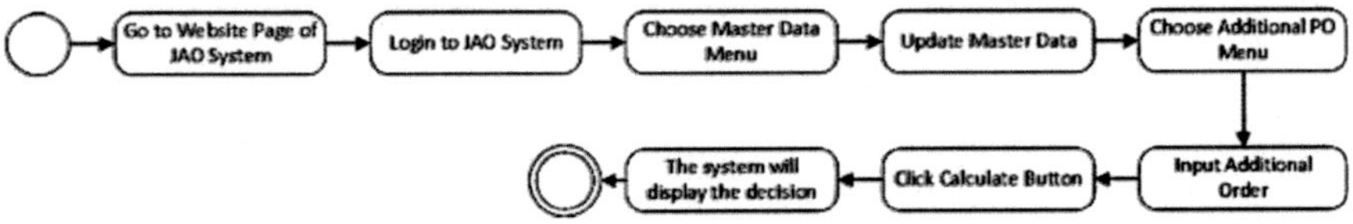

Fig. 5 Activity diagram JAO

4.4 User Interface Design

This section displays user interface of the JAO system that has been created. The displays can be seen as shown below.

Figure 6 is the first display when the decision maker accesses the website from the JAO system. Decision maker are asked to enter a username and password if they want to enter the system.

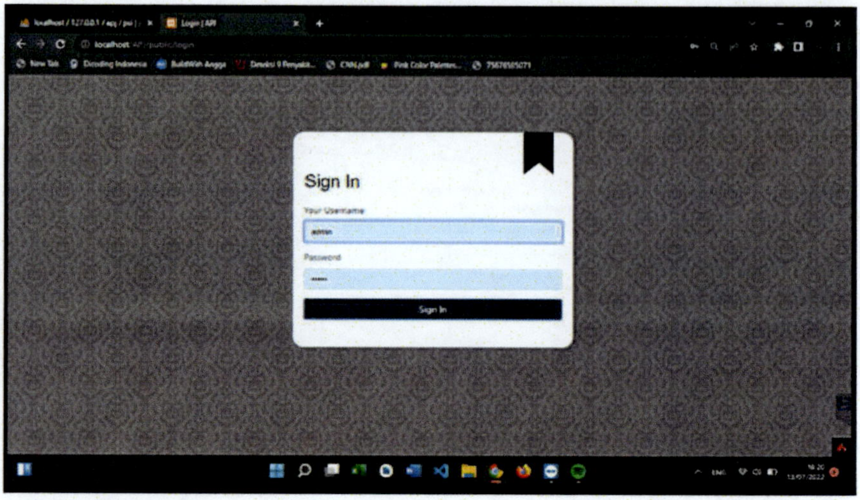

Fig. 6 Display of login page JAO

Figure 7 describes the main menu of JAO System. There are five menus of JAO System including Master data, Additional PO, Loading Machine, Judgement RM, and History.

Figure 8 describes some of the data obtained directly from the company's server. These data include forecasts, weight consume, flow process, cycle time, all stock raw materials, and stock each products or part numbers.

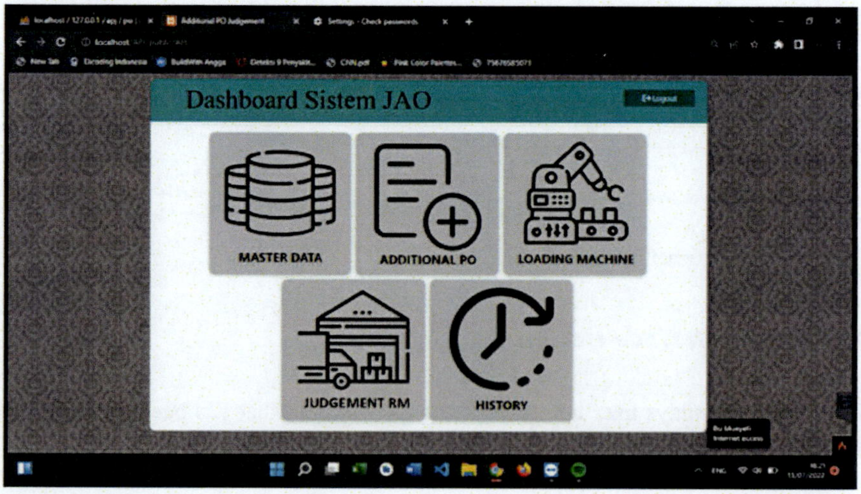

Fig. 7 Display of dashboard page JAO

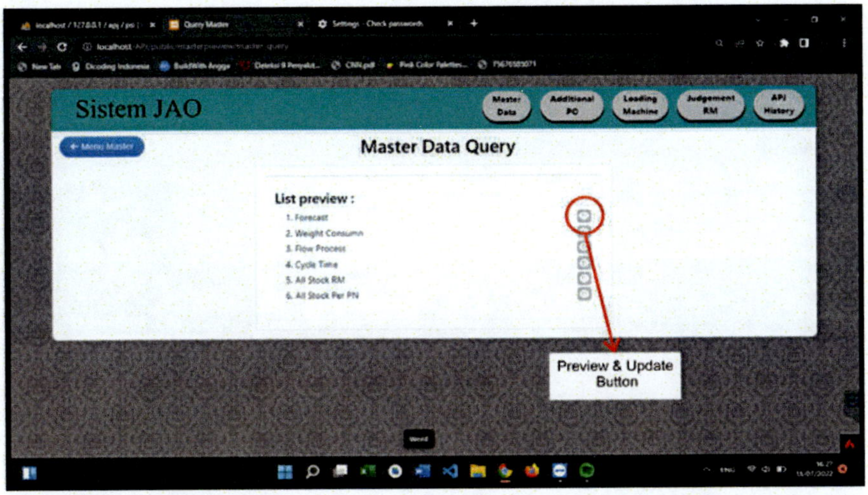

Fig. 8 Master data by query

Figure 9 describes the external data outside the company's server. This data include PSI Epson, ETA Raw Material, Balance SAP and Imported Parts, Running Machine and holiday list.

Figure 10 describes the display of JAO system when the decision maker wants to input an additional order request.

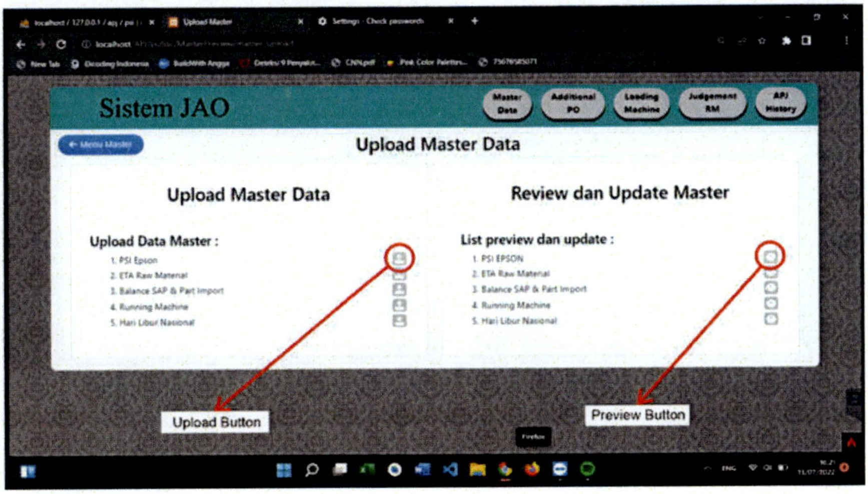

Fig. 9 Master data by uploaded

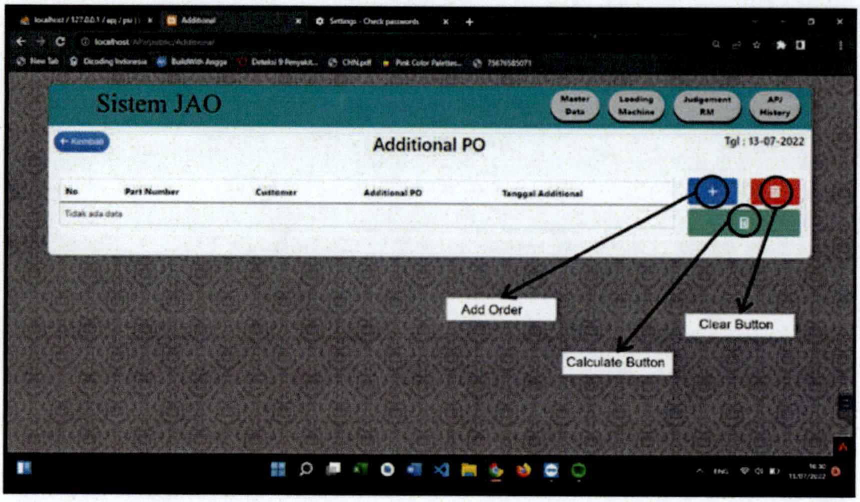

Fig. 10 Display of additional PO page

Figure 11 describes the decisions that have been made by the system. Two decisions are made based on processing time duration and the availability of materials.

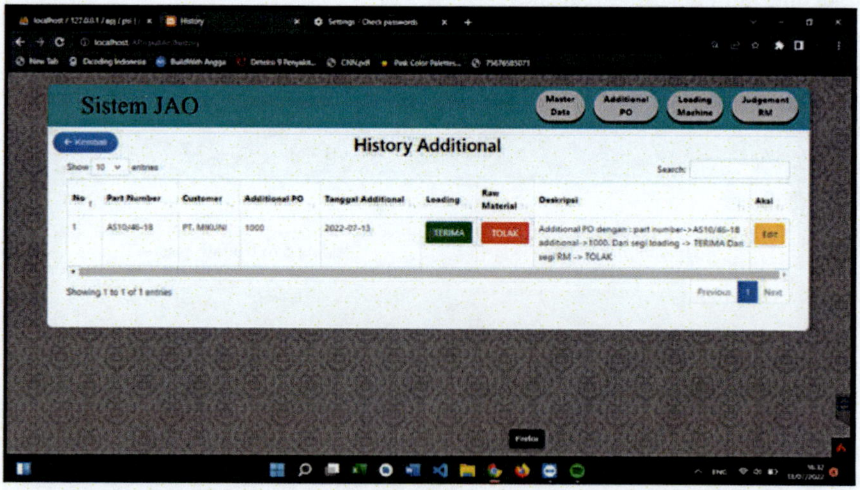

Fig. 11 Display of history additional

5 Conclusion

This research concluded that the development of a web-based decision support system to solve the JAO problem was successful. The JAO system successfully implemented in the company, besides that the data quality on this system is more real time because it is directly integrated to the company server. The test results on all functions of the system using black-box are successful and in accordance with expectations. The drawback of the system is no consideration regarding the defective products rate, besides that the forecast data used in the calculations only refers to one month ahead. The suggestions for further research, the system can be integrated with the material arrival prediction data.

References

1. Sodikin, Mashuri A (2012) Penjadwalan Produksi pada Sistem Manufaktur Repetitive Make to Order Flow Shop melalui Pendekatan Theory of Constraints 4(2)
2. Firzatullah RM (2019) Development of XYZ University's student admission site using waterfall method. J Mantik 3(2):10–19
3. Oluwatosin HS (2014) Client-server model. IOSR J Comput Eng 16(1):57–71. https://doi.org/10.9790/0661-16195771
4. Pribadi D, Saputra RA, Hudin JM, Gunawan (2020) Sistem Pendukung Keputusan
5. Wahid A (2020) Analisis metode waterfall untuk pengembangan sistem informasi. *J Ilmu-ilmu Inform dan Manaj STMIK* November:1–5
6. Heriyanti F, Ishak A (2020) Design of logistics information system in the finished product warehouse with the waterfall method: review literature. In: IOP conference series: materials science and engineering, vol 801, no. 1. https://doi.org/10.1088/1757-899X/801/1/012100
7. Bora, Ahmed S (2019) Mathematical modeling: an important tool for mathematics teaching. Int J Res Anal Rev 6(2):252–256
8. Raharjo B (2015) Self-taught learn codeigniter framework. Informatics Bandung, Bandung

Optimization Model for Perishable Product with Different Payment Schemes and Carbon Tax Regulation

Shafa Keysa Rinjani Hananta, Cucuk Nur Rosyidi, and Wakhid Ahmad Jauhari

Abstract Inventory is crucial in any business. Shortages or overstock may occur if it is not well planned. Companies that sell perishable goods may also experience this issue. When it comes to perishable goods, demand declines as they get older, resulting in frequent overstock. This extra inventory cannot be resold, resulting in a significant amount of food waste, which can increase carbon emissions and disposal cost. To solve this problem, an inventory optimization model in the form of Economic Ordering Quantity (EOQ) for perishable products is developed taking into account types of payments, carbon tax regulation, and the level of demand that depends on the age and price of the product. This study is built from the perspective of the retailer only. The model is developed for three payment schemes, which are cash payment, advance payment, and credit payment In the developed model, carbon emissions are assumed to be generated from three activities, namely, ordering, material handling, and storage. Additionally, a price reduction policy is implemented to increase the demand. The objective function of the model is to maximize the total profit. The proposed model is solved with analytical approach with the help of Mathematica 11.2 software. The model is applied to a numerical sample and a sensitivity analysis for validation. From the result of numerical example, it can be seen that the most profitable payment scheme is advanced payment.

Keywords EOQ · Payment schemes · Perishable product

1 Introduction

In a company, inventory is an idle resource but its existence cannot be avoided [1]. Therefore, inventory in a company must be optimized. The advantage of having inventory is that if there is a surge in demand at any time, the company can still meet

S. K. R. Hananta (✉) · C. N. Rosyidi · W. A. Jauhari
Department of Industrial Engineering, Universitas Sebelas Maret, Surakarta, Indonesia
e-mail: shafakeysa3@student.uns.ac.id

C. N. Rosyidi et al. (eds.), *Proceedings of the 6th Asia Pacific Conference on Manufacturing Systems and 4th International Manufacturing Engineering Conference*, Lecture Notes in Mechanical Engineering,
https://doi.org/10.1007/978-981-99-1245-2_25

the demand. However, if inventory is not carefully planned, there is a possibility of overstock, resulting in a buildup of raw materials, and storage costs will also increase.

Companies that manufacture or sell perishable goods also experience inventory issues. Products that perish and lose quality and value over time due to deterioration are referred to as perishables. As a product approaches its expiration date, demand for perishable goods gradually declines until it is almost zero. The company's production of a significant volume of food waste as a result of the excess inventory may result in disposal costs and have an adverse effect on the sustainability of the environment [2]. Controlling inventories for perishable goods is one method for reducing food waste. By calculating the ideal amount of orders for a product, inventory control can be accomplished.

Controlling a product's inventory can be done in a variety of ways. Economic Order Quantity (EOQ) is a widely used approach for determining the optimal order quantity of a product [3]. Harris [4] was the first to introduce EOQ. Furthermore, Ghare established the inventory model for perishable products in 1963 [5]. Then, Dobson et al. [6] created an EOQ model for perishable products that takes into account demand that varies depending on the product's age and level of freshness. Diaz et al. [7] also propose a model that takes into account the amount of demand, which varies depending on the product's age and selling price.

According to Dobson et al. [8], the level of product freshness must also be considered when developing models for efficient inventory control. As a product ages, its demand decreases linearly. The product's age can also influence the selling price [7]. In their study, Diaz et al. [7] made the assumption that the age of the product and the selling price affect the demand rate. The demand for the product will decline as it gets older. As a result, dynamic pricing should be adopted in order to boost product demand. One of the researchers who took payment into account when developing the EOQ model was Harris [4]. In the meantime, Goyal [8] developed an EOQ model by taking credit payments into account. Chang et al. [9] recently combined previous models to assess the impact of advance-cash-credit (ACC) payment on an inventory model. He et al. [10] studied the effects of carbon emission regulations, such as carbon taxes and cap-and-trade, on lot size and optimal emission. Shi et al. [11] then created an EOQ model for perishable products that takes into account payment aspects as well as carbon emissions.

The EOQ model has so far been developed by taking into account a wide range of factors, such as carbon emission, payment schemes, dynamic pricing, and demand rate that is dependent. However, there hasn't been any research on the development of an EOQ model for perishable goods that simultaneously takes into account the demand rate, which depends on the product's age and selling price, dynamic pricing, carbon emissions, and payment methods. These four factors were therefore taken into account in this development of an EOQ model for perishable goods based on studies by Shi et al. [11], Dobson et al. [6], and Diaz et al. [7]. The position of this research can be seen in Table 1.

Table 1 Research position

	Diaz et al. [7]	Dobson et al. [6]	Shi et al. [11]	This paper
Delay in payment	–	–	✓	✓
Cash payment	–	–	✓	✓
Advance payment	–	–	✓	✓
Carbon tax	–	–	✓	✓
Demand dependant on age	✓	✓	–	✓
Demand dependant on selling price	✓	✓	–	✓
Selling price dependant on age	✓	✓	–	✓
Optimum cycle time	✓	–	✓	✓
Objective function	Profit maximization	Profit maximization	Cost minimization	Profit maximization

2 Notation and Assumption

The following notation are used in the entire paper (Table 2).

The following assumptions are derived to formulate the inventory model. (1) No shortage is allowed, (2) the selling price of products depends on the age of the product, (3) product demand rate depends on the price and age of the product, (4) at the end of the cycle, all products are sold out, and (5) replenishment cycle time is instantaneous and lead time is negligible.

3 Mathematical Model

The demand rate function in this model depends on the age and selling price. We assume a constant price elasticity of demand α > 0 and thus, the demand rate at selling price P and age x is given by:

$$D(p, x) = D_o\left[\frac{P}{P_o}\right]^{-\alpha}\left[1 - \frac{x}{m}\right], \ \alpha > 0 \tag{1}$$

Regarding the price discount policy, we propose that the price reduction due to aging should follow a similar pattern to the demand reduction, using a paramater $\gamma \geq 0$ to control the speed of the price reduction, that is

Table 2 Notation used in developing the model

Notation	Meaning
Parameter	
c_h	Holding cost (excluding interest) (IDR/kg)
S	Amount of carbon emissions in inventory (kg)
c_p	Purchasing cost (IDR/kg)
E	Amount of carbon emission associated per Kg unit purchased (kg)
D_o	Demand (kg)
P_o	Selling price (IDR/kg)
I_c	Interest charged per unit time
I_e	Interest earned per unit time
K	Ordering cost (IDR)
G	Amount of carbon emission per order (kg)
L	Advanced paymnet period (month)
m	Maximum shelf life of the product, $m > 0$ (month)
M	Credit period of time granted, $M > 0$ (month)
N	Number of equal installments in an advanced payment
r_2	Rate of purchasing price discount for an advanced payment (%)
T	Tax paid on each carbon unit emitted (IDR/kg)
X	Age of the product (month)
Function	
Q	Order quantity (kg)
CT	Total carbon tax (IDR)
$I(x)$	The inventory level at age x (kg)
$D(p,x)$	Demand rate at price p and age x (kg)
$D(x)$	Demand rate function (kg)
$P(x)$	Product selling price at age x (IDR/kg)
TC	Total relevant cost (IDR/month)
TR	Total revenue (IDR/month
Decision variable	
T	Optimum cycle time (month)
Objective function	
TP	Total profit (IDR/month

$$P(x) = P_o\left[1 - \frac{x}{m}\right]^{\gamma}, \gamma \geq 0 \quad (2)$$

Replacing (2) in (1), we arrive at the following demand function:

$$D(x) = D_o\left[1 - \frac{x}{m}\right]^{1-\alpha\gamma} \quad (3)$$

From function (3), when $\gamma = 1/\alpha$, the demand is constant for any age and γ is not allowed to exceed $1/\alpha$ because in this scenario the demand will increase as the product is older where it requires an extreme discount to be applied.

The order quantity per replenishment cycle time T is as follows:

$$Q = \int_0^T D_o\left[1 - \frac{x}{m}\right]^{1-\alpha\gamma} dx \tag{4}$$

The inventory level I(x) at age x during the replenishment cycle [0, T] is given by:

$$I(x) = \int_t^T D_o\left[1 - \frac{x}{m}\right]^{1-\alpha\gamma} dx \tag{5}$$

The ordering cost per cycle is:

$$OC = K \tag{6}$$

The holding cost per cycle time T is given by:

$$HC = c_h \int_0^T I(x)dt = c_h \int_0^T \int_t^T D_o\left[1 - \frac{x}{m}\right]^{1-\alpha\gamma} dxdt \tag{7}$$

The total amount of carbon emissions per replenishment cycle includes the fixed carbon emission associated with placing an order, the variable amount of carbon emission associated with each unit purchased, and the amount of carbon emission associated with the storage of each unit held. Therefore, the total amount of carbon emissions per replenishment cycle time T is given by:

$$CT = \tau\left(G + E.\int_0^T D_o\left[1 - \frac{x}{m}\right]^{1-\alpha\gamma} dx + S\int_0^T \int_t^T D_o\left[1 - \frac{x}{m}\right]^{1-\alpha\gamma} dxdt\right) \tag{8}$$

The total revenue, TR, for this problem can be defined as:

$$TR = \int_0^T P_o D_o\left[1 - \frac{x}{m}\right]^{1-\alpha\gamma+\gamma} \tag{9}$$

3.1 The Case of Cash Payment

In this case, *retailer* pays for the item upon receipt of them. Hence, the total acquisition cost (AC_1) is as follow:

$$AC_1 = c_p Q = c_p \int_0^T D_o\left[1 - \frac{x}{m}\right]^{1-\alpha\gamma} dx \tag{10}$$

The interest charged IC_1 per cycle is given by:

$$IC_1 = I_c c_p \int_0^T \int_t^T D_o\left[1 - \frac{x}{m}\right]^{1-\alpha\gamma} dxdt \tag{11}$$

Combining the result from (6) to (11), the total profit per unit time is given by:

$$\begin{aligned} TP_1 = \frac{1}{T}\Bigg(&\int_0^T PoDo\left[1 - \frac{x}{m}\right]^{1-\alpha\gamma+\gamma} dx \\ &-\left((K + \tau G) + (c_p + \tau E)\int_0^T Do\left[1 - \frac{x}{m}\right]^{1-\alpha\gamma} dx\right. \\ &\left.+(c_h + \tau S + I_c c_p)\int_0^T \int_t^T D_o\left[1 - \frac{x}{m}\right]^{1-\alpha\gamma} dxdt\right) \end{aligned} \tag{12}$$

3.2 The Case of Advanced Payment

In this case, the retailer will pay the acquisition cost (AC_2) in n equal installments over L months prior to delivery. In general, suppliers will give a price discount r_2 to retailers as an incentive to get them to pay in advance. Hence, the total acquisition cost (AC_2) is given by:

$$AC_2 = (1 - r_2)c_p Q = (1 - r_2)c_p \int_0^T D_o\left[1 - \frac{x}{m}\right]^{1-\alpha\gamma} dx \tag{13}$$

The capital cost CC_1 per cycle is as follow:

$$CC_1 = \frac{1+n}{2n} I_c L(1-r_2)c_p \int_0^T \int_t^T D_o\left[1-\frac{x}{m}\right]^{1-\alpha\gamma} dxdt \tag{14}$$

The interest charged IC_1 per cycle is given by:

$$IC_2 = I_c(1-r_2)c_p \int_0^T \int_t^T D_o\left[1-\frac{x}{m}\right]^{1-\alpha\gamma} dxdt \tag{15}$$

Combining the results from (6) to (9) and (13) to (15), the total profit per unit time is given by:

$$\begin{aligned} TP_2 = \frac{1}{T}\Bigg[& \int_0^T PoDo\left[1-\frac{x}{m}\right]^{1-\alpha\gamma+\gamma} dx - (K+\tau G) \\ & - \left[(1-r_2)c_p + \tau \mathrm{E} + \frac{1+n}{2n}\mathrm{I}_c L(1-r_2)c_p\right] \int_0^T Do\left[1-\frac{x}{m}\right]^{1-\alpha\gamma} \\ & dx - \left[c_h + \tau \mathrm{S} + \mathrm{I}_c(1-r_2)c_p\right] \int_0^T \int_t^T D_o\left[1-\frac{x}{m}\right]^{1-\alpha\gamma} dxdt \Bigg] \end{aligned} \tag{16}$$

3.3 The Case of Credit Payment

In this case, retailer receives a fixed credit period (M) from supplier. It is clear that the acquisition cost per replenishment cycle time T is:

$$AC_3 = c_p Q = c_p \int_0^T D_o\left[1-\frac{x}{m}\right]^{1-\alpha\gamma} dx \tag{17}$$

In this payment scheme, there are two potential cases based on the value of credit period (M) and replenishment cycle time (T).

Sub-case 1 (M ≤ T)

In this case, retailer needs to pay interest for the items in stock after time M. Hence, the interest charged per cycle is as follows:

$$IC_{31} = I_c c_p \int_M^T \int_t^T D_o \left[1 - \frac{x}{m}\right]^{1-\alpha\gamma} dxdt \tag{18}$$

From age 0 until M, retailer could sell their product without paying anything to supplier. Hence, the interest earned per cycle is as follow:

$$IE_{31} = I_e \int_0^M PoDo(M - t)\left[1 - \frac{x}{m}\right]^{1-\alpha\gamma} dx \tag{19}$$

The total profit for this sub-case is:

$$\begin{aligned} TP_{31} = \frac{1}{T}\Bigg[& \int_0^T PoDo\left[1 - \frac{x}{m}\right]^{1-\alpha\gamma+\gamma} dx \\ & -(K + \tau G) - (c_p + \tau E) \int_0^T Do\left[1 - \frac{x}{m}\right]^{1-\alpha\gamma} dx \\ & - (c_h + \tau S) \int_0^T \int_t^T D_o\left[1 - \frac{x}{m}\right]^{1-\alpha\gamma} dxdt \\ & - I_c c_p \int_M^T \int_t^T D_o\left[1 - \frac{x}{m}\right]^{1-\alpha\gamma} dxdt \\ & + I_e \int_0^M PoDo\left[1 - \frac{x}{m}\right]^{1-\alpha\gamma+\gamma} (M - x)dx \Bigg] \end{aligned} \tag{20}$$

Sub-case 2 (M ≥ T)

In this case, the replenishment cycle time (T) is shorter than the credit period. Hence, the interest charged per cycle (IC_{32}) is 0. While the interest earned per cycle is given by:

$$IE_{32} = I_e \int_0^T PoDo(T - x)\left[1 - \frac{x}{m}\right]^{1-\alpha\gamma+\gamma} dx - I(0)(M - T) \tag{21}$$

The total profit for this sub-case is:

$$
\begin{aligned}
TP_{32} = \frac{1}{T}\Bigg[& \int_0^T PoDo\left[1-\frac{x}{m}\right]^{1-\alpha\gamma+\gamma} dx - (K+\tau G) \\
& - \left(c_p+\tau E\right)\int_0^T Do\left[1-\frac{x}{m}\right]^{1-\alpha\gamma+\gamma} dx \\
& - \left(c_h+\tau S\right)\int_0^T\int_t^T D_o\left[1-\frac{x}{m}\right]^{1-\alpha\gamma} dxdt \\
& + I_e\int_0^T Po\left[1-\frac{x}{m}\right]^{\gamma}\left(\int_0^T Do(T-x)\left[1-\frac{x}{m}\right]^{1-\alpha\gamma} dx + I(0)(M-T)\right)\Bigg]
\end{aligned}
\tag{22}
$$

4 Numerical Example and Analysis

In this section, we consider the following examples to illustrate the proposed model. The optimal solution obtained by using numerical example run by Mathematica 11.2. The parameter values in the numerical example were taken from Shi et al. [11] and Andini [12].

Example 1 For a cash payment, consider an inventory system where c_h = IDR100/kg, $S = 2.5$/kg, c_p = IDR4000/kg, $E = 5$/kg, D_o = 1000 kg, $\alpha = 1$, $\gamma = 0.5$, P_o = 8000/kg, $I_c = 10\%$, $K = 100.000$, $G = 250$ kg, $m = 2$ months, $\tau = 30$/kg. By using Mathematica 11.2, the optimal solution obtained for this model are $T = 0.249955$ month, $Q = 242$ kg, and TP = IDR2,983,720.

Example 2 For an advanced payment, in addition to those parameters in Example 1, we assume $L = 0.17$, $n = 3$, and $r_2 = 50\%$. By using Mathematica 11.2, the optimal solution obtained for this model are $T = 0.238562$ month, $Q = 232$ kg, and TP = IDR4,922,820.

Example 3 For a credit payment with sub-case where $M \le T$, in addition to those parameters in Example 1, we assume $I_e = 5\%$ and $M = 0.3$ month. By using Mathematica 11.2, the optimal solution obtained for this model are $T = 0.422589$ months, $Q = 400$ kg, and TP = IDR2,514,840.

Example 4 For a credit payment with sub-case where $M \geq T$, the numerical example used is the same with Example 3. By using Mathematica 11.2, the optimal solution obtained for this model are $T = 0.259542$ month, $Q = 251$ kg, and $TP =$ IDR3,048,960.

5 Sensitivity Analysis

In this section, we will discuss the influence of changes in the key parameters of the system on the optimum cycle time (T), order quantity (Q), and total profit (TP). Computational results are presented in Tables 3 and 4.

From the computational results shown in Table 2, we observe that the optimum cycle time (T) in both models will reduce as demand rate (D_o), selling price (P_o), and speed of the price reduction (γ) increase, while it will increase when tax (τ), maximum shelf life (m), and price elasticity of demand (α) gradually increase. Meanwhile, as the selling price (P_o) and speed of the price reduction (γ) rise, the order quantity (Q) will decline. However, it will rise in response to the rising of tax (τ), maximum shelf life (m), and price elasticity of demand (α). Additionally, as tax (τ) and speed of the price reduction (γ) increase, the overall profit (TP) will decrease. On the other hand, as the demand rate (D_o), selling price (P_o), tax (τ), and price elasticity of demand (α) increase, the overall profit will increase as well.

For the model with credit payment, where $M \leq T$, it can be seen from the computational results in Table 3 that the optimum cycle time (T) will increase when tax (τ), maximum shelf life (m), and price elasticity of demand (α) increase, while it will decrease when demand rate (D_o), selling price (P_o), and speed of the price reduction (γ) decrease. Meanwhile, as the selling price (P_o) and speed of the price reduction (γ) rise, the order quantity (Q) will decline. However, it will rise in response to the rising of tax (τ), maximum shelf life (m), and price elasticity of demand (α). Additionally, as tax (τ), maximum shelf life (m), and speed of the price reduction (γ) increase, the overall profit (TP) will decrease. On the other hand, as the demand rate (D_o), maximum shelf life (m), and price elasticity of demand (α) increase, the overall profit will increase as well.

Meanwhile, for the case when $M \geq T$, we can see the optimum cycle time (T) will decrease as demand rate (D_o), selling price (P_o), and price elasticity of demand (α) increase, whereas it will increase as tax (τ), maximum shelf life (m), and speed of the price reduction (γ) increase. The order quantity (Q) will decrease as the selling price (P_o) and speed of the price reduction (γ) increase It will, however, rise in response to increase in tax (τ), maximum shelf life (m), and price elasticity of demand (α). The overall profit (TP) will also decline as tax (τ) and price elasticity of demand (α) rise.On the other hand, as the demand rate (D_o), selling price (P_o), maximum shelf life (m), and speed of the price reduction (γ) rise, so will the overall profit (TP).

Table 3 Sensitivity analysis for cash and advanced payment model

Parameters	Changes	Cash payment			Advance payment		
		T	Q	TP	T	Q	TP
D_o	700	0.2996	201.69	1,971,230	0.285565	192.58	3,322,900
	800	0.2800	214.94	2,305,820	0.266962	206.28	3,853,190
	900	0.2637	229.32	2,643,490	0.25157	219.14	4,386,670
	1000	0.2500	241.97	2,983,720	0.238562	231.29	4,922,820
	1100	0.2382	254.01	3,326,140	0.22738	242.87	5,461,250
	1200	0.2279	265.51	3,670,450	0.217633	253.92	6,001,650
	1300	0.2188	276.54	4,016,430	0.209039	264.52	6,543,770
P_o	5600	0.3115	299.03	750,085	0.289491	278.75	2,679,750
	6400	0.2861	275.62	1,490,430	0.269035	259.77	3,423,970
	7200	0.2662	257.11	2,235,280	0.252416	244.28	4,171,880
	8000	0.2500	241.97	2,983,720	0.238562	231.29	4,922,820
	8800	0.2364	229.28	3,735,120	0.226778	220.22	5,676,320
	9600	0.2249	218.44	4,489,020	0.216593	210.62	6,432,000
	10,400	0.2149	209.02	5,245,060	0.207674	202.18	7,189,600
τ	21	0.2476	239.78	3,039,010	0.236315	229.19	4,978,500
	24	0.2484	240.51	3,020,570	0.237068	229.9	4,959,930
	27	0.2492	241.24	3,002,140	0.237816	230.6	4,941,370
	30	0.2500	241.97	2,983,720	0.238562	231.29	4,922,820
	33	0.2507	242.7	2,965,300	0.239304	231.99	4,904,280
	36	0.2515	243.42	2,946,900	0.240043	232.69	4,885,750
	39	0.2520	243.9	2,934,630	0.240779	233.38	4,867,220
m	1.4	0.2147	206.28	2,840,830	0.202662	195.14	4,762,380
	1.6	0.2275	219.2	2,897,720	0.215538	208.11	4,826,090
	1.8	0.2392	231.04	2,944,480	0.227455	220.11	4,878,610
	2	0.2500	241.97	2,983,720	0.238562	231.29	4,922,820
	2.2	0.2600	252.12	3,017,220	0.248974	241.79	4,960,680
	2.4	0.2693	261.59	3,046,210	0.25878	251.67	4,993,550
	2.6	0.2780	270.46	3,071,590	0.268051	261.02	5,022,410
α	0.7	0.2422	232.51	2,952,380	0.230949	222.82	4,875,520
	0.8	0.2447	235.55	2,962,690	0.233223	224.26	4,890,990
	0.9	0.2473	238.7	2,973,130	0.234656	226.95	4,906,760
	1	0.2500	241.97	2,983,720	0.238562	231.3	4,922,820
	1.1	0.2527	245.37	2,994,450	0.242686	235.9	4,939,210
	1.2	0.2556	249.91	3,005,340	0.247048	240.79	4,955,930
	1.3	0.2586	252.59	3,016,380	0.251672	245.97	4,973,020

(continued)

Table 3 (continued)

Parameters	Changes	Cash payment			Advance payment		
		T	Q	TP	T	Q	TP
γ	0.35	0.2627	251.34	3,010,560	0.244153	234.32	4,942,890
	0.4	0.2583	248.1	3,010,560	0.242261	233.31	4,936,180
	0.45	0.2540	244.98	2,997,060	0.240397	232.3	4,929,490
	0.5	0.2500	241.97	2,983,720	0.238562	231.3	4,922,820
	0.55	0.2460	239.07	2,970,540	0.234975	229.01	4,909,560
	0.6	0.2423	236.26	2,957,510	0.234656	228.62	4,906,760
	0.65	0.2387	233.54	2,944,620	0.233223	228.34	4,902,970

6 Conclusions

In this study, we proposed an EOQ model for perishable products with age- and price-dependent demand rates under various payment schemes and carbon tax regulation. The objective is to maximize profit, and the decision variable is replenishment cycle time. The order quantity can also be determined with this model. According to the numerical example, the advance payment scheme is more profitable than the other payment schemes. This is because the retailer obtained a price reduction from the supplier as an incentive in this scenario. As a result, compared to the other payment option, the retailer's overall relevant cost is lower.

Future research may expand this study in various directions. Other types of carbon-tax regulation (e.g., cap-and-trade regulation, carbon cap regulation, etc.) and the use of multi-criteria decision analysis to reduce both total relevant cost and carbon emission simultaneously could be an immediate viable extension. Finally, while the proposed model is developed solely from the retailer's perspective, this research can be expanded in the future to include both the supplier and retailer perspectives.

Table 4 Sensitivity analysis for credit payment (Model 3.1 and Model 3.2)

Parameters	Changes	M ≤ T			M ≥ T		
		T	Q	TP	T	Q	TP
D_o	700	0.454461	299.3	1,686,840	0.306912	206.37	2,020,450
	800	0.441434	332.88	1,962,100	0.288274	222.1	2,360,690
	900	0.431057	366.23	2,238,170	0.272746	236.9	2,703,680
	1000	0.422589	399.41	2,514,840	0.259542	250.93	3,048,960
	1100	0.415545	432.46	2,791,980	0.248131	264.29	3,396,210
	1200	0.40959	465.41	3,069,480	0.238137	277.08	3,745,150
	1300	0.404488	498.28	3,347,280	0.229287	289.36	4,095,570
P_o	5600	0.533534	496.19	392,560	0.326388	312.68	818,213
	6400	0.487235	456.23	1,092,290	0.298582	287.14	1,557,040
	7200	0.451436	424.91	1,800,290	0.276973	267.15	2,300,990
	8000	0.422589	399.41	2,514,840	0.259542	250.93	3,048,960
	8800	0.398652	378.07	3,234,760	0.245089	237.42	3,800,210
	9600	0.378345	359.84	3,959,200	0.232847	225.93	4,554,170
	10,400	0.360817	344.02	4,687,510	0.222302	216	5,310,450
τ	21	0.403849	382.72	2,625,270	0.2573	248.84	3,103,960
	24	0.410189	388.38	2,588,140	0.25805	249.54	3,085,620
	27	0.416434	393.94	2,551,340	0.2588	250.24	3,067,290
	30	0.422589	399.41	2,514,840	0.25954	250.93	3,048,960
	33	0.428658	404.79	2,478,630	0.26028	251.62	3,030,640
	36	0.434645	410.1	2,442,710	0.26102	252.31	3,012,330
	39	0.440552	415.32	2,407,050	0.26175	252.99	2,994,030
m	1.4	0.25624	244.13	2,760,020	0.223447	217.08	2,902,700
	1.6	0.301504	286.82	2,705,810	0.236711	229.56	2,961,030
	1.8	0.356893	338.57	2,621,630	0.248681	240.78	3,008,890
	2	0.422589	399.41	2,514,840	0.259542	250.93	3,048,960
	2.2	0.498488	469.08	2,391,100	0.269442	260.15	3,083,080
	2.4	0.584383	547.22	2,254,600	0.2785	268.57	3,112,510
	2.6	0.680066	633.44	2,108,390	0.286818	276.27	3,138,190
α	0.7	0.410744	382.62	2,488,150	0.273381	261.03	3,091,454
	0.8	0.414572	388.02	2,497,441	0.268584	257.56	3,077,113
	0.9	0.418518	393.61	2,506,084	0.263974	254.19	3,062,952
	1	0.422589	399.41	2,514,840	0.259542	250.93	3,048,960
	1.1	0.426792	403.14	2,523,708	0.255279	247.77	3,035,138
	1.2	0.431132	411.66	2,532,696	0.251177	244.7	3,021,475
	1.3	0.435619	418.15	2,541,797	0.247228	241.73	3,007,971

(continued)

Table 4 (continued)

Parameters	Changes	M ≤ T			M ≥ T		
		T	Q	TP	T	Q	TP
γ	0.35	0.439897	407.58	2,546,287	0.250564	240.21	3,015,156
	0.4	0.433931	404.8	2,535,711	0.253446	243.64	3,026,258
	0.45	0.428165	402.08	2,525,231	0.256437	247.21	3,037,519
	0.5	0.422589	399.41	2,514,840	0.259542	250.93	3,048,960
	0.55	0.417193	396.8	2,504,537	0.262768	254.8	3,060,570
	0.6	0.411969	394.23	2,494,322	0.266125	258.84	3,072,359
	0.65	0.406910	391.72	2,484,187	0.269619	263.06	3,084,348

References

1. Nasution AH, Prasetyawan Y (2008) Perencanaan dan Pengendalian Produksi. Yogyakarta: Graha Ilmu
2. Lebersorger S, Schneider F (2014) Food loss rates at the food retail, influencing factors and reasons as a basis for waste prevention measure. Waste Manage 34(11):1911–1919
3. Dutta D, Pavan K (2012) Fuzzy inventory model without shortage using trapezoidal fuzzy number with sensitivity analysis. Int Org Sci Res-J Math 4:32–37
4. Harris FW (1913) How many parts to make at once. Factory : the magazine of management, 135–136, and 152
5. Ghare P (1963) A model for an exponential decaying inventory. J Ind Eng 14:238–243
6. Dobson G, Pinker EJ, Yildiz O (2017) An EOQ model for perishable goods with age-dependent demand rate. Eur J Oper Res 25:84–88
7. Díaz BA, Lozano S, Palacio A (2017) Effects Of Dynamic pricing of perishable products on revenue and waste. Appl Math Model 45:148–164
8. Goyal SK (1985) Economic order quantity under condition of permissible delay in payment. J Operat Res Soc 36:335–338
9. Chang CT, Ouyang LY, Teng JT, Lai KK, Cárdenas-Barrón LE (2019) Manufacturer'S pricing and lot-sizing decisions for perishable goods under various payment terms by a discounted cash flow analysis. Int J Prod Econ 218:83–95
10. He P, Zhang W, Xu X, Bian Y (2015) Production lot-sizing and carbon emissions under cap-and-trade and carbon tax regulations. J Clean Prod 103:241–248
11. Shi Y, Zhang Z, Chen SC, Cárdenas-Barrón LE, Skouri K (2019) Optimal replenishment decisions for perishable products under cash, advance, and credit payments considering carbon tax regulations. Int J Prod Econ 223:1–16
12. Andini SH (2021) Pengembangan model optimasi penentuan kuantitas pemesanan ekonomis untuk perishable product dengan mempertimbangkan faktor donasi. Thesis. Universitas Sebelas Maret, Surakarta

Segmentation of Potential Sustainable Tourism Based on Support System Facility Perspective

Riska Yanu Fa'rifah, Fandi Achmad, Lutfia Septiningrum, and Iwan Inrawan Wiratmadja

Abstract This study aims to determine the segmentation of sustainable tourism development planning based on support system facilities in Rembang Regency using statistical analysis methods. By testing the Kaiser-Meyer-Olkin (KMO) test using 20 tourism objects as a sample, the results of the KMO assumption have been fulfilled so that the segmentation analysis of the potential for sustainable tourism objects can use the K-Means Cluster for the selection of alternative segmentation analysis. The results obtained from this research segmentation of tourism potential are divided into 3 groups, namely the Sustainable Development Support segment in cluster 1, which consists of 12 tourism objects, which are tourist attractions whose routes are spread in all directions from the city center. The Popular Local Tourism segment in cluster 2, as many as 4 tourism objects, is a group whose all routes are to the south of the city center. While the last segment of new local tourism potential, as many as 4 tourism objects, is a group of tourist objects whose routes all go south from the city center. From each segmentation group, a sustainable tourism area development planning strategy will be made.

Keywords Segmentation tourism potential · K-Means · Support system facilities

R. Y. Fa'rifah (✉) · F. Achmad
Telkom University, Bandung, Indonesia
e-mail: riskayanu@telkomuniversity.ac.id

L. Septiningrum
University of Yogyakarta, DI Yogyakarta, Indonesia

I. I. Wiratmadja
Bandung Institute of Technology, Bandung, Indonesia

C. N. Rosyidi et al. (eds.), *Proceedings of the 6th Asia Pacific Conference on Manufacturing Systems and 4th International Manufacturing Engineering Conference*, Lecture Notes in Mechanical Engineering,
https://doi.org/10.1007/978-981-99-1245-2_26

1 Introduction

Increasing tourism destinations and investment, making the tourism sector a key factor in both state and regional income, job creation, business development, and infrastructure. This has prompted several regions to start looking to develop tourism areas as one of the leading sectors for improving the regional economy. The current development of tourism areas cannot be separated from the concept of sustainable development, which has become the global agenda of every development process [1]. Tourism takes full account of the current and future economic, social, and environmental impacts, responding to the needs of visitors, the tourism industry, and the environment. The development of tourism areas requires proper planning from several alternative planning options, and of course, these alternatives need planning priorities. The Evaluation of Sustainable Tourism Strategy by involving government stakeholders in determining priorities for evaluating strategic tourism procedures [2]. The priority of tourism development on the island of Lombok to determine potential tourism sector areas by referring to the document data of the West Nusa Tenggara Province government [3]. Based on the description above, this study conducted a potential segmentation analysis based on the Support System Facilities (SSF) factor. According to [4] Supporting Systems and Facilities are services developed either directly or indirectly to support and encourage tourism potential. The Facility Support System affects the long-term success of the tourism industry because the needs of tourists for carrying capacity and facilities reflect the development of tourism in the area [5, 6]. Support systems and facilities in tourism are related to individual motivation to visit a tourist destination [5, 7]. This is done with the aim of supporting the government of Rembang Regency in order to increase the potential of sustainable tourism. Considering tourism is one of the important sectors of economic growth in Rembang [8]. This tourism potential segmentation analysis uses the k-means clustering method. The k-means algorithm is one of the most popular clustering methods used because this method is the easiest to implement, and researchers can determine the number of clusters themselves [9]. In addition, k-means clustering is a non-hierarchical algorithm that gives better results when compared to hierarchical clustering [10, 11].

2 Methodology

The type of research used in this research is quantitative research. Quantitative research aims to determine and explain the results of clustering using K-Means in evaluating support system facilities for determining potential tourism segmentation. In the process of collecting data, the methods used are field surveys and literature studies. The field survey method was carried out by means of structured observations of 20 tourist attractions as a sample in April to July 2022. Observations were carried out through interviews with the managers of tourist attractions and the tourism office

Table 1 Research variable

Variable	Information	Scale	Description
X1	Operational hour	Interval	The period in which the tourist attractions operate every day
X2	Distance from City center	Interval	The distance between the city center and tourist attractions is seen on google maps
X3	Directions from City center	Nominal	Tourist destinations that indicate the location of tourist attractions. The central point is in the town square. The directions in question are East, North, South, and West
X4	Rating	Interval	A standardized assessment of tourist attractions, the rating is obtained from Google Rating
X5	Average length of visit	Interval	The average time tourists use to enjoy tourist attractions is obtained from interviews with tourist attractions

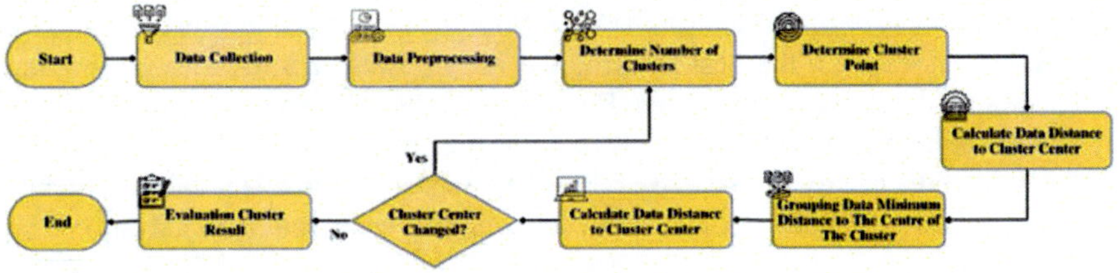

Fig. 1 Research flow evaluation support system facilities

of the Rembang district. In evaluating the variables in Table 1 used in this study, carried out with the steps that can be seen in Fig. 1.

The research variables used in this study can be described in Table 1.

K-means clustering is a prototype-based partitioning or data object clustering method that aims to group a number of clusters so that all members in a cluster are close to the appropriate prototype [12, 13]. K-means clustering begins by randomly selecting k cluster centre's or centroids. Each point in the dataset is assigned to the nearest centroid. Then recalculate the positions of all cluster centre's until no point can move from one cluster to another [9, 12, 14].

3 Result and Discussion

Segmentation analysis of tourism potential in this study uses the k-means clustering method based on the support system facility factor, which consists of operating hours of tourist objects, the distance of tourist objects from the city centre, direction from the city centre, rating results, and average visiting time. Before performing clustering, the assumption that must be fulfilled is that the sample taken must represent the population. In other words, the sample size of this study is sufficient for k-means clustering. Test the adequacy of the sample using the Kaiser-Meyer-Olkin (KMO) test with the results according to Table 2.

The KMO value obtained from all the variables used for the analysis there is only 1 variable whose value is below 0.5. Based on the KMO test, these variables do not meet the assumption of sample adequacy. However, the overall KMO test showed that the results were above 0.5 Based on Alabdulkarim [15], if the overall KMO test is above 0.5, then the sample size fulfilled. So that it can be said that overall, the sample used was sufficient to be used in the study and could represent the population, namely all tourism objects in Rembang Regency. The KMO assumption has been fulfilled so that a segmentation analysis of the potential for tourism objects using the k-means algorithm can be carried out. This study uses 2–6 number of clusters to obtain optimum segmentation results. The results of the segmentation analysis can be seen in Table 3.

The results of the optimum segmentation analysis are based on 3 measurements, namely the largest Dunn index [16], and the largest silhouette index [17–19]. In addition to these two criteria, it can also be seen from the smallest k-means connectivity [20]. Based on Table 3, the best result is the use of 3 clusters in the segmentation analysis using the k-means algorithm. Because the largest Dunn index and silhouette are in the use of these 3 clusters with index values of 0.43 and 0.26, respectively. In addition, connectivity on the use of 3 clusters is also the smallest value, which is 15.78. Based on the results of data processing, the best segmentation analysis can be visualized as shown in the following Fig. 2.

Based on Fig. 2, cluster 1 consists of tourist objects that mostly go east from the city center, namely Lasem District, with a distance of more than 3 h. The area is currently under development and is referred to as the "Kota Pusaka Lasem". Cluster 2 is the closest tourist attraction to the city center, with a distance of less than 3 km, and is located on the west side of the city center. The last cluster is a newly developed group of attractions located on the east side of the city center. So, it can

Table 2 KMO test

Variable	X_1	X_2	X_3	X_4	X_5	Overall
KMO test	0.53	0.51	0.52	0.48	0.53	0.51

Table 3 Results of K-means clustering

Validation measures	The number of clusters				
	2	3	4	5	6
K-means connectivity	16.44	15.78	17.25	26.82	27.57
Dunn	0.26	0.43	0.43	0.36	0.42
Silhouette	0.16	0.26	0.24	0.22	0.23

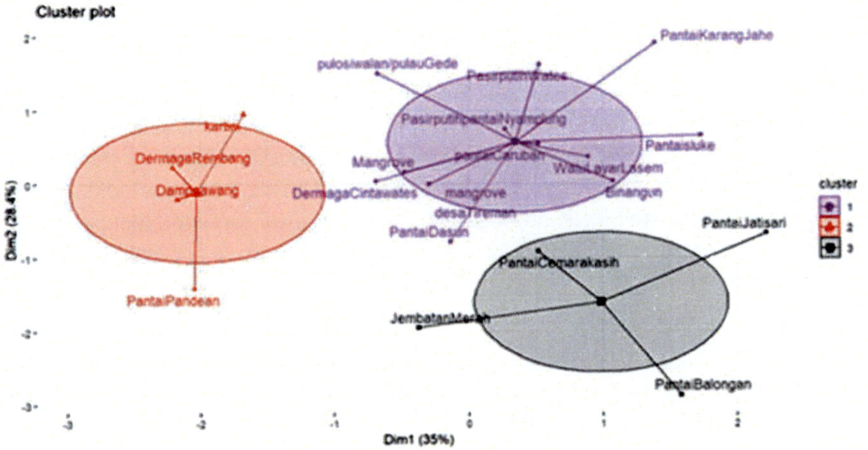

Fig. 2 Cluster plot of the best segmentation tourism potential analysis

be concluded that from the three clusters 3 segments are formed, namely cluster 1 is a sustainable development support segment consisting of 12 tourist objects, cluster 2 is a popular local tourism segment consisting of 4 tourist objects, and the last cluster is a segment of new local tourism potential. There are several criteria of sustainable tourism. These include sustainable tourism destination management, economic use for local communities, cultural preservation for the community and visitors, and environmental conservation [15]. Details of attractions can be seen in Table 4.

The results of this study are aimed at the tourism potential in Rembang Regency. This is different from the research conducted by Adianta, Ali, and Nugeraha [21] which conducted clustering to determine the segmentation of tourist visits to determine the profile and purpose of the tourist visit. Meanwhile, this study aims to explore the tourism potential in the Rembang Regency so that it can be developed based on existing resources to help improve the economy.

Table 4 Membership of segmentation tourism potential analysis using 3 clusters

Segment	Name of segment	Amount	Tourism
1	Sustainable development support	12	Watu Layar Lasem, Pantai Karang Jahe, Pantai Caruban, Pasir Putih Wates, Pulo Siwalan/Pualu Gede, Pasir Putih Pantai Nyamplung, Mangrove, Dermaga Cinta Wates, Binangun, Mangrove Desa Tireman, Pantai Sluke, Pantai Dasun
2	Popular local tourism	4	Dampo Awang, Dermaga Rembang, Pantai Pandean, Kartini
3	New local tourism potential	4	Pantai Jatisari, Pantai Cemara Kasih, Pantai Balongan, Jembatan Merah

4 Conclusion

In improving the regional economy through segmentation of sustainable tourism potential in Rembang district as a tourist destination, it can be concluded that 20 tourism objects can be divided into 3 segmentation groups, Cluster 1 sustainable development support consisting of 12 tourist objects, Cluster 2 popular local tourism consisting of 4 tourist objects, and Cluster 3 is a new local tourism potential which consists of 4 tourist objects. From each segmentation group, a sustainable tourism area development planning strategy will be made in the next research.

Acknowledgements This research was supported by Telkom University which assisted in providing the time and opportunity to collect data in Rembang Regency, Central Java, Indonesia. We would also like to thank the Tourism Office and the management of tourist attractions in Rembang Regency who have provided information related to the data needed in the research.

References

1. Badan Perencanaan Pembangunan Dearah Kabupaten Rembang (2022) Wisata Rembang [online]. Available: https://bappeda.rembangkab.go.id/tag/wisata-rembang/. Last accesed 26 Aug 2022
2. García-Melón M, Gómez-Navarro T, Acuña-Dutra S (2012) A combined ANP-delphi approach to evaluate sustainable tourism. Environ Impact Assess Rev 34:41–50. https://doi.org/10.1016/j.eiar.2011.12.001
3. Hanafi MM, Halim A (2009) Analysis of financial statements. UPP.STIM YKPN, Yogyakarta
4. Septiningrum L, Rizana AF, Soesanto RP, Rumanti AA (2022) Success factors for elevating indonesian tourism area potential. In: IEEE international conference on industrial engineering and engineering management (IEEM)
5. Lohmann G, Netto AP (2016) Tourism theory: concepts, models, and systems. Cabi
6. Lee-Ross D, Lashley C (2010) Entrepreneurship and small business management in the hospitality industry. Routledge
7. Page SJ (2014) Tourism management. Routledge
8. Rembang Regency Government (2022) Pemkab Rembang Terus Dorong Desa Kembangkan Potensi Wisata [online], available: https://rembangkab.go.id/berita/pemkab-rembang-terus-dorong-desa-kembangkan-potensi-wisata/. Las accessed: 24 Aug 2022
9. Hozumi Y, Wang R, Yin C, Wei G (2021) UMAP-assisted K -means clustering of large-scale SARS-CoV-2 mutation datasets. Comput Biol Med 131:1–14. https://doi.org/10.1016/j.compbiomed.2021.104264
10. Xie H, Zhang L, Lim CP, Yu Y, Liu C, Liu H (2019) Improving K-means clustering with enhanced Firefly Algorithms. Appl Soft Comput J 84:1–22. https://doi.org/10.1016/j.asoc.2019.105763
11. Sharma V, Bala M (2020) An improved task allocation strategy in cloud using modified K-means clustering technique. Egypt Inf J 21(4):201–208. https://doi.org/10.1016/j.eij.2020.02.001
12. Tan PN, Steinbach M, Kumar V (2016) Introduction to data mining. Pearson Education India
13. Patel E, Kushwaha DS (2020) Clustering cloud workloads: K-means vs gaussian mixture model. Proced Comput Sci 171:158–167
14. Gbadoubissa JEZ, Ari AAA, Gueroui AM (2020) Efficient k-means based clustering scheme for mobile networks cell sites management. J King Saud Univ-Comput Inf Sci 32(9):1063–1070

15. Sulistyadi Y, Eddyono F, Entas D (2019) Indikator Perencanaan Pengembangan Pariwisata Berkelanjutan. Aura Bandar Lampung
16. Kotary DK, Nanda SJ (2020) Automatic determination of K in distributed K-means clustering automatic determination of K in distributed K-means. Proced Comput Sci 165(2019):556–564. https://doi.org/10.1016/j.procs.2020.01.050
17. Naghizadeh A, Metaxas DN (2020) Condensed silhouette: an optimized optimized filtering filtering process process for cluster cluster selection in selection in K-Means. Proced Comput Sci 176:205–214. https://doi.org/10.1016/j.procs.2020.08.022
18. Clayman CL, Srinivasan SM, Sangwan RS, Clayman CL, Srinivasan SM, Sangwan RS (2020) K-means clustering and principal components analysis of microarray data L1000 components landmark genes K-means clustering and of principal analysis of microarray data of L1000 landmark genes. Proced Comput Sci 168:97–104. https://doi.org/10.1016/j.procs.2020.02.265
19. Xu H, Croot P, Zhang C (2021) Discovering hidden spatial patterns and their associations with controlling factors for potentially toxic elements in topsoil using hot spot analysis and K-means clustering analysis. Environ Int Col 151:1–14
20. Guo W, Wang W, Zhao S, Niu Y, Zhang Z, Liu X (2022) Density peak clustering with connectivity estimation. Knowl-Based Syst 243:108501. https://doi.org/10.1016/j.knosys.2022.108501
21. Adianta A, Ali S, Nugeraha P (2019) Analisis Segmentasi Demografisdan Perilaku pada Bisnis Pariwisata. J Perspekt Bisnis 2(2):10–18

Study of Learning Factories Concept for Production Laboratories in Bandung Polytechnic of State

Fachri Koeshardono, Sutrimo, Aris Suryadi, Muhammad Syahrain Basirisa, Lili Awalia, and Mugo Muchjayadi

Abstract Learning Factories are the place to test the new technologies and could also integrated the work and education environment. Polytechnic State of Bandung as the educational institution could adapt the concept of learning factories. The production laboratories in current state only utilized as learning places while it has the potential to become learning factory. The research in this study consists of identification to get the primary data for the laboratories. There are 3 labs that could be potentially become learning factory on its own or these labs could be collaborated as network of learning factories. The production capacity in every lab is different based on the products developed. The total for each lab in producing the parts for screw turbine: Fabrication Laboratory is 52 min; Machining Laboratory is 137 min, and Welding Laboratory is 289 min. By utilizing 8 h/day and 60 min per hour the number of productions for every lab is: 9 unit/day Fabrication Lab, 4 unit/day for Machining Lab and 2 unit/day for Welding Laboratory. As for cost estimation in producing screw turbine is 2.294.448 rupiah. The production laboratories in Bandung Polytechnic of State already have the capabilities to become learning factories. The mindset and the concept for each lab need to be developed.

Keywords Learning factory · Machining · Welding · Fabrication · Computer aided process planning

1 Introduction

In the era of Industry 4.0, engineering education institution need to prepare their student with the updated required skills. Besides that, the development of technology should be matched with the demand of the industries. Learning Factory is the concept that is rooted in manufacturing industry and education in the training of the labor.

F. Koeshardono (✉) · Sutrimo · A. Suryadi · M. S. Basirisa · L. Awalia · M. Muchjayadi
Bandung Polytechnic of State, Bandung Barat, Indonesia
e-mail: fachri.koeshardono@polban.ac.id

C. N. Rosyidi et al. (eds.), *Proceedings of the 6th Asia Pacific Conference on Manufacturing Systems and 4th International Manufacturing Engineering Conference*, Lecture Notes in Mechanical Engineering,
https://doi.org/10.1007/978-981-99-1245-2_27

It could be classified by its operation model, purpose and targets, process, setting, products, didactics and learning factory metrics [1]. Baena in EAFIT University transformed the learning process in their manufacturing workshop. The aim is to give the learner experience-based knowledge in real production environment [2]. B. Schallock in his paper explained the importance of soft skill beside hard skill for Learning Factory in preparing production staff in the Industrie 4.0 era. The highlighting skills are decision making, group work and performance monitoring skill. It is proved to be successful in Germany [3]. Management competencies regarding workers' participation could be also developed in Learning Factory and it is proven to be worked in Ruhr-Universität Bochum as reported by Wagner [4]. Reconfigurable Manufacturing System is essential in today's manufacturing world to meet the demand of the market. The market tends to be highly volatile, and the products should come in big varieties with the ability to be customized. Problem Based Learning in Learning Factory environment for such condition is explained by Andersen [5]. The design of Learning Factory also could be approached by Flexible Manufacturing Cell as reported by Mourtzis [6]. It is also needed to be addressed by educational institute that the learning process need to be in conjunction with the business world. The framework of collaboration between school and small medium enterprise studied by Matt [7]. The study about the utilization of Digital Twins, Virtual Reality and Augmented Reality in Learning Factory has promising future as explained by Tvenge [8]. Elbestawi developed the SEPT Learning Factory by utilizing Cyber-Physical Systems, Internet of things and Industrial Internet of Things. It is heavily designed to encourage the collaboration between human and robots [9].

Bandung Polytechnic of State as the educational institution could adapt the concept. These production laboratories in the current state only utilized as learning places. Learning Factory as mentioned before should be functioned at least in 2 areas. Hence, the development create learning factories should be studied intensively.

2 Materials and Methods

The research in this study consists of identification to get the primary data for the laboratories. There are 3 labs that could be potentially become learning factory on its own or collaborated as network of learning factories. The aim of this learning factory is to produce the good that could be beneficial for society.

The screw turbine was chosen since its rising using as renewable energy equipment in rural area. It is also relatively simple in its construction and sufficient as object for several learning subjects. The plan of screw turbine production then presented as Operation Process Chart (OPC). In the chart, the process time is tested by experimenting in the real environment. Besides that, it is also simulated in Computer Aided Manufacturing (CAM). From the result of experimentation and simulation, the time for each process is determined. From that time study, the calculation of the machine's capacities and the number of operators. The cost of product could be calculated to see if it is economical based on the capital and operational expenditure. The summary of research methodology is presented in Fig. 1.

Several production machines and auxiliaries are identified and sufficient in producing screw turbines. The first data that must be known are the layout of the machines in 3 laboratories which are: Fabrication Laboratory (Fig. 2), Machining Laboratory (Fig. 3) and Welding Laboratory (Fig. 4). Each lab is treated as one

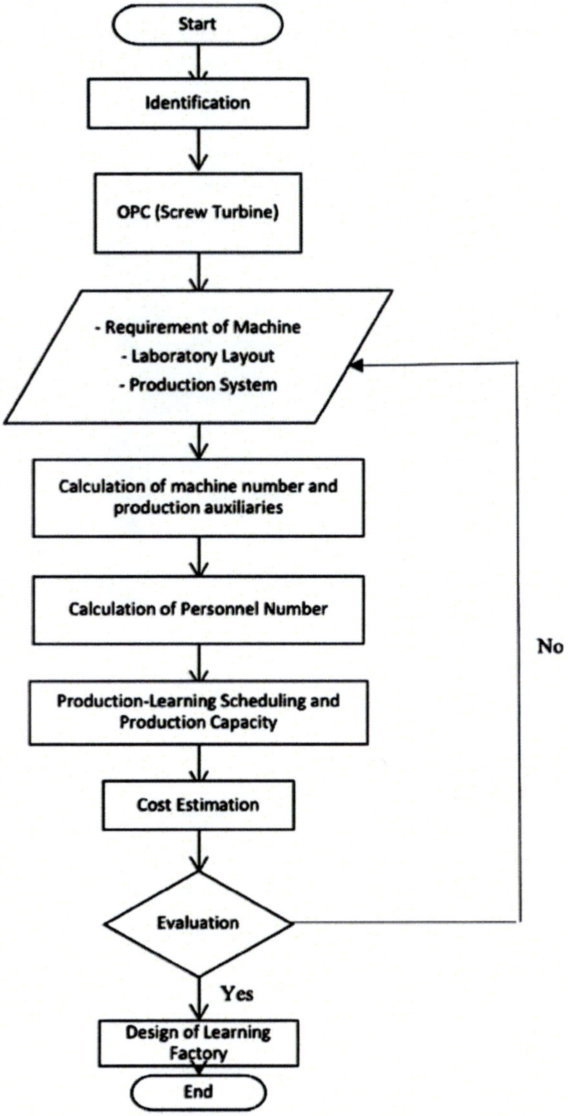

Fig. 1 Research methodology

machining center in the case of producing parts of screw turbine. Every lab has its own part in producing the parts for the product.

The section is written corresponding to the section number. The product as object of learning factory in Polytechnic State of Bandung is Screw Turbine (Fig. 5). The variation of manufacturing process is sufficient and could covered several study subjects and covered in several semesters.

In the perspective of market, the product has several advantages including its low cost in production while at the same time it is less harmful for the ecosystem since its low spinning speed so the fish could go through.

The Operation Process Chart (OPC) is developed by breaking down into subassemblies. It is consisting of: Screw, Turbine Axle, Cover and Frame. It is to be

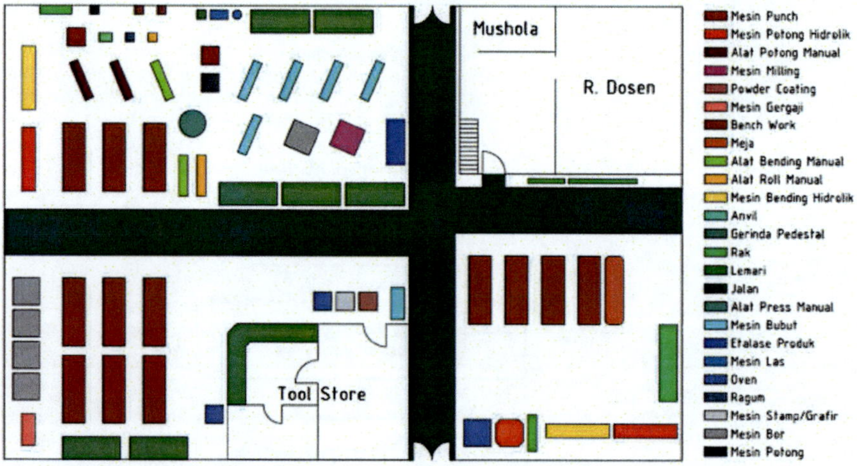

Fig. 2 Fabrication laboratory layout

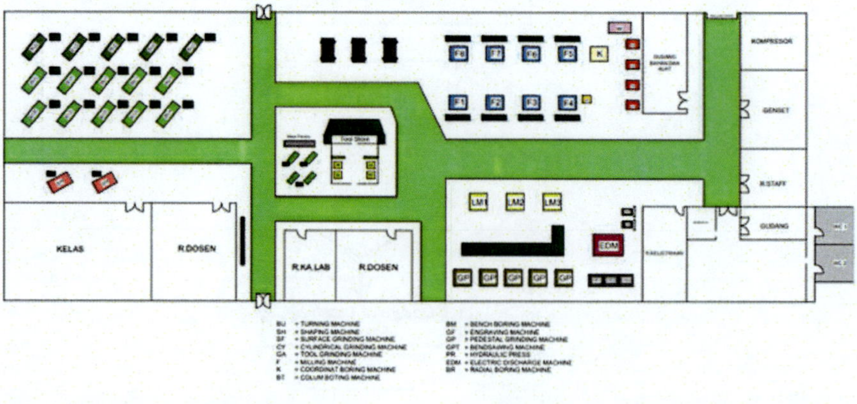

Fig. 3 Machining laboratory layout

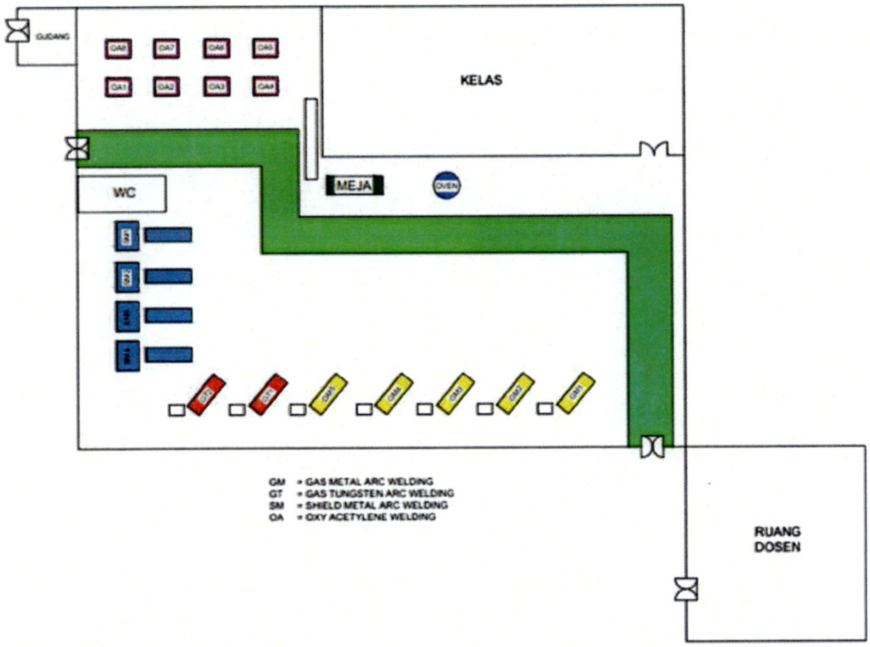

Fig. 4 Welding laboratory layout

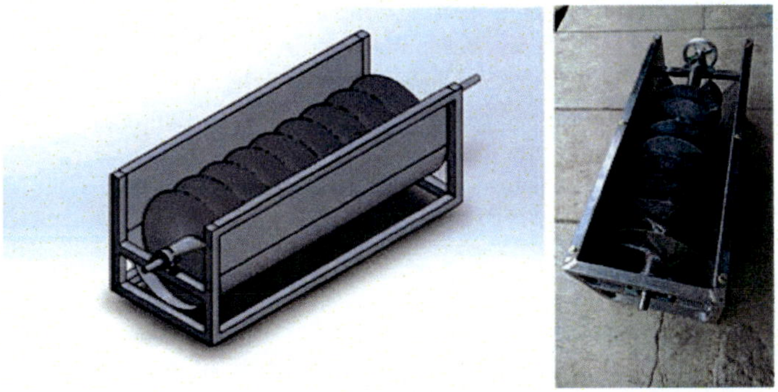

Fig. 5 Screw turbine

noted that every process always started and ended by measurement to get the good quality of the product. The time for these inspections is essential in making the OPC.

The time for each process is already including the setting time for each machine. The turbine axle subassembly that welded together with the screw is the most challenging, so it is worked by the lab operator instead of students. However, the other parts such as making the screw is a good example for students in learning process of shear machine and blanking machine (Fig. 6).

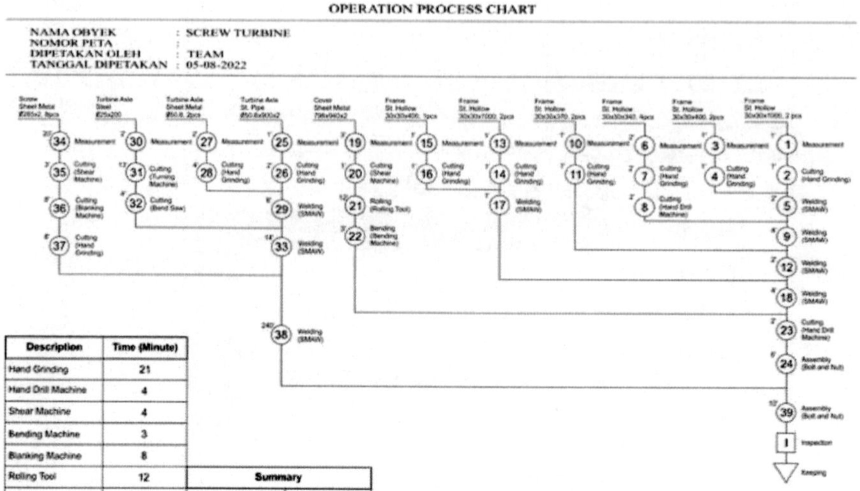

Description	Time (Minute)
Hand Grinding	21
Hand Drill Machine	4
Shear Machine	4
Bending Machine	3
Blanking Machine	8
Rolling Tool	12
Bend Saw	4
Turning Machine	13
SMAW	273
Assembly	16
TOTAL	358

Summary	
Act	Qty
○ Operation	39
□ Inspection	1
▽ Keeping	1
TOTAL	41

Fig. 6 Operation process chart screw turbine

The process of production the screw turbine could be simulated to get the balance in assembly. The simulation itself itself is done by Tecnomatix. There are 3 laboratories simulated by planning its worked parts. Frame, Cover, and Screw are produced in Fabrication Laboratory (Fig. 7) while Turbine Axle are produced in Machining Laboratory. The assembling process is worked at Welding Laboratory including the assembly process other than the welding itself.

3 Results

The production capacity in every lab is different based on the products developed. The total for each lab in producing the parts for screw turbine: Fabrication Laboratory is 52 min; Machining Laboratory is 137 min, and Welding Laboratory is 289 min. By utilizing 8 h/day and 60 min per hour the number of productions for every lab is: 9 unit/day Fabrication Lab, 4 unit/day for Machining Lab and 2 unit/day for Welding Laboratory.

As for cost estimation in producing screw turbine is 2,294,448 rupiah.

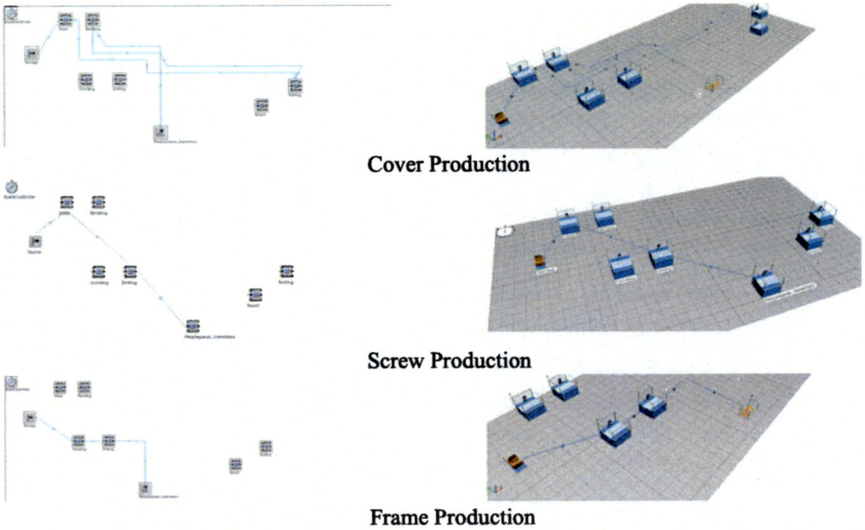

Fig. 7 Simulation of fabrication laboratory

Table 1 Machine and labor requirement for developing learning factory based on screw turbine

No	Machine	Time (minute)	Number	Lab	Labor
1	Hand grinding	21	1	Fabrication	2
2	Hand drill	12	1		
3	Shear	4	1		
4	Bending	3	1		
5	Rolling tool	12	1		
6	Turning	133	2	Machining	2
7	Bend saw	4	1		
8	SMAW	273	2	Welding	2

4 Conclusion

Production Laboratories in Bandung Polytechnic of State has the potentials as Learning Factories. The labs could be utilized as production plant without intervening with the learning process (less than 10% of facilities is used). The product itself could be used as the learning object of students. The lab organization should be reconstructed if the concept is adapted in Bandung Polytechnic of State.

References

1. Schuh G, Prote J-P, Dany S, Cremer S, Molitor M (2017) Classification of a hybrid production infrastructure in a learning factory morphology. Proced Manuf 9:17–24. https://doi.org/10.1016/j.promfg.2017.04.007
2. Baena F, Guarin A, Mora J, Sauza J, Retat S (2017) Learning factory: the path to industry 4.0. Proced Manuf 9:73–80. https://doi.org/10.1016/j.promfg.2017.04.022
3. Schallock B, Rybski C, Jochem R, Kohl H (2018) Learning factory for industry 4.0 to provide future skills beyond technical training. Proced Manuf 23:27–32. https://doi.org/10.1016/j.promfg.2018.03.156
4. Wagner P, Prinz C, Wannöffel M, Kreimeier D (2015) Learning factory for management, organization and workers' participation. Proced CIRP 32:115–119. https://doi.org/10.1016/j.procir.2015.02.118
5. Andersen A-L, Brunoe TD, Nielsen K (2019) Engineering education in changeable and reconfigurable manufacturing: using problem-based learning in a learning factory environment. Proced CIRP 81:7–12. https://doi.org/10.1016/j.procir.2019.03.002
6. Mourtzis D, Angelopoulos J, Dimitrakopoulos G (2020) Design and development of a flexible manufacturing cell in the concept of learning factory paradigm for the education of generation 4.0 engineers. Proced Manuf 45:361–366. https://doi.org/10.1016/j.promfg.2020.04.035
7. Matt DT, Rauch E, Dallasega P (2014) Mini-factory—a learning factory concept for students and small and medium sized enterprises. Proced CIRP 17:178–183. https://doi.org/10.1016/j.procir.2014.01.057
8. Tvenge N, Ogorodnyk O, Østbø NP, Martinsen K (2020) Added value of a virtual approach to simulation-based learning in a manufacturing learning factory. Proced CIRP 88:36–41. https://doi.org/10.1016/j.procir.2020.05.007
9. Elbestawi M, Centea D, Singh I, Wanyama T (2018) SEPT learning factory for industry 4.0 education and applied research. Proced Manuf 23:249–254. https://doi.org/10.1016/j.promfg.2018.04.025

Applying Importance Performance Analysis (IPA) to Measure the Customer Satisfaction on Food Safety Practices in Restaurant

Gusti Fauza, Aisyah Rizky Millenia, Asri Nursiwi, Dwiningtyas Padmaningrum, Setyaningrum Ariviani, and Hari Prasetyo

Abstract The growing number of restaurants required players in this business to improve the quality and safety of their foods. Supervision is carried out to ensure that each served product has high quality and safety that leads to customer satisfaction and loyalty to maintain business stability and continuity. In this study, customer satisfaction of an observed restaurant was evaluating using importance performance analysis (IPA) method. Twenty-five attributes sourcing from literatures and food safety practices guideline for restaurant are developed to evaluate food quality, service quality and environment and facilities of the restaurant. Involving 153 respondents the average satisfaction is 6.28 showing that customers satisfy on restaurant food safety practices. However, this performance is not in line with the microbial testing results that identifies the size of S. aureus in animal-based side dishes and E. coli in iced tea exceeded food safety limits. Further, the IPA diagram reveals that, seven attributes need improvement. Those attributes are from restaurant environment and facilities and food quality dimension.

Keywords IPA · Food safety · Restaurant · Customer satisfaction

G. Fauza (✉) · A. R. Millenia · A. Nursiwi · S. Ariviani
Department of Food Science and Technology, Faculty of Agriculture, Universitas Sebelas Maret, Surakarta, Indonesia
e-mail: gustifauza@staff.uns.ac.id

D. Padmaningrum
Department of Agricultural Extension and Communication, Faculty of Agriculture, Universitas Sebelas Maret, Surakarta, Indonesia

H. Prasetyo
Department of Industrial Engineering, Universitas Muhammadiyah Surakarta, Surakarta, Indonesia

C. N. Rosyidi et al. (eds.), *Proceedings of the 6th Asia Pacific Conference on Manufacturing Systems and 4th International Manufacturing Engineering Conference*, Lecture Notes in Mechanical Engineering,
https://doi.org/10.1007/978-981-99-1245-2_28

1 Introduction

Practical lifestyles, various food delivery services as a result of technological advance, and dining out habits are some of the reasons for the increasing number of conventional and fast-food restaurants. The significant growth in the number of restaurants increases the competition in this business as well. In foodservice business, customer satisfaction is a crucial factor for success [1, 2]. Some studies in this field reported that there is a strong relation between satisfaction and loyalty [3, 4]. Having loyal customers are the important aspect of business sustainability, hence identifying attributes that drive customer satisfaction is needed.

Food safety is a vital issue in the implementation of restaurant businesses, given a relatively high number case related to foodborne disease in Indonesia. In 2019, the National Agency of Drug and Food Control (BPOM) reported that there were 6205 cases of foodborne disease in Indonesia [5]. Nonetheless, the actual number is predicted to be greater. The WHO estimated that only one in a hundred foodborne outbreaks is reported, while the other 99 cases are underreported [6]. Meanwhile, [7] predicted that the costs incurred for foodborne cases (reported or underreported) ranged from 5.4 to 17 billion USD. The serious burden of foodborne disease in Indonesia and the great losses due to the incident have urged the government and food industry to pay more attention to food safety guarantees.

Restaurants, particularly full-service ones, have a potential to cause foodborne disease. Based on the CDC report [8], approximately 60% of foodborne diseases occurred in 2006 in the United States came from restaurants. From the business perspective, food safety is one of the important factors in consumers restaurant selection [9–13]. Consumers assessments of the cleanliness of the place and the safety of the food as well as the beverage are among the considerations to determine satisfaction and repurchase behavior [14]. In general, consumers use visual appearances to observe food safety and hygiene, such as the appearance of the served food and beverages, rooms, toilet, dining tables, utensils, and work staff [15]. Since the food safety practices in a restaurant is one of the keys driven the customer preferences hence the customer satisfaction on this dimension need to be assessed.

Studies on evaluating consumers satisfaction on restaurant has been done recently such as in [16–26]. However, limited research using consumer perception on food safety implementation in restaurant to evaluate the satisfaction. In this research the customer satisfaction is evaluated using customer perception on restaurant food safety practices. The Importance Performance Analysis (IPA) method is applied to measure and evaluate performance of attributes or parameters based on the consumer's point of view. The obtained data would show the level of consumer satisfaction to specific attributes of a product or service and the results can be used as a reference to improve the culinary business management. Further, this study also evaluates the microbial profile of some menu in restaurant. The evaluation is aimed to see whether the level of customer satisfaction on food safety practices in a restaurant is in line with its food safety status that represented by the microbial profile of some menu in the restaurant.

2 Method

2.1 Questionnaires Design

The questionnaire is consisting of three parts. The first part contains questions about the identity and demographic profile of respondents, second part contains questions about the importance of attributes, and third part contains questions about the performance of attributes. The attributes in questionnaire are constructed by referring related literatures which then are evaluated for content validity using Lawshe method [27]. Based on the literatures such as in [12, 15, 28–32], a total of 33 attributes from three customer satisfaction dimension e.g., food quality, service quality, and environmental conditions and facilities of restaurant, were obtained. Six panelists (lecturers from food science and technology department at Universitas Sebelas Maret and the restaurant owner) are asked to evaluate the attributes. Using Lawshe methods, eight attributes are invalid and results in 25 attributes used in this study as can be seen in Appendix A.

2.2 Data Collection

The survey is conducted in a restaurant which is located in Surakarta. It is a fully service restaurant where food is prepared once it is ordered by a customer. The questionnaire is distributed to customers who voluntarily agree to fill it. Filling out the questionnaire was done by giving an assessment of the importance and performance of attributes using a 7-point Likert scale ranging from not at all important/completely dissatisfied (1) to extremely important/completely satisfied (7) 0.155 questionnaires are received but 2 of them are invalid due to incompletion issue, hence only 153 respondents are used for evaluation.

2.3 Data Analysis

The survey data are analyzed using IPA method. IPA method was employed to identify service attributes and product attributes based on the level of performance and importance of these attributes from the consumer perception. The results of the IPA test are presented in a Cartesian diagram with the y-axis showing the level of importance and the x-axis showing the level of performance. The IPA Cartesian diagram is divided into four quadrants (Fig. 1). To construct this diagram, the average of performance rating and importance rating of attributes are calculated.

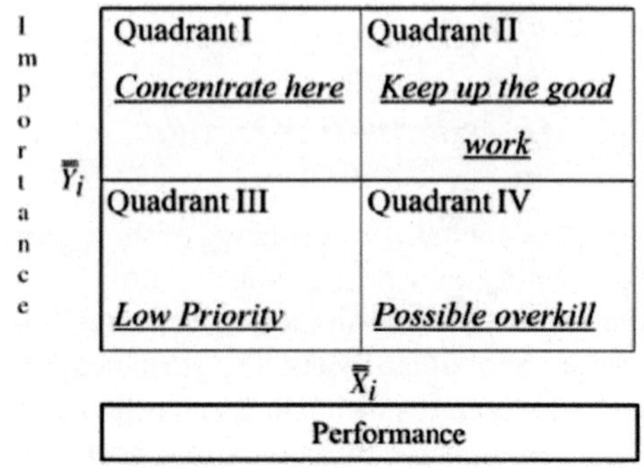

Fig. 1 A Cartesian diagram of IPA

2.4 Sample Analysis

The microbiological test was done to some of restaurant menus. It was conducted to see whether the level of customer satisfaction was in line with the results of microbial test. The restaurant providing many food and beverage menus. To select the representative samples, firstly, the variants on the menu were grouped into five categories, namely beverages, animal-based side dishes, vegetable side dishes, processed vegetables, and main courses. Further, the best seller menus in each category with has potential impact on foodborne disease were selected as the samples for microbial testing. It is to ensure that the samples for microbial testing were representative. Five menus, namely, iced tea, grilled chicken, stir fried water spinach, fried tempe and rice were selected for microbial testing. The test is done to detect TPC, *Escherichia coli, Staphylococcus aureusI, Salmonella* sp and *Bacillus cereus.*

3 Results and Discussion

3.1 Demographic Profile of Respondents

Based on the survey data, the demographic profile of the 153 respondents is summarized and presented in Table 1. As presented in Table 1, the majority respondents were female (67%) and 84% respondents have visited the restaurant at least twice. The restaurant is located near university.

3.2 Validity and Reliability Test

Testing the validity of the instrument is done by calculating the value of the Pearson Correlation coefficient, *r*. The value of *r* table is 0.1587. As seen in Table 2, the

Table 1 Demographic profile of respondents

No	Characteristics of respondents	n	Percentage (%)
1	*Gender*		
	Male	50	33
	Female	103	67
2	*Frequency of eating at the restaurant each week*		
	Once	24	16
	Twice—4 times	53	34
	5—7 times	35	23
	8—10 times	21	14
	> 10 times	20	13
3	*Education level*		
	Senior high school	6	4
	Diploma/undergraduate	109	71
	Postgraduate	35	23
	Others	3	2

validity test indicates that *r* value of the 25 attributes, both the importance and performance level, is higher than 0.1587 with a range of 0.410–0.833. Therefore, all attributes are declared valid.

The reliability testing was carried out by calculating the value of Cronbach's alpha [33]. Tests were carried out for each attribute dimension i.e., food quality, service quality, environmental conditions, and facilities. Dimensions are declared reliable if the Cronbach's alpha value is higher than 0.7 [34].

As can be seen from Table 3, the results of the reliability test revealed that the importance and performance levels of the three attributes ranged from 0.810 to 0.907 or higher than 0.7. It indicates that the research instrument is able to produce consistent data.

3.3 *Importance Performance Analysis*

The IPA diagram was constructed by calculating the average of performance and importance level of each attribute. The values can be seen in Table 4, where the diagram is presented by Fig. 2. As can be seen in Table 4, customers perceive that the performance of restaurant quality food quality, service quality, and environment and facility, is satisfying as the value above 6 with a grand average 6.282 (represented by the bold values in the Performance column). Further, customers think that those attributes are essential since the grand average is nearly 6.5 as shown by the bold values at the Importance column.

Table 2 The results of validity test

Attributes		Importance level			Performance level		
		r	*p value*	Result	*r*	*p value*	Result
Food quality (M)	1	0.725	0.000	Valid	0.775	0.000	Valid
	2	0.716	0.000	Valid	0.809	0.000	Valid
	3	0.721	0.000	Valid	0.732	0.000	Valid
	4	0.718	0.000	Valid	0.709	0.000	Valid
	5	0.710	0.000	Valid	0.753	0.000	Valid
	6	0.724	0.000	Valid	0.680	0.000	Valid
Service quality (P)	1	0.410	0.000	Valid	0.564	0.000	Valid
	2	0.648	0.000	Valid	0.750	0.000	Valid
	3	0.731	0.000	Valid	0.714	0.000	Valid
	4	0.761	0.000	Valid	0.608	0.000	Valid
	5	0.629	0.000	Valid	0.816	0.000	Valid
	6	0.711	0.000	Valid	0.675	0.000	Valid
	7	0.462	0.000	Valid	0.748	0.000	Valid
	8	0.722	0.000	Valid	0.716	0.000	Valid
	9	0.681	0.000	Valid	0.702	0.000	Valid
Environment and facility (L)	1	0.678	0.000	Valid	0.754	0.000	Valid
	2	0.763	0.000	Valid	0.790	0.000	Valid
	3	0.697	0.000	Valid	0.688	0.000	Valid
	4	0.643	0.000	Valid	0.641	0.000	Valid
	5	0.705	0.000	Valid	0.674	0.000	Valid
	6	0.792	0.000	Valid	0.833	0.000	Valid
	7	0.702	0.000	Valid	0.775	0.000	Valid
	8	0.605	0.000	Valid	0.678	0.000	Valid
	9	0.724	0.000	Valid	0.755	0.000	Valid
	10	0.742	0.000	Valid	0.793	0.000	Valid

Table 3 The results of reliability test

Attribute	Importance		Performance	
	Cronbach's alpha	Result	Cronbach's alpha	Result
Food quality (M)	0.810	Reliable	0.831	Reliable
Service quality (P)	0.821	Reliable	0.868	Reliable
Environment and facility (L)	0.885	Reliable	0.907	Reliable

Table 4 The average of performance and importance level

Attributes		Average	
		Performance	Importance
Food quality (M)	1	6.431	6.627
	2	6.477	6.771
	3	6.458	6.686
	4	6.261	6.660
	5	6.399	6.699
	6	6.523	6.712
Subsection average score		**6.425**	**6.693**
Service quality (P)	1	6.333	6.641
	2	6.020	6.222
	3	6.340	6.288
	4	6.157	6.065
	5	6.307	6.582
	6	6.307	6.366
	7	6.464	6.673
	8	6.137	5.758
	9	6.503	6.539
Subsection average score		**6.285**	**6.348**
Environment and facility (L)	1	6.196	6.614
	2	6.157	6.536
	3	6.288	6.301
	4	6.268	6.523
	5	6.163	6.458
	6	6.131	6.542
	7	6.078	6.471
	8	6.471	6.667
	9	6.098	6.510
	10	6.085	6.516
Subsection average score		**6.194**	**6.514**
Grand average		**6.282**	**6.497**

Table 4 shows that in overall, the customers satisfied on the restaurant's food safety practices as the grand average of restaurant performance is above 6. Further, Table 4 also presents that consumers consider food quality dimension that related to food safety e.g., fresh food and beverage (M1), no foreign object found in food (M2), no strange or typical smells found (M3), food should safe from flies or other insect (M4), fresh ingredients used (M5), and consumers are safe from illness after consuming food (M6), is the most important attributes among others. The result is

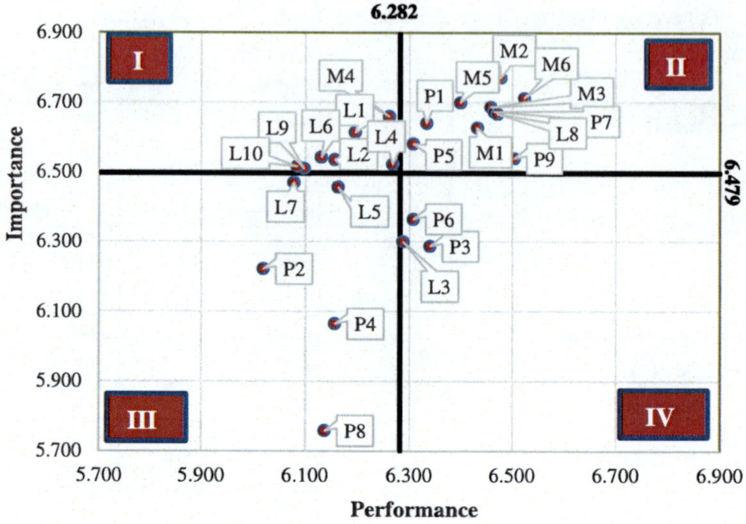

Fig. 2 IPA's diagram of the observed restaurant

similar to [15] but quite different from [35] who found that the students in Ghana overlooked on food freshness.

Despite the customers consideration on the importance of the food quality's attributes, one of those attributes (M4) falls in Quadrant 1 along with six attributes from environment and facility condition, as can be seen in Fig. 2. The attributes that are included in Quadrant I are considered by customer as important but the perceived level of their performance is low, Therefore, these attributes require immediate attempts to improve the performance or improvement should be prioritized in this quadrant [36]. The value of attribute importance is relatively high, indicating that the quality of the attribute has significant effect on the level of service quality. Principally, suboptimal performance on the attributes in Quadrant I will simply reduce the level of consumer satisfaction in overall [39].

Still in line with [15], customers of the observed restaurant also consider the clean and well maintain service area and facility such as kitchen, dining area (dine table and chairs, cloth used to clean the table) and toilettes, as the important dimension after food quality. However, from the customer perception, the performance of those attributes although quite satisfying but still below the expectation since they fall in Quadrant 1. They are clean and well-maintained kitchen (L1), clean and maintain service area of restaurant (L2), no pests and rodents are presented (L4), clean and tidy garbage disposal area (L6), toilet with adequate numbers and facility (L9), and clean and well-maintained toilet (L10).

Further, in this study, microbial profile of the five samples of the restaurant menus were tested and the result is presented by Table 5. The results of microbiological testing showed that the microbial status of the sample iced tea (at third replication) and grilled chicken (all replication) was not met the requirement. Meanwhile, the categories of vegetable side dishes (fried tempe), processed vegetables (stir fried water spinach), and main course (white rice) were acceptable based on the applicable critical limits.

Table 5 demonstrates that on the third replication, the average MPN value is 9.2 MPN/ml. It exceeds the critical limit (< 3 MPN/ml) (BPOM, 2012). Pathogenic *Escherichia coli* strains produce endotoxins that may cause mild to severe diarrhea [36]. Among the factors allegedly contribute to the contamination of iced tea by *E. coli* in the observed restaurant are the personal hygiene of the employees and the food safety of materials (water form refillable gallon bottles, ice cubes, steeped tea, and sugar water). Personal hygiene has a significant effect on *E. coli* bacterial contamination in food products. A comparison was done on foods produced by good personal hygiene and poor personal hygiene, showing that respectively, 5.3 and 58.3% of food samples were contaminated with *E. coli* [37]. The materials of the iced tea potentially contribute to *E. coli* contamination in the sample. The water from refillable gallon bottles used to steep tea and the ice cubes are also at risk for being contaminated, moreover, the steeped tea should be single use to minimize potential cross-contamination [38].

In addition to the MPN value of *E. coli* in a sample of iced tea, the number of *Staphylococcus aureus* in all samples of grilled chicken was classified unacceptable based on food safety standard. The hazard is caused by the enterotoxigenic strains produced by *S. aureus*. In general, the human body has a rapid response to the

Table 5 The results of microbiological test

Sample	Type of test	Unit	Critical limit	Replication		
				1	2	3
				Result	Result	Result
Iced tea	TPC	Colonies/ml	1×10^5	2.6×10^3	1.2×10^3	1.2×10^3
	Escherichia coli	MPN/ml	< 3	< 3	< 3	**9.2**[a]
Grilled chicken	TPC	Colonies/g	1×10^5	1.6×10^3	9.4×10^3	3.5×10^2
	Staphylococcus aureus	Colonies/g /25 g	1×10^2	**2.2×10^2**[a]	**1.3×10^4**[a]	**7.1×10^3**[a]
	Salmonella sp		Negative	Negative	Negative	Negative
Stir fried water spinach	TPC	Colonies/g /25 g	1×10^5	1.1×10^3	4.2×10^3	9.9. $\times 10^2$
	Salmonella sp	Colonies/g /25 g	Negative	Negative	Negative	Negative
Fried tempe	TPC	Colonies/g	1×10^5	2.5×10^2	5.0×10^2	6.2×10^2
Rice	TPC	Colonies/g	1×10^5	10^1	10^1	10^1
	Bacillus cereus	Colonies/g	1×10^3	Undetected	Undetected	Undetected

[a]The value over the critical limit

enterotoxigenic strains of *S. aureus*, with symptoms such as stomach cramps and vomiting. Staphylococcal food poisoning can be caused the consumption of foods containing at least 1 μg of *S. aureus* [36] Unlike others, enterotoxigenic *S. aureus* strains are resistant to heat, cold, and proteolytic enzymes [39].

S. aureus, as shown in Table 4, was presented in all samples of grilled chicken served by observed restaurant and exceeded the critical limit (10^2 colonies/g) (BPOM, 2012), with the highest average of 1.3×10^4 colonies/g of the sample. However, the analysis indicated that the size of *S. aureus* in the grilled chicken samples was insufficient to produce toxins. In humans, intoxication occurs due to the presence of at least 1 μg of *S. aureus* in food, while this amount can only be produced by at least 10^5 colonies/g of the food [36]. Meanwhile, different studies reported the ability of *S. aureus* to produce enterotoxins when the size reaches 10^6 colonies/g [39–41]. The presence of *S. aureus* in the grilled chicken samples was possibly due to cross-contamination from food handlers' bare hand during food processing and less appropriate seasoning or grilling process to kill bacteria in chicken meat. Moreover, the high nutrient content in chicken meat also accelerates the growth rate of bacteria in foods.

Despite the estimation that the size of colonies found in the samples is less to produce enterotoxins in sufficient quantities, it exceeds the critical limit for food safety, indicating the necessity to evaluate the grilled chicken production processes. The evaluation involves the bacterial doubling time (DT) and the influence of pathogenic bacteria on the sensory and biochemical properties of the product. Gibson et al. [42] reported that *S. aureus* grew at two doubling times (DT), namely in vivo (1.87 h/1 h 52 min) and on laboratory condition (DT 0.4 h/24 min). Meanwhile, the highest average size of *S. aureus* in the grilled chicken sample was 1.3×10^4 colonies/g, indicating the possibility for threatening food safety if consumed after being stored ± 6 h after without any cooling or heating. In addition to the DT, physical changes in foods due to *S. aureus* contamination also become a concern. In fact, *S. aureus* is not a type of spoilage bacteria. Although the size of colonies exceeds the food safety limit, these bacteria are not able to significantly change the physical, biochemical, and sensory properties of foods. It indicates the inability to identify the presence of *S. aureus* solely by visual observation [43].

From a business perspective, even though some employee personnel hygiene (P2) has relatively low performance and could cause food borne disease none of those attributes causes of the consumer dissatisfaction to the observed restaurant. However, based on a food safety perspective, as indicated by the findings on samples of iced tea and grilled chicken, it is important to immediately improve and monitor any potential violation against food safety standard. The management is required to pay attention on this issue. Monitoring and improvement are not only aimed at maintaining the safety of the products served by the restaurant, but also a part of marketing strategies. Customer perceived value may change over time. Access to information and personal knowledge potentially affect consumer perspective on food safety. Personal knowledge plays a role in one's point of view on a thing or phenomenon [44]. Therefore, it is expected that monitoring and continuous improvement would advance the performance of products or services in line with the development of information related to food safety in the community. As a result, it will reduce the gap between the performance of the observed restaurant and the customer expectation based on both information and personal knowledge.

4 Conclusion

The customer satisfaction measurement on restaurant using the food safety perspective have been done. Based on findings obtained from this study, it can be concluded that the restaurant under study has good and satisfactory performance in relation to food safety practices based on customer perception. However, a thorough monitoring is required regarding the employee personnel hygiene and material quality, considering the results of microbiological testing that reveals the presence of *E. coli* in one sample of iced tea and *Staphylococcus aureus* in all samples of grilled chicken.

Acknowledgements The authors thankfully acknowledge the financial support provided by Ministry of Research, Technology and Higher Education Indonesia on this research with contract no. 1035.1/UN27.22/PT.01.03/202.

Appendix A. List of Attributes

Attributes code		Description
Food quality (M)	1	Only fresh food and beverages are served to customers
	2	No foreign object found in the food and beverages
	3	Food and beverages do not emit any strange or typical smells
	4	Food and beverages must be covered from any flies or other insects
	5	All fresh food ingredients and food additives shall be used under good conditions
	6	Consumers do not suffer from illness or food poisoning after consuming food and/or beverages
Service quality (P)	1	Clean, carefully washed utensils to serve food and beverages
	2	Employees do not use bare hands to handle and serve food and beverages. They are required to wear gloves or tongs

(continued)

(continued)

Attributes code		Description
	3	Employees wear clean and neat uniforms or clothes
	4	Employees have short, clean nails without any polish
	5	Employees come to work in good health
	6	Employees use appropriate personal protective equipment (i.e., aprons, gloves, masks)
	7	Employees do not cough and sneeze over food or beverages
	8	Employees are not allowed to wear jewellery or accessories, except watch
	9	All employees, without exception, are not allowed to smoke during the food processing and serving
Environment and facilities (L)	1	The kitchen is clean and well-maintained
	2	Service area and all surfaces of the restaurant are clean and well-maintained
	3	The terrace, the yard, the parking lot, and the building are clean and tidy
	4	No pests and rodents are present around the restaurant
	5	The restaurant is located in a safe location, not exposed to any source of contamination
	6	The restaurant has a clean and tidy garbage disposal area
	7	The cleanliness of the dining table and the cloth used to clean the table
	8	The restaurant provides proper hand washing sinks
	9	The availability of toilets with adequate number and facilities (running water, tissue, hand dryer, soap, and sink)
	10	The toilet is clean and well-maintained

References

1. Andaleeb SS, Caskey A (2007) Satisfaction with food services. J Foodserv Bus Res 10(2):51–65. https://doi.org/10.1300/J369v10n02
2. Saad Andaleeb S, Conway C (2006) Customer satisfaction in the restaurant industry: an examination of the transaction-specific model. J Serv Mark 20(1):3–11. https://doi.org/10.1108/08876040610646536
3. Szymanski DM, Henard DH (2001) Customer satisfaction: a meta analysis of the empirical evidence. J Acad Mark Sci 29(1):16–35
4. Fornell C, Johnson MD, Anderson EW, Cha J, Bryant BE (1996) The American customer satisfaction index: nature, purpose, and finding. J Mark 60(4):7–18
5. BPOM (2019) Laporan Tahunan Pusat Data dan Informasi Obat dan Makanan Tahun 2019," 2019. [Online]. Available: https://www.pom.go.id/new/admin/dat/20200817/Laporan_Tahunan_2019_Pusat_Data_dan_Informasi_Obat_dan_Makanan.pdf
6. Romanda F, Priyambodo P, Risanti ED (2017) Hubungan Personal Hygiene dengan Keberadaan Escherichia Coli pada Makanan di Tempat Pengolahan Makanan (Tpm) Buffer Area Bandara Adi Soemarmo Surakarta. Biomedika 8(1). https://doi.org/10.23917/BIOMEDIKA.V8I1.2899
7. On W, Rahayu S, ahayu WP (2022) Estimates for the burden and costs of foodborne diarrheal illness in Indonesia. Asia-Pacific J Food Saf Secur 3(1):3–16, 2017, Accessed: 22 Jan 2022. [Online]. Available: https://researcharchive.lincoln.ac.nz/handle/10182/9510
8. Jacob CJ, Powell DA (2009) Where does foodborne illness happen-in the home, at foodservice, or elsewhere-and does it matter? Foodborne Pathog Dis 6(9):1121–1123. https://doi.org/10.1089/FPD.2008.0256
9. Ali J, Nath T (2013) Factors affecting consumers' eating-out choices in india: implications for the restaurant industry. J Foodserv Bus Res 16(2):197–209. https://doi.org/10.1080/15378020.2013.782243
10. Auty S (1992) Consumer choice and segmentation in the restaurant industry. Serv Ind J 12(3):324–339. https://doi.org/10.1080/02642069200000042
11. Duarte Alonso A, O'neill M, Liu Y, O'shea M (2013) Factors driving consumer restaurant choice: an exploratory study from the southeastern United States. J Hosp Mark Manag 22(5):547–567. https://doi.org/10.1080/19368623.2012.671562
12. Aksoydan E (2007) Hygiene factors influencing customers' choice of dining-out units: findings from a study of university academic staff. J Food Saf 27(3):300–316. https://doi.org/10.1111/J.1745-4565.2007.00081.X
13. Cullen F (2004) Factors influencing restaurant selection in Dublin. J Foodserv Bus Res 7(2):53–85. https://doi.org/10.1300/J369v07n02_05
14. Ryu K, Shawn Jang S (2008) Dinescape: a scale for customers quality perception of dining environments. J Foodserv Bus Res 11(1):2–22. https://doi.org/10.1080/15378020801926551
15. Bai L, Wang M, Yang Y, Gong S (2019) Food safety in restaurants: the consumer perspective. Int J Hosp Manag 77:139–146. https://doi.org/10.1016/J.IJHM.2018.06.023
16. Karki D (2018) How food quality, price, ambiance and service quality effects customer satisfaction: a study on nepalese restaurants in 2018
17. Haghighi M (2012) Evaluation of factors affecting customer loyalty in the restaurant industry. African J Bus Manag 6(14):5039–5046. https://doi.org/10.5897/ajbm11.2765
18. Waly GZ et al (2021) Quality of service evaluation based on importance performance analysis method and the kano model. J Phys Conf Ser 1764(1):012199. https://doi.org/10.1088/1742-6596/1764/1/012199
19. Tuncer I, Unusan C, Cobanoglu C (2021) Service quality, perceived value and customer satisfaction on behavioral intention in restaurants: an integrated structural model. J Qual Assur Hosp Tour 22(4):447–475. https://doi.org/10.1080/1528008X.2020.1802390
20. Loo PT, Khoo-Lattimore C, Boo HC (2021) How should I respond to a complaining customer? A model of cognitive-emotive-behavioral from the perspective of restaurant service employees. Int J Hosp Manag 95(January):102882.https://doi.org/10.1016/j.ijhm.2021.102882

21. Shin YH (2020) The influence of quality of physical environment, food and service on customer trust, customer satisfaction, and loyalty and moderating effect of gender: an empirical study on foreigners in South Korean Restaurant. Int J Adv Cult Technol 8(3):172–185
22. Cha JM, Borchgrevink CP (2019) Customers' perceptions in value and food safety on customer satisfaction and loyalty in restaurant environments: moderating roles of gender and restaurant types. J Qual Assur Hosp Tour 20(2):143–161. https://doi.org/10.1080/1528008X.2018.1512934
23. Berliansyah RA, Suroso A (2018) The influence of food and beverage quality, service quality, place, and perceived price to customer satisfaction and repurchase intention. J Res Manag 1(1). https://doi.org/10.32424/jorim.v1i1.18
24. Zhong Y, Moon HC (2020) What drives customer satisfaction, loyalty, and happiness in fast-food restaurants in china? Perceived price, service quality, food quality, physical environment quality, and the moderating role of gender. Foods 9(4):460. https://doi.org/10.3390/foods9040460
25. Gabrow RY (2021) Evaluation of customer satisfaction and service quality using Servqual model: the case of fast-food restaurants in Iraq. Period Eng Nat Sci 9(1):336–345. https://doi.org/10.21533/PEN.V9I2.1915.G823
26. DiPietro RB, Levitt J (2019) Restaurant authenticity: factors that influence perception, satisfaction and return intentions at regional american-style restaurants. Int J Hosp Tour Adm 20(1):101–127. https://doi.org/10.1080/15256480.2017.1359734
27. Lawshe CH (1975) A quantitative approach to content validity. Pers Psychol 28:563–575
28. Barber N, Goodman RJ, Goh BK (2011) Restaurant consumers repeat patronage: a service quality concern. Int J Hosp Manag 30(2):329–336. https://doi.org/10.1016/J.IJHM.2010.08.008
29. Lee LE, Niode O, Simonne AH, Bruhn CM (2012) Consumer perceptions on food safety in Asian and Mexican restaurants. Food Control 26(2):531–538
30. Liu P, Lee YM (2018) An investigation of consumers' perception of food safety in the restaurants. Int J Hosp Manag 73:29–35. https://doi.org/10.1016/J.IJHM.2018.01.018
31. Park H, Almanza BA, Miao L, Sydnor S, (Shawn) Jang SC (2016) Consumer perceptions and emotions about sanitation conditions in full-service restaurants 19(5):474–487. https://doi.org/10.1080/15378020.2016.1189740
32. Ungku Fatimah UZA, Boo HC, Sambasivan M, Salleh R (2011) Foodservice hygiene factors-the consumer perspective. Int J Hosp Manag 30(1):8–45. https://doi.org/10.1016/j.ijhm.2010.04.001
33. Tzeng GH, Chang HF (2011) Applying importance-performance analysis as a service quality measure in food service industry, vol 6, no 3. Universidad Alberto Hurtado. Facultad de Economía y Negocios, pp 106–115
34. van Griethuijsen RALF et al (2015) Global patterns in students' views of science and interest in science. Res Sci Educ 45(4):581–603. https://doi.org/10.1007/s11165-014-9438-6
35. Adam I, Hiamey SE, Afenyo EA (2014) Students' food safety concerns and choice of eating place in Ghana. Food Control 43:135–141. https://doi.org/10.1016/j.foodcont.2014.03.005
36. SNI, "SNI 7388:2009 Batas maksimum cemaran mikroba dalam pangan," Standar Nas. Indones, p 17
37. Yunus SP, Pinontoan O, Hubungan Personal Higiene dan Fasilitas Sanitasi dengan Kontaminasi Escherichia Coli Pada Makanan di Rumah Makan Padang Kota Manado Dan Kota Bitung Relationship Personal Hygiene and Sanitation Facilities with Escherichia Coli Contamination Food in Padang Restaurant in Manado and Bitung City
38. Imanuel LA (2019) Identifikasi Bakteri Escherichia coli pada Es Teh di Pasar Malam Kampung Solor Kota Kupang. Karya Tulis Ilm pp 1–59
39. Yildirim T, Sadati F, Kocaman B, Siriken B (2019) Staphylococcus aureus and Staphylococcal enterotoxin detection in raw milk and cheese origin coagulase positive isolates. Int J Sci Lett 1(1):30–41
40. Grispoldi L et al (2019) Study on the growth and enterotoxin production by Staphylococcus aureus in canned meat before retorting. Toxins (Basel) 11(5). https://doi.org/10.3390/toxins11050291

41. Paper S (2008) Staphylococcal Gastroenteritis. Mod Food Microbiol 20(3):545–566. https://doi.org/10.1007/0-387-23413-6_23
42. Gibson B, Wilson DJ, Feil E, Eyre-Walker A (1880) The distribution of bacterial doubling times in the wild. Proc R Soc B Biol Sci 285:2018. https://doi.org/10.1098/rspb.2018.0789
43. Pisestyani H, Ramadhani NN, Sudarwanto M, Lukman DW, Wicaksono A (2021) Sanitation and hygienic practices of ready-to-drink milk seller based on total of coliform and staphylococcus aureus. J Med Vet 4(1):14. https://doi.org/10.20473/jmv.vol4.iss1.2021.14-22
44. Hasell A, Stroud NJ (2019) The differential effects of knowledge on perceptions of genetically modified food safety. Int J Publ Opin Res. 32(1):111–131. https://doi.org/10.1093/ijpor/edz020

Natural-Based Tourism Potential Identification from Indonesia in Post Pandemic Situation

Rayinda P. Soesanto, Afrin F. Rizana, and Artamevia S. Rizaldi

Abstract Indonesia has many areas with tourism potential. It possesses many beautiful natural areas that has potential to be a tourist destination. However, many areas that has potential to become tourism area has not been optimally cultivated. The identification of tourism potential is a complex problem and requires the involvement of a well-defined set of criteria. There has always been some risk involved in choosing potential places for development of tourism. Due to this challenge, this study aimed to propose tourism potential identification for natural-based tourism. The identification process was performed using SECI: socialization, externalization, combination, and internalization. SECI was used to convert the tacit knowledge within the stakeholder and combine it with the information from literature study to gain more information about the indicators and the sub-indicators of tourism potential from stakeholder. Six indicators and 17 sub-indicators were obtained from SECI. Moreover, a multi-level AHP was employed to investigate the level of importance of each indicator and sub-indicator. It is found that the most important aspect that determine the tourism potential area is the accessibility of that area.

Keywords AHP · Indonesia · Potential · Tourism · SECI

1 Introduction

The COVID-19 pandemic has altered both the environment of the travel and tourism sector as well as consumer and tourist behavior, which has adjusted to the new normal. In order for the actors in the tourism industry to survive and prosper, they must be able to quickly adjust to the new normal "disruption" [1]. One of the industries most impacted by this pandemic is the tourist industry [2]. Tourism is considered as critical component of the country's economic development because it has been one of the

R. P. Soesanto (✉) · A. F. Rizana · A. S. Rizaldi
Telkom University, Bandung 40257, Indonesia
e-mail: raysoesanto@telkomuniversity.ac.id

C. N. Rosyidi et al. (eds.), *Proceedings of the 6th Asia Pacific Conference on Manufacturing Systems and 4th International Manufacturing Engineering Conference*, Lecture Notes in Mechanical Engineering,
https://doi.org/10.1007/978-981-99-1245-2_29

main sources of foreign exchange earnings. Therefore, the government of Indonesia depends heavily on the tourism industry [3]. Tourism has been seen as a chance to develop a destination and create jobs, offering amenities that locals and tourists can both enjoy, fostering economic growth and development.

The tourist sector can be thought of as a medium of economic development for developing nations that do not necessitate significant investment, such as Indonesia. One of the key sources of funding for the growth of tourism is natural and cultural attractions, which are currently available for Indonesia [4]. Natural-based tourism (NBT) has grown to play a significant role in the tourist industry and economy as a result of the increasing number of natural tourism locations and travellers [5].

One of the most crucial elements of regional and local development is competitiveness. Increasing destination competitiveness has taken center stage in discussions on tourism policy, both in academic literature and in government discussions. The strength of this industry, its ability to draw large numbers of tourists, and eventually its capability to create money are all determined by how competitive each country's tourism sector is. There has always been some risk involved in choosing potential places for development of tourism. Due to this challenge, this study aimed to propose tourism potential identification for natural-based tourism.

The identification of tourism potential is a complex problem and requires the involvement of a well-defined set of criteria. The determination of the method, criteria, and instrument used in tourism potential identification are crucial since the result will heavily depend on the criteria used. There are many means in tourism potential identification. Previous study regarding the tourism potential identification is the study from Rivero et al. [6] that evaluated tourism potential using an item response theory by considering internal and external aspects. The study from Yan et al. [7] employ a multi-criteria decision-making approach i.e. weighted sum model to assess tourism potential. The study of Hoang et al. [8] combined GIS, AHP, and PCA to assess tourism potential in one area.

2 Methodology

Natural-based tourism (NBT) is defined as the part of tourism that occurs in natural destinations, including activities centered around natural features and attractions [9, 10]. Nature-based tourism definitions range from the location where a specific tourism service is provided in nature to normative definitions outlining how services and activities should be provided. According to Fossgard and Fredman [9] NBT planning focuses on place and product way. NBT planning must consider the resources and capabilities of suitable places to implement recreational programs. The NBT context shows how outdoor activities in natural settings contribute to tourists' experiences, demonstrating the importance of natural attractions. Tourism requires recreation resources in addition to natural beauty. Pull factors require accessible natural habitats and resources. NBT destinations cannot be built in remote locations, highlighting the importance of access to natural places and resources.

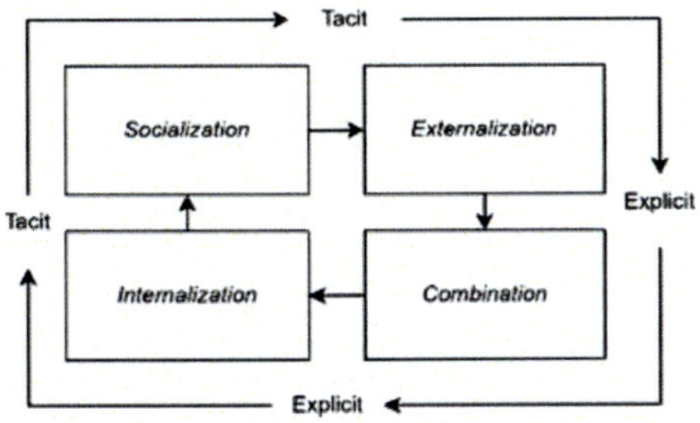

Fig. 1 SECI method

To identify the indicators of tourism potential for NBT, knowledge conversion using SECI is used. Takeuchi and Nonaka [11] contend that knowledge conversion serves as the impetus for the entire process of knowledge generation. Through these channels, individual experiences are shared and amplified throughout the organization. Figure 1 depicts SECI. In this study, SECI is used to convert tacit stakeholder knowledge and combine it with literature study information to gain more information about tourism potential indicators and sub-indicators. Previous study that is used the knowledge conversion is the study from [12–16].

The first phase of SECI is socialization. In this phase, respondents were interviewed face-to-face. Tourism indicators and measuring instruments have been developed from implicit to tacit knowledge. The next phase is externalization, this phase involves the transformation of implicit knowledge into explicit knowledge. This is accomplished by identifying the outcomes of the socialization stage as parameters and indicators that are utilized to calculate weights when employing AHP. The AHP approach, one of the multi-criteria decision analyses, can be used to solve issues involving several choice criteria [17]. Due to its use of pairwise comparisons for measurement and its reliance on expert opinions to create priority scales, AHP is appropriate for multicriteria decision making [18, 19]. The combination stage converts implicit knowledge into explicit knowledge by combining the prior phase's measuring devices and NBT potential indicators. Tourism experts combine each statistic and indicator. Internalization converts explicit to tacit knowledge. After the combination stage, NBT potential indicators and measuring instruments are shared with stakeholders.

3 Result and Discussion

In carrying out the socialization phase of SECI, stakeholder are needed to obtain tacit knowledge about indicators and measuring tools for NBT potential. Transforming the tacit knowledge from stakeholder to another tacit knowledge is done by conducting interviews with stakeholder. The stakeholder that taken the interview are the representative from tourism office, and two manager of tourist area. Table 1 shows the result of the interview.

Table 1 Interview result

Respondent	Interview result
1	Explanation of indicators for the development of natural tourism potential and measuring tools
2	Explanation of indicators and measuring tools for nature tourism based on 6A (Attraction, Amenities, Accommodation, Activity, Accessibilities, and Ancillary Services)
3	Explanation of indicators and measuring tools for nature tourism based on 6A (Attraction, Amenities, Accommodation, Activity, Accessibilities, and Ancillary Services)

In the externalization phase, the tacit to explicit knowledge switch occurs at the externalization step. The outcomes of three resource persons' interviews at the Socialization stage will be documented in explicit knowledge because they will be expressed in terms that are easier to understand in the form of indicators and measurement tools. Table 2 shows the externalization example.

In the combination phase, the documented and compiled explicit knowledge from each stakeholder at the externalization stage, namely in the form of indicators, will be merged with additional explicit knowledge linked to measurement tools and indicators of chosen NBT potential that may be utilized to determine NBT potential. Table 3 shows the example of combination phase.

Following the merger, the weighing will be done with the AHP method that was recommended by a tourist specialist. Six parameters with 17 indications and 65 measuring equipment were obtained from the combined results. There is a hierarchy of parameters and indicators for determining natural tourism potential in Fig. 2.

Following the merging of indicators and measurement tools, weighting is performed. In the stage that follows, the weighing will now be carried out using the AHP approach. Table 4 shows the consistency result from the AHP.

In cases where the value of CR is less than or equal to 0.1, expert judgment is reliable [23]. From Table 4, it is known that the CR value is 0.085 so that conclude the result is acceptable and consistent. According to Saaty [18], the consistency ratio cannot be determined using a 2×2 matrix because the value of RI for this matrix is 0. The result of indicator and weight for this study is shown in Table 5.

Internalization is achieved by discussing indicators and measurement devices. It also discusses how AHP weighted each parameter and indicator. Parameter

Table 2 Externalization phase result

Parameter	Indicator	Measuring instrument
Attraction	There is a peculiarity in tourist attractions	Tourists can enjoy flora and fauna in national parks/wildlife reserves
	The cleanliness of tourist destinations	There are trash cans at every tourist location
	Comfort in tourist attractions	Clean and cool air in tourist areas

Table 3 Combination phase result

Parameter	Indicator	Measuring instrument		Combined results
		Combined results of respondents	Other sources	
Attraction	There is a peculiarity in tourist attractions	Tourists can enjoy flora and fauna in national parks/ wildlife reserves	There are flora that can be enjoyed or in national parks in tourist areas/there is fauna that can be enjoyed by tourists or wildlife reserves/there are beaches/there is sea/ there are mountains/ there are forests, there are waterfalls/there are lakes/there are rivers in tourist areas [20]	There are flora that can be enjoyed or in national parks in tourist areas/there is fauna that can be enjoyed by tourists or wildlife reserves/there are beaches/there is sea/ there are mountains/ there are forests, there are waterfalls/there are lakes/there are rivers in tourist areas
	The cleanliness of tourist destinations	There are trash cans at every tourist location	There are closed waste locations consisting of organic and inorganic waste locations at each tourist location [21, 22]	There are closed waste locations consisting of organic and inorganic waste locations at each tourist location

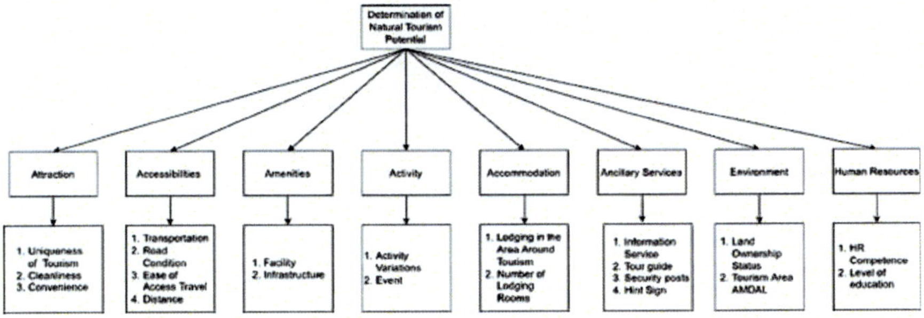

Fig. 2 Hierarchical structure

Table 4 Consistency result

Consistency	Value
Λmax	6.529
Consistency index (CI)	0.106
Random index (RI)	1.240
Consistency ratio (CR)	0.085

Table 5 Parameter and indicator for NBT potential

Parameter	Weight	Indicator	Weight
Attraction	0.100	The uniqueness of nature tourism	0.067
		Cleanliness in tourist attractions	0.467
		Convenience in tourist attractions	0.467
Accessibilities	0.436	Transportation	0.067
		Road conditions	0.427
		Ease of access	0.427
		Tourist spot distance	0.079
Amenities	0.168	Facilities	0.125
		Infrastructure	0.875
Activity	0.149	Activities at tourist attractions	0.167
		Event	0.833
Accommodation	0.100	Lodging	0.833
		Number of rooms	0.167
Ancillary services	0.047	Tourist information service	0.147
		Tour guide	0.624
		Security posts	0.156
		Sign	0.074

and indicator weights determine natural tourism potential. The discussion aims to explore measuring techniques, relative weights of each parameter and indicator, and transform explicit knowledge into tacit knowledge about natural tourism potential indicators. Discussion results show tacit knowledge was delivered as expected.

4 Conclusion

The result of natural tourism potential identification by using SECI and AHP show that there are six important aspects that should be considered in determining tourism potential of one area i.e. attraction, accessibility, amenities, activity, accommodation, ancillary services. Among those aspects, accessibility was the most important, followed by amenities, activities, accommodations, natural attractions, and ancillary services. In tourism development, stakeholders should prioritize road condition, transportation, and access. Without proper access, it's hard to develop a tourist destination. Other than that, the existence of supporting infrastructure and facilities also important. This research has a limitation only until the determination of the weight of each parameter and indicator. For further study, it would be good if there's a framework regarding how to measure each sub-indicator. Moreover, this study only focus on Natural-based tourism. It would be more beneficial if future study also consider other type of tourism such as cultural tourism, artificial tourism, etc.

References

1. Kementerian Pariwisata dan Ekonomi Kreatif, "Tren Pariwisata 2021," 2021.
2. Kristiana Y, Pramono R, Brian R (2021) Adaptation strategy of tourism industry stakeholders during the COVID-19 pandemic: a case study in Indonesia. J Asian Financ Econ Bus 8(4):0213–0223. https://doi.org/10.13106/jafeb.2021.vol8.no4.0213
3. Ikasari H (2021) Strategies to improve the attractiveness of Sam Poo Kong Temple As cultural tourism site of Semarang city. Central Java , Indonesia 9(10):1–9. https://doi.org/10.21776/ub.jitode.2021.009.01.01
4. Aliansyah H, Hermawan W (2021) Peran Sektor Pariwisata Pada Pertumbuhan Ekonomi Kabupaten/Kota Di Jawa Barat. Bina Ekon 23(1):39–55. https://doi.org/10.26593/be.v23i1.4654.39-55
5. Mäntymaa E, Ovaskainen V, Juutinen A, Tyrväinen L (2018) Integrating nature-based tourism and forestry in private lands under heterogeneous visitor preferences for forest attributes. J Environ Plan Manag 61(4):724–746. https://doi.org/10.1080/09640568.2017.1333408
6. Sánchez Rivero M, Sánchez Martín JM, Rengifo Gallego JI (2016) Methodological approach for assessing the potential of a rural tourism destination: an application in the province of Cáceres (Spain). Curr Issues Tour 19(11):1084–1102. https://doi.org/10.1080/13683500.2014.978745
7. Yan L, Gao BW, Zhang M (2017) A mathematical model for tourism potential assessment. Tour Manag 63:355–365. https://doi.org/10.1016/j.tourman.2017.07.003
8. Hoang HTT, Truong QH, Nguyen AT, Hens L (2018) Multicriteria evaluation of tourism potential in the central highlands of Vietnam: combining geographic information system (GIS), analytic hierarchy process (AHP) and principal component analysis (PCA). Sustain 10(9):1–20. https://doi.org/10.3390/su10093097
9. Fossgard K, Fredman P (2019) Dimensions in the nature-based tourism experiences cape: an explorative analysis. J Outdoor Recreat Tour 28(2018):0–1. https://doi.org/10.1016/j.jort.2019.04.001
10. Rezvani M, Nickravesh F, Darban Astaneh A, Kazemi N (2022) A risk-based decision-making approach for identifying natural-based tourism potential areas. J Outdoor Recreat Tour 37(December 2021):100485. https://doi.org/10.1016/j.jort.2021.100485
11. Nonaka I, Takeuchi H (1995) Knowledge-creating company. Knowledge-Creating Co. https://doi.org/10.1016/S0969-4765(04)00066-9
12. Kurniawati A, Andrawina L, Soesanto RP (2016) E-Learning Pada Kegiatan Maintenance Mesin Berdasarkan Knowledge Conversion Dengan Metode SECI. J Rekayasa Sist Ind 1:137–140
13. Atma and Hediyanto (2017) Best Practice Kegiatan Corrective Maintenance untuk Kerusakan Bearing pada Mesin Millac 5H 6P Berdasarkan Knowledge Conversion. Pros SNTI dan SATELIT 2017, vol 2017, no July 2018, pp C1-7
14. Mahatmavidya PA, Soesanto RP, Kurniawati A, Andrawina L (2018) Characteristic and competency measurement instrument development for maintenance staff of mechanical expertise with SECI method: a case of manufacturing company. In: IOP conference series: materials science and engineering, vol 319, no 1. https://doi.org/10.1088/1757-899X/319/1/012008
15. Kusumastuti D, Soesanto RP, Kurniawati A, Kurniawan MT (2020) E-learning content design using ADDIE and SECI: case of shelving activity in research organization. Int J Adv Sci Eng Inf Technol 10(3):1072–1077. https://doi.org/10.18517/ijaseit.10.3.10841
16. Ananto BD, Kurniawati A, Rizana AF, Athari N (2021) Perancangan Konten e-learning Pada Kegiatan Pembuatan Dan Pemasangan Kubah masjid di Perusahaan Tiga Berlian Menggunakan Metode Seci Dan Addie. J@ti Undip J Tek Ind 16(2)

17. Dirpan A (2018) Combining an analytic hierarchy process and TOPSIS for selecting postharvest technology method for Selayar citrus in Indonesia. In: IOP Conference Series: Earth and Environmental Science, vol 156, no 1. https://doi.org/10.1088/1755-1315/156/1/012031
18. Saaty TL (2008) Decision making with the analytic hierarchy process. Int J Serv Sci 1(1):83–98. https://doi.org/10.1504/ijssci.2008.017590
19. Soesanto RP, Tripiawan W, Darmawan I (2020) Design of Multicriteria decision making tools for IT project selection: a case from software house. In: IOP Conference Series: Materials Science and Engineering, vol 847, no 1. https://doi.org/10.1088/1757-899X/847/1/012013
20. Isdarmanto, Dasar Dasar Kepariwisataan dan Pengelolaan Destinasi Pariwisata. Yogyakarta: Gerbang Media Aksara (2016)
21. Fitri WY, Wibowo AW, Ariyanto DB (2020) Kebijakan Pengelolaan Sampah Di Daerah Utama Tujuan Wisata. J Kebijak Publik 11(2):105. https://doi.org/10.31258/jkp.11.2.p.105-112
22. Kementrian Pariwisata Dan Ekonomi Kreatif, *Peraturan Menteri Pariwisata Dan Ekonomi Kreatif Nomor 2 Tahun 2021*. Jakarta, 2021
23. Fatimah HA, Trisminingsih R (2018) Analyzing success factors of enterprise resource planning adoption using analytical hierarchy process. Int J Innov Enterp Syst 2(01):45–49. https://doi.org/10.25124/ijies.v2i01.16

Development of Framework for Flexible Job Shop Scheduling Based on Digital Twin to Tackle Disturbing Events

Muhammad Usman Nisar, Anas Ma'ruf, and Abdul Hakim Halim

Abstract Due to dynamism in job shop scheduling (JSS), the manufacturing environment is highly complex, dynamic, and unpredictable, where disturbing events such as new job arrival, machine breakdown, etc. are unavoidable and change the initial state of the manufacturing environment. To address these issues, this study employs the notion of a flexible shop floor that can meet specific product demands while also adapting to disturbing occurrences in real time. Under new job arrival, machine failure, order cancellation, and rush orders, flexible job shop scheduling (FJSS) is addressed for this purpose, and an efficient framework based on digital twin technology is provided, capable of recognizing and capturing the disturbing factors in real time and triggering timely re-scheduling to mitigate the negative impact in a cost-effective manner. This research could tackle disturbing factors effectively and efficiently in real time.

Keywords Flexible job shop scheduling · Disturbing factors · Proactive–reactive rescheduling · Digital twin

1 Introduction

1.1 Background

Job shop scheduling (JSS) and flexible job shop scheduling (FJSS), two of these categories of shop scheduling, are NP-hard issues. They are one of the most well-known combinatorial optimization problems and have drawn a lot of interest from

M. U. Nisar · A. Ma'ruf · A. H. Halim (✉)
Institut Teknologi Bandung, Bandung 40132, Indonesia
e-mail: ahakimhalim@mail.ti.itb.ac.id

M. U. Nisar
e-mail: 33421701@mahasiswa.itb.ac.id

C. N. Rosyidi et al. (eds.), *Proceedings of the 6th Asia Pacific Conference on Manufacturing Systems and 4th International Manufacturing Engineering Conference*, Lecture Notes in Mechanical Engineering,
https://doi.org/10.1007/978-981-99-1245-2_30

scholars because they (1) reflect real-world applications and (2) have great scientific research value. An ideal JSS is a method of assigning n jobs (each job consists of a sequence of operations) to m machines (each operation can only be processed by a specific machine) with a purpose to find an effective schedule which is an allocation of the operations at a specific time t, in order to optimize some objective function (makespan, tardiness, flow time etc.) under some constraints (scheduling efficiency, stability, cost etc.) [1–3].

Flexible job shop scheduling problem (FJSS) is a common relaxation and generalization of JSS, providing more flexibility in terms of allocating the job-operations on available candidate machines. Due to flexibility of machining the same job with available candidate machines, FJSS has to make two decisions simultaneously: (1) machine assignment (the selection of an appropriate machine for each operation from its available candidate machines), and (2) operation sequencing (to determine the order of processing the allocated jobs in each machine) [3, 4].

2 Research Gap

FJSS has shown and expanded its worth in a variety of ways; nonetheless, there are a number of shortcomings related with FJSS that must be recognized and addressed effectively. These shortcomings are detailed below;

1. The job processing route is unexpected, and each processing machine is not unique, providing greater flexibility to the company while raising the complexity of resolving it [5]. As a result, it is critical to record both, the job processing route and the processing machine, in real time [4].
2. The majority of FJSS literature assumes that the production environment is static (that all relevant data is accessible at time zero and does not change over time) [1]. Furthermore, a number of suppositions are made when studying static scheduling, one of which is that machines are always available in the scheduling horizon (i.e., never break down or require planned maintenance) [6], and that job characteristics, such as the release date, due date, and processing time, are predetermined [1, 5]. However, these assumptions are very restrictive in many real-world settings [7].
3. As FJSS's components are more synchronized, connected, and interdependent than typical manufacturing systems are [8], it is unavoidable that it may experience unpredictable disturbing factors [6, 9, 10], such as new job arrival, order cancellations, rush order, random machine breakdowns, tool breakage, etc. Table 1 summarizes research done for these disturbing factors (machine failure being the most examined and rush order and order cancellation being the least studied) with various resolution methods (metaheuristic mainly employed and digital twin less used) (see Fig. 1).

To address these issues, companies' production systems must be more intelligent, responsive, and flexible [28] allowing them to quickly identify these disturbing

Table 1 Existing studies for number of issues versus resolution methods used

	New job arrival	Machine failure	Rush order	Job cancellation	Others
1-Digital twin		[6, 11]	[6]		
2-Approximate	[4, 8, 12–18]	[4, 8, 12–15, 17–20]	[16]	[13, 16]	[4, 14, 15, 21, 22]
3-Exact	[23, 24]	[23–26]			[27]
This article: combo of 1&2	✓	✓	✓	✓	

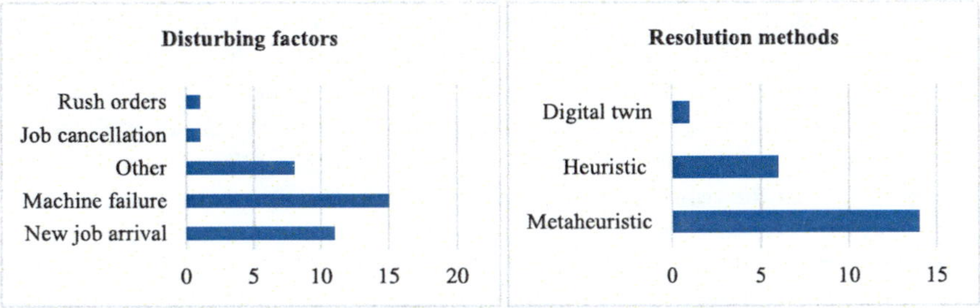

Fig. 1 Literature studies for different disturbing factors

factors in real time and provide a warning as soon as possible for prompt rescheduling in a cost-effective manner [29]. Developing effective method for dealing with the challenges of such practical environments is the major motivation behind this research. Thus, the goal of this research is to develop a framework using a mix of digital twin technology and a metaheuristic algorithm to identify, capture, and mitigate the negative impact of machine failure, new job arrival, job cancellation, and rush orders.

3 Proposed Framework

Numerous industrial practices demonstrate that digital twin technology can enable real-time tracking of production processes that are subject to unpredictability and hidden influence [11]. In DT, by using the concepts of internet of things (IoT) and cyber-physical systems (CPSs) etc., detailed information about machines, orders (or jobs), operations, and equipment, etc. can be collected in real-time [6] and can be made available to use for decision making. It effectively performs real-time mapping and bi-directional interaction of data between the physical and information worlds, and thus deeply integrating operational data, environmental changes, dynamic disturbing factors, and other information in the physical production line with data in the virtual space [5].

In general, a physical job shop and a virtual job shop mutually cooperate through a cyber-physical system (CPS) [30]. In a physical job shop, dynamic disturbances can be detected in time, the type of dynamic events is identified, and the updated data information is passed to the virtual job shop and subsequently, the virtual job shop reschedules the unfinished operations in the previous simulation plan immediately after a dynamic event occurs, by executing the built-in algorithm, and an optimal scheduling result is fed back to the shop floor control system in the physical job shop [30]. With this type of real-time connection, production systems can be managed, monitored, and controlled and industrial processes can be modified to accomplish the aims of self-adaptation, quick responsiveness, and intelligence [2]. The DT-based approach is considered superior to the traditional dynamic scheduling method, since they can provide accurate prediction, and detect disturbance based on a dynamic reference [1].

Figure 2 demonstrates a framework based on digital twin technology for addressing disturbing factors in flexible job shop scheduling. The initial schedule is developed with the following factors in mind: the history of the same process, delays and distortions, potential dynamic occurrences, and periodic rescheduling information. It should be emphasized that the schedule is believed to be established with some degree of freedom (flexibility) to accommodate and respond to disturbing factors based on the proactive rescheduling strategy. Moreover, a periodic policy reschedules the facility on a regular basis and applies the schedules on a rolling time horizon [31]. In rolling time horizon, the entire process is divided into several time periods according to the actual processing conditions and each period is a cycle and the system automatically performs another rescheduling for the remaining unprocessed operations after completing one period until the end of the entire process [14].

The initial schedule is then subsequently input into the company's control system, and job processing begins based on that schedule, while rescheduling continues in line with the cycles stated in the initial plan. It is worth noting that the company's control system is integrated with digital twin technology in order to map real time data, allowing for bidirectional interaction between the physical and information worlds. Using this data, tracking, monitoring, and management of the manufacturing process is done, and simulation makes it easier to detect the presence of disturbing factors (new job arrival, machine failure, rush order and order cancellation) in real time.

4 Conclusion

Due to the unpredictability of customer demands for customized products, manufacturing companies are seeing a change in their market. To address this issue, industries are using technological advancements to increase the competitiveness of their production tools. One such technology in this regard is digital twin technology. Through the use of digital twin technology, it is possible to follow, oversee, and manage the production process as well as to detect and record dynamic events like the arrival

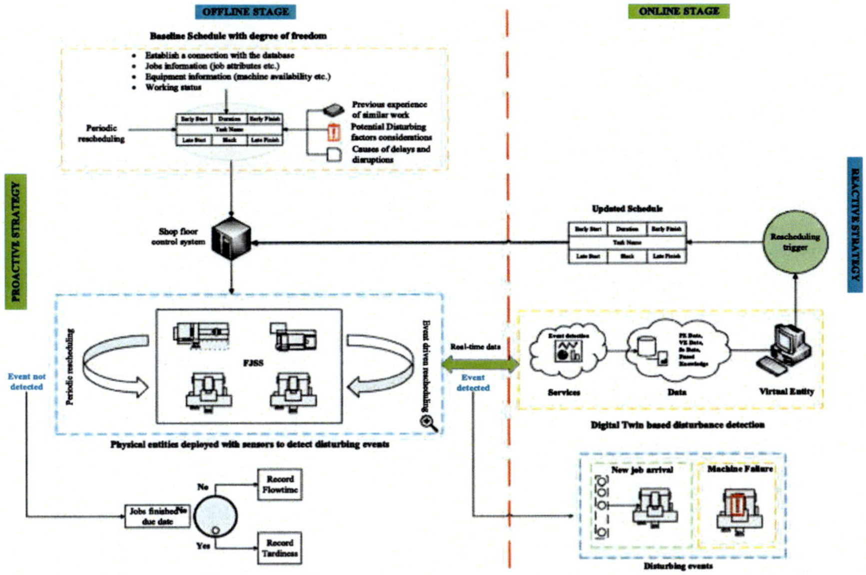

Fig. 2 Digital twin technology-based framework for FJSS to capture real time disturbing factors to kick-off timely rescheduling in cost-effective manner

of new jobs, equipment breakdown, rush order and job cancellation, in real time to prompt cost-effective and on-time rescheduling. According to studies, digital twin technology can effectively handle such unpredictable dynamic situations. A framework based on digital twins has been developed for this purpose, using proactive reactive rescheduling method using hybrid policy to initiate rescheduling periodically as well as whenever a disturbing event occurs. Future study will concentrate on applying digital twin technologies in a real-world situation while taking into account a variety of disturbing factors.

References

1. Zhang M, Tao F, Nee AJJoMS (2021) Digital twin enhanced dynamic job-shop scheduling 58:146–156
2. Saqlain M et al (2022) A monte-carlo tree search algorithm for the flexible job-shop scheduling in manufacturing systems, p 1–24
3. Zhang F et al (2021) Collaborative multifidelity-based surrogate models for genetic programming in dynamic flexible job shop scheduling
4. Fangfang Z, Yi M, Mengjie Z (2019) A new representation in genetic programming for evolving dispatching rules for dynamic flexible job shop scheduling. EvoCOP
5. Lilan L et al (2022) Digital twin-driven adaptive scheduling for flexible job shops. Sustainability
6. Ghaleb M, Taghipour S, Zolfagharinia HJJOMS (2021) Real-time integrated production-scheduling and maintenance-planning in a flexible job shop with machine deterioration and condition-based maintenance 61:423–449

7. Erich T (2022) Types of flexible job shop scheduling: a constraint programming experiment. ICAART
8. Liping Z et al (2012) Scheduling flexible job shop in dynamic environment based on a memetic algorithm. In: 2012 IEEE 11th international conference on cognitive informatics and cognitive computing
9. Singh MR (2014) A study on flexible flow shop and job shop scheduling using meta-heuristic approaches
10. Donya R et al (2013) A new approach to reducing the effects of stochastic disruptions in flexible flow shop problems with stability and nervousness. Int J Manage Sci Eng Manage
11. Zhang F et al (2022) Digital twin data-driven proactive job-shop scheduling strategy towards asymmetric manufacturing execution decision 12(1):1–19
12. Sicheng Z, Tak Nam W, Wong TN (2017) Flexible job-shop scheduling/rescheduling in dynamic environment: a hybrid MAS/ACO approach. Int J Prod Res
13. Ahmad Shahrizal M, Zakaria Z, Deris S (2016) Rescheduling for JSSP and FJSSP using clonal selection principle approach—a theory
14. Wang L, Luo C, Cai JJJOAT (2017) A variable interval rescheduling strategy for dynamic flexible job shop scheduling problem by improved genetic algorithm
15. Wang Y, Han JJIJOA, Propagation (2021) A FJSSP method based on dynamic multi-objective squirrel search algorithm
16. Ferreira IC et al (2020) Artificial bee colony algorithm applied to dynamic flexible job shop problems. In: International conference on information processing and management of uncertainty in knowledge-based systems. Springer
17. Ning G, Cao D (2021) Multi-step genetic algorithm for solving dynamic flexible job shop scheduling problem. In: 2021 IEEE international conference on advances in electrical engineering and computer applications (AEECA). IEEE
18. Tian S et al (2019) An energy-efficient scheduling approach for flexible job shop problem in an internet of manufacturing things environment, 7:62695–62704
19. Zhao M et al (2019) A random forest-based job shop rescheduling decision model with machine failures, p 1–11
20. Yonemoto R, Suwa HJIJOAT (2021) Reactive scheduling based on adaptive manipulator operations in a job shop configuration with two machines. 15(5): 661–668
21. Álvarez-Gil N et al (2021) A discrete firefly algorithm for solving the flexible job-shop scheduling problem in a make-to-order manufacturing system 29(4):1353–1374
22. Andrade-Pineda JL et al (2020) Scheduling a dual-resource flexible job shop with makespan and due date-related criteria 291(1): 5–35
23. Mokhtari H, Dadgar MJC, Research O (2015) Scheduling optimization of a stochastic flexible job-shop system with time-varying machine failure rate. 61:31–45
24. Chen Q, Deng L, Wang HJIJOSM (2018) Optimization of multi-task job-shop scheduling based on uncertainty theory algorithm 17(3):543–552
25. Xiaoqiang W, Songling T, Lei Z (2019) The internet of things enabled shop floor scheduling and process control method based on petri nets. IEEE Access
26. Wang L et al (2021) Dynamic job-shop scheduling in smart manufacturing using deep reinforcement learning 190:107969
27. Cao Z et al (2019) An adaptive scheduling algorithm for dynamic jobs for dealing with the flexible job shop scheduling problem 61(3): 299–309
28. Tliba K et al (2022) Digital twin-driven dynamic scheduling of a hybrid flow shop. J Intell Manuf
29. Tighazoui A, Christophe S, Nathalie S (2020) New efficiency-stability criterion in a rescheduling problem with dynamic jobs weights. In: 2020 7th international conference on control, decision and information technologies (CoDIT)
30. Yan Q et al (2022) Digital twin-enabled dynamic scheduling with preventive maintenance using a double-layer Q-learning algorithm 144:105823
31. Vieira GE, Herrmann JW, Lin EJJOS (2003) Rescheduling manufacturing systems: a framework of strategies, policies, and methods. 6(1):39–62

Online Transportation Order Matching Considering Passenger Priority Using Recency Frequency Monetary Method

Andreas Handojo, Nyoman Pujawan, Budi Santosa, and Moses Laksono Singgih

Abstract With the development of transportation technology, online transportation is one of the choices of public transportation which is currently growing rapidly. Passengers can book transportation services through the app. The application will appoint a driver to fulfill the order. It takes a system that can prioritize service to loyal customers, to get priority to be served first. This is important in retaining loyal customers. In this research, the Recency Frequency Monetary (RFM) method was used to classify loyal customers. Based on this classification, drivers are appointed to fulfill orders from passengers (order matching). The method used to pair drivers and passengers is the Hungarian method. The simulation of this system will be tested using the Google Maps API. Based on the tests carried out, the results obtained that the system made was able to prioritize the fulfillment of orders from passengers based on its Recency Frequency Monetary value.

Keywords Recency frequency monetary · Hungarian · Online transportation · Customer segmentation

1 Introduction

Technology developments make all areas of life more advanced, including in the field of transportation. This is indicated by the emergence of several online transportation companies such as Uber, Lyft, Didi [1, 3]. In the online transportation business, passengers can place orders through the application. The online transportation provider will fulfill the order by appointing a driver (order matching). One of the

A. Handojo (✉) · N. Pujawan · B. Santosa · M. L. Singgih
Industrial Engineering Department, Sepuluh Nopember Institute of Technology, Surabaya, Indonesia
e-mail: handojo@petra.ac.id

A. Handojo
Informatics Department, Petra Christian University, Surabaya, Indonesia

C. N. Rosyidi et al. (eds.), *Proceedings of the 6th Asia Pacific Conference on Manufacturing Systems and 4th International Manufacturing Engineering Conference*, Lecture Notes in Mechanical Engineering,
https://doi.org/10.1007/978-981-99-1245-2_31

problems faced by this business is the condition of uncertainty in meeting emerging demands. This problem is caused because passengers can place orders anytime and anywhere. This condition certainly causes uncertainty in meeting demand [6, 10].

To fulfill incoming orders, online transportation companies need to prioritize which orders will be served first. In order to determine the matching assignment between passengers and drivers. This is important to build good relationships with application users and improve service to loyal customers. Loyal customers who use the service need to be prioritized to get the best and fastest service. Therefore, it is necessary to classify existing customers.

This research aims to classify loyal customers in the online transportation business so that online transportation provider can prioritize these passengers in the services provided. In this research, those who are categorized as loyal customers are passengers who recently used the application, frequently used online transportation services, and spent a lot of money using the service. Based on the priority level of the passengers, the system will match the available passengers and drivers based on the closest pick-up distance.

In determining the priority of passengers in this study, the Recency Frequency Monetary (RFM) method will be used to segment passenger activity. The RFM method was chosen because it is a method that can perform simple but effective customer segmentation [5, 9]. Meanwhile, the process of matching orders between drivers and passengers in this study will be carried out using the Hungarian method by considering the pick-up distance and passenger RFM score. The Hungarian method is used because it is one of the methods that is widely applied to assignment problems because the calculation time is fast and relatively accurate [8, 11].

2 Literature Review

The RFM method was first introduced by Stone and Bob in 1989. The RFM method is used as a simple but effective segmentation method. This method is applied in market segmentation [4]. RFM analysis is widely used to determine customer ratings based on customer transaction history information that has been recorded in the past. This method has been used in various applications that involve large numbers of customers, such as online purchases, retailing, and others [5].

In this method, customers will be grouped based on 3 dimensions, namely Recency (R), Frequency (F), Monetary (M). Where the recency value is obtained from the number of days between the last transaction made by the customer and the current day. While the frequency value is obtained from the number of transactions made by the customer. While the monetary value is the total money that has been spent by the customer. The RFM score will be processed using a numerical scale based on the quantile value. For example, a number from 5 to 1 can be applied, where the top quantile is assigned a score of 5, and the lowest quantile is assigned a score of 1. After obtaining a score for each recency, frequency, and monetary, customers will be grouped based on their total score.

Research in order matching on drivers and passengers has been done before. For example, in a study conducted by Diamant and Baron [7], which made the installation of two queues by considering the patience of the customer. This study conduct research on order fulfillment system with a combinatorial optimization method to maximize driver acceptance rate for each order without reducing customer satisfaction. The advantage of this research is that drivers can accept or reject tasks that have been assigned to the machine, so as a precaution, the machine will pair jobs with more than one suitable driver [12]. There is research with order fulfillment system by combining Hungarian and Greedy methods. This research focuses on combining the two methods to produce a more effective order fulfillment system. The simulation is carried out statically where passengers will wait at existing stations, and the calculated factors are the distance and waiting time of the passengers [2]. As for customer segmentation research using RFM, there are studies that prioritize customers based on customer transaction data at the company. This study compares the K-Means and Fuzzy-C-Means methods to create customer priority segmentation that has previously been calculated using the RFM method [5]. Based on the studies above, there is no research that applies the customer segmentation method with RFM to pair drivers and passengers (order matching).

3 Research Method

The research method in this study can be seen in Fig. 1. The main process is reading the required data from the database, performing the RFM process, calculating the factors that are used as factors to be considered, installing the Hungarian method, and displaying the results obtained into the application. Figure 1 shows the system process.

The application will generate passenger and driver data on the map using the Google Map API. Driver and passenger locations that have been randomly generated will be depicted on a map using the Google Maps API. Passengers that appear are counted as passengers who place orders. Meanwhile, the drivers that appear are counted as available drivers. Transaction history data from passengers in the past will be considered using the RFM method. The score from the RFM is used as one of the values in the calculation of the Hungarian method. In addition to the RFM score, the Hungarian method will consider the mileage factor from the driver to the pickup distance from the passenger. The order matching results obtained are then depicted

Fig. 1 System process

on the map in the form of a pickup route. The generation of passenger and driver appearances is carried out repeatedly, the number of occurrences of passengers and drivers will be determined randomly. Each occurrence is called a batch. Passengers and drivers who fail to find a partner in one batch will go to the next batch to be processed again.

4 Implementation

The RFM calculations are carried out on transaction data from passengers. For example, on Table 1 show as example from 5 passenger transaction data. From the transaction data, the recency, frequency, and monetary values of each passenger will be calculated. For example, the process was executed on 02/05/2022. The recency value is calculated by finding the number of days since the passenger last transacted. For example, in Passenger 1, the value is zero because the last transaction was the day when the data was processed (02/05/2022). Frequency is calculated by finding the number of times a passenger has made a transaction. For example, Passenger 1 gets a value of 14 because it has made 14 transactions on the entire transaction data. Monetary is calculated by finding how many fees have been paid by the passenger in the transaction. For example, in Passenger 1, the value of 248,100 is obtained from the total price that has been paid by Passenger 1. The results of the total transaction data from each passenger can be seen in Table 2.

After getting the recency, frequency and monetary data for each passenger, the RFM value is calculated. Giving the RFM value is done by determining the category grouping of the existing data by using quantile division (Table 3). In the example,

Table 1 Passenger transaction data (example)

Id passenger	Transaction date	Price
1	02/05/2022	27,500
2	29/04/2022	20,250
3	01/04/2022	13,750
4	23/03/2022	20,250
5	02/02/2022	10,000

Table 2 Result of Total RFM Transaction Data

Id passenger	Recency	Frequency	Monetary
1	0	14	248,100
2	3	14	308,500
3	27	10	198,600
4	40	6	99,750
5	67	4	49,000

Table 3 Quantile on RFM

Quantile	Recency	Frequency	Monetary
Q1	14	10	100,000
Q2	30	7	80,000
Q3	70	5	50,000
Q4	90	3	20,000

this research will use quantile calculations on the recency, frequency, and monetary values to obtain the RFM category they have. The division of quantile values can be determined. For example, like other RFM scoring, the recency category will be divided into 5 categories (category 5 to category 1), with the division of the value limit being less than 14 days, 30, 70, and above 90 days. The first quantile category will have the best RFM value (5 points), while the fifth quantile category will have the worst RFM value (1 point).

For example, in Passenger 1 (Table 2), the Recency Factor will be categorized as 5 (RFM score 5) because the passenger made a transaction less than 14 days ago. Meanwhile, the frequency factor will be categorized as category 5 (RFM score 5) because the passenger has transacted more than 10 times. The monetary factor will be included in category 5 (RFM score 5) because the passenger has spent more than 100,000. The RFM results that have been obtained are then used to calculate the total RFM score for each passenger by adding up the recency, frequency, and monetary categorization scores. The higher the RFM score obtained, the higher the priority level of the passenger. The results of the calculation of the RFM score can be seen in Table 4.

To test the result from RFM score and Hungarian method, this research conduct 60 batches can be seen in Table 5. Each batch will bring up drivers and passengers randomly, both in terms of number and location of appearance. Each batch running in duration 1 min. This testing conduct with weight for each factor: distance 10%, duration 10%, RFM score 80%. From the test results obtained, it can be concluded that the application of the RFM method in the order matching process on online transportation can run well with a standard deviation pickup distance of 1285 km.

To visualize the position of drivers, passengers and routes from the results of the assignment, a visualization of driver and passenger markers and the route taken on the Google Maps API is used. Drivers who are available are visualized using a blue

Table 4 RFM score

Id passenger	R score	F score	M score	RFM score
1	5	5	5	15
2	5	5	5	15
3	4	4	5	13
4	2	3	4	9
5	2	2	2	6

Table 5 Testing result

	RFM + Hungarian Result
Number of batches	60
Number of assignments	328
Min pick-up time	0.033 min
Max pick-up time	9.893 min
Avg pick-up time	4.368 min
Standard deviation pickup time	2.452 min
Total pick-up time	1433.883 min
Min pick-up distance	0.016 km
Max pick-up distance	4.487 km
Avg pick-up distance	1.275 km
Standard deviation pick up distance	1.285 km
Total pick-up distance	424.927 km

marker icon, while passengers order location use a red marker icon. The red route is used to visualize the pickup route from the driver's starting point to the passenger's starting point. The visualization image on the map can be seen in Fig. 2.

Fig. 2 Order matching result

5 Conclusion

With the development of the online transportation business. Online transportation service providers need a system that can provide better service to loyal customers. One of them is by prioritizing the fulfillment of orders to customers. In this research, customer classification was carried out using the RFM method which classifies customers based on factors recently using the service (recency), how often they use the service (frequency), how much money has been spent in using the service (monetary). The results of this RFM score are used as one of the determinants of order fulfillment. the order matching process between the driver and passenger is carried out using the Hungarian method by considering the pickup distance factor. From the tests carried out, the results obtained that customer priorities with the RFM method on online transportation can run well. Where, testing using 60 batches (with a duration of 1 min/batch) resulted in a standard deviation pickup distance of 1285 km.

Future research can be developed by including additional factors in the calculation of the assignment method, such as the driver's rating, the total distance traveled by the driver, and others. Research can also be developed by performing different weighting of the factors in the calculation of the assignment method. In addition, research can be developed, such as by doing the same thing with drivers. So that in matching orders, both drivers and passengers are prioritized.

References

1. Alonso-mora J, Samaranayake S, Wallar A, Frazzoli E, Rus D (2017) On-demand high-capacity ride-sharing via dynamic trip-vehicle assignment. Proc Natl Acad Sci
2. Babicheva T, Burghout W, Andreasson I, Faul N (2018) The matching problem of empty vehicle redistribution in autonomous taxi systems. Procedia Computer Science 130:119–125. https://doi.org/10.1016/j.procs.2018.04.020
3. Bai R, Li J, Atkin JAD, Kendall G (2014) A novel approach to independent taxi scheduling problem based on stable matching. J Oper Res Soc 65(10):1501–1510. https://doi.org/10.1057/jors.2013.96
4. Birant D (2011) Data mining using RFM analysis. In: Tech, pp 91–108
5. Christy AJ, Umamakeswari A, Priyatharsini L, Neyaa A (2018) RFM ranking—an effective approach to customer segmentation. J King Saud Univ Comput Inf Sci. https://doi.org/10.1016/j.jksuci.2018.09.004
6. Cohen T (2018) Being ready for the next Uber: can local government reinvent itself? Eur Transp Res Rev 10:57. https://doi.org/10.1186/s12544-018-0330-8
7. Diamant A, Baron O (2019) Double-sided matching queues: priority and impatient customers. Oper Res Lett. https://doi.org/10.1016/j.orl.2019.03.003
8. Hou L, Li D, Zhang D (2018) Ride-matching and routing optimization: models and a large neighborhood search heuristic. Transp Res Part E 118:143–162
9. McCarty JA, Hastak M (2007) Segmentation approaches in data-mining: a comparison of RFM, CHAID, and logistic regression. J Bus Res 60(6):656–662. https://doi.org/10.1016/j.jbusres.2006.06.015
10. Miao F, Han S, Hendawi AM, Khalefa ME, Stankovic JA, Pappas GJ (2017) Data-driven distributionally robust vehicle balancing using dynamic region partitions. In: Proceedings of

the 8th international conference on cyber-physical systems—ICCPS '17, pp 261–271. https://doi.org/10.1145/3055004.3055024
11. Munkres J (1957) Algorithms for the assignment and transportation problems. J Soc Ind Appl Math 5(1):32–38. https://doi.org/10.1137/0105003
12. Zhang L, Hu T, Min Y, Wu G, Zhang J, Feng P, Ye J (2017) A taxi order dispatch model based on combinatorial optimization. In: Proceedings of the 23rd ACM SIGKDD international conference on knowledge discovery and data mining. https://doi.org/10.1145/3097983.3098138

Manufacturing Losses Analysis of Ring Spinning Machine Based on Overall Equipment Effectiveness Evaluation: A Textile Case Study

Hasna Khairunnisa, Annas Toat, Ahmad Darmawi, and Agus Ardiyanto

Abstract Ring spinning machine is the largest number of machines in yarn manufacturing production floor. This condition leads to a higher amount of machine downtime in compared to other machines. Downtime is considered as manufacturing losses and have to be reduced. Manufacturing losses on ring spinning machines in PT Z were analyzed based on Overall Equipment Effectiveness (OEE) evaluation to measure the current machines conditions. The results showed 87% of availability rate, 94% of performance rate, 97% of quality rate, and 80% of OEE value, which were all considered low compared to the ideal values. Manufacturing losses were then computed and divided into six big losses, where idling and minor losses held the biggest loss with 43% of overall time loss. Those losses were examined using pareto and 5 whys approach to find the possible cause of each problems. The frequent cause of problems was mainly the irregular doffing time, worker (operator, doffer, feeder) problems, and machine problems. Improvement actions on each of the losses and failure were finally proposed based on the critical cause of loss.

Keywords Ring spinning · OEE · Six big losses · Yarn manufacturing

1 Introduction

The yarn manufacturing process consists of several steps and machines based on the type of yarn processed. Among those machines, the ring spinning machine is considered as critical section since rope-like fiber strands become a fine yarn through a high amount of attenuation [1]. Ring spinning machines generally holds the largest number of machines compared to others in yarn manufacturing, as also happened in PT Z. Hence the ring spinning machine downtime contributes more to overall

H. Khairunnisa (✉) · A. Toat · A. Darmawi · A. Ardiyanto
Yarn Manufacturing Engineering Departement, Akademi Komunitas Industri Tekstil dan Produk Tekstil Surakarta, Surakarta, Indonesia
e-mail: hasna@ak-tekstilsolo.ac.id

C. N. Rosyidi et al. (eds.), *Proceedings of the 6th Asia Pacific Conference on Manufacturing Systems and 4th International Manufacturing Engineering Conference*, Lecture Notes in Mechanical Engineering,
https://doi.org/10.1007/978-981-99-1245-2_32

production loss compared to other yarn manufacturing machines. Unit AB of PT Z has 156 units of ring spinning machines with 516 spindles each. Those 156 machines produce different variants of product based on the fiber type and yarn number, which determine the machine settings like spindle speed and twist per inch. One of the yarn variants is TR 30 KT 65/35 yarn where its monthly production did not meet the machine efficiency target of 96% due to machine downtime. Downtime data showed that most of the downtime were caused by machine breakdown and spare part change.

Therefore, machine maintenance is really crucial on ensuring a good machine performance. One of the steps in implementing Total Productive Maintenance is the measurement cycle where Overall Equipment Effectiveness (OEE) is calculated. OEE is a quantitative metric launched by Nakajima to measure productivity of individual equipment in a factory [2]. OEE is the main metric which incorporates three basic indicators of equipment performance and reliability, which are availability (stop losses), performance (speed losses), and quality (defect losses) [3]. Studies on machine performance analysis were widely performed using OEE in various industry, for instance cork industry [4], food industry [5] automobile industry [6], and paper manufacturer [7]. Ring spinning machine was researched by [8] who analyzed its TPM implementation in Indonesian company and [1] who performed TPM implementation in Bangladesh company. Stoppage losses of this machines were examined by [9] in Bangladesh. There are still limited researches found on machine performance evaluation in the textile industry, specifically in ring spinning machine. Problem solving on daily operations in the spinning mill are generally based on production managers judgement and trial-and-errors. This statement corresponds to [10] saying textile industry in general is facing varied operational issues, such as process inefficiency and ill-defined maintenance routines. Hence this research aims to analyze manufacturing loss of the ring spinning machines, and look closely on each causes of the problems by giving real case study on spinning mills factory in Indonesia.

2 Research Method

This research was carried out on ring spinning machines of one of the spinning unit in PT Z which produces Teteron Rayon (TR) yarn. It consists of 65% of polyester fibers and 35% of rayon fibers (TR 65/35). Data are calculated based on OEE evaluation as the basis for further analysis. Since there was no recorded detailed data on machine stoppage from the company, data is manually collected for three days with three shifts each on total of 6708 spindles from 13 ring spinning machines (A14-A26). Those 13 of ring spinning machines were selected since they produced the same type of yarn with the same yarn number (30 Ne), twist per inch (17.35), and spindle speed. There are eight hours per shift, hence the total running time for all 13 machines in observation is 1440 min with no planned downtime making the loading time for each of 13 machines is also 1440 min.

Availability rate, performance rate, and quality rate were calculated based on the collected data to determine the OEE value. To calculate the availability rate, breakdown stoppage data for each machine are needed to determine the loading time before being compared with the operation time of each machine. Performance rate was calculated by determining the total production, ideal cycle time, and operation time. The total production was calculated in kilograms based on the standard cop weight and determined by Eq. (1), which includes unit conversion of yarn count from Ne (hank/lbs) to Nm (meter/gram) showed by constant of 0,59. The unit conversion is needed since the Ne is commonly used as the yarn count while the counter shown on the machines are in meters and the cop weight are in grams. Total stopped spindle are also considered since not all of 516 spindles of the machines were running at the same time due to machine breakdown or idles. This will provide more detailed calculation.

$$\text{Total Production (unit)} \frac{\frac{\text{Counter} \times 0{,}59}{\text{Yarn Count (Ne)}} \times \text{Total running spindle}}{\text{Cop Weight (gr)}} - \text{Total stopped spindle} \tag{1}$$

The quality rate was calculated using the total of defective ring cops compared with the total production in unit. After determining availability rate, performance rate, and quality rate, therefore the OEE can be determined using Eq. (2).

$$\text{OEE} = \text{availability rate} \times \text{performance rate} \times \text{quality rate} \tag{2}$$

3 Result and Discussion

3.1 OEE and Six Big Losses

The calculated OEE of the 13 ring spinning machines varies from 69 to 92% with average value of 80% presented on Table 1. There is no standard of OEE percentage that is considered bad or good, but OEE of 85% or higher should be pursued as an ideal goal of equipment [11]. Compared to the generally accepted worldwide goal for each factor [12], the overall ring spinning performance were considered low, where the availability rate is less than 90%, performance rate is under 95%, quality rate is below 99%, and overall OEE is less than 85%.

Machine A24 has the biggest OEE value while machine A17 has the lowest one. Machine A17 has the longest time of stoppage breakdown and among those whose defect cop are more than 100. Whereas machine A24's stoppage breakdown was only one-third of machine A17. These data speak that machine idle time plays a very important role on the overall OEE value. The OEE value of all of the 13 ring spinning machines varies so much with 23% difference that the highest OEE value is 92% (A24) while the lowest OEE value is only 69% (A17). All of those 13 machines

Table 1 Overall equipment effectiveness (OEE)

No. M/C	Availability rate (%)	Performance rate (%)	Quality rate (%)	OEE (%)
A14	82	93	98	75
A15	84	86	97	70
A16	83	94	97	76
A17	79	90	97	69
A18	76	95	97	70
A19	86	97	98	82
A20	96	92	99	87
A21	91	90	99	81
A22	94	99	98	91
A23	88	91	96	77
A24	93	102	97	92
A25	91	94	96	82
A26	92	100	98	90
Avg	**87**	**94**	**97**	**80**

produce the same type of yarn with the same yarn number and produced in the same production batch, yet the effectiveness of each machines shows lack of balance. Looking closer to each of the OEE component, which are availability rate (87% avg), performance rate (94% avg), and quality rate (97% avg), gives insight that the quality of the yarn cop product actually does not have as much problem as the high machine idle time shown by a low rate of availability. In general, most of the ring spinning machines does not meet the ideal OEE value [12], only 4 machines out of the 13 machines has OEE value more than 85%. This shows that the production process of the ring spinning in this company should be improved and the balance on each of machines should be met. Six big losses analysis on the next part will dig deeper about this in detail.

Activity that caused stoppage losses of the ring spinning machines during three days observations were further computed based on the six big losses category. Figure 1 shows that doffing holds the biggest total of time loss compared to other activities, followed by process change. Doffing is an obliged activity of ring spinning process to release the finished product of yarn in cops into the next station which are the winding machines. But irregular doffing time of each machines and while there are 156 ring spinning machines running at the same time make this problem to be more complex.

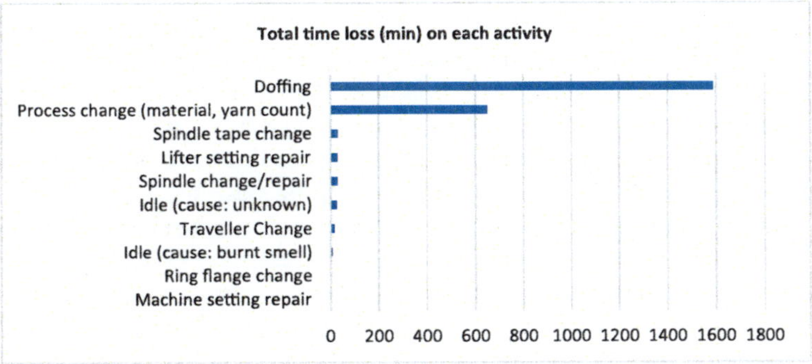

Fig. 1 Total time loss for each activity

3.2 *Losses Analysis*

After grouping each of activity into six big losses, analysis was then performed to look closer on each of the losses and how they compared to each other using pareto chart shown on Fig. 2. The most time loss was idling and minor losses, which has spent a total of 1615 min, corresponding to 43.45% of all time lost. It can be suggested from the chart that the idling and minor losses which includes doffing is the problem to be solved first in order to increase the machine performance therefore improve the OEE value. While reduced speed losses, setup and adjustment losses, defect losses, and equipment failure losses are the next priorities on solving problems, respectively. Reduced speed losses which includes reduced spindle speed and stopped spindle accounts for 25.54% of all losses or 949.13 min. Setup and adjustment losses, defect losses, and equipment failure losses are accounts for 17.70%, 10.35%, and 2.96% respectively. Whereas there is no recorded data for reduced yield losses.

To understand the problem better and to find the possible causes of each of the losses, root cause analysis is then performed using 5 whys approach, as shown in

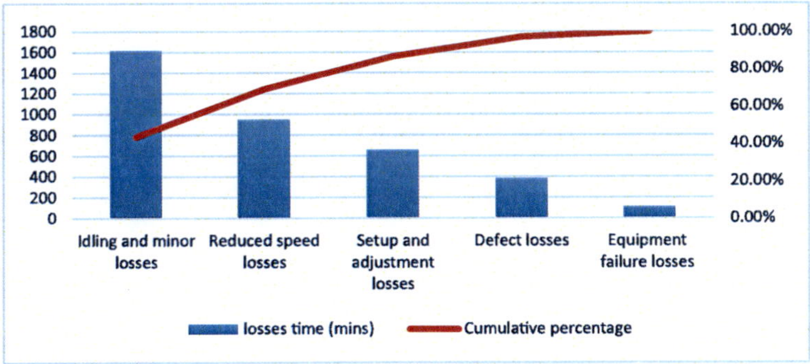

Fig. 2 Pareto chart

Table 2 for idling and minor, Table 3 for reduced speed, Table 4 for setup and adjustment, Table 5 for defect, and Table 6 for equipment failure. The five whys analysis gives insight on causes of each failures. The frequent causes of all failures were the irregular doffing time, worker, and machine problems. Irregular doffing time leads to many other problems that follow like doffer problem, higher doffing time, and cop defect. Worker/man on the ring spinning machine like operator, doffer, feeder, and maintenance technician also contribute a lot on the smooth production process. Their lack of skill, coordination, and awareness on the process gives impact to production therefore causing losses. Thus, actions from company management holds big roles on solving this worker problem. While machine problem (spindle tape, spindle, traveler, apron, etc.) and breakdown shows lack of machine maintenance and control.

Table 2 Idling and minor losses analysis

Failure	Why 1	Why 2	Why 3	Why 4	Why 5
Machine on idle without any breakdown	Operators purposely let the machine on idle	Operators only focused on connecting the yarn break	Lack of the operators concern on the machine condition	Lack of training on solving machine problems	Lack of employee training and skill development
High doffing time	No doffers available when doffing ready	Delay of doffers when doffing	Doffers unaware of doffing status	Lack information on doffing status	No communication of operators and doffers
	Overwhelmed doffers	Same doffing time on different machines	No time interval of each doffing	No scheduled doffing time	Different machine counter on each process

Table 3 Reduced speed losses analysis

Failure	Why 1	Why 2	Why 3	Why 4	Why 5
Low machine performances	High slip factor on spindle tape	No spindle tape change before broken	Lack of information on the part lifetime	No records on part changes	
	Stopped spindle	Spindle break	Lack of machine maintenance and long repair time	Shorter time for repairs because of late report	High number of machines: longer machine checking time for QC
		No traveler			
		Apron break			
		Late sliver feeding	Feeders are unavailable on time	Feeders not concerned on feeding time	No communication of feeders and operators

Table 4 Setup and adjustment losses analysis

Failure	Why 1	Why 2	Why 3	Why 4	Why 5
Too long machine off for process change	No set-up workers ready	Varied & uncertain times of process change	No scheduled process change time	Lack of communication between QC, maintenance, production	No awareness that waiting is a waste

Table 5 Defect losses analysis

Failure	Why 1	Why 2	Why 3	Why 4	Why 5
Cop weight is not up to standard	Lapping of the material	Operators are too late on connecting the yarn break	Lack of operator's skill to connect the yarn breaks	Overwhelmed operators with large areas of ring spinning machines	Lack of employee training and skill development
	Feeding materials (sliver) out of stocks	Feeders do not aware of the material stock status	Feeders are unavailable on time	Feeders are not concerned about feeding time	No communication of feeders and operators
	Perform early doffing	No interval between each machines' doffing time	Difficulty on determining doffing time	Different machine counter for each process	No clear interval of each machine doffing
Ring cut yarn	Too big yarn diameter in the cop	Yarn in the cop is rubbed with ring flange and traveler	Lifter setting problem	Lack of machine setting control	

Table 6 Equipment failure losses analysis

Failure	Why 1	Why 2	Why 3	Why 4	Why 5
Spindle tape change	Spindle tape is stuck, cannot rotate	Spindle tape break/ disconnected	Use of obsolete of spindle tape	No data of spindle tape lifetime	No records on part changes
Spindle change	Vibrated/ trembled spindle	Broken part	Worn-out spindle	Fibers are stick to the spindle	Lack of machines cleanliness
Machine off caused by burnt smell	Friction of metal objects with fly waste	Too many fly wastes on the machines	Lack of machines cleanliness		
Setting lifter repair	Incorrect lifter setting	Lack of machine setting control			

3.3 Proposed Improvement

Based on the root-cause analysis on each of the losses, actions of improvements are proposed for each of the failure and cause of loss. Detailed improvement alternatives are hoped to provide clear description and instruction on how to overcome and prevent the same losses on the ring spinning machine in the future. The proposed improvements are presented on Table 7. In order to tackle the man problems, employee training and re-training program are suggested mainly to increase their skill, communication and awareness of the machine and process condition. The maintenance process of the machine also has to be improved specifically to have a correct part lifetime data and better part-change implementation. There also should be a particular improvement on doffing activity to reduce not only the overall doffing time but also the complexity of arranging the doffing schedule and intervals. In general, the current SOPs must be applied well and new SOPs for tackling common problems also must be generated to accomplish a better process thus minimize the manufacturing losses.

4 Conclusion

This research conducted manufacturing losses analysis on the ring spinning machines based on a case study on one of the biggest yarn manufacturing company in Central Java. OEE value was evaluated for the sample machines to find out the current status on the machines. Results showed that all parameters: 87% of availability rate, 94% of performance rate, 97% quality rate, and 80% of final OEE value were all below the ideal values and need to be improved. The OEE values on each of the observed ring spinning machines were also highly varied which shows unbalanced conditions of each machine. Six big losses of the machines were examined using pareto analysis which revealed that idling and minor losses are causing the highest time loss of 43%. Amongst of the idling loss were mainly caused by doffing activity which held 66% of overall time loss. The possible cause of each losses was analyzed using the 5 whys approach to understand the problems and their possible causes. Finally, proposed improvement for each of the losses were established as useful inputs to the company and also for further study on implementing the TPM not only on the ring spinning machines but also on the other machines used in the yarn manufacturing process.

Table 7 Proposed improvements

Losses	Failure	Cause of loss	Proposed improvements
Idling and minor losses	Machine on idle without any break-down	Lack of employee training and skill development	Arrange operator training program to study more on the SOPs, increase awareness' on the machine condition and to increase the response speed when the machine is in trouble
	High doffing time	Lack of communication between operators and doffers	Increase awareness on the doffing activity, improve communication between operators and doffers on machine status
		Different machine counter on each process	Balance machine counter in order to determine the doffing time and doffing intervals from each machine easier
Reduced speed losses	Low machine perfor-mances	No records on part changes	Take detailed recording on part changes history to determine the lifetime of each part and make a scheduled part change based on their lifetime
		High number of machines: longer machine checking time for QC	Make and utilize information board(s) containing critical machine conditions, status and any machine breakdowns. The information board can be filled by both operators or QC
		Lack of communication between feeders and operators	Increase awareness on feeding activity, improve communication of operators and feeders on the sliver stock status and feeding time
Setup and adjustment losses	Machine being off for too long in process change	Lack of awareness that waiting is also considered as production waste	Arrange and perform employee training program to increase their awareness on manufacturing waste and strengthen communication about the production line and production status
Defect losses	Cop weight is not up to standard	Lack of employee training and skill development	Perform evaluation for operators' skill on handling and operating the machines, set up retraining program for low performers of operators
		Lack of communication between feeders and operators	Increase awareness on the feeding activity, improve communication of operators and feeders on sliver stock status and feeding time
		No clear interval of machine doffing	Determine machine doffing interval by balancing process and each of the machine counters

(continued)

Table 7 (continued)

Losses	Failure	Cause of loss	Proposed improvements
	Ring cut yarn	Lack of machine setting control	Enhance machine setting control, improve record keeping on machine breakdowns or troubles, and suggest regular control by operators
Equipment failure losses	Spindle tape change	No records on part changes	Carefully record part changes history so valid part lifetime can be determined and it can be used as the basis for the next part changes
	Spindle change	Lack of control on the cleanliness of the machines	Conduct a regular cleaning not only on fly waste but also on roving waste, keep all the waste produced from the process on the designated place, fast response on any uncleanness found
	Machine off caused by burnt smell		

References

1. Ahmad N, Hossen J, Ali SM (2018) Improvement of overall equipment efficiency of ring frame through total productive maintenance: a textile case. Int J Adv Manuf Technol 94(1):239–256
2. Muchiri P, Pintelon L (2008) Performance measurement using overall equipment effectiveness (OEE): literature review and practical application discussion. Int J Prod Res 46(13):3517–3535
3. Marinhoa P, Pimentela D, Casaisa R, Silvaa FJG, Sáa JC, Ferreiraa LP (2021) Selecting the best tools and framework to evaluate equipment malfunctions and improve the OEE in the cork industry. Int J Ind Eng Manage 12(4):286
4. Sousa E, Silva FJG, Ferreira LP, Pereira MT, Gouveia R, Silva RP (2018) Applying SMED methodology in cork stoppers production. Procedia Manuf 17:611–622
5. Tsarouhas P (2018) Improving operation of the croissant production line through overall equipment effectiveness (OEE): a case study. Int J Product Perform Manag 68(1):88–108
6. Gupta P, Vardhan S (2016) Optimizing OEE, productivity and production cost for improving sales volume in an automobile industry through TPM: a case study. Int J Prod Res 54(10):2976–2988
7. Li X, Liu G, Hao X (2021) Research on improved OEE measurement method based on the multiproduct production system. Appl Sci 11(2):490
8. Martomo ZI, Laksono PW (2018) Analysis of total productive maintenance (TPM) implementation using overall equipment effectiveness (OEE) and six big losses: a case study. In: AIP conference proceedings. AIP Publishing LLC, p 030026
9. Hossen J, Ahmad N, Ali SM (2017) An application of Pareto analysis and cause-and-effect diagram (CED) to examine stoppage losses: a textile case from Bangladesh. J Text Inst 108(11):2013–2020
10. Mahmood A (2020) Smart lean in ring spinning—a case study to improve performance of yarn manufacturing process. J Text Inst 111(11):1681–1696
11. Junior ML, de Paula Santos LC, Chamie NRG, Pierezan R, Loures ER, dos Santos EP, da Costa SEG, de Lima EP (2020) Automated system gains in lean manufacturing improvement projects. Procedia Manuf 51:1340–1347
12. Gupta AK, Garg RK (2012) OEE improvement by TPM implementation: a case study. Int J IT Eng Appl Sci Res 1(1):115–124

Improving the Efficiency of the Coal Car Body Assembly Process Using Work Standards at PT INKA

Farhana Mazaya Putri and Bambang Suhardi

Abstract PT INKA (Industri Kereta Api) is a train manufacturer company and has become one of Indonesia's proud strategic industries. PT INKA's production floor experienced problems, such as the existence of non-value-added activities and uneven load distribution between workstations. It is necessary to identify the waste that occurs in the car body assembly production line and repair it using the work standardization method with consideration of the operator's workload. The data processing in this study are the calculation of each work element time, identify VA, NNVA, VA, and waste, proposed improvements with work standardization, determination of the number of proposed operators, comparison of working ratio operators, and comparison of operators' workload. Waste in the form of non-value-added work activities was found in several car body assembly stations. Policy steps that can be taken to improve operational performance in the car body assembly department are optimizing the working ratio and operator workload through standardization of work elements.

Keywords Work standards · Workload · Lean manufacturing

1 Introduction

PT INKA (Industri Kereta Api) is the first and largest State-Owned Enterprise manufacturing integrated rail and transportation facilities in the region of Southeast Asia. PT INKA has 40 years of experience in developing rail facilities as a transportation solution, by offering various products and after-sales services to ensure consumers get quality transportation [1]. Domestically, PT INKA (Persero) strives to continue to encourage the development of an integrated rail-based transportation system.

F. M. Putri · B. Suhardi (✉)
Department of Industrial Engineering, Faculty of Engineering, Universitas Sebelas Maret, Surakarta 57126, Indonesia
e-mail: bambangsuhardi@staff.uns.ac.id

C. N. Rosyidi et al. (eds.), *Proceedings of the 6th Asia Pacific Conference on Manufacturing Systems and 4th International Manufacturing Engineering Conference*, Lecture Notes in Mechanical Engineering,
https://doi.org/10.1007/978-981-99-1245-2_33

Despite having tremendous opportunities, PT INKA (Persero) is currently facing important challenges to maintain and increase its share in the international market. In terms of competing and surviving in the global market, PT INKA (Persero) must offer high-quality products on time. For this reason, the industry has no choice but to reduce workstation cycle times, minimize production lead times, and increase efficiency in every line of work. It is realized that reducing cycle times and/or increasing functional efficiency will help PT INKA (Persero) to gain a competitive advantage in the form of responsiveness and cost advantage. Based on observations in the car body assembly department, there are several problems such as the existence of non-value-added activities and uneven load distribution between work stations.

In the process of assembling the car body of the KKBW coal train, waste is still found in several production lines. This waste is in the form of activities that do not add value to the product and the unequal workload by each operator. Therefore, this study aims to identify the waste that occurs in the car body assembly production line and then repair it using the work standardization method with consideration of the operators' workload. This improvement is expected to increase the efficiency of the production process in the car body assembly department.

2 Literature Review

The measurement of working time was first introduced by Frederick W. Taylor. The downtime measurement method is suitable for repetitive and short work. Measurement of time will produce output in the form of standard time where this time will be used for standard workers in completing a job. Standard time is the time required by a worker who has an average level of ability to complete a job [2]. The standard time is not established directly but needs to be added such as leeway and adjustment factors. This needs to be done because the ability of each person is different in completing a job [3].

Lean manufacturing is a series of activities to reduce waste, increase value-added activities, and reduce the occurrence of activities that do not provide added value (non-value added). The main concept of lean manufacturing is to reduce waste or waste. Waste is defined as all activities along the value stream mapping that do not provide added value in the process of input into output [4]. The waste should be eliminated as it will result in corporate productivity such as on-time delivery of products, increases resource utilization, reduced production costs, and the number of WIPs [5].

In lean manufacturing, there are several tools that can be used to reduce waste such as the 5S and Yamazumi chart [6], ECRS method [7], VSM (Value Stream Mapping) [8], and standardization research [9]. Previous research has applied the lean concept to the furniture industry which is able to reduce waste and create work standardization [9]. Work standardization is the basis for efficient and effective work methods. Work standardization is a tool to produce quality products on the basis of operator movements arranged in the right order so as not to cause non-value-added activities. Work standardization includes many tools, namely standardized worksheet (standardized work chart), work combination sheet (work combination table), process capacity sheet, operator work instruction (work instruction, operator instruction), and cycle balance chart (stack chart, yamazumi chart) [10].

3 Methodology

3.1 Data Collecting

Data collection was carried out in the car body assembly department of the KKBW coal train. The data was collected in February 2022. All observations and recordings were carried out in the first shift from 07.00 to 16.00 to maintain data consistency because the operators who work are always the same in each shift. In addition, this is also done to avoid variations in the adjustment factor due to differences in operator skills in carrying out their work. The data taken are in the form of work elements, processing time, adjustment factors, allowances, and the number of workers.

3.2 Data Testing

Data testing is in the form of uniformity tests and data adequacy tests. The data uniformity test was carried out to determine whether the data collected was uniform or not. Data is said to be uniform if it is between the upper control limit and the lower control limit. Meanwhile, the data adequacy test is carried out to ensure that the data collected is objectively sufficient. Testing the data using a 95% confidence level with a value of $k = 2$ and $s = 0.05$.

The data uniformity test for each work element is carried out in several stages: calculating the average value of processing time, calculating the standard deviation, calculating the value of the upper control limit and lower control limit, and determining whether the data collected per element is uniform or not. Based on the calculation, the processing time data on each work element is said to be uniform because the data that has been collected is between the upper control limit and the lower control limit.

The data adequacy test for each work element is carried out by determining the value of N' or the amount of theoretical data, then determining whether the data collected per element is sufficient or not by comparing the amount of observational data and the amount of theoretical data. The data is said to be sufficient if the amount of observational data is greater than the amount of theoretical data. Based on the calculations, the processing time data on each work element is said to be sufficient because the amount of observation data is greater than the amount of theoretical data.

3.3 Data Processing

The data processing in this study begins with the calculation of the cycle time of each work element, the calculation of adjustments using the Shumard method, the calculation of adjustments using the objective method, the calculation of the normal time of each work element, the calculation of allowances, the calculation of the standard time of each work element. Table 1 is a recap of the calculation of the standard time for each work element in the car body assembly department of the KKBW coal train.

The next data processing is the classification of work elements, calculation of operator working ratio, calculation of operator workload, proposed improvement with standardization, determination of the number of proposed operators, calculation of operator working ratio after repair, calculation of operator workload after repair, and calculation working ratio and operator workload after repair.

4 Results and Discussions

4.1 Identify VA, NNVA, VA, and Waste

The work elements at each station are grouped into three categories, namely Value-Added Activity (VA), Necessary but Non-Value-Added Activity (NNVA), and Non-Value-Added Activity (NVA). In addition, the types of waste generated by activities with non-value added are also identified. Table 2 is a recap of activities and waste for each operator and work element in the KKBW coal car body assembly department.

4.2 Proposed Improvements with Work Standardization

Improvements with standardization of work are carried out after calculating the working ratio and workload of workers. The proposed standardization of work is to eliminate non-value-added work elements so that the work carried out by operators is only value-added and necessary but non-value-added work elements. Tables 3, 4, 5, 6, 7, 8, 9 and 10 are tables of proposed work standards for each operator.

Table 3 shows the proposed work standard for the welder operator station 1 where there is one non-value-added work element that is eliminated, namely the underframe reforming work element.

Table 4 shows the proposed work standard for the welder operator station 2 where there is one non-value-added work element that is eliminated, namely the work element of reforming the cover underframe and side wall.

Table 5 shows the proposed work standards for the welder operator station 4 where the welder operator station 4 does not change the job description. Therefore,

Table 1 Time of each work element

Work station	No.	Work element	Cycle time (s)	Adjust ment	Normal time (s)	Allowance (%)	Standard time (s)
1	1	Taking underframe	289.33	1.33	383.37	23	471.54
	2	Placing the front end on underframe	288.33	1.24	356.57	25	445.72
	3	Placing chamber on jig	453.00	1.05	473.39	33	629.60
	4	Reforming underframe	2361.33	1.02	2402.66	38	3315.67
	5	Taking front end	114.00	1.41	161.12	23	198.18
	6	Placing the underframe on jig 1	79.00	1.50	118.63	29	153.03
	7	Placing the clamp on front end	138.33	1.53	211.65	26.5	267.74
	8	Placing front end retaining plate using weld	524.00	1.44	754.56	28.5	969.61
	9	Placing turn buckle using weld (1)	208.67	1.67	347.43	29.5	449.92
	10	Tack welding (1)	321.00	1.30	415.70	40	581.97
	11	Taking rear end	143.67	1.41	203.05	23	249.75
	12	Placing rear end on underframe	94.00	1.50	141.16	29	182.09
	13	Placing clamp on rear end	152.67	1.53	233.58	26.5	295.48
	14	Placing rear end retaining plate using weld	555.67	1.44	800.16	28.5	1028.21
	15	Placing turn buckle using weld (2)	461.33	1.67	768.12	29.5	994.72
	16	Tack welding (2)	264.67	1.30	342.74	40	479.84

(continued)

Table 1 (continued)

Work station	No.	Work element	Cycle time (s)	Adjust ment	Normal time (s)	Allowance (%)	Standard time (s)
	17	Taking side wall 1	269.00	1.24	332.66	23	409.18
	18	Placing side wall 1 on underframe	207.67	1.24	256.81	29	331.29
	19	Placing turn buckle using weld (3)	401.33	1.67	668.22	28.5	858.66
	20	Tack welding (3)	139.33	1.30	180.44	40	252.61
	21	Taking side wall 2	155.00	1.24	191.68	23	235.77
	22	Placing side wall 2 on underframe	233.67	1.24	288.97	29	372.77
	23	Placing turn buckle using weld (4)	540.67	1.67	900.21	28.5	1156.77
	24	Tack welding (4)	2413.33	1.30	3125.27	40	4375.37
	25	Moving sub-assembly to jig 2	205.67	0.97	199.84	23	245.80
2	26	Placing sub-assembly to jig 2	59.67	1.50	89.60	26	112.90
	27	Total welding	11,526.00	1.22	14,109.75	48.5	20,952.97
	28	Welding on the lower pillar	1532.33	1.48	2267.85	44.5	3277.05
	29	Taking cover underframe dan side wall	70.67	0.99	69.96	24	86.75
	30	Reforming cover underframe and side wall	361.33	1.04	374.28	35	505.28
	31	Placing underframe cover and side wall using weld	1277.33	1.39	1772.30	25.5	2224.24

(continued)

Table 1 (continued)

Work station	No.	Work element	Cycle time (s)	Adjust ment	Normal time (s)	Allowance (%)	Standard time (s)
	32	Taking accessories	256.67	0.99	254.10	22	310.00
	33	Placing accessories	1098.33	1.48	1625.53	30	2113.19
	34	Cutting turn buckle	1928.00	1.30	2496.76	29	3220.82
	35	Total grinding	9345.00	1.41	13,199.81	48.5	19,601.72
	36	Moving sub-assembly to jig 3	247.33	1.41	349.56	23	429.96
3	37	Placing sub-assembly to jig 3	82.00	1.41	115.89	25.5	145.45
	38	Dimensional reforming	4510.00	1.32	5945.68	50	8918.53
	39	Recording the final car body dimensions	2938.00	1.31	3838.99	22	4683.56
4	40	Waiting for the car body after the primer painting	616.00	1.18	725.85	14	827.47
	41	Taking car body after the primer painting	325.67	1.41	460.28	22	561.54
	42	Placing car body on repair jig	184.33	1.24	227.96	25	284.95
	43	Total repair	23,182.00	1.41	32,744.58	51	49,444.31
Total			70,555.33		94,926.71		136,651.97

the standard of work applied to the operator of the 4th station welder is the same as what has been done before.

Table 6 shows the proposed work standards for helper operator station 1 where the helper operator station 1 does not change the job description. Therefore, the work standards applied to station 1 helper operators are the same as those that have been done before.

Table 7 shows the proposed work standards for helper operator station 2 where the helper operator station 2 does not change the job description. Therefore, the work

Table 2 Identification of activities and waste car body assembly

Operator	Work element	Working category			Type of waste
		VA	NNVA	NVA	
Welder station 1	Reforming underframe			✓	Overprocessing
	Placing front end retaining plate using weld	✓			–
	Placing turn buckle using weld (1)	✓			–
	Tack welding (1)	✓			–
	Placing rear end retaining plate using weld	✓			–
	Placing turn buckle using weld (2)	✓			–
	Tack welding (2)	✓			–
	Placing turn buckle using weld (3)	✓			–
	Tack welding (3)	✓			–
	Placing turn buckle using weld (4)	✓			–
	Tack welding (4)	✓			–
Welder station 2	Total welding	✓			–
	Welding on the lower pillar	✓			–
	Reforming cover underframe and side wall			✓	Overprocessing
	Placing underframe cover and side wall using weld	✓			–
	Placing accessories	✓			–
	Cutting turn buckle	✓			–
	Total grinding	✓			–
Welder station 4	Total repair	✓			–
Helper station 1	Taking underframe	✓			–
	Placing the underframe on jig 1	✓			–
	Placing chamber on jig	✓			–
	Taking front end	✓			–
	Placing the front end on underframe	✓			–
	Placing the clamp on front end	✓			–
	Taking rear end	✓			–

(continued)

Table 2 (continued)

Operator	Work element	Working category			Type of waste
		VA	NNVA	NVA	
	Placing rear end on underframe	✓			–
	Placing clamp on rear end	✓			–
	Taking side wall 1	✓			–
	Placing side wall 1 on underframe	✓			–
	Taking side wall 2	✓			–
	Placing side wall 2 on underframe	✓			–
	Moving sub-assembly to jig 2		✓		–
Helper station 2	Placing sub-assembly to jig 2		✓		–
	Taking cover underframe dan side wall		✓		–
	Taking accessories		✓		–
	Moving sub-assembly to jig 3		✓		–
Helper station 3	Placing sub-assembly to jig 3		✓		–
	Recording the final car body dimensions	✓			–
Helper station 4	Waiting for the car body after the primer painting			✓	Waiting
	Taking car body after the primer painting		✓		–
	Placing car body on repair jig		✓		–
Reformer	Dimensional reforming	✓			–

standards applied to the helper operator station 2 are the same as those that have been done before.

Table 8 shows the proposed work standard for the helper operator station 3 where the helper operator station 3 does not change the job description. Therefore, the work standards applied to the station 3 helper operators are the same as those that have been done before.

Table 9 shows the proposed work standard for station 4 helper operators where there is one non-value-added work element that is eliminated, namely the work element waiting for the car body after the primary painting is available.

Table 3 Table of work standards for welder station 1

Operator	Work element	Standard time (s)
Welder station 1	Placing front end retaining plate using weld	969.61
	Placing turn buckle using weld (1)	449.92
	Tack welding (1)	581.97
	Placing rear end retaining plate using weld	1028.21
	Placing turn buckle using weld (2)	994.72
	Tack welding (2)	479.84
	Placing turn buckle using weld (3)	858.66
	Tack welding (3)	252.61
	Placing turn buckle using weld (4)	1156.77
	Tack welding (4)	4375.37
Total time		11,147.68

Table 4 Table of work standards for welder station 2

Operator	Work element	Standard time (s)
Welder station 2	Total welding	20,952.97
	Welding on the lower pillar	3277.05
	Placing underframe cover and side wall using weld	2224.24
	Placing accessories	2113.19
	Cutting turn buckle	3220.82
	Total grinding	19,601.72
Total time		51,389.99

Table 5 Table of work standards for welder station 4

Operator	Work element	Standard time (s)
Welder station 4	Total repair	49,444.31
Total time		49,444.31

Table 10 shows the proposed work standards for reformer operators where the reformer operators have not changed their job descriptions. Therefore, the standard of work applied to the reformer operator is the same as before.

4.3 Determination of the Number of Proposed Operators

The number of operators in the car body assembly department needs to be adjusted considering that the workload between operators is not evenly distributed. At station 1 forming the underframe and setting jig, there is a change in the number of operators,

Table 6 Table of work standards for helper station 1

Operator	Work element	Standard time (s)
Helper station 1	Taking underframe	471.54
	Placing the underframe on jig 1	445.72
	Placing chamber on jig	629.60
	Taking front end	198.18
	Placing the front end on underframe	153.03
	Placing the clamp on front end	267.74
	Taking rear end	249.75
	Placing rear end on underframe	182.09
	Placing clamp on rear end	295.48
	Taking side wall 1	409.18
	Placing side wall 1 on underframe	331.29
	Taking side wall 2	235.77
	Placing side wall 2 on underframe	372.77
	Moving sub-assembly to jig 2	245.80
Total time		4487.94

Table 7 Table of work standards for helper station 2

Operator	Work element	Standard time (s)
Helper station 2	Placing sub-assembly to jig 2	112.90
	Taking cover underframe dan side wall	429.96
	Taking accessories	86.75
	Moving sub-assembly to jig 3	310.00
Total time		939.61

Table 8 Table of work standards for helper station 3

Operator	Work element	Standard time (s)
Helper station 3	Placing sub-assembly to jig 3	4683.56
	Recording the final car body dimensions	145.45
Total time		4829.01

Table 9 Table of work standards for helper station 4

Operator	Work element	Standard time (s)
Helper station 4	Placing car body on repair jig	284.95
	Taking car body after the primer painting	827.47
Total time		1112.42

Table 10 Table of work standards for reformer

Operator	Work element	Standard time (s)
Reformer	Dimensional reforming	8918.53
Total time		8918.53

the transfer of 1 welder to station 4 so that the welder at station 1 becomes 2 people. At station 2 welding and installation of accessories, there is a change in the number of operators, namely the transfer of 2 welders to station 4 so that the welder at station 2 becomes 6 people. At station 3 reforming there is a change in the number of operators, namely the addition of three helpers so that the helper at station 3 becomes five people. In addition, at station 3 there is also a change in the reformer operator, which was originally 4 people, then increased to 8 people. At the final repair station 4 there is a change in the number of operators, namely the addition of 8 welders to station 4 so that the welder at station 4 becomes 20 people.

4.4 Comparison of Working Ratio Operators

The calculation of the working ratio of each operator is carried out to determine the ratio of the value-added, necessary but non-value-added, and non-value-added activities carried out by the operator. Work elements that do not provide added value will be eliminated later so that the car body assembly process becomes more efficient.

The working ratio of each operator in the car body assembly department before the standardization of work can be visualized in a graph as shown in Fig. 1. Meanwhile, the working ratio of each operator in the car body assembly department after the proposed repair can be visualized in a graph as shown in Fig. 2.

Improvements using work standardization make each operator work optimally, as seen from the high ratio of activities that provide added value. Welder station 1 previously carried out 76.7% value-added activity and 23.3% non-value-added activity. After repair, the station 1 welder does 100% of the value-added work. Welder station 2 previously carried out 99.1% value-added activity and 0.9% non-value-added activity. After repair, the station 2 welder does 100% of the value-added work. Welder station 4 has not changed because it has 100% done value-added work.

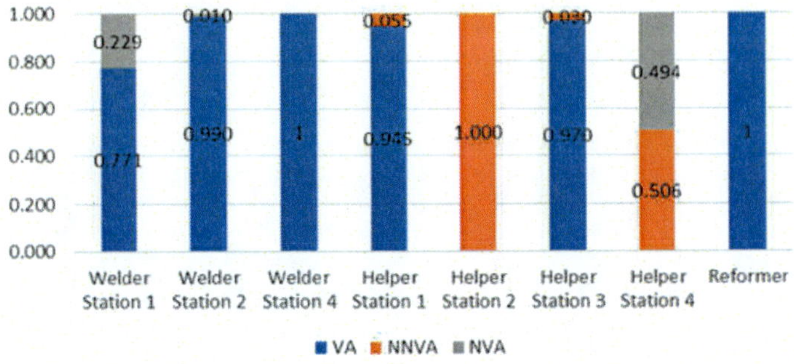

Fig. 1 The working ratio of operators before the standardization of work

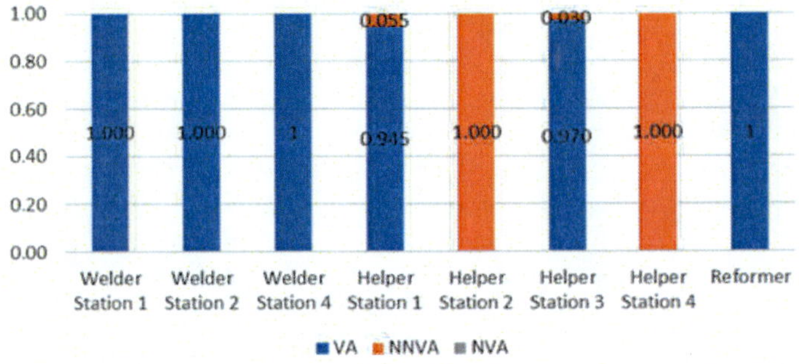

Fig. 2 The working ratio of operators after standardization of work

Station 1 helper did not change because 94.7% had done value-added work and 5.3% had important work even though it did not provide added value. Station 2 helpers previously performed 58.7% necessary but non-value-added activity and 41.3% non-value added activity. After repairs, station 2 helpers do 100% of the essential work even though it's not value-added. Station 3 helper has not changed because 96.8% have done value-added work and 3.2% important work even though it does not provide added value. Station 4 helpers previously performed 51.7% necessary but non-value-added activity and 48.3% non-value-added activity. After repairs, the station 4 helper does 100% of the essential work though it's not value added. Reformers have not changed because they have done 100% value-added work.

4.5 Comparison of Operators' Workload

Calculation of the workload of each operator is carried out to determine the amount of workload of each operator. The workload of each operator in the car body assembly department can be visualized in a graph as shown in Fig. 3. Meanwhile, the workload

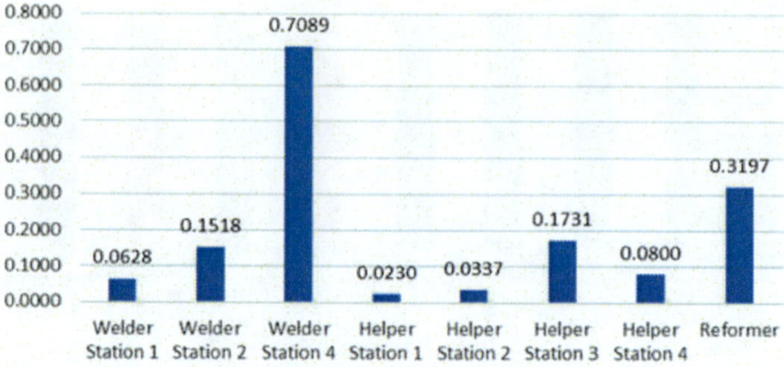

Fig. 3 The workload of operators before the standardization of work

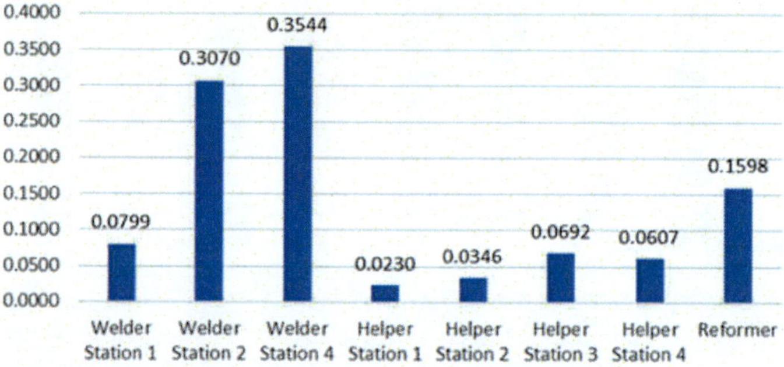

Fig. 4 The workload of operators after standardization of work

of each operator in the car body assembly department after the proposed repair can be visualized in a graph as shown in Fig. 4.

Improvements with changes in the number of operators made the workload of each operator almost evenly distributed even though there were still operators with slightly higher workloads.

5 Conclusion

Waste in the form of non-value-added work activities was found in several car body assembly stations. At station 1 there is waste in the form of overprocessing, namely the underframe reforming work element of 13.262 s. At station 2 there is waste in the form of overprocessing, namely the work elements of reforming cover underframe and sidewall of 2.021 s. At station 4 there is waste in the form of waiting, namely the work element waiting for the car body after the primary painting is available for 3.309 s.

Policy steps that can be taken to improve operational performance in the car body assembly department are optimizing the working ratio and operator workload through standardization of work elements. Standardization of work elements is done by eliminating work elements that do not provide added value. In addition, it is necessary to improve the quality control performance of each part that will enter the car body assembly department so as to minimize the excessive reforming process. Expert supervision is also required during the assembly process. PT. INKA (Persero) also needs to provide sufficient training to employees involved in the production process, especially new employees, so that the skills among employees are balanced and adequate. Training should be carried out together with more experienced employees to avoid skills gaps.

The limitation of this research is that there are no economic factors considered in the proposed change in the number of workers. Considerations from the economic side in the future need to be carried out as input for managerial. Suggestions for further research in reducing waste in the car body assembly process of PT INKA (Persero) is to perform a line balance analysis to reduce the waiting time that occurs so that the solutions provided can be more optimal.

References

1. Company profile of PT INKA (Persero) (2021) http://www.inka.com/corporation/7. Accessed 30 Aug 2022
2. Wignjosoebroto S (2000) Ergonomi, Studi Gerak dan Waktu Teknik Analisis untuk Meningkatkan Produktivitas Kerja. PT. Gunawidya, Jakarta
3. Zadry HR, Susanti L, Yuliandra B, Jumeno D (2015) Analisis dan Perancangan Sistem Kerja. Andalas University Press, Padang
4. Gasperz V, Fontana A (2011) Lean six sigma for manufacturing and service industries waste elimination and continuous cost reduction. Vinchristo Publication, Bogor
5. Woehrle SL, Abou-Shady L (2010) Using dynamic value stream mapping and lean accounting box scores to support lean implementation. Am J Bus Educ 3:8
6. Rahani AR, Ashraf AM (2012) Production flow analysis through value stream mapping: a lean manufacturing process case study. Procedia Engineering 41:1727–1734
7. Miranda FAA (2011) Application of work sampling and ECRS (eliminate, combine, re-layout, and simplify) principles of improvement at TO1 assembly. In: 21st ASEMEP national technical symposium. Manila, pp 1–7
8. Helleno AL, Isaias de Moraes AJ, Simon AT (2017) Integrating sustainability indicators and lean manufacturing to assess manufacturing processes: application case studies in Brazilian industry. J Cleaner Prod 153:405–416
9. Suhardi B, Sari AP, Laksono PW (2016) Implementation of standardization work to improve productivity in Indonesian furniture industry. J Eng Appl Sci 11(12):2602–2606
10. Liker JK, Meier D (2007) The Toyota way fieldbook. Erlangga Group, Jakarta

The Sustainability-Oriented Total Cost of Ownership Model of the Electric Motorcycle

Yuniaristanto, Sayyidah Maulidatul Afraah, Wahyudi Sutopo, and Muhammad Hisjam

Abstract Indonesia is one of the largest motorcycle markets in the world. This condition has an impact on high air pollution and greenhouse gas emissions. Migration from fossil fuel to electric motorcycles can reduce greenhouse gas emissions. Many consumers are not aware of the operational cost savings of using EM. These savings can be calculated using the total cost of ownership model (TCO), where the motorcycle owner will know all the costs involved during the lifetime of the motorcycle. This study aims to propose a sustainability-oriented TCO model for electric motorcycles (EMs) and conventional motorcycles (CMs). This TCO model considers consumer, social and environmental aspects. This proposed model can provide information on TCO for EMs and CMs to motorcycle owners for consideration in buying motorcycles in the future.

Keywords Electric motorcycle · The total cost of ownership · Sustainability · Conventional motorcycle

1 Introduction

In this decade, various problems related to energy scarcity, air pollution, and greenhouse gas emissions are sweeping the world [1]. One of the main causes is the high energy consumption from fuel oil in transportation activities, which impacts high carbon emissions in the world [2]. One of the efforts to reduce carbon emissions is migrating energy consumption from fossil energy to alternative energy. This effort can be realized using electric vehicles (EVs), where Evs can be charged from the electricity network at home or public charging stations [3]. Thus, EV is an environmentally friendly, energy-efficient transportation solution and saves operational and maintenance costs [4].

Yuniaristanto (✉) · S. M. Afraah · W. Sutopo · M. Hisjam
Universitas Sebelas Maret, Jalan Ir. Sutami 36A 57126, Surakarta, Indonesia
e-mail: yuniaristanto@ft.uns.ac.id

C. N. Rosyidi et al. (eds.), *Proceedings of the 6th Asia Pacific Conference on Manufacturing Systems and 4th International Manufacturing Engineering Conference*, Lecture Notes in Mechanical Engineering,
https://doi.org/10.1007/978-981-99-1245-2_34

The number of vehicles in Indonesia continues to increase, were in 2021, it has reached 143,797,227 units, dominated by 121,209,304 motorcycles and 16,903,094 passenger cars [5]. Very rapid growth is found in the type of motorcycle, where sales in 2021 will reach 5,057,516 units [6]. These vehicles' ownership and sales certainly impact the high carbon emissions produced. So it is necessary to migrate from conventional vehicles (CV) to electric vehicles (EV), which are more environmentally friendly. This solution was strengthened by the issuance of Presidential Regulation no. 55 of 2019 concerning the Acceleration of the Battery-Based Electric Motor Vehicle Program for Road Transportation as one of the government's efforts to reduce the disposal of GHG or Greenhouse Gas.

However, until March 2022, the number of new EV owners reached 16,060 units consisting of three wheels, passenger cars, motorcycles, buses, and freight cars [7]. This fact shows that the EV adoption rate is still very low. The low level of EV adoption is caused by the high purchase price of Evs when compared to conventional vehicles (CV) [8, 9]. Another reason is that consumers have not realized the magnitude of operational cost savings by using electric vehicles [10, 11]. In addition, many factors still play a role in consumer decisions in purchasing Evs [12–14]. These factors can be classified into monetary (such as purchase price, taxes, operational costs, parking fees, and other costs) and non-monetary (such as mileage, size, weight, attitude, charging time, and charging station). The monetary group is usually approached with the Total Cost of Ownership (TCO) model. TCO is a philosophy aimed at calculating the actual cost of buying a product such as a motorcycle [15]. The main objective of TCO analysis is to make better purchasing decisions by considering the issue of costs beyond price [15], which are hidden and ongoing before the product is implemented [16].

Many studies have discussed the TCO model on Evs to compare the TCO value between various types of vehicles and brands between CV and EV. Zirogiannis et al. compared the TCO between conventional and electric cars to assess the effect of the I standard or Corporate Average Fuel Economy [17]. Nian et al. [18] compared the TCO of 17 car brands consisting of conventional and electric to create a new business model for promoting EV adoption in Singapore. Vanhaverbeke et al. [19] compared the TCO between passenger and private cars in Belgium to maximize the RES (Renewable Energy Sources) used, guarantee cost status, and minimize costs. Similar research has also developed the TCO model by adding a social cost orientation to compare conventional and electric cars in Indonesia. The study also concluded that the exemption of PPnBM and the increase in fossil fuel prices could result in a lower TCO EV than CV [20].

Most previous studies took the object of study as electric vehicles (Evs). There are very few studies that examine the object of electric motorcycles. In addition, previous research is still focused on consumer orientation. There is no research has been found that considers sustainable concept vehicles. Thus, this study aims to propose a TCO model for electric motorcycles by considering sustainability.

This research will focus on the electric motorcycle (EM) for several reasons: (i) the number of conventional motorcycles (CM) in Indonesia is far more than the number of conventional cars, reaching almost eight times [17], while the CO_2 emissions produced are similar [18]; (ii) Indonesia is a country that has a very high density of motorcycles [9]; and (iii) motorcycles are a fundamental mode of transportation for the Indonesia people which are widely used for activities with high mobility, such as motorcycle taxis, working, school, and travel.

Based on the background of these problems, this study proposes a TCO model for CM and EM by considering a more comprehensive consumer, social and environmental orientation according to the concept of sustainable transportation. The TCO model is expected to be used to accelerate EM adoption in Indonesia. The urgency of this research is to help accelerate the adoption of EM in the Indonesian market so that the government's target in 2025 is to adopt about 2.1 million EM owners. In addition, this research is an effort to conserve fossil fuels and reduce the high carbon emissions in big cities in Indonesia.

2 Model Development

The development of the TCO model was based on research by Riyanto et al. [20] and Danielis et al. [21]. The total cost of ownership or TCO consists of three components: consumer-oriented TCO, social-oriented TCO, and environmental-oriented TCO. The consumer-oriented TCO is composed of investment costs (motorcycle purchase price, motorcycle resale price, battery purchase price, and battery resale price) and operational costs (taxes, insurance, discounts, energy, and maintenance costs). The social orientation of TCO is composed of the loss of refueling time, the loss of maintenance time, and the impact of emissions on health. Environmental orientation TCO is composed of air quality, greenhouse gas emissions, and noise.

This model has not considered the ease of charging, the availability of charging stations, and the motorcycle's performance, such as speed and torque. The model also assumes that the value of motorcycles and batteries decreases constantly or linearly, while energy costs and maintenance costs increase yearly following the inflation rate.

2.1 Consumer Oriented TCO

Consumer-oriented TCO can be formulated as Eq. 1.

$$TCO_C = PP + \sum_{t=1}^{j} \frac{C_{opex}}{(1+r)^t} - RV(1+r)^{-j} + BR \qquad (1)$$

The first term describes the motorcycle purchase price, while the second describes the operational costs considering the discount rate over the ownership. The third term describes the resale value of motorcycles, while the last describes the battery replacement costs. Each term in Eq. (1) will be explained below.

a. Motorcycle Purchase Price (PP)

The purchase price is the OTR (on the road) price which includes the net purchase price of the vehicle, including taxes and other administrative costs. This price will also decrease over time, such as the effect on the lifetime of a motorcycle.

b. *Operational Costs (C_{opex})*

C_{opex} is the sum of all cash payments, which is shown in the formula below:

$$C_{opex} = C_{EC} + C_{MR} + C_{IN} + C_{TA} + C_{OF} + C_{IP} \tag{2}$$

C_{opex} covers the cost of energy consumption (C_{EC}) which increases every year, maintenance or repair costs (C_{MR}), insurance costs (C_{IN}) paid at the beginning of the purchase, annual tax costs (C_{TA}), parking fees (C_{OF}), and loan interest costs (C_{IP}).

c. *Discount rate I*

This discount rate is an important component in the calculation because it affects the high and low prices. Thus, the calculation results can be drawn and adjusted to the present value at a certain time. The discount rate in this study is obtained from the sum of the average inflation rate with the bank's average interest rate in the last 12 months.

d. *Resale value (RV)*

The resale value formula refers to Kong et al. [22]. The resale value is calculated separately between the battery and the motorcycle. The percentage of vehicle depreciation used in the study will refer to the survey data.

e. *Battery replacement cost (BR)*

This term applies only to the TCO model for EM. The economic life of the battery in this study is 1000 cycles, where the battery capacity decreases to 70% and requires replacement. Battery replacement costs use the following formula:

$$BR = \frac{f(BP - CB)}{(1 + r)^i} \tag{3}$$

The cost of a battery replacement (*BR*) is obtained by subtraction of the battery purchase price (*BP*) from the battery resale price (*CB*), which takes into account the discount rate in the year of replacement. The battery replacement period (i) is obtained by dividing the economic life of the battery by the charging frequency. At the same time, the frequency of battery replacement (f) is obtained by dividing the length of ownership by the battery replacement period (i).

2.2 Social Oriented TCO

This model assumes an increase in emissions from conventional motorcycles by 6% per year [23], so it can be formulated as follows.

$$TCO_S = \sum_{t=1}^{j} \frac{C_{MT} + C_{FT} + C_{HI}}{(1+r)^t} \quad (4)$$

The model notation is described below.

a. *Cost of time loss for maintenance*

C_{MT} is the cost of time loss for maintenance. The cost of time loss for maintenance in year t is calculated by multiplying the frequency of repairs and maintenance for a year (f_M), the time required for one maintenance (T_M), and monetized by external costs (e_w). The C_{MT} formula can be written as follows.

$$C_{MT} = f_M x T_M x e_w \quad (5)$$

b. *Cost of time loss for refueling*

C_{FT} is cost of time loss for refueling and can be written as Eq. 6. C_{FT} can be calculated by multiplying the frequency of charging for a year (f_F), the time required for a single charge (T_F), and monetized by external costs (e_w).

$$C_{FT} = f_F x T_F x e_w \quad (6)$$

c. *Health impact emission cost*

C_{HI} is health impact emission cost (see Eq. 7). C_{HI} can be obtained by multiplying the sigma of the level of the emission produced from each motorcycle $\left(P_{HI-p}\right)$, external cost (e_p), and annual mileage (*AKT*). Each type of emission will increase *k* % every year because it is influenced by the lifetime of the motorcycle (t).

$$C_{HI} = (100\% \ + k(t-1))x\left(\left(\sum_P P_{HI-p} x e_p\right) x AKT\right) \quad (7)$$

2.3 Environmentally Oriented TCO

Environmentally oriented TCO can be formulated as Eq. 8.

$$TCO_E = \sum_{t=1}^{j} \frac{(C_{NOISE} + C_{GHG} + C_{AQ})}{(1+r)^t} \tag{8}$$

a. *Greenhouse gas cost*

The cost of greenhouse gas emissions (C_{GHG}) is obtained by multiplying the sigma of greenhouse gas emissions (P_{GHG-p}) generated from each motorcycle, each external cost (e_p) and annual mileage (*AKT*) [24]. Each type of emission will increase *k* % every year because it is influenced by the lifetime of the motorcycle (t).

$$C_{GHG} = (100\% \ + k(t-1))\left(\sum_P (f_p x P_{GHG-p}) x e_p x AKT\right) \tag{9}$$

b. *Cost of local air quality emissions*

The cost of local air quality emissions (C_{AQ}) is obtained by multiplying the sigma of emissions generated from each motorcycle (P_{AQ-p}), external cost (e_p) and annual mileage (*AKT*) [24]. Each type of emission will increase *k* % every year because it is influenced by the lifetime of the motorcycle (t).

$$C_{AQ} = (100\% \ + k(t-1))\left(\sum_P P_{AQ-p}) x e_p x AKT\right) \tag{10}$$

c. *Noise cost*

The annual noise cost (C_{NOISE}) is the multiplication between the user's annual mileage (AKT) [21] and noise cost (see Eq. 11). Noise cost is the sum of the multiplication between urban noise costs (C_{NU}) and the percentage of trips in urban areas ($\%\ NU$) and the multiplication between suburban noise costs (C_{NSU}) and the percentage of trips in suburban areas (*%NSU*).

$$C_{Noise} = AKTx((\%\ NUxC_{NU}) + (\%\ NSUxC_{NSU})) \tag{11}$$

2.4 TCO Main Model

The TCO model is the sum of the total cost of ownership, which is oriented toward consumer (TCO_C), social (TCO_S), and environmental (TCO_E). The TCO EM model can be formulated as in Eq. 12.

$$TCO_{Tot} = TCO_C + TCO_S + TCO_E \tag{12}$$

All secondary data from stakeholders and primary data from survey results are used as input values in the TCO model for electric motorcycles.

3 Research Methodology

This paper describes proposing a consumer, social and environmental oriented TCO model. The next stage of this research is collecting data. The data collected are primary and secondary data. Primary data was collected through a survey by distributing questionnaires to respondents who were users of conventional motorcycles and electric motorcycles. The priority of respondents consists of respondents with the province domiciled with the highest level of motorcycle ownership in Indonesia. This primary data consists of data that is directly related to user behavior consisting of the brand, payment system, performance, mileage, downtown and suburban trips, type of fuel, filling time, service frequency, service fee, service time, types of services, parking fees, and insurance costs. Secondary data can be obtained from documents of stakeholders, namely the government, manufacturers, motorcycle dealers or distributors, as well as data from previous studies. Secondary data include the purchase price of electric motorcycles and batteries, components of energy consumption costs, prices of components for routine repairs and services, emissions, and external social and environmental costs.

The data is then processed using the TCO model that has been developed. TCO calculation results are analyzed by comparing EM and CM, including investment, operational, social, and environmental costs. Based on the comparison of EM and CM, the parameters and attributes that influence the TCO value can be seen. These results can be used as material for consideration in formulating policies and strategies to accelerate the diffusion of EM in Indonesia.

4 Conclusion

The level of EV adoption in Indonesia is still very low due to the high purchase price of EVs compared to CVs. But on the other hand, many consumers have not realized the amount of savings from operating costs using an EV compared to a CV. The TCO model can approximate the calculation of these savings. The majority of research on TCO is conducted on vehicles or cars, while motorcycles are still rare. In fact, motorcycles are a mode that dominates the number of vehicles in Asia, especially in developing countries. In addition, much TCO research focuses on consumer orientation and has not comprehensively considered the sustainability aspect. This research proposes a TCO model for CM and EM that considers a more comprehensive consumer, social and environmental orientation following sustainable transportation.

The next stage of this research is data collection, both primary data and secondary data. Primary data was collected through a survey by distributing questionnaires to respondents who are CM and EM owners. Secondary data can be obtained from stakeholder documents, namely the government, manufacturers, motorcycle dealers, or distributors, and data from previous research. Based on these data, the TCO value of EM and CM can be calculated and compared every year during the ownership period. This comparison shows which cost components cause waste, and savings when using EM or CM can be found. Motorcycle owners can certainly use the findings from this research in deciding the purchase of a motorcycle in the future. This is expected to motivate CM owners to migrate to EMs. The results of this research can also be considered in formulating policies and strategies for accelerating EM diffusion in Indonesia.

Acknowledgements This research was funded by the Institution of Research and Community Services of Universitas Sebelas Maret through Fundamental Research Grant number: 254/UN27.22/PT.01.03/2022

References

1. Sang Y-N, Bekhet HA (2015) Modelling electric vehicle usage intentions: an empirical study in Malaysia. J Clean Prod 92:75–83. https://doi.org/10.1016/j.jclepro.2014.12.045
2. Tanwir NS, Hamzah MI (2020) Predicting purchase intention of hybrid electric vehicles: evidence from an emerging economy. World Electr Veh J 11. https://doi.org/10.3390/WEVJ11020035
3. Ahmadi L, Croiset E, Elkamel A et al (2015) Effect of socio-economic factors on EV/HEV/PHEV adoption rate in Ontario. Technol Forecast Soc Change 98:93–104. https://doi.org/10.1016/j.techfore.2015.06.012
4. Sutopo W, Astuti RW, Purwanto A, Nizam M (2013) Commercialization model of new technology lithium ion battery: a case study for smart electrical vehicle. In: Proceedings of the 2013 joint international conference on rural information and communication technology and electric-vehicle technology, rICT and ICEV-T
5. Badan Pusat Statistik Jumlah Kendaraan Bermotor Menurut Provinsi dan Jenis Kendaraan (unit). 2021. https://www.bps.go.id/indikator/indikator/view_data_pub/0000/api_pub/V2w4dFkwdFNLNU5mSE95Und2UDRMQT09/da_10/1. Accessed 5 Apr 2022
6. AISI Statistic Distribution. https://www.aisi.or.id/statistic/. Accessed 16 Apr 2022
7. Rahayu IRS (2022) Jumlah Kendaraan Listrik di Indonesia Capai 16.060 Unit. In: Kompas. https://money.kompas.com/read/2022/03/21/143726326/jumlah-kendaraan-listrik-di-indonesia-capai-16060-unit. Accessed 12 Apr 2022
8. Egbue O, Long S (2012) Critical issues in the supply chain of lithium for electric vehicle batteries. EMJ Eng Manage J 24:52–62. https://doi.org/10.1080/10429247.2012.11431947
9. Rousseau A, Halbach S, Michaels L et al (2014) Electric drive vehicle development and evaluation using system simulation. IFAC
10. Krause RM, Carley SR, Lane BW, Graham JD (2013) Perception and reality: public knowledge of plug-in electric vehicles in 21 US cities. Energy Policy 63:433–440
11. Zhang X, Wang K, Hao Y et al (2013) The impact of government policy on preference for NEVs: the evidence from China. Energy Policy 61:382–393
12. Berkeley N, Bailey D, Jones A, Jarvis D (2017) Assessing the transition towards battery electric vehicles: a multi-level perspective on drivers of, and barriers to, take up. Transp Res Part A Policy Pract 106:320–332. https://doi.org/10.1016/j.tra.2017.10.004

13. Coffman M, Bernstein P, Wee S (2017) Electric vehicles revisited: a review of factors that affect adoption. Transp Rev 37:79–93. https://doi.org/10.1080/01441647.2016.1217282
14. Liao F, Molin E, van Wee B (2017) Consumer preferences for electric vehicles: a literature review. Transp Rev 37:252–275. https://doi.org/10.1080/01441647.2016.1230794
15. Ellram LM, Siferd SP (1998) Total cost of ownership: a key concept in strategic cost management decisions. J Bus Logist 19:55–84
16. Bailey JT, Heidt SR (2003) Why is total cost of ownership (TCO) important? Darwin Mag
17. Zirogiannis N, Duncan D, Carley S et al (2019) The effect of CAFE standards on vehicle sales projections: a total cost of ownership approach. Transp Policy 75:70–87. https://doi.org/10.1016/J.TRANPOL.2019.01.006
18. Nian V, Hari MP, Yuan J (2019) A new business model for encouraging the adoption of electric vehicles in the absence of policy support. Appl Energy 235:1106–1117. https://doi.org/10.1016/j.apenergy.2018.10.126
19. Vanhaverbeke L, Schreurs D, Clerck Q De et al (2017) Total cost of ownership of electric vehicles
20. Riyanto R, Riyadi SA, Nuryakin C, Gerald Massie NW (2019) Estimating the Total Cost of Ownership (TCO) of Electrified Vehicle in Indonesia. ICEVT 2019—Proceeding 6th International Conference Electrical Vehicle Technology 2019:88–99. https://doi.org/10.1109/ICEVT48285.2019.8994030
21. Danielis R, Giansoldati M, Scorrano M (2019) Consumer-and society-oriented cost of ownership of electric and conventional cars in Italy. Work Pap SIET Soc Ital di Econ dei Trasp e della Logist 19_3:1–21
22. Kong D, Xia Q, Xue Y, Zhao X (2020) Effects of multi policies on electric vehicle diffusion under subsidy policy abolishment in China: a multi-actor perspective. Appl Energy 266:114887. https://doi.org/10.1016/j.apenergy.2020.114887
23. Hassani A, Hosseini V (2016) An assessment of gasoline motorcycle emissions performance and understanding their contribution to Tehran air pollution. Transp Res Part D Transp Environ 47:1–12. https://doi.org/10.1016/j.trd.2016.05.003
24. Mitropoulos LK, Prevedouros PD, Kopelias P (2017) Total cost of ownership and externalities of conventional, hybrid and electric vehicle. Transp Res Procedia 24:267–274. https://doi.org/10.1016/J.TRPRO.2017.05.117

Supplier Selection and Order Allocation: Systematic Literature Review

Enty Nur Hayati, Cucuk Nur Rosyidi, Wakhid Ahmad Jauhari, Pringgo Widyo Laksono, and Retno Wulan Damayanti

Abstract In the primary buying problem, supplier selection and order allocation (SSOA) are two strategic choices. This paper's evaluation focuses on the topic of order allocation and supplier selection from 2014 to 2022. Our goal is to give a categorization of these works over a nine-year period as part of a systematic assessment of the scientific literature. By integrating all published research, the systematic review employs a keyword search strategy to improve the quality of the entire study and reduce bias. Use the Cochrane approach to look for publications on supplier selection and order distribution. Finally, there are numerous avenues for future research in the problem of supplier selection and order allocation, including taking into account multiple sources of uncertainty at once, considering supplier selection and order allocation in the service industry, and creating new techniques for pandemic situations brought on by Covid-19.

Keywords Supplier selection · Order allocation · Multi-criteria decision making

1 Introduction

In order to increase customer competitiveness, the supplier selection and order allocationproblem requires the design of techniques and models for evaluating and assessing the performance of a group of providers. Due to the wide range of quantitative and qualitative factors used in the evaluation and decision-making process, this decision is more complicated. To address this issue, several decision-making strategies have been put out in the literature, particularly multi-criteria analysis combining both quantitative and qualitative data. Two key strategic considerations in purchasing are supplier selection and order allocation. Supplier selection and order allocation

E. N. Hayati (✉) · C. N. Rosyidi · W. A. Jauhari · P. W. Laksono · R. W. Damayanti
Department of Industrial Engineering, Faculty of Engineering, Universitas Sebelas Maret, Surakarta, Indonesia
e-mail: enty@student.uns.ac.id

C. N. Rosyidi et al. (eds.), *Proceedings of the 6th Asia Pacific Conference on Manufacturing Systems and 4th International Manufacturing Engineering Conference*, Lecture Notes in Mechanical Engineering,
https://doi.org/10.1007/978-981-99-1245-2_35

difficulties are covered in a number of taxonomy and literature review publications, including [1–5]. There have been more intriguing studies on supplier allocation and order selection in recent years, but few of them have been the subject of comprehensive reviews. In order to classify these works across a nine-year period, our goal is to give a comprehensive evaluation of the scientific literature. In order to connect existing quantitative methods with empirical research, this study suggests a new structure and classification of existing research streams, as well as different methods used for SSP and order allocation. It also identifies gaps in the state-of-the-art and outlines potential future research directions. In order to classify these works across a nine-year period, our goal is to give a comprehensive evaluation of the scientific literature. In order to connect existing quantitative methods with empirical research, this study suggests a new structure and classification of existing research streams, as well as different methods used for SSP and order allocation. It also identifies gaps in the state-of-the-art and outlines potential future research directions.

This paper's major contribution is to give a thorough analysis of the literature on supplier selection by order allocation during the previous nine years, which reveals a marked growth in the number of published research papers in this field. Additionally, to improve researchers' comprehension of the subject matter of choosing suppliers and allocating orders as well as to make it simpler for researchers to locate useful data on this topic. In addition to highlighting any gaps that may be filled by additional study using MCDM techniques and optimization models, this paper aims to detect emerging trends in the supplier selection and order location fields.

Finally, the goal of this study is to make it easier for researchers to identify relevant information in this enormous body of literature. This study evaluates research published in international scientific journals with a specific focus on this topic, taking into account the field's ongoing evolution, and develops a useful framework for categorizing the most publications produced between 2014 and 2022. As a result, the researcher offers a bibliography of 21 works that have been published. To respond to the following inquiries: Which journal devotes the most pages to this problem? Which techniques for making multi-criteria decisions are frequently used in the supplier selection process? Which mathematical approach is frequently used to the challenges of supplier selection and order allocation?

2 Methodology

In this article, we outline a systematic review to categorize publications on this issue that were published between 2014 and 2022. Research issues concerning the difficulty of supplier selection and order allocation were examined using a systematic review. By integrating all published research, the systematic review employs a keyword search strategy to improve the quality of the entire study and reduce bias. The Cochrane method, one of the most used systematic review techniques, has a few phases designed to guarantee the caliber of the articles under investigation [6]. Using the Cochrane approach, look for papers on supplier selection and order allocation by doing the following procedures [7]:

a. Finding relevant keywords

Supplier selection and order allocation are the important phrases employed in this study. Researchers who study supplier choice and order allocation are also known by these terms. For a more precise search, this study employs Boolean logic: AND, OR. (" ") The research's key terms are "supplier selection" OR "order allocation," respectively ("supplier selection" AND "order allocation").

b. Database search using keywords

The study's scientific database used to search the literature was published, and it is titled Use Google Scholar.

c. Criteria for selecting scientific articles;

Step 1: 2014 to 2022 sees the publication of chosen papers.
Step 2: English-language articles must be published online.
Step 3: Articles from reviewed publications.
Step 4: Abstracts, keywords, and citations are downloaded with Publish Or Perish software. To visually display paper search results using the Vos Viewer application. To save and manage articles using the Mendeley software.
Step 5: The chosen article is one that addresses the selection of suppliers and order allocation.

Eight articles on supplier selection and order allocation were the most recently published in 2020, while there were none in 2016.

3 Result

a. Publication Year

Of the 21 selected articles published in the time interval between 2014 and 2022. Eight articles on supplier selection and order allocation were the most recently published in 2020, while there were none in 2016.

b. List of Publications

Papers on order distribution and supplier selection have been published in these journals. "Computer and Industrial Engineering", "Sustainable Production and Consumption", "Int. J. Production Economics", "Computers and Chemical Engineering", "European Journal of Operation Research", have published several papers in this field (Table 1).

Table 1 List of published journals

Journal name	Amount	Journal name Journal name	Amount
Computer and Industrial Engineering	5	Journal of Enterprise Information Management	1
Sustainable Production and Consumption	1	European Journal of Operation Research	1
MDPI	2	International Journal of Production Research	1
Int. J. Production Economics	2	Mersin University of Maritime Faculty	1
Journal of Cleaner Production	1	Business Review	1
Journal of Manufacturing Systems	1	Springer	1
Computers and Chemical Engineering	1	etc	2

c. Most Popular Applications

Model applications have been categorized. Case studies have been explored by several authors. In the area of supplier selection and order distribution, Automotive Industry is a well-liked program (Table 2).

d. Most Popular Sources of Uncertainty

Demand, capacity, and cost are the three most common sources of uncertainty, as seen in Fig. 1. Demand has received the most attention among them (50%), compared to other elements. This variable often has a major impact on how orders are distributed and is regarded as one of the optimization model's constraints.

e. Classification of Mathematical Models

Multi-objective optimization is a branch of mathematical optimization including the simultaneous optimization of several objective functions while also combining the use of many criteria for decision making. A linear program multi-objective model,

Table 2 Most popular application

Application	Article
Automotive industry	[8–13]
Electronics/computer company	[14, 15]
Food industry	[16]
Health industry/hospital	[17–20]
Medical device industry	[21–23]
Space product industry	[24]
Rubber industry	[25]
Domestic industry	[26–28]

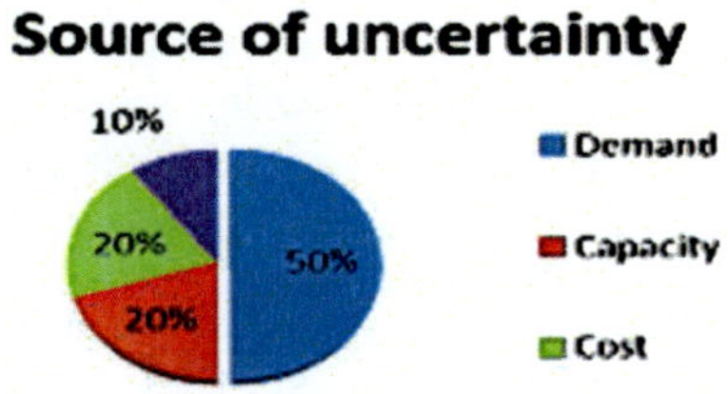

Fig. 1 Sources of uncertainty

Table 3 Classification of mathematical models

Multi-source model	Mathematical model	Number of articles
Multi purpose	Multi-objective linear program	5
	Multi-objective non-linear	4
	ε-constraint	2
	Mix integer non linear programming	1
Bi-objective	Integer program	2

Table 4 Optimization of the model based on the number of elements

Model optimization based on number of elements	
Multi	References
Period	[8–10, 12, 14, 16, 18, 19, 21, 26–30]
Product	[8–11, 14–16, 18, 19, 21, 28–30]
Supplier	[8–12, 14, 17–22, 25–31]
Scenarios	[16, 27]

a non-linear multi-objective model, and finally the constraint model make up the mathematical framework for multi-objectives (Tables 3 and 4).

4 Discussion

The talk that follows addresses each study topic on choosing suppliers and allocating orders:

Question 1: Which multi-criteria decision-making techniques are frequently used in the selection of suppliers ?

Based on the results of the literature review, there are several journals that address the issue of supplier selection and order allocation. The journal that most discusses this issue is the journal "Computer & Industrial Engineering", namely from 21 articles cited, as many as 5 articles were published in the journal.

Question 2: Which methods of multi-criteria decision-making are typically employed while choosing providers? According to a review of the literature on supplier selection and order allocation that was conducted using papers from 2014 to 2022, or for 9 years, it is clear that other variables, such uncertainty, have a significant impact on this issue. The uncertainty variables covered in these articles include product complaints, cost, capacity, and uncertainty in demand. Compared to other uncertainty components, demand uncertainty has the strongest influence on supplier choice and order allocation.

Question 3: Which mathematical approach is frequently used to the challenges of supplier selection and order allocation? Based on Table 2, product availability, suppliers, times, and scenarios must all be considered throughout the model's optimization depending on the number of elements, supplier selection, and order allocation. The topic of supplier selection and the distribution of orders for several suppliers, multiple periods, and multiple goods is covered in the publications that have been evaluated. Single-objective or multi-objective or bi-objective issues must also be considered in this issue, so as to create an appropriate mathematical model.

5 Conclusion

Order distribution and supplier choice can provide the business a competitive edge. Cost, quality, and timing are the three main factors to consider when choosing a provider. Recently, more ecological and environmental issues, such reducing carbon emissions, have been taken into account when choosing suppliers and allocating orders.

Future study on challenges related to order allocation and supplier selection has several potential areas. Among them are: (i) Taking into account several sources of uncertainty at once. (ii) Take into account supplier allocation and order placement in service sectors like healthcare systems (eg hospitals). (iii) Future study may focus on creating new techniques for scenarios like those that arise during a pandemic caused by Covid-19. (iv) Supplier selection and order allocation articles have not examined a number of aspects, such as cost, capacity, and demand, which may be approximated using sophisticated forecasting approaches like machine learning and deep learning.

References

1. Suryawanshi P, Dutta P (2022) Optimization models for supply chains under risk, uncertainty, and resilience: a state-of-the-art review and future research directions. Transp Res Part E Logist Transp Rev 157:102553
2. Naqvi MA, Amin SH (2021) Supplier selection and order allocation: a literature review. J Data Inf Manag 3(2):125–139

3. Saputro TE, Figueira G, Almada-Lobo B (2022) A comprehensive framework and literature review of supplier selection under different purchasing strategies. Comput Ind Eng 167(January):108010
4. Aouadni S, Aouadni I, Rebaï A (2019) A systematic review on supplier selection and order allocation problems. J Ind Eng Int 15(s1):267–289
5. Mukherjee K (2014) Supplier selection criteria and methods; past, present and future Supplier selection criteria and methods: past, present and future. Int J Oper Res x(1)
6. Higgins J, Green S (2008) Overview of reviews. Cochrane handbook for systematic reviews of interventions. Cochrane Database Systematic Reviews, pp 187–235
7. Damayanti RW, Hartono B, Wijaya AR (2019) The complexity of megaprojects in developing countries: a literature review. IEEE Int Conf Ind Eng Eng Manage 2019(December) 1366–1370
8. Moheb-Alizadeh H, Handfield R (2019) Sustainable supplier selection and order allocation: a novel multi-objective programming model with a hybrid solution approach. Comput Ind Eng 129:192–209
9. Kaur H, Prakash Singh S (2021) Multi-stage hybrid model for supplier selection and order allocation considering disruption risks and disruptive technologies. Int J Prod Econ 231
10. Li F, Wu CH, Zhou L, Xu G, Liu Y, Tsai SB (2021) A model integrating environmental concerns and supply risks for dynamic sustainable supplier selection and order allocation. Soft Comput 25(1):535–549
11. Ayhan MB, Kilic HS (2015) A two stage approach for supplier selection problem in multi-item/multi-supplier environment with quantity discounts. Comput Ind Eng 85:1–12
12. Kellner F, Utz S (2019) Sustainability in supplier selection and order allocation: combining integer variables with Markowitz portfolio theory. J Clean Prod 214:462–474
13. Moheb-Alizadeh H, Handfield R (2018) An integrated chance-constrained stochastic model for efficient and sustainable supplier selection and order allocation. Int J Prod Res 56(21):6890–6916
14. Baek SH, Kim JS (2020) Efficient algorithms for a large-scale supplier selection and order allocation problem considering carbon emissions and quantity discounts. Mathematics 8(10):1–16
15. You SY, Zhang LJ, Xu XG, Liu HC (2020) A new integrated multi-criteria decision making and multi-objective programming model for sustainable supplier selection and order allocation. Symmetry (Basel) 12(2)
16. Islam S, Amin SH, Wardley LJ (2021) Machine learning and optimization models for supplier selection and order allocation planning. Int J Prod Econ 242
17. Kumcu S, Özyörük B (2020) Selection of the best supplier with 0–1 goal programming by integrating ahp in the healthcare industry. Mersin Univ J Marit Fac
18. Franco C, Alfonso-Lizarazo E (2020) Optimization under uncertainty of the pharmaceutical supply chain in hospitals. Comput Chem Eng 135
19. Imran M, Kang C, Babar Ramzan M (2018) Medicine supply chain model for an integrated healthcare system with uncertain product complaints. J Manuf Syst 46:13–28
20. Yazdani M, Torkayesh AE, Chatterjee P (2020) An integrated decision-making model for supplier evaluation in public healthcare system: the case study of a Spanish hospital. J Enterp Inf Manag 33(5):965–989
21. Bektur G (2020) An integrated methodology for the selection of sustainable suppliers and order allocation problem with quantity discounts, lost sales and varying supplier availabilities. Sustain Prod Consum 23:111–127
22. Stević Ž, Pamučar D, Puška A, Chatterjee P (2020) Sustainable supplier selection in healthcare industries using a new MCDM method: measurement of alternatives and ranking according to Compromise solution (MARCOS). Comput Ind Eng 140
23. Ghadimi P, Ghassemi Toosi F, Heavey C (2018) A multi-agent systems approach for sustainable supplier selection and order allocation in a partnership supply chain. Eur J Oper Res 269(1):286–301
24. Hoseini SA, Fallahpour A, Wong KY, Mahdiyar A, Saberi M, Durdyev S (2021) Sustainable supplier selection in construction industry through hybrid fuzzy-based approaches. Sustainability 13(3):1–19

25. Srinual N, Mehnen J, Udomsub R, Multi criteria supplier selection from social aspects in thai tyre rubber industry
26. Hamdan S, Cheaitou A (2017) Dynamic green supplier selection and order allocation with quantity discounts and varying supplier availability. Comput Ind Eng 110:573–589
27. Sarkis J, Dhavale DG (2015) Supplier selection for sustainable operations: a triple-bottom-line approach using a Bayesian framework. Int J Prod Econ 166:177–191
28. Lamba K, Singh SP, Mishra N (2019) Integrated decisions for supplier selection and lot-sizing considering different carbon emission regulations in Big Data environment. Comput Ind Eng 128(April 2018):1052–1062
29. Hosseini ZS, Flapper SD, Pirayesh M (2022) Sustainable supplier selection and order allocation under demand, supplier availability and supplier grading uncertainties. Comput Ind Eng 165(September 2021):107811
30. Amadea B, Dewi K, Rosyidi CN, Aisyati A, An optimisation model of drug order quantity and distribution using continuous review approach by considering secondary suppliers
31. Pamucar D, Torkayesh AE, Biswas S (2022) Supplier selection in healthcare supply chain management during the COVID-19 pandemic: a novel fuzzy rough decision-making approach. Ann Oper Res

A Literature Review on the Energy Efficiency Issue in Operating Room Planning and Scheduling

Ahmad Kholid Alghofari, Cucuk Nur Rosyidi, Wakhid Ahmad Jauhari, and Pringgo Widyo Laksono

Abstract This paper provides a review of the literature on operating room planning and scheduling issues. This paper investigates the most recent consideration in Operating Room (OR) planning and scheduling in order to comprehensively classify the issues studied. This Article reviewed that OR planning considers the efficiency energy issue. The literature published between 2015 and 2022 explicitly mentions OR planning, scheduling, and efficiency energy. The literature review starts with a collection of papers on OR planning, scheduling, and energy efficiency. Then, the theme's efficiency is determined using bibliometric analysis. Furthermore, future research opportunities in operating room planning and scheduling with efficiency in mind are identified. This Article identify several papers that have made efforts to minimize energy use in the operating room. This provides an opportunity for further research on operating room scheduling by considering energy efficiency and uncertainty factors.

Keywords Energy efficiency · Operating room planning · Scheduling

1 Introduction

Excellent Operating Room (OR) performance is one of the most critical and costly resources for efficiency and productivity in most hospitals. OR services are the hospital's most significant source of income and patient satisfaction [1, 2]. Furthermore, according to many studies, OR is the most important expense and pay (up to 60%) of Hospitals [3]. The performance of OR utilization is heavily influenced

A. K. Alghofari (✉)
Department of Industrial Engineering, Muhammadiyah University of Surakarta, Surakarta, Indonesia
e-mail: ahmad.kholid@ums.ac.id

A. K. Alghofari · C. N. Rosyidi · W. A. Jauhari · P. W. Laksono
Department of Industrial Engineering, Universitas Sebelas Maret, Surakarta, Indonesia

C. N. Rosyidi et al. (eds.), *Proceedings of the 6th Asia Pacific Conference on Manufacturing Systems and 4th International Manufacturing Engineering Conference*, Lecture Notes in Mechanical Engineering,
https://doi.org/10.1007/978-981-99-1245-2_36

by hospital management's planning and scheduling policies [4]. Furthermore, many parties are involved in the OR scheduling process, including management, surgeons, patients, and surgical facilities. A mathematical model must be created and optimized to solve these problems based on limited resources and the desired objective function [5]. The objective functions are cost and overtime minimization and OR utilization maximization [6].

From various reviews, much research has been done on OR planning and scheduling. Several literature reviews have been conducted to demonstrate the evolution of research on the OR theme. Cardoen [7], who reviewed 247 papers from 1950 to 2008, considered six factors which are: patient characterization, performance measurement, decision category, research methods, uncertainty, and research application. Samudra [8] continued to use review data until 2014. A review of the planning and scheduling of OR research on elective and non-elective types of patients has been carried out by van Riet & Demeulemeester [9]. Hof [10] reviewed Case Mix Planning (CMP) as part of medium and long-term planning for tactical and strategic levels. Wang (2021) conducted a research review with a special concentration on the type of patient, both inpatient and outpatient [3].

Based on the previous studies, OR planning and scheduling are more focused on efforts to utilize OR utilities. The availability of patients, surgeons, nurses, operating rooms, and operating facilities are usually the primary considerations in planning and scheduling. On the other hand, energy efficiency is an important issue that is increasingly emerging and becoming a trend in the future [11]. Hospitals have also been designed with energy efficiency in minimal conditions [12]. In the OR, each supporting facility has equipment that uses different energy consumption for each location. Therefore, efforts to find references related to OR planning and scheduling considering uncertainty and energy efficiency are carried out to to respond to the following issues:

i. RQ1. What is the most recent factor to consider in OR planning and scheduling?
ii. RQ2. How are energy efficiency issues related to operating room planning and scheduling considerations?
iii. RQ3. Which areas of research should be investigated further to support OR planning and scheduling? Taking into account energy efficiency variables?

2 Methods

A Systematic literature Review Methodology (SRM) is used to investigate the research theme of operating room planning and scheduling while taking uncertainty and energy efficiency into account. SRM as a method for determining, quantifying, and analyzing the contribution of scientific journal articles to answering research questions [13]. The following steps in the systematic review procedure identify keyword search identification. The terms operating room planning, operating room scheduling, energy efficiency healthcare, and energy efficiency scheduling were used

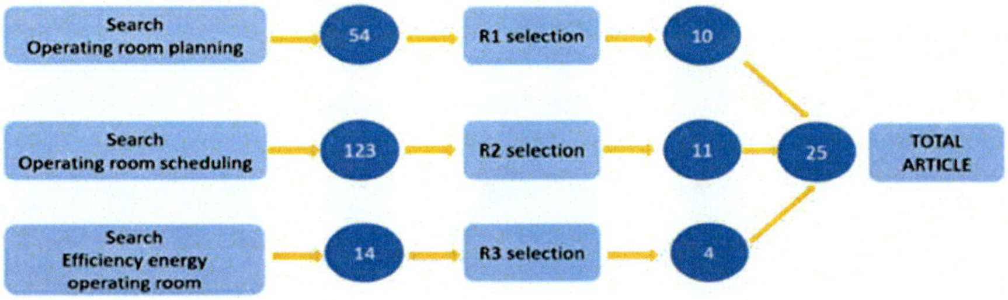

Fig. 1 Flow of systematic review

in this study. Furthermore, researchers in operating room planning and scheduling use these terms.

The analysis was carried out on libraries taken from the Scopus database and with the snowball search method. Data is taken using "publish and perish" software. Database selected based on the suitability of the theme. The article should be written in English and available for full paper. Full articles with abstracts, keywords, and citations are downloaded and stored in Mendeley software. The articles will be map visualization using Vos Viewer to show the research gap. There are overall 191 articles collected; after being selected by a number of citations and relevance to the topic, there are 25 articles. Following a thorough examination, only 23 articles are suitable for further analysis (Fig. 1).

3 Result and Discussion

3.1 Bibliometric Analysis

Bibliometrics is a quantitative data analysis method that employs mathematical and statistical tools to assess the interdependence and impact of publications in a given field of study [14]. This approach can provide a high-level overview of a large body of academic literature while also trying to identify influential studies, authors, journals, organizations, and countries over time [15]. Based on an examination of the journals and research areas of the 23 titles chosen, the following conclusions were reached. The keywords in the 191 publications evaluated in this study were examined using the VOS viewer (Fig. 2). Across all articles, nineteen keywords appeared more than ten times in the title and abstract fields. These keywords were found 204 times in publications categorized as OR scheduling, energy efficiency, optimization, and OR Planning. Figure 1 depicts high-frequency keyword density, with lighter colors indicating more key term occurrences.

There are 23 articles published in international journals that discuss operating room planning and scheduling as well as energy efficiency. The selected articles are shown in Table 1.

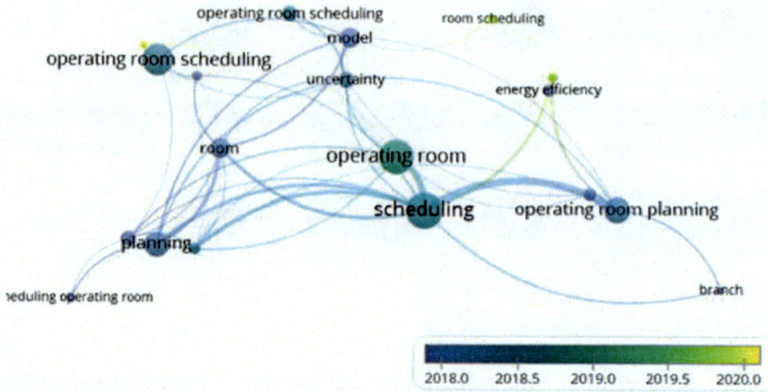

Fig. 2 Map of co-occurrence network visualization of titles and keywords abstract and Visualization map of keyword frequency evolution over time

3.2 The Energy Efficiency Issues Related to the Operating Room

Based on the results obtained, five papers are relevant to energy efficiency (Table 2). The five papers discuss how to consider energy efficiency in the operating room. Some technical engineering is done to increase efficiency. Jones [16] created the Fundamental Use of Surgical Energy (FUSE) plan to close a knowledge gap in the best practice use of energy devices for all health care professionals and staff who work in the operating room, procedural and interventional suites. Tejero [17] studies energy use optimization in operating room ventilation during inactivity periods. Sukarno [18] states that a dual duct air system may decrease the power requirement of an operating room HVAC system when compared to a well-controlled conventional two-position system. The proposed dual duct system is more expensive to install but will provide a favorable payback for operating room applications.

The following is a discussion to answer each research question about energy efficiency in operating room planning and scheduling.

RQ1. What is the most recent consideration in RO planning and scheduling?

There are three types of OR planning and scheduling, according to the findings of the literature review. According to an analysis of performance criteria, utilization and waiting time are the most prominent among researchers. Patient delay, preferences, and completion time, on the other hand, have received less attention than other criteria [19, 20]. Uncertainty mathematical modeling and multi-objective functions are more appealing to scholars since they are more realistic, whereas duration uncertainty is the most interesting area among scholars [2, 21].

RQ2. How are energy efficiency issues related to operating room planning and scheduling considerations?

Based on the literature review of journals related to energy efficiency in the operating room, most of the research on energy efficiency is related to efforts to save

Table 1 Selected article OR planning and scheduling (author, title, decision level, performance criteria, and uncertainty)

Author (year)	Title	Decision level	Performance criteria	Uncertainty
Aringhieri et al. (2015)	A two level metaheuristic for the operating room scheduling and assignment problem	S, T, O	U, FA	LOS, RU
Saadouli et al. (2015)	A stochastic optimization and simulation approach for scheduling operating rooms and recovery beds in an orthopedic surgery department	O	OT, U, FA	D, RU
Addis et al. (2016)	Operating room scheduling and rescheduling: a rolling horizon approach	T, O	WT, OT, U	RO, D, Ar
Xiang et al. (2015)	An ant colony optimization approach for solving an operating room surgery scheduling problem	O	WT, MS, U, OT, P	D, Ar, LOS
Molina-Pariente, Fernandez-Viagas, et al. (2015)	Integrated operating room planning and scheduling problem with assistant surgeon dependent surgery durations	T, O	U, WT, OT	D, RU
Guido and Conforti (2017)	A hybrid genetic approach for solving an integrated multi-objective operating room planning and scheduling problem	S, T	U, MS	D, RU
Doulabi et al. (2016)	A constraint-programming-based branch-and-price-and-cut approach for operating room planning and scheduling	T, O	MS, OT	RU,
Dios et al. (2015)	A Decision Support System for Operating Room scheduling	O	U, VR	D, RU
Roshanaei et al. (2017)	Collaborative operating room planning and scheduling	S, O	P, WT, U	D
Jebali and Diabat (2017)	A Chance-constrained operating room planning with elective and emergency cases under downstream capacity constraints	T, O	WT, OT, RR, U	D, RU, LOS
Molina-Pariente et al. (2018)	A stochastic approach for solving the operating room scheduling problem	T, O	OT, WT, U	D, Ar, RU
Li et al. (2017)	Scheduling elective surgeries: the tradeoff among bed capacity, waiting patients and operating room utilization using goal programming	T, O	WT, U, RR,	D, Ar, RU

(continued)

Table 1 (continued)

Author (year)	Title	Decision level	Performance criteria	Uncertainty
Kroer et al. (2018)	Planning and scheduling operating rooms for elective and emergency surgeries with uncertain duration	S, O	OT, U, FA	D, Ar

Decision Level: Strategic (S), Tactical (T), Operational (O)
Performance Criteria: Waiting time (WT), Utilization (U), Overtime (OT), Make span (MS), Recovery room availability (RR), Variable resource requirement (VR), Preference (P), Financial Asset (FA)
Uncertainty: Duration (D), Arrivals (Ar), Length of Stay (LOS), Resource Uncertainty (RU)

Table 2 Paper related to energy efficiency in OR

Author (year)	Title	Journal	Objective
Jones (2017)	Fundamental use of surgical energy (FUSE): an essential educational program for operating room safety	The Permanente Journal	The FUSE program was developed to fill a knowledge gap in the best-practice use of energy devices by providing a standardized educational tool
Tejero-González (2021)	Optimization of energy use in operating room ventilation during inactivity periods	Building Research and Information	The study compares setback ventilation in operating rooms using two approaches: two operating rooms with setback ventilation and two operating rooms with normal ventilation and heat recovery system operation
Sukarno (2021)	A multi-stage heat-pipe heat exchanger for improving the energy efficiency of a hospital operating room's HVAC system	International Journal of Low-Carbon Technologies	Conducting research on engineering HVAC systems in operating rooms for more efficient energy use
Bartholomew (2015)	Energy-efficient approach for operating rooms	ASHRAE Journal	This research will show how a dual duct HVAC system version saves significantly more operational energy than a standard system
Tejero-González (2022)	Safety and energy implications of setback control in operating rooms during unoccupied periods	Applied Sciences (Switzerland)	Research provides evidence of the savings and safety of this energy efficiency measure in operating rooms

energy used in the operating room. As stated by Chen [12], efforts to save energy use in OR can significantly save hospital expenses and reduce carbon emissions due to energy use.

The hospital has been designed with energy efficiency in mind by using efficient Air Conditioning (AC) [12]. The same considerations regarding the efficiency of using AC in the Operating Room (RO) have also been studied by Chien [22].

RQ3. Which areas of research should be investigated further to support OR planning and scheduling?

Based on the discussion of RQ1 and RQ2, the following gaps must be addressed in future studies: (1) Energy reduction efforts are carried out to reduce carbon use based on the development of increasing energy efficiency. Trianni's data (2018) shows that from 2008–2017, several 828 research titles on energy efficiency are growing. (2) RO planning and scheduling usually consider performance criteria in operations planning (RQ1). However, previous researchers have never considered energy efficiency in OR planning and scheduling. At the same time, OR requires equipment that mostly requires an energy source (RQ2). The distributed energy in the operating room opens opportunities for optimization and considering performance criteria in operating room planning and scheduling.

4 Conclusions

The issue of OR planning and scheduling is critical in healthcare management. This paper analyzed and thoroughly reviewed the field. The Scopus database was used for the analysis, which covered the years 2015 to 2022. OR scheduling issues are classified into three decision levels: strategic, tactical, and operational. Following the review were performance criteria, uncertainty, strategy, solution, and methods. Energy efficiency in various fields also affects the use of the operating room. Several researchers have made efforts to minimize energy use in the operating room. This opens opportunities for further research on operating room scheduling by considering energy efficiency and uncertainty factors. Future research can focus more on OR planning and scheduling by considering the uncertainty factor and energy efficiency.

References

1. Denton BT (ed) (2013) Handbook of healthcare operations management methods and applications, vol 1, 1st ed. Springer. https://doi.org/10.1007/978-1-4614-5885-2
2. Azar M, Carrasco RA, Mondschein S (2022) Dealing with uncertain surgery times in operating room scheduling. Eur J Oper Res 299(1):377–394. https://doi.org/10.1016/j.ejor.2021.09.010
3. Wang L, Demeulemeester E, Vansteenkiste N, Rademakers FE (2021) Operating room planning and scheduling for outpatients and inpatients: a review and future research. Oper Res Health Care 31. Elsevier B.V. https://doi.org/10.1016/j.orhc.2021.100323

4. Bello C, Urman RD, Andereggen L, Doll D, Luedi MM (2022) Operational and strategic decision making in the perioperative setting: meeting budgetary challenges and quality of care goals, best practice and research: clinical anaesthesiology. Bailliere Tindall Ltd. https://doi.org/10.1016/j.bpa.2022.04.003
5. Kahraman C, Topcu YI (2018) Operations research applications in health care management, vol. 262. Springer International Publishing, Cham. https://doi.org/10.1007/978-3-319-65455-3
6. Zhu S, Fan W, Yang S, Pei J, Pardalos PM (2019) Operating room planning and surgical case scheduling: a review of literature. J Comb Optimization 37(3). Springer, New York, pp 757–805. https://doi.org/10.1007/s10878-018-0322-6
7. Cardoen B, Demeulemeester E, Beliën J (2010) Operating room planning and scheduling: a literature review. Eur J Oper Res 201(3):921–932. https://doi.org/10.1016/j.ejor.2009.04.011
8. Samudra M, van Riet C, Demeulemeester E, Cardoen B, Vansteenkiste N, Rademakers FE (2016) Scheduling operating rooms: achievements, challenges and pitfalls. J Sched 19(5):493–525. https://doi.org/10.1007/s10951-016-0489-6
9. van Riet C, Demeulemeester E (2015) Trade-offs in operating room planning for electives and emergencies: a review. Oper Res Health Care 7:52–69. https://doi.org/10.1016/j.orhc.2015.05.005
10. Hof S, Fügener A, Schoenfelder J, Brunner JO (2017) Case mix planning in hospitals: a review and future agenda. Health Care Manag Sci 20(2):207–220. https://doi.org/10.1007/s10729-015-9342-2
11. Worrell E, Bernstein L, Roy J, Price L, Harnisch J (2015) Industrial energy efficiency and climate change mitigation. Energy Effic. Accessed 21 Dec 2015. [Online]. Available: https://doi.org/10.1007/s12053-008-9032-8
12. Chen Y (2018) Study on energy-saving design and operation of hospital purification air conditioning system. Int J Low-Carbon Technol 13(2):184–190. https://doi.org/10.1093/ijlct/cty013
13. Staples M, Niazi M (2007) Experiences using systematic review guidelines. J Syst Softw 80(9):1425–1437. https://doi.org/10.1016/j.jss.2006.09.046
14. Gil M, Wróbel K, Montewka J, Goerlandt F (2020) A bibliometric analysis and systematic review of shipboard decision support systems for accident prevention. Safety Science 128. Elsevier B.V. https://doi.org/10.1016/j.ssci.2020.104717
15. van Eck NJ, Waltman L (2010) Software survey: VOSviewer, a computer program for bibliometric mapping. Scientometrics 84(2):523–538. https://doi.org/10.1007/s11192-009-0146-3
16. Jones SB (2017) Fundamental use of surgical energy (FUSE): an Essential educational program for operating room safety. Perm J 21. https://doi.org/10.7812/TPP/16-050
17. Tejero-González A (2021) Energy use optimization in ventilation of operating rooms during inactivity periods. Build Res Inf 49(3):308–324. https://doi.org/10.1080/09613218.2020.1817723
18. Sukarno R (2021) Multi-stage heat-pipe heat exchanger for improving energy efficiency of the HVAC system in a hospital operating room. Int J Low-Carbon Technol 16(2):259–267. https://doi.org/10.1093/ijlct/ctaa048
19. Zhu S, Fan W, Liu T, Yang S, Pardalos PM (2020) Dynamic three-stage operating room scheduling considering patient waiting time and surgical overtime costs. J Comb Optim 39(1):185–215. https://doi.org/10.1007/s10878-019-00463-5
20. Park J, Kim BI, Eom M, Choi BK (2021) Operating room scheduling considering surgeons' preferences and cooperative operations. Comput Ind Eng 157. https://doi.org/10.1016/j.cie.2021.107306
21. Khaniyev T, Kayış E, Güllü R (2020) Next-day operating room scheduling with uncertain surgery durations: exact analysis and heuristics. Eur J Oper Res 286(1):49–62. https://doi.org/10.1016/j.ejor.2020.03.002
22. Chien TY (2020) Comparative analysis of energy consumption, indoor thermal-hygrometric conditions, and air quality for HVAC, LDAC, and RDAC systems used in operating rooms. Appl Sci (Switzerland) 10(11). https://doi.org/10.3390/app10113721

Economic Order Quantity Model for Perishable Product by Considering Donation Factor

Cucuk Nur Rosyidi and Syania Hilda Andini

Abstract The aim of this research is to develop an optimization model which can be used to determine the economic order quantity of perishable product by considering the donation factor. The objective function of the model was to maximize the company profit. The demand of the product was influenced by the selling price and the amount of donation. The price will have negative effect on the price. The higher the price, the lower the demand. Conversely, the donation has positive impact on the demand. The higher the donation, the higher the demand. Donation has been acknowledged as an activity which will expand product demand, increase brand reputation, and obtain subsidies from the government. This model also considers the consumer willingness to purchase which assume to be linearly decrease following the age of the product. A numerical example and sensitivity analysis were given to respectively show the applicability of the model and study the impact of certain input parameters to both decision variables and objective function.

Keywords Donation · EOQ · Perishables product · Government subsidies

1 Introduction

Product inventory has to be managed to avoid both product shortage and excess inventory. Product shortage will make a company fail to meet certain customer demand, while excess inventory will make the company has to bear unnecessary cost. Most of perishable products are fresh products such as fruit, meat, vegetables, and dairy products with short shelf life and age-dependent [1]. Hence, in the case of perishable products, excess inventory will decrease the willingness to buy of the products and as the products achieve its shelf life, they have to be disposed with a certain cost of disposal [2]. The excess inventory of perishable products will also increase the

C. N. Rosyidi (✉) · S. H. Andini
Universitas Sebelas Maret, Surakarta, Indonesia
e-mail: cucuknur@staff.uns.ac.id

C. N. Rosyidi et al. (eds.), *Proceedings of the 6th Asia Pacific Conference on Manufacturing Systems and 4th International Manufacturing Engineering Conference*, Lecture Notes in Mechanical Engineering,
https://doi.org/10.1007/978-981-99-1245-2_37

food waste [3]. This issue becomes one of the important global issues that must be addressed by many countries including Indonesia. Based on a report by [4], Indonesia became the second-highest country in the world as a producer of food waste after Saudi Arabia. The decomposition of organic substance of the perishable products results methane gas (CH_4). The gas has a global warming potential of 25 times greater than CO_2 [5]. When food waste is unavoidable, a company may consider to donate their product to food banks due to the high level of food insecurity even in developed countries like the United States [6]. Donations are a form of corporate social responsibility and with the donation, the company has fulfilled its social responsibility to the community, increased brand reputation, and will increase product demand [7].

The main problem that are often faced by companies is how to determining the optimal order quantity. The widely used model to determine the optimal order quantity is the Economic Order Quantity (EOQ) [8]. The EOQ model was first developed by F. W. Haris by assuming deterministic product demand and constant quality. Perishable products do not conform with the latter assumption. Their quality is time dependent and will decrease with certain rate. Deterioration can occur in the form of spoilage, damage, obsolescence, or loss of marginal value resulting in a decrease in the value of the product [9]. According to [10], deteriorated products have three category, namely obsolescence, perishable, and physical depletion or decay. Obsolescence is a product that becomes obsolete due to the development of technology or the emergence of new products, such as cosmetics, accessories, and fashion products. Perishables are products that perish or have a short life, such as fruits and vegetables. Physical depletion is a product that is easily damaged by time, such as flour, chemicals, and gasoline. Hence, the traditional EOQ cannot be used to solve cases of perishable products.

Inventory depletion has been discussed in [11] by considering the exponentally decline of demand. Research [12] developed an inventory model of perishable products for both deterministic and stochastic demand for single and multiple items. Muriana [13] developed a model for perishable product considering stochastic demand. Diaz et al. [2] developed an EOQ model of perishable product by considering the willingness to purchase from the client-side considering the cost of product disposal at the end of product life cycle. Wang et al. [7] conducted a study on the closed-loop supply chain model by considering product recovery and donations. A model is needed to determine the optimal order quantity and price in order to maximize the annual profit of the company. This study refers to the model of [2] regarding the EOQ model for perishable products with linear willingness to purchase. This study also refers to the research of [7] for the product recovery and donation. Based on those two studies, in this research we develop an optimization model for determining the optimal EOQ for perishable products by considering donations to maximize the company's annual profit.

2 Model Assumptions and Notations

2.1 Assumptions

The following assumptions are used in this research:

1. Demand is assumed to be dependent on the price and the number of donated products.
2. Shortage is not allowed.
3. The willingness to purchase decreases linearly.
4. Unsold products are donated at the end of their shelf life.
5. The donated products are safely consumable.
6. The government provides subsidies for the donations.
7. Donation activities increase reputation revenue.

2.2 Notations

The objective function of this model is to maximize the total profit with two decision variables, namely order quantity (Q) and product selling price (p). The model notations are as the following:

α	Market size
β	Price sensitivity
p	Selling price per kilogram (IDR/Kg)
γ	Donation sensitivity coefficient
d	Donation quantity (Kg)
D	Annual demand quantity (Kg)
a	Number of working days
W	Product life cycle (days)
r	Demand per unit time (Kg/day)
b	Reputation revenue from donated product (IDR/Kg)
k	Donation subsidies from the government (IDR/Kg)
Cm	Annual holding cost per kilogram (IDR/Kg)
Co	Ordering cost (IDR)
h	Cost of donation activity per kilogram of product (IDR/Kg)
$s\ (t, i)$	Sales quantity at time t of cycle i (Kg)
$S\ (t, i)$	Cumulative sales quantity at time t of cycle i (Kg)
$I\ (t, i)$	Inventory at time t of cycle i (Kg)
$\bar{I}(i)$	Average inventory at cycle i (Kg)
Q	Order quantity (Kg)
T	Cycle Time (days)
TC	Total cost (IDR)
TR	Total revenue (IDR)
TP	Total profit (IDR)

3 Model Formulation

The annual demand in this model depends on the selling price of the product per kilogram and the total number of products donated per cycle. The annual demand function is expressed in Eq. (1).

$$D = \alpha - \beta p + \gamma \sum_{i}^{n} d(i) \tag{1}$$

The annual demand is divided by the number of working days to yield the daily demand.

$$r = \frac{D}{a} \tag{2}$$

The willingness to purchase is assumed to be linearly decreased and reaches zero at the end of its life. Hence, the number of products sold at time t during cycle i is the product of the probability of the product sold with the daily demand as shown in Eq. (3).

$$s(t, i) = r\left[1 - \frac{t}{W}\right] \tag{3}$$

The inventory level is the difference between the number of products ordered and the cumulative number of products sold at time t cycle i:

$$I(t, i) = Q - S(t, i) \tag{4}$$

Hence, the average inventory in each cycle can be found by the following equation:

$$E(\bar{I}(i)) = Q\left[\frac{1}{2} + \frac{Q}{6rW}\right] \tag{5}$$

The number of products sold per cycle can be found by Eq. (6).

$$E(S(T, i)) = Q - \frac{Q^2}{2rW} \tag{6}$$

The number of products donated at the end of each cycle is the difference between the number of products ordered and the cumulative number of products sold at the end of each cycle as shown in Eq. (7).

$$E(d(i)) = \frac{Q^2}{2rW} \tag{7}$$

The total cost is obtained as the sum of the ordering cost, the average inventory cost, and the value of donation cost. Ordering cost is the product of the ordering cost per order by the annual demand divided by the order quantity. The average inventory cost is the product of the annual inventory cost with the average inventory. The total cost equation is determined as follow:

$$TC = Co\frac{\alpha - \beta p + \gamma \frac{Qa}{2W}}{Q} + \frac{CmQ}{2} + \frac{CmQ^2a}{6\left(\alpha - \beta p + \gamma \frac{Qa}{2W}\right)W} + h\frac{Q^2a^2}{4W^2} \quad (8)$$

Total revenue is found as the sum of revenue from product sales and income from donation activities. The total revenue consists of revenue from product sales, donation activities and subsidies from the government as well as reputation income from donation activities. Equation (9) shows the total revenue.

$$TR = p\left(\alpha - \beta p + \gamma \frac{Qa}{2W}\right) - \frac{pQa}{2W} + (b + k)\frac{Qa}{2W} \quad (9)$$

Total annual profit is the difference between total revenue and total cost as shown in Eq. (10).

$$TP = p\left(\alpha - \beta p + \gamma \frac{Qa}{2W}\right) + (b + k)\frac{Qa}{2W} - \frac{pQa}{2W} - Co\frac{\alpha - \beta p + \gamma \frac{Qa}{2W}}{Q} - \frac{CmQ}{2} - \frac{CmQ^2a}{6\left(\alpha - \beta p + \gamma \frac{Qa}{2W}\right)W} - h\frac{Q^2a^2}{4W^2} \quad (10)$$

4 Numerical Example

The following parameters are used in the numerical example of this research: $\alpha =$ 50,000; $\beta = 2$; $\gamma = 1$; $a = 360$ days; $W = 30$ days; $b =$ IDR 6000/kg; $k =$ IDR 2000/Kg; $Cm =$ IDR 4000/Kg; $Co =$ IDR 100,000/order; $h =$ IDR 100/Kg. The optimal solutions were found as follows:

Q^* = 71 kg
p^* = IDR 13.204.19
TP = IDR 262.797.320

The optimal donated product for the numerical example was found to be 426 kg.

5 Sensitivity Analysis

Sensitivity analysis was performed to study the effect of some key parameters toward both decision variable and objective function. In this research, we investigate the effects of holding cost, ordering cost, market size price sensitivity coefficient, and shelf life. The sensitivity was done by change those parameters value from − 30% until 30% with the increment of 10%. The results of sensitivity analysis are shown in Figs. 1, 2, and 3. Figure 1 shows the results of sensitivity analysis of holding cost, ordering cost, price sensitivity coefficient, and shelf life on the value of optimal order. From the figure we can see that holding cost and price sensitivity coefficient has no effect on the optimal order. The ordering cost is relatively not sensitive to the optimal order since the increase of parameter values slightly increase the optimal order. The shelf life has significant effect on optimal order quantity and hence the order quantity sensitive to the change of shelf life.

Figure 2 shows the effect of parameters changes on the optimal price. The figure shows that holding cost has no effect on optimal price. The ordering cost and shelf life have a very small effect on the optimal price. The price sensitivity has a significant impact on optimal price since this parameter directly related to the price as the decision variable.

The effect of the parameters changes on the objective function is shown in Fig. 3. The figure shows that holding cost, ordering cost, and shelf life have very small

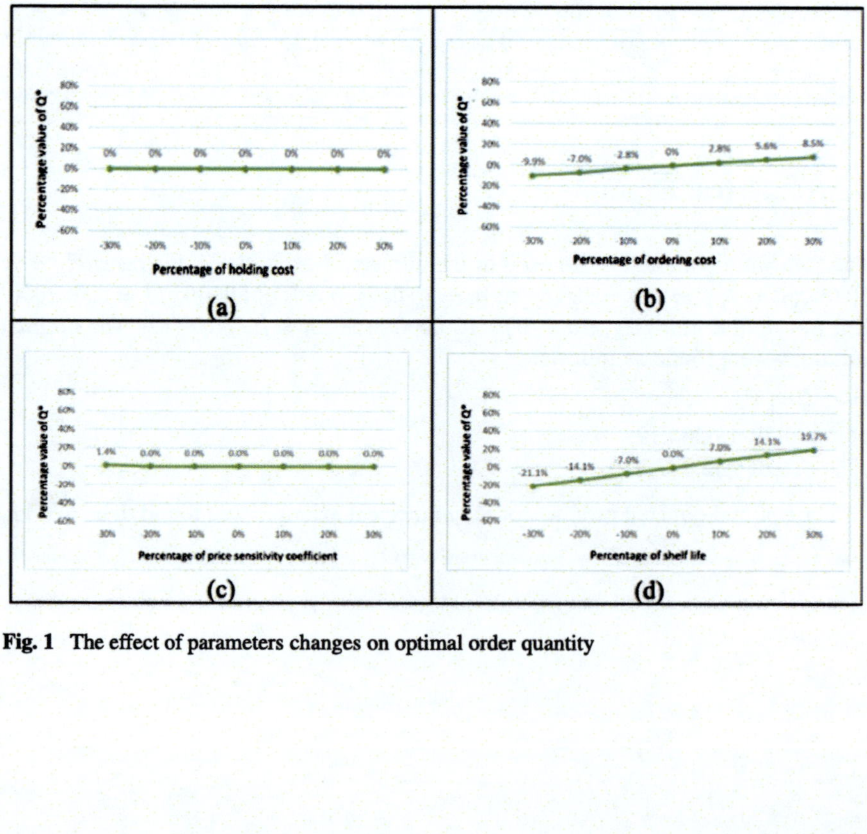

Fig. 1 The effect of parameters changes on optimal order quantity

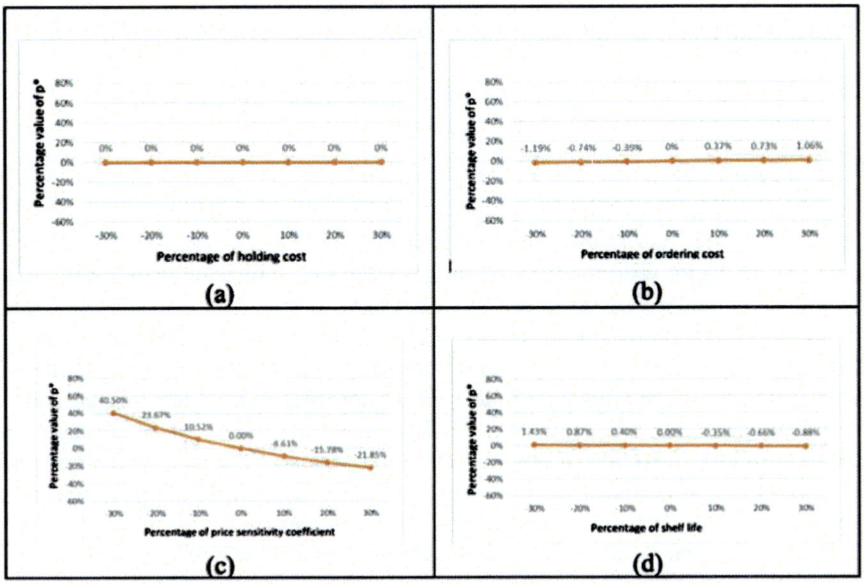

Fig. 2 The effect of parameters changes on optimal price

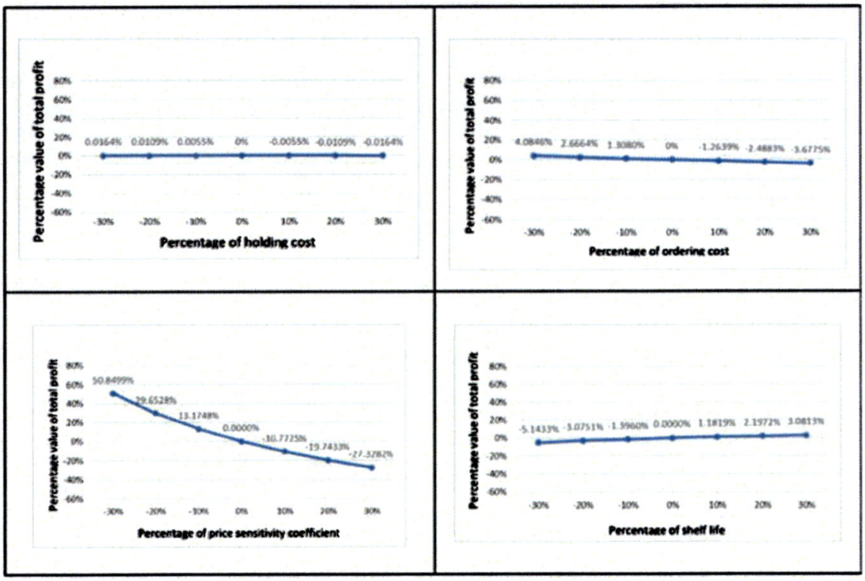

Fig. 3 The effect of parameters changes on total profit

impact to the profit, while the price sensitivity coefficient has a significant impact on the total profit.

6 Conclusions

An EOQ model for perishable product was developed in this research by considering the donation factor. The main objective of this study is to determine the optimal order quantity and selling price that maximize the expected total annual profit under the assumed conditions. The sensitivity analysis showed that shelf life has a significant impact on optimal order while and price sensitivity coefficient have significant impacts on both optimal price and total profit. Those results give an important information for the decision maker to be carefully maintain the product shelf life and keep the product insensitive to the price by efficiently manage the supply and demand to make the customer insensitive to the price of the product. For further research, it is necessary to consider several factors in the model such as product shortage, multi-item products, and the involvement of some other parties in the system.

References

1. Li R, Lan H, Mawhinney JR (2010) A review on deteriorating inventory study. J Serv Sci Manag 03(01):117–129. https://doi.org/10.4236/jssm.2010.31015
2. Díaz RDS, Paternina-Arboleda CD, Martínez-Flores JL, Jimenez-Barros MA (2020) Economic order quantity for perishables with decreasing willingness to purchase during their life cycle. Oper Res Perspect 7:7100146. https://doi.org/10.1016/j.orp.2020.100146
3. Lebersorger S, Schneider F (2014) Food loss rates at the food retail, influencing factors and reasons as a basis for waste prevention measures. Waste Manage 34(11):1911–1919. https://doi.org/10.1016/j.wasman.2014.06.013
4. Green A (2022) Food loss and waste. https://impact.economist.com/perspectives/sustainability/food-sustainability-index-2016/infographic/food-loss-and-waste. Accessed 6 Oct 2022
5. Shahid K, Hittinger E (2021) Techno-economic optimization of food waste diversion to treatment facilities to determine cost effectiveness of policy incentives. J Clean Prod 279:122634. https://doi.org/10.1016/j.jclepro.2020.122634
6. Roe BE, Qi D, Bender KE (2020) Some issues in the ethics of food waste. Physiol Behav 219:112860. https://doi.org/10.1016/j.physbeh.2020.112860
7. Wang Y, Wang Z, Li B, Liu Z, Zhu X, Wang Q (2019) Closed-loop supply chain models with product recovery and donation. J Clean Prod 227:861–876. https://doi.org/10.1016/j.jclepro.2019.04.236
8. Dutta D, Kumar P (2012) Fuzzy inventory model without shortage using trapezoidal fuzzy number with sensitivity analysis. IOSR J Math 4(3):32–37. https://doi.org/10.9790/5728-0433237
9. Chiao HE, Wee HM, Yang PC (2008) A deteriorating inventory model with two storage facilities, partial backordering and quantity discount. In: Proceedings of the 7th international conference on machine learning and cybernetics. ICMLC 7:3903–3908. https://doi.org/10.1109/ICMLC.2008.4621085
10. Widyadana IGA (2013) Penentuan ukuran pemesanan material dengan memperhatikan decay Inventory dan Quantity Unit Discount 1(1):57–62

11. Ghare PM, Schrader GF (1963) A model for an exponential decaying inventory. J Ind Eng 14:238–243
12. Nahmias S (1982) Perishable inventory theory: a review. Oper Res 30(4):680–708. https://doi.org/10.1287/opre.30.4.680
13. Muriana C (2016) An EOQ model for perishable products with fixed shelf life under stochastic demand conditions. Eur J Oper Res 255(2):388–396. https://doi.org/10.1016/j.ejor.2016.04.036

Optimization of the Automotive Manufacturing Plant Using Systematic Layout Planning (SLP) with Modular Layout

Muhammed Nafis Bin Osman Zahid
and Muhammad Zulhilmi Bin Shamsul Amran

Abstract The project is a part of the improvement activities carried out in the automotive part manufacturing lines. Facilities layout planning become a very crucial task due to the nature of the industry that deals with various kinds of automotive component production. Dealing with the increasing demand and the variation of the components to be produced, the company faced limited space to locate their machines. Failure to properly set up their layout will cause waste in time, cost, space and production processes. Moreover, the quality of the components produced will also be affected and turned down the customers which mainly come from the international automotive brands. The research implements a prominent tool for layout plan called Systematic Layout Planning (SLP). The SLP tool mainly consists of several stages of analysis for facility planning. Apart from that, an extensive analysis is employed by using modular layout. The result indicates the reduction of the area used on the production floor which leads to the reduction of material travel distance and increases the production rates.

Keywords Systematic layout planning · Modular layout · Plant simulation analysis

1 Introduction

Facility layout planning is an important process in any manufacturing project. Waste and losses can be prevented with sufficient time spent on planning and arrangement before the equipment is installed [1]. Facilities can be arranged orderly in a logical sequence based on the dedicated component produced within the facilities. However,

M. N. B. O. Zahid (✉) · M. Z. B. S. Amran
Faculty of Manufacturing and Mechatronic Engineering Technology, Universiti Malaysia Pahang, 26600 Pekan, Pahang, Malaysia
e-mail: nafis@ump.edu.my

C. N. Rosyidi et al. (eds.), *Proceedings of the 6th Asia Pacific Conference on Manufacturing Systems and 4th International Manufacturing Engineering Conference*, Lecture Notes in Mechanical Engineering,
https://doi.org/10.1007/978-981-99-1245-2_38

this is not the case when the industry needs to deal with a large and variety of components at the same production line. Redesigning the layout each time the demand for the component changes will consume a lot of time and cost. For decades, Systematic Layout Planning has been established and used extensively. The flexibility of the layout of the plant would benefit the company and reduce the material handling costs by about 10 to 30% [2]. The facility layout of manufacturing involves several configurations that deal with the plant site with lines, buildings, major facilities, work areas, aisles and other pertinent features such as departmental boundaries.

Facility planning and the process of producing products are closely intertwined. In actuality, designing or redesigning a facility's layout will take a lot of time and money, especially for manufacturing companies, regardless of their sizes and capacity. This is due to the fact that poor facility design can result in a variety of issues, including operator disorganization, poor product quality, customer dissatisfaction, and more. One solution to the facility layout problem is systematic layout planning (SLP). Simple techniques and no need for advanced mathematics, algorithms, or computer software make it very simple to understand and master.

The work is carried out in one of a leading automotive component manufacturing plants. The company produces many types of automotive components for local and international brands. The production quantity is based on made-to-stock and the company relied on the previous order to project their operations. Due to the increasing demand and variation of components, the company faced limited space to locate new machines to cope with the ordered quantity. Since there is no established procedure for relayout in this company, thus it is initial exposure for them to implement the SLP methods due to the well establishment of this methods and a lot of project examples have been carried out. The objectives of this study are to investigate the layout problem based on the machine locations on island A3 in the manufacturing plant. Then, the SLP analysis is conducted and incorporated with the modular layout alternatives.

2 Systematic Layout Planning (SLP) and Modular Layout Design

There are some differences in layout planning between the manufacturing facilities and those of offices, retail centers and warehouses [3]. Facility layout is an important element that needs to be well planned to establish an efficient system [4]. Generally, up to 50% of the operating cost can be reduced with a well-designed manufacturing layout plant [5]. A methodical approach to layout design is called systematic layout planning (SLP) [1]. SLP comprises a collection of standards for recognizing and assessing various actions and options included in any layout design approach, as well as a framework of phases through which each layout design project must proceed. SLP aims to increase productivity and reduce production costs by employing a minimum workflow [5]. Moreover, the previous implementation also showed that the tool minimizes the material handling cost, reduces walking distance and increases line performance and efficiency [6]. Based on the general cost distribution, 20% to 50% of the total production cost is the material handling cost [2] whereas 10% to 30% of the total cost can be reduced by implementing an efficient facility layout.

Several works carry out using SLP have remarkably increased the performance of the manufacturing facilities. The implementation of SLP in the garment industry has led to the minimization of the material handling cost by up to 23% and material transfer time by 34% [7]. Another example of SLP implementation is recorded in the lean manufacturing improvement project conducted at the company producing cable trays, cable support systems and pole pipes. The outcomes of the analysis reduced the travel distance by 57% and 10% of the traveling time [8]. Another study extended the analysis of SLP and integrate it with multi-criteria decision-making tools. The proposed alternatives are evaluated using this methodology in comparison to the existing layout. The SLP analysis proposed an improved layout that enhanced the flow of the materials, space utilization and flexibility [9].

The modular design represents the standard interfaces where customization is realized through independent parts [10]. Modular design or modularity in design is an approach that can produce smaller parts called modules or skids when the system is subdivided. SLP design process does not have specific methods to deal with product mix segmentation and department planning. Thus, it is disadvantageous in dealing with manufacturing facilities that cater a variety of components. It is possible to solve this by taking an additional step to improve the SLP process by including production flow analysis into it. The modular layout is one of the procedures for enhancing SLP process. The ideas behind layout modules will go beyond the procedures and input data that are now needed for facility layout planning. The layout module also supports the need of new generation of facility layout beyond the traditional layout. A layout module is defined as a group of machines connected by a material flow network that exhibits a flow pattern characteristic of a specific type of layout, such as the Flowline, Cellular or Functional Layout [11]. The idea of layout modules offers an additional choice beyond the three conventional layouts (Flowline, Cellular or Functional Layout) that are still being researched and used in the industry. Discrete Event Simulation (DES) is a method for creating a digital model of an understudied system [12]. One of the effects of changing the manufacturing layout is that it is time-consuming and expensive. Discrete Event Simulation is the best method for creating and evaluating a wide range of scenario alternatives that aid the production manager in layout planning to minimise such consequences.

3 Methodology

Gathering or measuring the important data and helpful data to the project is one of the fundamental steps to moving on to the next level of analysis. The company provides all of the data, including the design of the present layout, capacity references product data, data history for sales quantity by product, and data from the production master plan, which are all crucial pieces of information for the project. The methodology of the study is derived from the processing steps illustrated in Fig. 1. Data analysis is the process of evaluating data either using analytical or logical methodologies. The five factors used as input data in systematic layout planning are P, Q, R, S, and T. P stands for "product," Q for "quantity," R for "route," S for "supporting services," and T for "time." P, Q, R, S, and T are crucial components of the systematic layout planning (SLP) used for this project since data must be available to arrange the plant layout in accordance with the process flow. The systematic layout planning will proceed as per the procedure based on the input data and an understanding of the responsibilities and relationships between the activities.

There are steps that must be followed in order to determine the alternatives layout. This heuristic process will propose the module layout for the company. Figure 2 displays the procedures of producing the layout module.

The layout's workspace is the component that must assess the space required of the machines. The space area of the aisle around the machines is also crucial in the layout

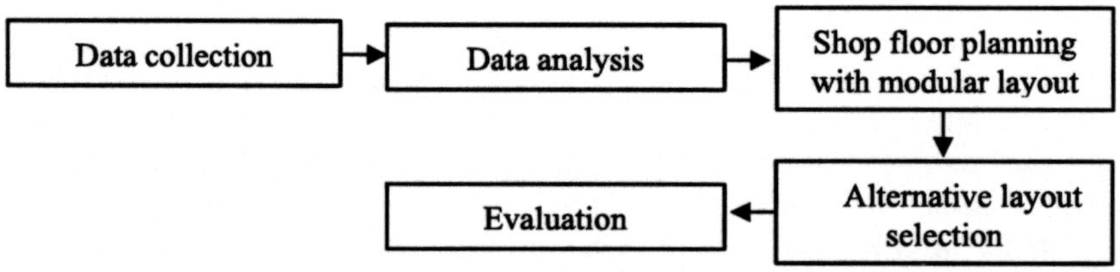

Fig. 1 Methodology process flowchart

Fig. 2 Modular layout procedure

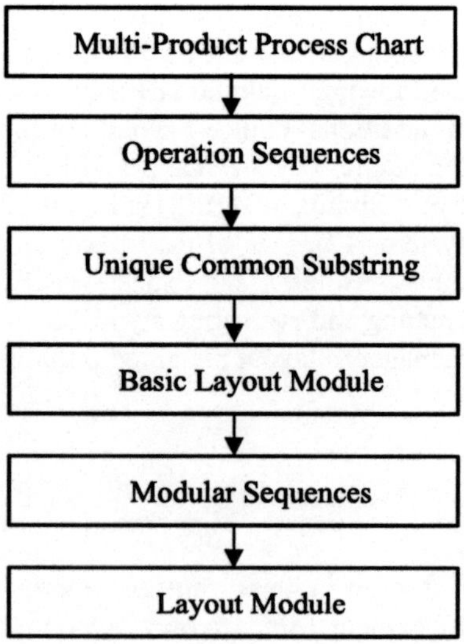

module, which uses the dimension of the machines to compute their area. This is due to the fact that workspaces must have an aisle for operators. Additionally, the distance between one machine and another or other requirements in the manufacturing line is crucial. The space between machines is primarily used to ensure smooth product material movement along the manufacturing line. The company has already presented the current plan, including the dimensions of the equipment and facilities. As a result, it is possible to determine how much room each equipment along the production line needs. According to the type of product, the computation was done by island and one product type at a time. In this manner, it is simpler to compare and determine the typical area requirement for the factory floor.

Some heuristic processes must be followed in order to determine the alternative layout. The automotive industry produced about 42 different product components for various brands of cars. Therefore, a multi-product process diagram is suitable for a production line that produces a variety of products. Figure 1 illustrates an example of a multi-product process chart that groups all product components or items onto a single table.

A layout module demonstrates the flow pattern feature and may include a product, a process, and sequences. By sharing comparable routing, layout modules can be constructed by identifying process sequences that are shared by two or more types of operation sequences. The common substrings among operations sequences are what this is known as. The term "common substring" refers to a group of process sequences that are shared by two or more types of operation sequences. These operation sequences are the superstrings of the substrings. In each superstring there is an operation that does not belong to the common substring that known as residual operation. The next step involves the analysis of dominant common substrings using the suggested heuristic approach based on Merger coefficients for all pairs of prominent common substrings. The relationship between common substring and the type of layout module can be categorized based on flowline module, branched flowline module, patterned floe module and functional layout module. If two layout modules have many common machines, they may be merged into a Function Layout module to reduce machine duplication. Next, the original part routings are replaced by the combination of residual machines and the model modules created by the above procedure.

The modular sequences are designed to incorporate into the module any residual system that exists in an adjacent model module to eliminate machine duplication. As a consequence, in the layout module, new strongly linked components arise and result in the Flowline, Branched Flowline or Patterned Flow being broken in that module. In order to determine the feasibility of duplicating machines in several modules, the total available capacity per machine of each type for the entire production period needs to be computed. Eventually, depending on the processes, potentially shared machines can be located. The total distance travelled to complete the process from beginning to end is what matters most. Shorter travel distances are preferable because they will cut down on processing time and prevent production line delays, which increase costs.

4 Results and Discussion

The input data consist of the layout design, product capacity references, data history (production rate), cycle time, and product sequences references. The information made available enables a thorough investigation of the production rates and sequences, as well as the connection between the machining process and alternative layout selection. The company has separated its production plant into six areas, which they refer to as Islands (A1, A2, A3, A4, A5, A6). Each island produces a variety of products that go through five or more processes. The outcome addressed in this article is only noted on island A3. Figure 3 indicates the operation process or sequence of two types of models which are denoted as model 1 and model 2 on four dedicated production lines. In total, one layout contains 8 separate processes. Each dot is marked with a number that indicates the process cycle time. A different line of manufacture is also shown by the varying colours of the dots.

Using the total distance travelled procedure, options are chosen and laid out. The distance travelled is calculated by measuring the of the materials travelled from one machine to others. This is based on the layout design supplied by the manufacturer in CAD format which replicate exactly the same line setup on the production floor. The layout module's adjustment comes after the modular sequences simplification of the remaining machines. The arrangement of the layout is based on many line layout strategies, including U-shape, I-shape, and L-shape layouts. The proposed alternatives are drafted in Fig. 4. The layout will then be chosen by calculating the entire

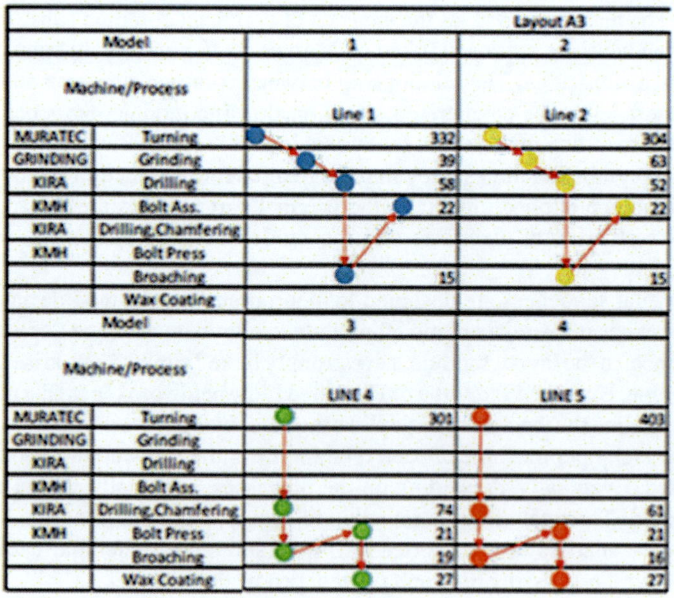

Fig. 3 Multi-product process chart on island A3

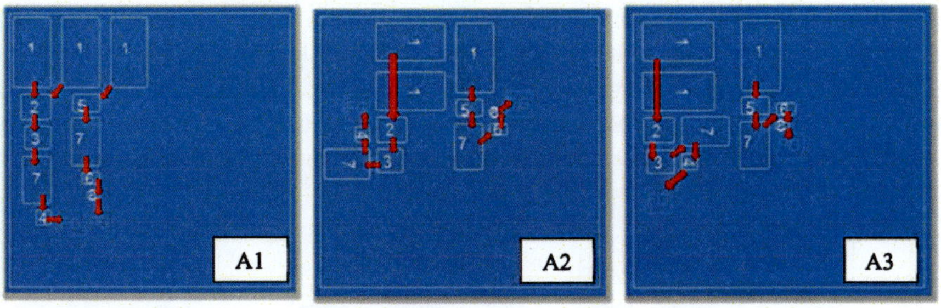

Fig. 4 Alternative 1, 2 and 3 layout for Island A3

Table 1 Total distance traveled and plant area optimization

Total distance travel between		
Alternative 1	Alternative 2	Alternative 3
28.61 m	34.34 m	38.42 m

Table 2 The comparison of existing and after optimization of layout in island A3

Plant area optimization (Island A3)		
Existing Layout (m^2)	SLP (m^2)	Reduction
183.721	95.206	48.18%

distance that the material will go, as described in the methodology section. Alternative 1 achieves the minimum total distance travel that is different in the range of 6 to 10 m compare to Alternative 2 and 3. Table 1 summarizes the total distance travelled and the area optimization achieved in this study. The spaced reduction recorded after implementing Alternative 1 is approximately 48%. Obviously, alternative layout 1 has the shortest material travel distance and consequently utilizes minimum plant area to equip with the facilities. Thus, island A3 should be arranged according to alternative layout 1, which is chosen as the best option.

Table 2 compares the island A3 before and after optimization utilising Systematic Layout Planning (SLP) with Modular layout. The spaced reduction recorded after implementing the Alternative 1 is approximately 48%. Obviously, the alternative layout 1 has the shortest material travel distance and consequently utilize minimum plant area to equip with the facilities. Thus, the island A3 should be arranged according to alternative layout 1, which is chosen as the best option.

4.1 Validation Using Plant Simulation Software

The machine process sequence for island A3 is simulated based on Alternative 1 using Tecnomatix Plant Simulation software (Fig. 5). The data which consist of process cycle time and percent availability are inserted in the setting of each process. According to resource statistics from the simulation as illustrated in Fig. 6 (example of model 1), the working percentage is still below optimal levels, and buffering and bottlenecks in the line can be noticed. Line balancing can be used to address this issue. But, the simulation carried out in this project is primary to forecast how many units can be produced in a day.

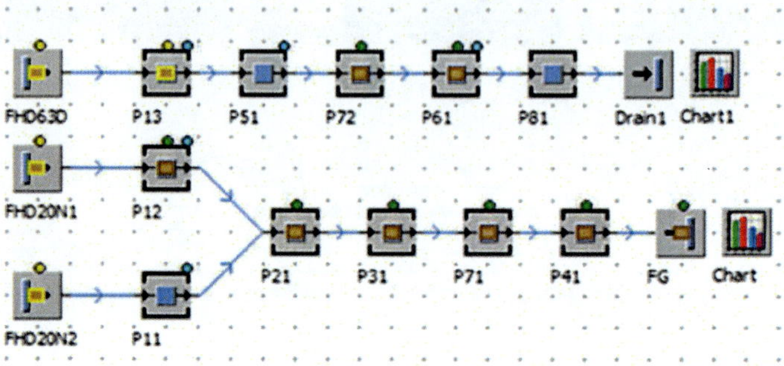

Fig. 5 Machine process flow in tecnomatix plant simulation software for layout A3

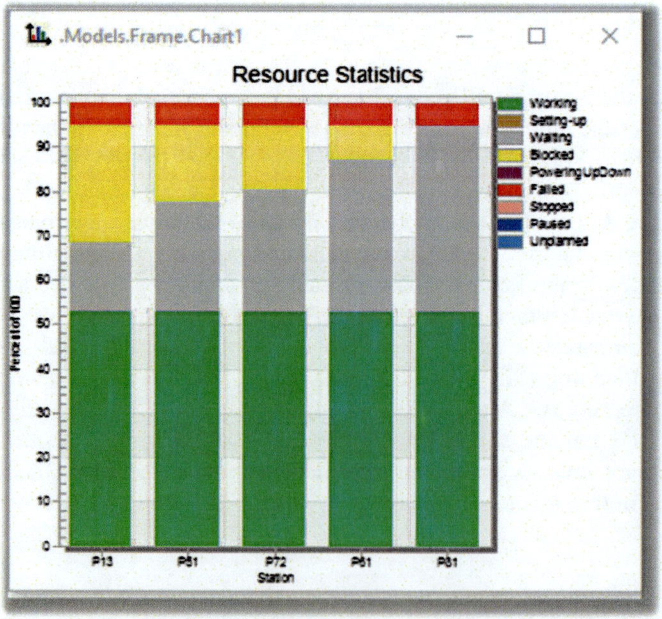

Fig. 6 Resources statistics for model FHD63D

Table 3 Production goal per day and quantity produce based on simulation software

Model	Quantity target (unit/day)	Quantity produced (unit/day)
Model 1	200	254
Model 2	325	357

There are two models produced in four production lines at island A3. The simulation outcomes indicate that the line setup can achieve the daily production goals (Table 3). The model 1 sequence is capable to produce 25% more that the targeted unit. Whereas the model 2 sequence can produce 10% extra units from the daily target production. The predicted production unit from the simulation strongly indicate the effectiveness of the proposed Alternative layout for Island A3.

5 Conclusion

Systematic Layout Planning (SLP) with Modular Layout is a practical solution to the issue of restricted space faced by the automotive part industry. The main factor that contributed to this issue are due to the rising demand on variety of products which lead to the increasing of machines brought into the manufacturing plant. The usage of a modular layout allows the simplification of the operation sequences, which can lower the amount of workspace needed by the workstation. The usage of a modular layout allows for the simplification of the operation sequences, which can lower the amount of workspace needed by the workstation. Based on the study, it is strongly advised to implement the plant A1 shown in Fig. 4 as the best alternate configuration in island A3. The choice will drastically lower the layout area and cut down the total distance travel of the parts. In addition, it is anticipated that production rates on the manufacturing lines will exceed the daily targeted unit that is derived based on the customer demand. In light of the project's continued advancement, it is critical to remember that observational activities are crucial to ensuring the veracity and correctness of the data gathered. Therefore, additional time is required to comprehend the process flow and gather the necessary information for analysis and simulation. After all, the suggested method was effectively used in the context of manufacturing automobile parts for a variety of car brands. It gives the company's technical team methodical instructions on how to handle facility expansion and re-layout projects that are necessary to meet market demand.

References

1. Richard Muther LH (1973) Systematic layout planning. 4th edn. Management & Industrial Research Publication
2. Tompkins JA, White JA, Bozer YA, Tanchoco JMA (2010) Facilities planning. Wiley

3. Amit N, Suhadak N, Johari N, Kassim I (2012) Using simulation to solve facility layout for food industry at XYZ Company. In: Paper presented at the 2012 IEEE symposium on humanities, science and engineering research
4. Chen C, Tran Huy D, Tiong LK, Chen IM, Cai Y (2018) Optimal facility layout planning for AGV-based modular prefabricated manufacturing system. Autom Constr. https://doi.org/10.1016/j.autcon.2018.08.008
5. Sisay Geremew Gebeyehu TB (2018) Systematic layout planning: an effective tool for minimizing work flow. Manuf Facil: Desk Rev 5 (1)
6. Liu Y-S, Tang L-N, Ma Y-Z, Yang T (2018) TFT-LCD module cell layout design using simulation and fuzzy multiple attribute group decision-making approach. Appl Soft Comput 68:873–888. https://doi.org/10.1016/j.asoc.2017.10.026
7. Suhardi B, Juwita E, Astuti RD (2019) Facility layout improvement in sewing department with systematic layout planning and ergonomics
8. Ramadhan D, Widodo L, Gozali L, Sukania IW, Daywi, FJ, Doaly CO (2020) Redesigning the facility layout with systematic layout planning method and lean manufacturing approach on the production floor at PT. Baruna Trayindo Jaya
9. EC M, Keraita JN (2018) Improvement of facility layout using systematic layout planning
10. Ameen W, AlKahtani M, Mohammed MK, Abdulhameed O, El-Tamimi AM (2018) Investigation of the effect of buffer storage capacity and repair rate on production line efficiency. J King Saud Univ Eng Sci 30(3):243–249. https://doi.org/10.1016/j.jksues.2018.03.001
11. Huang H (2003) Facility layout using layout modules. The Ohio State University
12. Chraibi A, El Cadi AA, Kharraja S, Artiba A (2016) Using discrete event simulation to evaluate operating theater layout. IFAC-PapersOnLine 49(12):95–98. https://doi.org/10.1016/j.ifacol.2016.07.556

Inventory Model for Reusable Items with Uniform Demand Arrival

Yosi Agustina Hidayat

Abstract This paper develops the ordering and reuse policies for inventory management of reusable items. The purpose of this research is to minimize the total cost of inventory, which includes cost of purchasing new inventory, cost of ordering new inventory, cost of storing new inventory and reused items, cost of the recovery process setup, and operational costs of the recovery process, in uniform arrival distribution of goods, both for new items and recoverable items, during the planning horizon, which is usually one year. The optimum decision variables are order lot size for new items (Q) for each time of purchase and batch quantity for recovering and reusing recoverable items (q). The proposed model shows that when using the assumption of uniform arrival, the economic order lot size for new items (Q) becomes smaller than the assumption of simultaneous arrivals of goods. Using a uniform arrival rate is a good starting point when estimating throughput capacity. Reducing the number of purchases of new items is a good strategic decision because it can indirectly support government policies to reduce the use of natural resources in the manufacture of these new products. This also supports reuse policy of maximum utilization of an item before its useful life expires.

Keywords Reuse items · New items · Recoverable items · Inventory · Uniform demand arrival · Sustainable production

Y. A. Hidayat (✉)
Institut Teknologi Bandung (ITB), Bandung, West Java 40132, Indonesia
e-mail: yosi@itb.ac.id

C. N. Rosyidi et al. (eds.), *Proceedings of the 6th Asia Pacific Conference on Manufacturing Systems and 4th International Manufacturing Engineering Conference*, Lecture Notes in Mechanical Engineering,
https://doi.org/10.1007/978-981-99-1245-2_39

1 Introduction

1.1 Background

The type of inventory managed by a company may take various forms, ranging from raw materials, auxiliary materials, parts, supplies, to the form of finished goods. Inventory is essentially idle resource in which the existence is waiting for further processing [1–3]. Further processes here can be in the form of production, marketing, or consumption activities. The nature of inventory, which is an idle resource, becomes a waste for the company. On the other hand, the existence of inventory is important to ensure the fulfillment of the needs of its users. This trade-off condition makes inventory management policies in a company becomes very important.

Inventory should be managed in such a way that it can meet the needs of its users or customers at the minimum possible total cost. Economic Order Quantity (EOQ) was developed by Ford Whitman Harris in 1913 as a way to manage inventory, but it has certain assumptions that may not align with real-world situations, such as deterministic demand and simultaneous arrival of ordered goods. These assumptions represent the ideal situation but do not represent the actual situation that can occur in the industry. For example, demand can be probabilistic in many cases, the price of goods are influenced by the number of goods purchased/ordered in the form of discounts, and goods can come in stages (uniform) when they are ordered. EOQ becomes the basis for the subsequent developed models with various new assumptions by many researchers. The limitations of EOQ to accommodate real assumptions that occur in the industrial world are complemented by the existence of Economic Production Quantity (EPQ). The factory's production capacity has limitations, so it takes the involvement of buyers and producers in determining the amount of orders at a limited level [4].

EPQ assumes that an item or inventory is only one time of usage, once its usefulness is exhausted, it will be discarded. However, if this is conducted continuously, then the available natural resources will eventually run out due to their limited number and continuously being used [4]. This is the concern of many people at this time. Awareness of the environment has made the government with its regulations begin to motivate industry players to start recovery and reuse activities. By implementing this, it is expected that the industry can participate in maintaining the availability of natural resources to support sustainable development. Reuse activities can reduce the use of new materials and this can certainly reduce production costs. Currently, a new era has emerged where supply chain management is concerned with repairing, remanufacturing, and recycling products that have expired and are collected from end consumers. Then the company reprocesses it into new products that can later be resold and generate greater profits because the capital becomes smaller. Examples of inventory control systems like this can be used in machine tools, medical and surgical instruments, photocopiers, motor vehicle parts, and office furniture [4]. The recovery of used products is known to be economically more attractive than their disposal [5]. Metal recycling, waste paper recycling and deposit systems for soft drink bottles are typical examples of product recovery that have been around for a long time. Increased concern for environmental issues has also increased concern for inventory reuse [6]. Environmental costs over the entire life cycle of industrial products play an important role in the calculation of total production costs [7]. This paper refers to the latest research from Mokhtari [4] regarding co-ordering and reuse policies for inventory management that can be reused (reuse items). This paper adds an assumption that the reused items to enter the production process are gradually and limited, since not all of them arrive instantly. Therefore, a uniform arrival assumption to both reused and new purchased items is developed.

1.2 Inventory System for Reusable Items

Initially, the company purchases new inventory. The inventory is then used until it is close to its useful life. Before its useful life expires, the inventory will enter the recovery process and then become reusable inventory. When the reused inventory has reached the end of its useful life, the reused inventory will be discarded and replaced by new inventory. The activities of buying new inventory, using it, recovering it, reusing it, and throwing it away when its useful life has expired, will be an activity that keeps on repeating itself. Before its useful life expires, the inventory can still be recovered for reuse. Recovering activities will occur in batches because they are waiting for inventory to be recovered at a certain amount point. In addition, it is also necessary to know some related terms, such as,

(a) Reusable items are inventory that can be used several times after the recovery process.
(b) Usable Items are inventory that is now ready for use (new inventory or used but restored inventory).
(c) Recoverable items are inventory that was used recently, and can still be recovered for reuse.

2 Literature Review

An initial study that aims to find the economic order quantity for newly purchased products, the optimal recovery lot size for recoverable products, and the optimal order of orders and setups is in [5]. Later on, a production and recovery inventory system (remanufacturing or repair) for a single product is assumed to have unlimited storage capacity, infinite planning horizon, constant demand level, no shortage and zero lead time [8]. The purpose of study in [8] is to find a solution to the remanufacturing problem of reverse logistics and to change the general assumptions considered in the production-remanufacturing inventory model literature.

According to [9], in the companies that can produce new products and recover the value of used products by remanufacturing and disassembling components, it is important to obtain the optimal policy parameters. The fee structure consists of setup fees, switching fees, holding fees, and non-EOQ fees. The purpose of this study is to determine the optimal production and remanufacturing policy that means finding the optimal number of production and remanufacturing lots of each part to minimize the total cost.

Our research objective is to determine the quantity of inventory to be ordered (Q) and the number of batches for the recovery and reuse process (q). This is to reduce the total cost of inventory, which includes the cost of purchasing new inventory, the cost of ordering new inventory, the cost of storing new inventory and reused items, the cost of the recovery process setup, and the operational cost of the recovery process. This paper extends research of Mokhtari [4] with the assumption development being carried out, namely the arrival of goods are uniform (in stages) at a certain speed, both for new items and used items that enter the batch for recovery.

3 Model Development

3.1 Model Component

1. Performance Criteria

The performance criterion which is the objective function of this paper is the minimization of the total inventory cost which includes the cost of purchasing new inventory, the cost of ordering new inventory, the cost of storing new inventory and reused items, the setup cost of the recovery process, and the operational costs of the recovery process during the planning horizon which is usually one year.

2. Decision Variables

There are two decision variables related to the determination of this reusable items inventory policy, namely:

- Order lot size for new items (Q) for each time you make a purchase.
- Batch quantity for the process of recovering and reusing recoverable items (q).

3. Notation

Some notations that will be used for later model formulation in Table 1. These notations also refer to the notations in Mokhtari [4] with the addition of R for the arrival speed of inventory, which represents the assumption of uniform demand arrival.

The inventory patterns for recoverable, usable, and recoverable items are shown in Fig. 1.

It is assumed that the holding costs for usable and re-coverable items are equivalent. Because inventory can be used $(m + 1)$ times, including one time when new and several times when it is recovered and reused, the formulation of the cost of purchasing new inventory (items) (PC) is shown in Eq. (1).

Table 1 List of model notations

Notation	Explanation	Unit	Notation	Explanation	Unit
D	Total annual demand	Unit/ year	T_R	Inventory cycle for single recovery and batch reuse q	Cycle
m	Number of times an item can be recovered and reused	Times	t_q	Inventory cycle for recovery and reuse of batch q until it reaches the end of its useful life	Cycle
C	Purchase cost per unit	IDR/ unit	Q	Order quantity for new items	Unit
A_u	Constant ordering cost	IDR/ order	q	Batch quantity for recovery	Unit
A_r	Constant recovery setup cost	IDR/ setup	p	Ratio of order quantity for new items Q to batch quantity for recovery q (Q = pq)	
h_u	Storage cost for one unit of goods that can be used per unit time	IDR/ unit	R_Q	Speed of arrival of new items	Unit/ period
h_r	Storage cost for one unit of used goods per unit time	IDR/ unit	R_q	Arrival speed of items to be recovered into batches	Unit/ period
r	Cost of recovery per unit	IDR/ unit	TC	Total system cost	IDR/ year
T_C	Time interval to fully use all items in order Q	Period	HC	Total holding cost	IDR/ year

$$PC = C\frac{D}{(m+1)} \tag{1}$$

where $D/(m + 1)$ represents the total quantity of goods to be purchased to meet the annual demand for D. The ordering cost of a new item is shown in Eq. (2).

$$OC = \text{Au}\frac{D}{(m+1)Q} \tag{2}$$

where $D/((m + 1)$ Q) indicates the total number of new items that must be ordered. Assuming $Q = pq$, then the formula is shown in Eq. (3).

$$OC = \text{Au}\frac{D}{(m+1)pq} \tag{3}$$

As we know, the number of times each item is restored in each *TR* interval is *m*. In addition, by assuming $Q = pq$, we can conclude that the number of recovery process implementations at each *TC* interval is *p x m*. Therefore, knowing the total

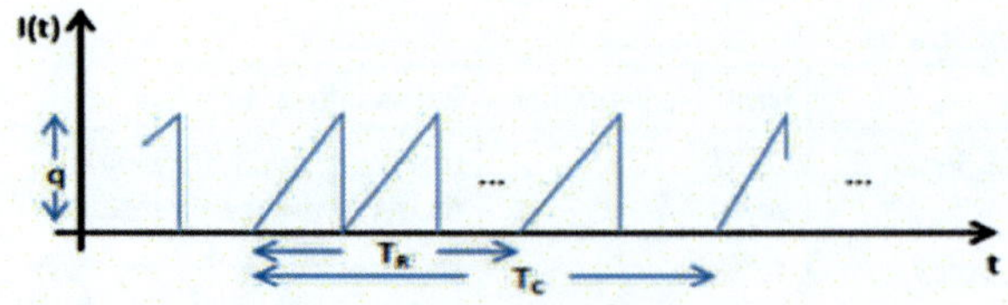

(a) Inventory pattern for recoverable items

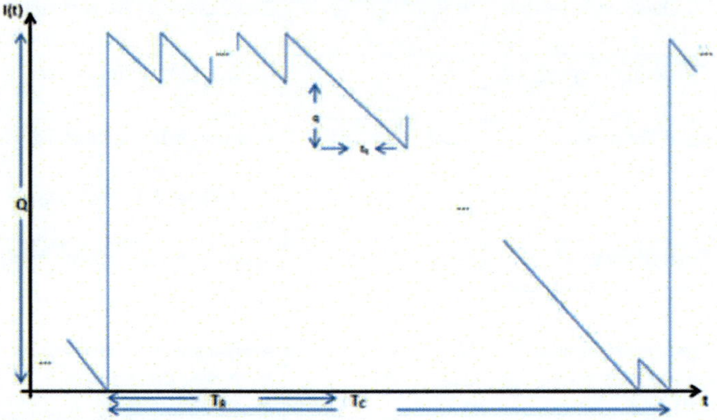

(b) Inventory curve for usable items

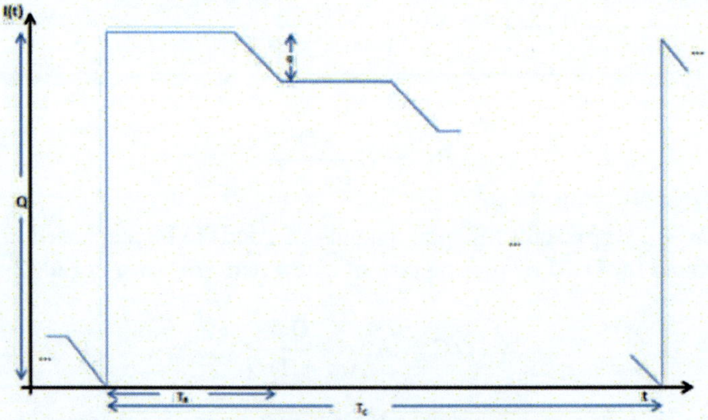

(c) Inventory curve for usable and recoverable items

Fig. 1 Inventory patterns

number of times a new item has to be ordered is $D/((m+1)\,Q)$, the total number of recoveries obtained is $pm \times D/((m+1)\,Q) = p(D/Q)(m/(m+1))$ per year, and the setup costs for the recovery process (*SC*) are calculated in Eq. (4).

$$SC = Arp\frac{D}{Q}\left(\frac{m}{m+1}\right) \tag{4}$$

The purchase of $D/(m+1)$ new items in one year, each of which is restored several times. Thus, the total number of times the carried out recovery process is $m\text{x}D/(m+1)$ per year. Therefore, the operational cost of the recovery process (*RC*), as a function of the number of times the carried out recovery process, as shown in Eq. (5).

$$RC = rD\left(\frac{m}{m+1}\right) \tag{5}$$

where r represents the cost of recovery per unit of goods. As a comparison, before presenting the model (especially related to storage costs) with the assumption of uni-form arrival of goods, the author first displays the model with the assumption of simultaneous arrival of goods. This model with simultaneous arrivals [4].

$$q = \frac{q}{D} \tag{6}$$

Since the item can be used $(m+1)$ times, the T_R interval is shown in Eq. (7).

$$TR = (m+1)t\text{q} \;= (m+1)\frac{q}{D} \tag{7}$$

$$Tc = pTR = (m+1)\frac{pq}{D} \tag{8}$$

The area under the inventory curve in Fig. 1c at the *TC* interval is shown in Eq. (9).

$$I = TR[Q + (Q-q) + (Q-2q) + \cdots + Q - (p-1)q] - p\frac{qtq}{2} \tag{9}$$

Or it can be simplified in Eq. (10).

$$I = \frac{pq^2}{2D}[(m+1)(p+1) - 1] \tag{10}$$

So that the average inventory $\overline{I} = I/Tc$ can be calculated in Eq. (11).

$$\overline{I} = \frac{q}{2}[(p+1) - 1/(m+1)] \tag{11}$$

Thus, the total holding cost (*HC*) where $HC = h.\overline{I}$, is formulated in Eq. (12).

$$HC = h\frac{q}{2}[(p+1) - 1/(m+1)] \tag{12}$$

The total cost is formulated in Eq. (13).

$$TC(p,q) = C\frac{D}{(m+1)} + Au\frac{D}{(m+1)pq} + Arp\frac{D}{Q}\left(\frac{m}{m+1}\right) + rD\left(\frac{m}{m+1}\right) + h\frac{q}{2}[(p+1) - 1/(m+1)] \tag{13}$$

The inventory model above is a generalization of the basic EOQ model. In the basic EOQ, there is no reuse option, and therefore, we have $m = 0$, $p = 1$ and $Q = q$. Therefore, the total cost of *TC* (*p*, *q*) is simplified in Eq. (14).

$$TC(Q) = CD + Au\frac{D}{Q} + h\frac{Q}{2} \tag{14}$$

To find the optimal inventory system design, we must set the partial derivation of the total cost of *TC* (*p*, *q*) with respect to *p* and *q* to zero and solve the resulting system of equations, as shown in Eqs. (15) and (16).

$$\frac{\partial TC(p.q)}{\partial p} = -\text{Au}\frac{D}{(m+1)p^2q} + h\frac{q}{2} = 0 \tag{15}$$

$$\frac{\partial TC(p.q)}{\partial p} = -Au\frac{D}{(m+1)p^2q} - \text{Ar}\frac{D}{q^2}\left(\frac{m}{m+1}\right) + h\frac{q}{2}\left[(p+1) - \frac{1}{(m+1)}\right] = 0 \tag{16}$$

By solving this system, the optimal solution form is in Eqs. (17) and (18).

$$p^* = \sqrt{\frac{1}{(m+1)}\left(\frac{Au}{Ar}\right)} \tag{17}$$

$$q^* = \sqrt{\frac{2ArD}{h}} \tag{18}$$

and, therefore, the optimal order quantity for the new item $Q^* = P^*q^*$ is obtained in Eq. (19).

$$Q^* = \sqrt{\frac{2AuD}{h(m+1)}} \tag{19}$$

When goods arrive uniformly, goods q and Q will be supplied during t at a rate of R (where $R > D$). During period t, there is also a demand for goods of $t.D$. Therefore, at the end of period t_1 the number of items in the warehouse is $q_{\max}.t$. Thus, in a state of uniform arrival supply with speed R, what changes is the average inventory level over time. This will result in changes in storage costs, while procurement costs remains the same. To calculate the total holding cost (*HC*), we must obtain the average inventory level. For this purpose, we need to calculate the level of inventory held over a time period. Because holding costs are the same for usable and recoverable items, inventory levels are used for this purpose. Before doing so, the three inventory cycles, t_q, T_R, and T_C are shown in Eqs. (20–22).

$$tq = \frac{q}{Rq} \tag{20}$$

Since the item can be used $(m + 1)$ times, the T_R interval can be achieved is shown in Eq. (21).

$$TR = (m + 1)tq = (m + 1)\frac{q}{Rq} \tag{21}$$

$$Tc = pTR = (m + 1)\frac{pq}{Rq} \tag{22}$$

The area under the inventory curve at the *TC* interval is shown in Eq. (23).

$$I = TR[Q + (Q - q) + (Q - 2q) + \ldots + Q - (p - 1)q] - p\frac{qtq}{2} \tag{23}$$

Or it can be simplified into Eq. (24).

$$I = \frac{pq^2}{2D}[(m + 1)(p + 1) - 1] \tag{24}$$

So that the average inventory $\overline{I} = I/Tc$ is shown in Eq. (25).

$$\overline{I} = \frac{q}{2}\left[(p + 1) - 1/(m + 1)\left(1 - \frac{D}{R}\right)\right] \tag{25}$$

Thus, the total holding cost (HC) where $HC = h.\overline{I}$, is formulated in Eq. (26).

$$HC = h\frac{q}{2}\left[(p + 1) - 1/(m + 1)\left(1 - \frac{D}{R}\right)\right] \tag{26}$$

The total cost is formulated in Eq. (27).

$$TC(p,q) = C\frac{D}{(m+1)} + Au\frac{D}{(m+1)pq} + Arp\frac{D}{Q}\left(\frac{m}{m+1}\right) + rD\left(\frac{m}{m+1}\right)$$
$$+ h\frac{q}{2}\left[(p+1) - 1/(m+1)\left(1 - \frac{D}{R}\right)\right] \quad (27)$$

The inventory model above is a generalization of the basic EOQ model. In the basic EOQ, there is no reuse option, and therefore, we have $m = 0, p = 1$ and $Q = q$. Therefore, the total cost of *TC* (*p*, *q*) is simplified in Eq. (28).

$$TC(Q) = CD + Au\frac{D}{Q} + h\frac{Q}{2}\left(1 - \frac{D}{R}\right) \quad (28)$$

To find the optimal inventory system design, we must set the partial derivation of the total cost of *TC* (*p*, *q*) with respect to p and q to zero and solve the resulting system of Eqs. (29) and (30).

$$\frac{\partial TC(p.q)}{\partial p} = -Au\frac{D}{(m+1)p^2q} + h\frac{q}{2}\left(1 - \frac{D}{R}\right) = 0 \quad (29)$$

$$\frac{\partial TC(p.q)}{\partial p} = -Au\frac{D}{(m+1)p^2q} - Ar\frac{D}{q^2}\left(\frac{m}{m+1}\right)$$
$$+ h\frac{q}{2}\left[(p+1) - \frac{1}{(m+1)}\left(1 - \frac{D}{R}\right)\right] = 0 \quad (30)$$

By solving this system, the optimal solution form is shown in Eqs. (31) and (32).

$$p^* = \sqrt{\frac{1}{(m+1)}\left(\frac{Au}{Ar}\right)} \quad (31)$$

$$q^* = \sqrt{\frac{2ArD}{h\left(1 - \frac{D}{Rq}\right)}} \quad (32)$$

And, therefore, the optimal order quantity for the new item $Q^* = P^*q^*$ is obtained as shown in Eq. (33).

$$Q^* = \sqrt{\frac{2AuD}{h(m+1)\left(1 - \frac{D}{RQ}\right)}} \quad (33)$$

Since it is clear, the above solution is reduced to the EOQ model, when $m = 0$.

4 Numerical Testing

For comparison, two numerical examples are presented, namely a numerical example for with the assumption of simultaneous arrival of goods, and the assumption of uniform arrival of goods. It is known that the data of a company taken from Mokhtari [4] on a planning horizon (1 year) related to its inventory management are as follows:

The calculation results of the two assumptions is shown in Table 2.

Based on Table 3, it can be seen that when using the assumption of simultaneous arrival of goods, namely the basic assumption in Wilson's EOQ model, it will produce a slightly higher total cost than when using the assumption of uniform arrival. This result is in line with the results of the Wilson model validity test, where when the uniform assumption is used, the total cost generated is also smaller than the assumption of simultaneous arrival of goods due to the decrease of holding cost. However, these results also show that when using the assumption of uniform arrival, the economic order lot size for new items (Q) becomes significantly smaller than the assumption of simultaneous arrivals of goods. In manufacturing industries, the uniform distribution is easy to estimate and comprise a helpful and accurate decision support tool for managers until they begin to learn about the distribution characteristics of the demand during the lead-time. Using a uniform arrival distribution is also a good starting point when estimating throughput capacity. Therefore, reducing the number of purchases of new items that must be made is a good strategic decision because it can indirectly support government policies to reduce the use of natural resources in the manufacture of these new products. This also supports item reuse activities if the item still be used further before its useful life expires (Table 3).

Table 2 Parameter setting

Parameter	Data	Unit	Parameter	Data	Unit
D	10,000	Unit/year	A_r	150,000	IDR/setup
R_Q	100,000	Unit/year	h_u	2000	IDR/unit
Rq	80,000	Unit/year	h_r	3000	IDR/unit
C	1,000,000	IDR/unit	m	2	Times
A_u	100,000	IDR/order			

Table 3 Numerical Results

Assumption	p^*	q^* [unit/year]	Q^* [unit/year]	$TC\ (p^*,q^*)$ [IDR/year]
Simultaneous demand arrival	0.577	1000	1732	3,871,071,367
Uniform demand arrival	0.569	1069	609	3,870,802,208

5 Conclusion and Future Research

The importance of inventory management in a company and the different forms that inventory can take, such as raw materials, finished goods, and supplies. While inventory is considered an idle resource and can lead to waste, it is necessary to meet the needs of customers. With growing environmental awareness, governments have started implementing regulations to encourage industries to engage in recovery and reuse activities. This shift has led to supply chain management focusing on repairing, remanufacturing, and recycling products to reduce the use of new materials and promote sustainability. This approach can also result in cost reductions and generate greater profits. The co-ordering and reuse policies for inventory management are beneficial for items that can be reused. It introduces the assumption that reused items enter the production process gradually and in limited quantities, rather than all at once. This development allows for a uniform arrival assumption for both reused and newly purchased items. To address the limitations of EOQ, the concept of Economic Production Quantity (EPQ) was introduced. EPQ takes into account the limited production capacity of a factory and involves the collaboration between buyers and producers to determine the order quantities. However, EPQ assumes that an item is used only once and then discarded, which raises concerns about the depletion of natural resources. We develops this research by adding the assumption of uniform arrival of goods, both for new items and recoverable items. The test was carried out by developing two calculation models, where the storage costs for new items and recoverable items are the same ($h = h_r = h_u$). For the future, the proposed model will be extended by considering different storage costs for new items and recoverable items, discounted price, shortage condition, and non-zero lead times.

References

1. Hadley G, Whitin T (1963) Analysis of inventory systems. Prentice-Hall, Englewood Cliffs
2. Buchan JE, Koenigsberg E (1963) Scientific inventory management. Prentice Hall of India New Delhi
3. Tersine RJ (1994) Principles of inventory and materials management, 4th edn. Prentice-Hall International, Inc., New Jersey
4. Mokhtari H (2018) Joint ordering and reuse policy for reusable items inventory management. Sustain Prod Consum 15:163–172. https://doi.org/10.1016/j.spc.2018.07.002
5. Choi D-W, Hwang H, Koh S-G (2007) A generalized ordering and recovery policy for reusable items. Eur J Oper Res 182:764–774
6. Fleischmann M, Bloemhof-Ruwaard JM, Dekker R, van der Laan E, van Nunen JAEE, van Wassenhove LN (1997) Quantitative models for reverse logistics: a review. Eur J Oper Res 103(1):1–17
7. Spengler Th, Püchert H, Penkuhn T, Rentz O (1997) Environmental integrated production and recycling management. Eur J Oper Res 97(2):308-326

8. El Saadany AMA, Jaber MY, Bonney M (2013) How many times to remanufacture? Int J Prod Econ 143(2):598–604
9. Kozlovskaya N, Pakhomova N, Richte K (2018) An EOQ inventory model with remanufacturing and dismantling for parts. Oper Res Develop Sector: 63–80. Part of the Asset Analytics book series (ASAN)

Introduction of AHP Decision Model for Prioritizing Critical Risk of WMSD in the Automotive Working Environment

Fazilah Abdul Aziz, Nik Mohd Zuki Nik Mohamed, and Ahmad Nasser Mohd Rose

Abstract Work-related musculoskeletal disorders (WMSD) is one of the essential occupational diseases that is a frequent and prevalent pain complaint by automotive workers. The importance of decisions in WMSD risk factors is related to the type of job task and working environment. This paper introduces the AHP decision model as an ergonomics assessment tool to determine relative measures of significance and assess priority weights for different risk categories of WMSD. This paper broadly categorizes the identified risk factors for WMSD into individual, organizational, physical, and psychosocial factors. Organizational factors indicated the most critical ergonomics risk factors influencing WMSD growth among automotive component manufacturer workers. Meanwhile, exposure to physical demand at work (EPD) and high workload (HWL) are the most critical risk factors. Since the various risk factors occurring under each category may vary significantly from context to context, a practical method for assessing and determining some significance level is required. AHP analysis does this in WMSD risk factors in the automotive working environment and is used as an ergonomics assessment tool.

Keywords Work-related musculoskeletal disorders (WMSD) · Analytic hierarchical process (AHP) · AHP decision model · Automotive industry

1 Introduction

Automotive workers are constantly exposed to various ergonomics risks due to the nature of their working environment, where they must deal with machines, heavy tools, and materials. Repetitive work tasks are more likely to be performed by automotive workers. According to a study, pushing and pulling are involved in 10% of

F. A. Aziz (✉) · N. M. Z. N. Mohamed · A. N. M. Rose
Faculty of Manufacturing and Mechatronic Engineering Technology, University Malaysia Pahang, 26600 Pekan, Pahang, Malaysia
e-mail: fazilahaa@ump.edu.my

C. N. Rosyidi et al. (eds.), *Proceedings of the 6th Asia Pacific Conference on Manufacturing Systems and 4th International Manufacturing Engineering Conference*, Lecture Notes in Mechanical Engineering,
https://doi.org/10.1007/978-981-99-1245-2_40

working operations in the automobile industry, and heavy object handling is required in more than 40% of cases [1]. Hence, the automotive industry has a significant incidence of work-related musculoskeletal disorders (WMSD).

Scholars commonly acknowledge the complexity of WMSDs [2]. The development of WMSD is impacted by the physical requirements of some tasks, work organisation risks, and psychosocial environments [3]. Furthermore, personal perceptions and other pertinent personality factors are equally important in risk management [4]. The focus of proactive ergonomics should be on taking proactive measures to prevent WMSDs by reducing workplace risk factors, early diagnosis, and future identification of relevant risk factors [5]. Thus, this study intends to prioritise the working environment's critical WMSD risk factors.

The AHP model is a problem-solving technique, and the ultimate hierarchy is to quantify risk levels and determine the relative significance of each module. Pairwise comparisons are used in the popular multi-criteria decision-making process known as AHP to assess the relative weights of various options [6]. Several previous studies of applying the AHP technique within the ergonomics risk context exist. The ergonomic risk associated with MSD was considered when ten processes were prioritized using the AHP, and workers using mobile elevating work platforms were considered [7]. An AHP was conducted by nine experts employing a pairwise comparison counting on the tasks for ergonomic interventions for the prevention of WMSD in a small manufacturing assembly line [8]. The ergonomics work tools are designed based on the AHP method to determine the factors that cause musculoskeletal complaints and the ergonomic risk level [9].

This study aims to develop an AHP decision model for prioritizing the critical risk of WMSD in the automotive working environment. Finally, this work will contribute to a body of knowledge needed to assist occupational safety and health (OSH) practitioners by producing practical WMSD risk assessment tools in the workplace for the worker's well-being, productivity, and organizational sustainability.

2 Develop AHP Decision Model

This paper applied a multi-criteria decision-making process in prioritizing the WMSD risk factors in the workplace. The combined case analyses and expert opinions created the hierarchical WMSD risk factors module. Figure 1 demonstrations the procedure of the AHP technique in classifying the critical WMSD risk factors.

This study applies to formulate an appropriate hierarchy of the AHP model consisting of the goal, factor and sub-factor. The hierarchy in Fig. 2 shows four major criteria that make up the risk factors for WMSD. In the second level, the goal is influenced by four factors: personal, organizational, physical, and psychosocial. The determination and evaluation of the sub-factors were included in the third and fourth levels.

In this study, verbal replies were translated into numerical figures that represented the values of a_{ij} using a 9-point linguistic scale [10]. Table 1 gives a description of

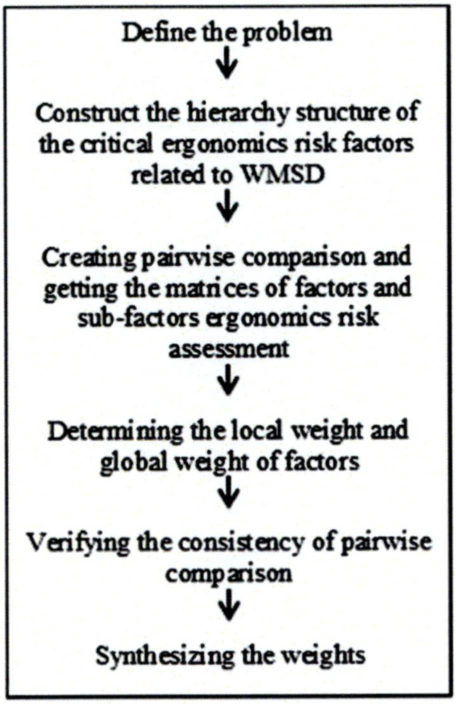

Fig. 1 Procedure of AHP decision model for critical WMSD risk factors prioritization

the scale. At the level of the primary criteria, the relative weights of the individual factors (IF), organizational factors (OF), physical factors (PYF), and psychosocial factors (PSF) are compared.

Equations 1–3 describe the consistency index and consistency ratio. In Table 2, the random index (RI) is displayed. It is considered acceptable if the consistency ratio is less than or equal to 0.10 (10%). Otherwise, the pairwise comparison must be repeated since the results are deemed incongruous.

$$\text{Consistency index (CI)}\frac{\lambda_{\max} - n}{n - 1} \tag{1}$$

$$\lambda_{\max} = \sum_{i=1}^{n}\left[\left(\sum_{i=1}^{n} GM_i\right)(w_j)\right] \tag{2}$$

where, $\lambda_{\max}$ = maximum eigenvalue and n is the number of elements

$$\text{Consistency ratio (CR)} = \frac{\text{Consistency index (CI)}}{\text{Random index (RI)}} \leq 0.10 \tag{3}$$

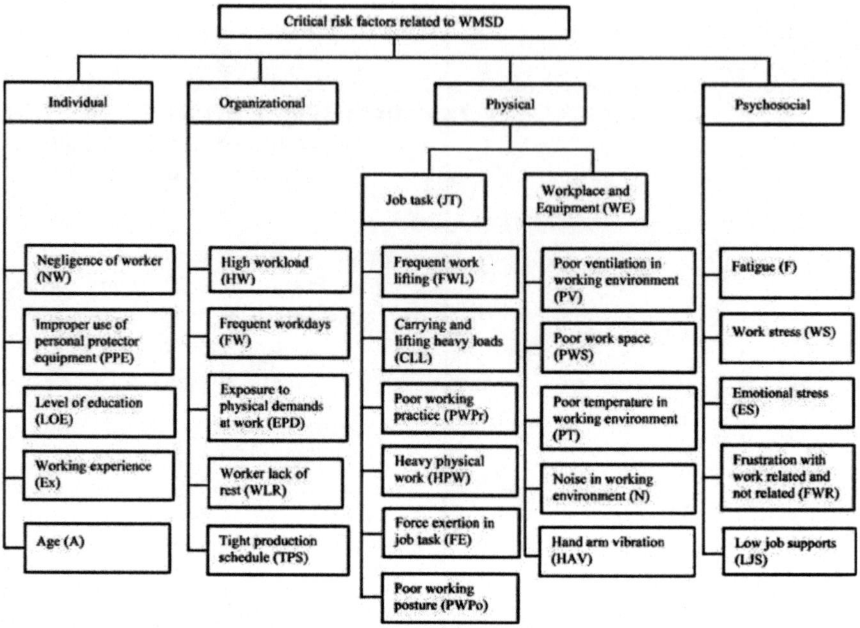

Fig. 2 AHP decision model of critical WMSD risks factors

Table 1 Pair-wise comparison scale used with AHP technique

Intensity of importance	Score	Definition	Explanation
1	1	Equal importance	Two elements contribute equally to the property
2	3	Moderate importance of one over another	Experience and judgment slightly favour one over the other
3	5	Essential or strong importance	Experience and judgment strongly favour one over another
4	7	Very strong importance	An element is strongly favoured and its dominance is demonstrated in practice
5	9	Extreme importance	The evidence favoring one element over another is one of the highest possible orders of affirmation
Reciprocals		When activity i compared to j assigned one of the above numbers, the activity j compared to i is assigned its reciprocal	
Rational		Ratios arising from forcing consistency of judgments	

Table 2 Random index [11]

n	1	2	3	4	5	6	7	8	9	10
RI	0	0	0.58	0.9	1.12	1.124	1.32	1.41	1.45	1.49

This paper employed a questionnaire survey to accumulate the data required to use the AHP technique. The survey was developed and led to prioritizing and assigning the importance of weightings to assess ergonomics risk factors. The WMSD risk factors list was in the questionnaire, as identified in the previous research review.

In order to assess the significance of each factor and sub-factor, a series of AHP-based questionnaire surveys are being used. The pair-wise comparison was used to conduct this AHP-based questionnaire survey. The questionnaire has 78 questions, 1 at the main factor level (level 6) and 4 at the sub-factor level (72).

3 Application of AHP Decision Model

For the created AHP-based questionnaire, a pilot study has been undertaken. Two academics and eight experienced professionals took part. The AHP-based questionnaire has been modified in response to ideas and comments from respondents. The revised questionnaire now uses the correct wording. Before utilizing the technology to perform the survey, elaborate on the studied risk variables. Because the nine-point scale frequently caused participants' confusion during pairwise comparisons, we reduced the point scale from 9 to 5 while keeping the score values of 1, 3, 5, 7, and 9.

The process of identifying decision-critical risk variables has started with the creation of an expert panel that will make the decisions. The AHP-based questionnaire survey for this study required recruiting fifteen production operation specialists from an automobile component manufacturer (ACM). These specialists anticipated having at least 10 years of professional experience and an in-depth understanding of ACM production processes. Microsoft Excel was used to perform the computations based on the AHP methodology.

Microsoft Excel was used to perform the computations based on the AHP methodology. Each comparison module's consistency ratio is acceptable because $CR < 0.1$. The results shown in Table 3 were determined qualitatively by the ACM's automotive industry specialists. The judgement is consistent in Table 3 because the inconsistency ratio ranges from 0.0000 to 0.0895.

The local weights of risk factors were identified in this investigation (expressed in normalized geometric means). The greater the local weights of risk factors, the greater their significance in determining how risky WMSD is. The risk factor connected to WMSD was given priority in the current investigation. The decision-maker will be assisted in gradual risk control by this prioritization. Tables 4, 5, 6, 7, 8, 9 and 10 provide local weights for risk factors and sub-factors related to ergonomics.

Table 3 Random index

Judgment matrix	CR	Consistency validation (CR < 0.1)
Ergonomics risk factors	0.0691	Accepted
Individual risk factors	0.0877	Accepted
Organizational risk factors	0.0108	Accepted
Physical risk factors	0.0000	Accepted
Job task risk factors	0.0712	Accepted
Workplace and equipment risk factors	0.0762	Accepted
Psychosocial risk factors	0.0257	Accepted

Table 4 Local weights of WMSD risk main factors

Risk factors	IF	OF	PhyF	PsyF	Local weights	%
IF	0.1311	0.2036	0.1032	0.0888	0.1317	13.17
OF	0.2344	0.3641	0.3246	0.5339	***0.3643***	***36.43***
PhyF	0.3308	0.2920	0.2604	0.1717	0.2637	26.37
PsyF	0.3036	0.1402	0.3119	0.2056	0.2404	24.04
Total					1	100

Table 5 Local weights of individual risk sub-factors

Risk sub-factors	NW	PPE	LOE	WE	A	Local weights	%
NW	0.2094	0.3682	0.3046	0.0860	0.2439	***0.2424***	***24.24***
PPE	0.0784	0.1379	0.1405	0.2087	0.1777	0.1486	14.86
LOE	0.2279	0.3253	0.3316	0.4863	0.2305	***0.3203***	***32.03***
WE	0.4214	0.1118	0.1180	0.1731	0.2729	0.2194	21.94
A	0.0629	0.0568	0.1053	0.0459	0.0749	0.0692	6.92
					Total	1	100

Table 6 Local weights of organizational risk sub-factors

Risk sub-factors	HW	EPD	TPS	FW	WLR	Local weights	%
HW	0.2241	0.1736	0.2442	0.2903	0.2343	0.2333	23.33
EPD	0.3345	0.2592	0.2630	0.2142	0.2240	***0.2590***	***25.90***
TPS	0.1518	0.1631	0.1655	0.1552	0.2013	0.1674	16.74
FW	0.1501	0.2354	0.2073	0.1945	0.1945	0.1963	19.63
WLR	0.1395	0.1687	0.1199	0.1458	0.1459	0.1440	14.40
					Total	1	100

Table 7 Local weights of physical risk sub-factors

Risk sub-factors	JT	WE	Local weights	%
JT	0.1943	0.1943	0.1943	19.43
WE	0.8057	0.8057	0.8057	80.57
		Total	1	100

Table 8 Local weights of physical job task risk sub-factors

	CLL	PWP	HPW	FWL	FE	PWP_O	Local weights	%
CLL	0.0972	0.2303	0.1587	0.0558	0.1088	0.0629	0.1189	11.89
PWP_r	0.0488	0.1157	0.1747	0.2380	0.1395	0.0965	0.1355	13.55
HPW	0.0844	0.0913	0.1378	0.1038	0.1881	0.1586	0.1273	12.73
FWL	0.1475	0.0412	0.1125	0.0847	0.0733	0.1086	0.0946	9.46
FE	0.2580	0.2395	0.2117	0.3340	0.2889	0.3379	***0.2783***	***27.83***
PWPo	0.3641	0.2821	0.2046	0.1836	0.2014	0.2355	0.2452	24.52
						Total	1	100

Table 9 Local weight of physical workplace and equipment risk sub-factors

	PV	PWS	PT	N	HAV	Local weights	%
PV	0.1218	0.0625	0.1285	0.1497	0.1187	0.1162	11.62
PWS	0.3403	0.1744	0.3489	0.3188	0.0953	0.2555	25.55
PT	0.1510	0.0796	0.1593	0.1584	0.2661	0.1629	16.29
N	0.1289	0.0867	0.1593	0.1584	0.1940	0.1454	14.54
HAV	0.2580	0.5968	0.2041	0.2147	0.3259	***0.3199***	***31.99***
					Total	1	100

Table 10 Local weights of psychosocial risk sub-factors

	F	WS	ES	FW	LJS	Local weights	Percent, %
F	0.2169	0.2127	0.2212	0.1941	0.2151	0.2120	21.20
WS	0.0979	0.0960	0.1009	0.1135	0.0810	0.0979	9.79
ES	0.1429	0.1385	0.1457	0.0668	0.2290	0.1446	14.46
FW	0.2016	0.1525	0.3173	0.1804	0.1369	0.1977	19.77
LJS	0.3407	0.4003	0.2150	0.4451	0.3379	***0.3478***	***34.78***
					Total	1	100

Referring to Table 4, the organizational risk component had the most significant value (36.43%), followed by the physical risk factor (26.37%), the psycho-social risk factor (24.04%), and the individual risk factor (13.17%).

To assess each sub-impact factors on the primary factor, the same process was used for all of them. Prioritizing sub-factors positioned beneath each factor in the hierarchy was done using pairwise comparisons. According to Table 5, this study discovered that level of education (LOE), which has a local weight of 32.03%, has a more significant impact than the other individual risk sub-factors. Furthermore, it was determined that the negligence of the workforce (NW) was a second crucial component (24.24%).

Referring to Table 6, the most critical organizational risk factor was identified as exposure to physical demand at work (EPD), which was followed by high workload (HW), frequent workdays (FW), tight production schedule (TPS), and worker lack of rest (WLR).

The physical sub-factors listed in Table 7 with a local weight of 80.57% are the workplace and equipment (WE), which have the most significant results.

The most significant job tasks physical risk sub-factor, as indicated in Table 8, is force exertion (FE), which has the highest weight of 27.83%. However, Table 9 reveals that among the sub-factors of the workplace and equipment, hand-arm vibration (HAV) has the highest weight, at 31.9%.

According to Table 10, the most important psychosocial risk factor was low job support (LJS), which had a local weight of 34.78%. Although Table 11 lists the overall importance and ranking of WMSD risk factors.

Based on calculations made using the AHP technique, it can be seen that organizational elements made up most of the decision-makers evaluated as first-priority risks. These findings demonstrate that the AHP decision model can generate significant WMSD risk factors and WMSD risk sub-factors based on expert team evaluation.

4 Discussions

This study has described the creation and use of the AHP decision model for ranking the most crucial WMSD risk factors. Critical risk factors and their sub-factors can be prioritised using the AHP decision paradigm. After physical, psychosocial, and personality factors, organisational issues were found to be the most crucial ergonomics risk factors promoting the growth of WMSD in workers. Employers must therefore assess the organisational practices and policies that may expose their employees to WMSD risk factors.

According to this study's findings, the most significant individual risk sub-factor affecting the WMSD is level of education (LOE). Higher educated personnel frequently hold positions in workplaces where they are less likely to be exposed to the high prevalence of WMSDs [12]. As opposed to shop floor employees in the automobile industry, who are exposed to physical risks that cause symptoms of musculoskeletal problems, [13]. There was a strong correlation between the variables' educational levels and the occurrence of musculoskeletal disorders.

The most important physical component has been identified as the work environment and equipment (WE). Worker musculoskeletal problems and symptoms are

Table 11 Global weights of WMSD risk factors

Main factors	Sub-factors	Global weight	Rank
IF	NW	3.19	14
	PPE	1.96	19
	LOE	4.22	11
	WE	2.89	16
	A	0.91	22
OF	HWL	***8.50***	***2***
	FW	***7.15***	***4***
	EPD	***9.43***	***1***
	WLR	***5.24***	***8***
	TPS	***6.10***	***6***
Phy-JT	CLL	*0.61*	*25*
	PWPr	*0.69*	*23*
	HPW	*0.65*	*24*
	FWL	*0.49*	*26*
	FE	1.43	20
	PWPo	1.26	21
Phy-WE	PV	2.47	17
	PWS	**5.43**	**7**
	PT	3.46	13
	N	3.09	15
	HAV	***6.80***	***5***
PsyF	F	***5.10***	***9***
	WS	2.35	18
	ES	3.47	12
	FW	4.75	10
	LJS	***8.36***	***3***

a result of poor workplace ergonomics [14]. To lower the risk of WMSD among workers, equipment with ergonomic interventions is required [15]. The most critical workplace and equipment physical risk sub-factor was hand-arm vibration (HAV). Technical machine operators' health and productivity are impacted by vibration exposure [16].

Moreover, the most critical job task physical risk sub-factor was force exertion (FE). Workers who perform physically demanding work activities develop WMSD [5]. Low job support (LJS) was the most important psychosocial risk factor in this study. A similar finding in the previous study is that low job support is the most frequent risk factor thmay affect WMSD [17] such as back region disorders [18] and upper-extremity disorders [19].

However, the most critical risk factors are excessive workloads (HWL) and exposure to physical demands at work (EPD). WMSD concerns have impacted most industrial workers who engage in demanding physical activity [20]. In order to reduce WMSD risk factors in the workplace in the manufacturing industry, it is necessary to prevent excessive workloads [21]. As a result, while designing ergonomic interventions for the workplace, occupational safety and health practitioners must prioritize these risk factors.

5 Conclusions

This study can identify the critical WMSD risk factors in the working environment using qualitative multi-criteria decision-making based on AHP models of parts of this study area. The AHP model can manage the risk factors that need to be detoured and stopped through ergonomic intervention. The AHP decision model could improve workplace ergonomics evaluation in automotive production plant operational situations. This paper has offered insightful information regarding the actual data collection and use of the AHP decision model.

Acknowledgements A special thanks to local automotive component manufacturer workers for their assistance in data collection. Also, the Universiti Malaysia Pahang provides financial support for this paper (RDU200749).

References

1. Argubi-Wollesen A, Wollesen B, Leitner M, Mattes K (2017) Human body mechanics of pushing and pulling: analyzing the factors of task-related strain on the musculoskeletal system. Saf Health Work 8:11–18. https://doi.org/10.1016/j.shaw.2016.07.003
2. Mossa G, Boenzi F, Digiesi S, Mummolo G, Romano VA (2016) Productivity and ergonomic risk in human based production systems: a job-rotation scheduling model. Int J Prod Econ 171:471–477. https://doi.org/10.1016/j.ijpe.2015.06.017
3. Oakman J, Rothmore P, Tappin D (2016) Intervention development to reduce musculoskeletal disorders: Is the process on target? Appl Ergon 56:179–186. https://doi.org/10.1016/j.apergo.2016.03.019
4. Aziz RA, Rebi MAT, Rani A, Rohani JM (2014) Work-related musculoskeletal disorders among assembly workers in Malaysia. J. Occup. Saf. Heal. 11:33–38
5. Chander DS, Cavatorta MP (2018) Multi-directional one-handed strength assessments using anybody modeling systems. Appl Ergon 67:225–236. https://doi.org/10.1016/j.apergo.2017.09.015
6. Liu Y, Eckert CM, Earl C (2020) A review of fuzzy AHP methods for decision-making with subjective judgements. Expert Syst Appl 161:113738. https://doi.org/10.1016/j.eswa.2020.113738
7. Supattananon N, Yawara P, Youngwilai A, Arkararungruangkul R (2022) Productivity improvement through a work study with ergonomic risk assessment: A case study of a high-voltage line distribution operation. Eng Appl Sci Res 49:688–695

8. Shin W, Park M (2017) Ergonomic interventions for prevention of work-related musculoskeletal disorders in a small manufacturing assembly line. Int J Occup Saf Ergon 3548:1–13. https://doi.org/10.1080/10803548.2017.1373487
9. Widodo L, Ariyanti S, Octavia J (2019) Peningkatan produktifitas UKM Produk Sugarwax melalui Intervensi Ergonomi di Stasiun Kerja. Peningkatan Produktifitas UKM Prod. Sugarwax melalui Interv. Ergon. di Stasiun Kerja.
10. Saaty TL (1980) The analytic hierarchy process. Decis Anal 1–17
11. Saaty TL (1994) How to make a decision: the analytic hierarchy process. Interfaces (Providence) 24:19–43. https://doi.org/10.1287/inte.24.6.19
12. Yarandi MS, Ghasemi M, Ghanjal A (2022) The relationship between individual, physical and psychosocial risk factors with musculoskeletal disorders and related disabilities in flight security personnel. Int J Occup Saf Ergon ISSN 28:369–397. https://doi.org/10.1080/10803548.2020.1760528
13. Falck AC, Örtengren R, Rosenqvist M (2014) Assembly failures and action cost in relation to complexity level and assembly ergonomics in manual assembly (part 2). Int J Ind Ergon 44:455–459. https://doi.org/10.1016/j.ergon.2014.02.001
14. Boyle S, Fitzgerald C, Conlon BJ, Vijendren A (2022) A national survey of workplace - related musculoskeletal disorder and ergonomic practices amongst Irish otolaryngologists. Irish J Med Sci (1971-) 191:623–628. https://doi.org/10.1007/s11845-021-02642-y
15. Kee D, Haslam R (2019) Prevalence of work-related musculoskeletal disorders in agriculture workers in Korea and preventative interventions. Work 64:763–775. https://doi.org/10.3233/WOR-193038
16. Mościcka-Teske A, Sadłowska-Wrzesińska J, Butlewski M, Misztal A, Jacukowicz A (2017) Stressful work characteristics, health indicators and work behavior: the case of machine operators. Int J Occup Saf Ergon 23:510–518. https://doi.org/10.1080/10803548.2016.1197577
17. Das SK, Mukhopadhyay S, Koilakuntla M (2015) Analytic hierarchy process to rate risk factors associated with WMSDs in general physicians. J Health Manag 17:241–247. https://doi.org/10.1177/0972063415575813
18. Zhang Q, Dong H, Zhu C, Liu G (2019) Low back pain in emergency ambulance workers in tertiary hospitals in China and its risk factors among ambulance nurses: a cross-sectional study. BMJ Open 9:e029264
19. Martimo KP, Shiri R, Miranda H, Ketola R, Varonen H, Viikari-Juntura E (2010) Effectiveness of an ergonomic intervention on the productivity of workers with upper-extremity disorders—a randomized controlled trial. Scand J Work Environ Heal 36:25–33. https://doi.org/10.5271/sjweh.2880
20. Ziaei M, Choobineh A, Abdoli-Eramaki M, Ghaem H (2018) Individual, physical, and organizational risk factors for musculoskeletal disorders among municipality solid waste collectors in Shiraz. Iran Ind Health 56:308–319. https://doi.org/10.2486/indhealth.2018-0011
21. Moussavi SE, Zare M, Mahdjoub M, Grunder O (2019) Balancing high operator's workload through a new job rotation approach: application to an automotive assembly line. Int J Ind Ergon 71:136–144. https://doi.org/10.1016/j.ergon.2019.03.003

Design Thinking Process for Solving Supplier Selection and Risk Management Problem: A Case Study

Tiara Priscilla Nandyasari and Yosef Daryanto

Abstract A case study is performed in a small-sized enterprise that manufactures custom cellphone cases. The company uses plain cases as the main raw material. This business is characterized by fast-changing cellphone case models or types. A concern was raised because some consumer orders were canceled due to delays in fulfilling the order on time. It occurs due to the unavailability of the plain case both in the company's and supplier's stock. Based on the existing problem, the company plans for supplier evaluation and determining future supplier priorities. The target is to reduce delays or cancellations of orders by customers. This study uses the design thinking process to solve the problem. Analytic Hierarchy Process method was used for supplier selection. Three alternative suppliers were available and analyzed using six criteria of price, availability, quality, delivery, warranty & claims, and service. In addition, supplier risk management is proposed to anticipate potential risks that may occur in every activity in the business process. The proposed risk management is developed using the House of Risk model. As a result, three preventive actions can be taken, namely conducting joint performance evaluations with suppliers as an action with priority, reconfirming orders to suppliers, and finally building cooperation contracts with suppliers.

Keywords Design thinking process · Supplier selection · AHP · House of risk · Risk management

1 Introduction

Supplier is one of the critical aspects of the sustainability of a company's operations [1]. The performance of the suppliers determines the availability of materials or supplies for the company's operations. Unfortunately, complex market conditions

T. P. Nandyasari · Y. Daryanto (✉)
Universitas Atma Jaya Yogyakarta, Yogyakarta 55281, Indonesia
e-mail: yosef.daryanto@uajy.ac.id

C. N. Rosyidi et al. (eds.), *Proceedings of the 6th Asia Pacific Conference on Manufacturing Systems and 4th International Manufacturing Engineering Conference*, Lecture Notes in Mechanical Engineering,
https://doi.org/10.1007/978-981-99-1245-2_41

and uncertainty due to various disruptions increase the risk of supply failures [2]. Hence, a company can incorporate risk criteria in the supplier selection process [2–4]. Alternatively, the company may prepare a supplier risk assessment strategy to properly identify potential hazards and minimize them [4]. A supplier risk assessment can proactively assess a supply disruption in advance, or reactively take action after a detrimental event occurs [5].

Company X is a small-sized business that produces custom cellphone cases with the main raw material of plain cases. This business is characterized by fast-changing cellphone case models or types. Based on the interviews with the owner and staff, it is known that the company experienced some difficulties in fulfilling customer orders. On some occasions, their suppliers do not have the stock of specific cellphone cases. Suppliers need times to update or renew stock that follows the release of new types of cellphones. The unavailability of raw materials from the main supplier will cause delays in processing customer orders, causing lateness from what was promised to the customer, that in some cases make the consumers cancel their orders. The purpose of this study is to determine the main supplier that is suitable and meets the company's criteria to reduce delays or cancellations of orders. Then, to increase the company's readiness in anticipating the emergence of supply problems, this study also proposes risk management plan that is needed by the company.

2 Research Method

As a problem solving research, this study is divided into five stages following the Design Thinking Process, including Empathize, Define, Ideate, Prototype, and Test as shown in Fig. 1. Design thinking process is a solution-focused problem solving approach. It helps problem solver or design team produce creative and innovative future results [6], by clarifying the problem, generating ideas, and modelling [7]. It is a rational problem-solving approach by understanding stakeholder needs related to a problem, reframing the problem on a human-centric way, creating various alternative solutions in brainstorming sessions, and adopting a hands-on approach to prototyping and testing [8]. This problem solving approach is best taught to engineering students [7].

The Empathize stage ensures that the solution can meet the needs of the stakeholders. Hence, the owner and the staff were interviewed. It was identified that sometimes, their supplier does not have the stock of specific cellphone cases, especially for new types of cellphones. This unavailability will cause delays in processing customer orders, causing lateness from what was promised to the customer, that in some cases make the consumers cancel their orders. A more reliable supplier was needed.

At the Define stage, further information is collected to get a clear picture of the system and the problem. The business process starts when consumers specify the design for their cellphone case. Customers can choose a design that the company already has or propose a new design on their own (custom design). Then, Company X

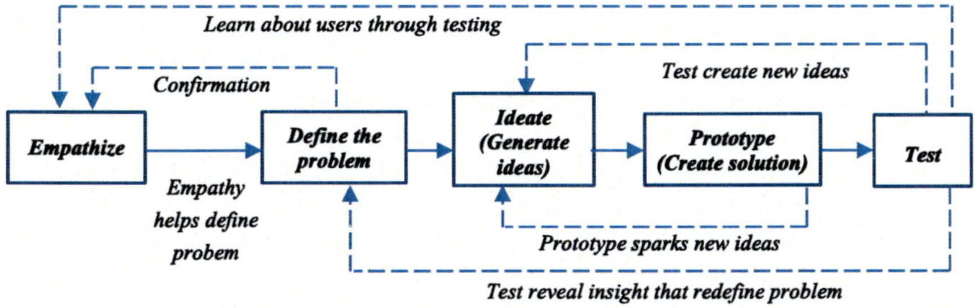

Fig. 1 Design thinking process for problem-solving

will process customer orders. The ordered product will usually reach the consumer in two to three days. If Company X has the case with the appropriate type in its stock, then the design/custom process for the case can be carried out immediately. However, if the stock of cellphone cases is not available, the company will order the plain cases from the supplier. It is done immediately on the same day as the arrival of customer orders. One batch of cellphone cases is ten units. But sometimes the stock is less than 10 or even the supplier has no stock, especially for new cellphone types. The supplier is located in the same city. Currently, the company has one supplier who acts as the main supplier. The supplier has lower prices when compared to other suppliers. The supplier also provides good and honest service. It can be identified that the company experienced a supply problem that cause a delay in customer order fulfillment. The current supplier is unable to ensure the availability of the required material, especially the case for new cellphone types. Confirmation and discussions were held, means the Empathize and Define stages are iterative processes to clearly understand stakeholders' needs.

Then, in the Ideate phase, alternative solutions were generated. Originally, there are two alternative solutions, (1) keeping the current supplier and adding another new supplier, and (2) reselecting one main supplier (supplier priority). However, from the next discussion, the owner only wants raw material supply from a single source. Hence, the alternative solution of determining supplier priority was chosen to solve the problem.

To perform a supplier selection, several factors or criteria need to be considered. The criteria set in selecting suppliers need to match the interests and needs of the company [9]. The more criteria the company sets for selecting suppliers, the more complicated and complex the supplier selection process will be, so a method is needed to make decisions in supplier selection. Methods that are often used in making decisions to choose suppliers include Analytic Hierarchy Process (AHP), Technique for Order Preference by Similarity to Ideal Solution (TOPSIS), MOORA [10]. Among these methods, the AHP method considers consistency in the assessment used to determine priorities. The AHP method can solve the problem based on the judgment of several decision-makers with different opinions to achieve one goal [11]. This method has a simple model that will be easier to apply in small and medium enterprises. The MOORA method is considered difficult when used in deciding group assessments (unifying judgments or opinions from more than one party), while TOPSIS must involve accurate data and does not have a comparison index as an indicator [12]. The method is selected through a discussion with the owner.

For further Supplier Relationship Management (SRM), this study proposes a risk management plan. SRM has an important role in building better relationships with suppliers [13]. SRM can improve the quality of suppliers' performance [14]. Tavana et al. [15] studied an integrated supply chain risk–benefit assessment and supplier selection for establishing a strong and mutually beneficial long-term supply chain relationship. The risk management planning is carried out using the House of Risk (HOR) model. HOR consists of two parts in the outline. The first HOR is the risk identification section, while the second HOR is the risk management process [16, 17].

Next is the Prototype phase which is the stage to design and realize the chosen solution to solve the existing problem, namely by using the AHP method and followed by the House of Risk. After data collection, the step is the determination of criteria and sub-criteria. The process is carried out together with the company. Several criteria and sub-criteria obtained from references were proposed to the company. Eights references were used as shown in Table 1. The criteria and sub-criteria are then adjusted to the company's needs in selecting suppliers. Six criteria are finally selected, namely Price, Availability, Delivery, Quality, Warranty & Claim Policy, and Service. In addition to this, sub-criteria are set for each of the predefined criteria. A summary of the determination of criteria and sub-criteria is shown in Table 1.

Then the alternative suppliers were determined. There are three alternative suppliers, those are suppliers who are currently or have worked with Company X. The alternatives were B-Cell, which is the current main supplier, also A-Cell and

Table 1 Summary of criteria and sub-criteria

Number	Criteria	Sub-criteria	Code
1	Price (Ardhy & Dahda [18], Astanti et al. [19], Lukmandono et al. [20])	Low price	P1
		Additional cost	P2
2	Availability (Afandi [21], Mahendra [22])	Ability to fulfill the quantity	A1
		A complete type of cellphone case	A2
		Speed of updating the latest type	A3
3	Quality (Mahendra [22], Taufik et al. [23], Pratama et al. [24])	Good quality	Q1
		Provide goods without defects	Q2
4	Delivery (Astanti et al. [19], Mahendra [22], Taufik et al. [23])	Delivery time	D1
		Correctness of the type and quantity	D2
5	Warranty & Claim Policy (Taufik et al. [23], Kurniawan et al. [24])	Warranty time terms	W1
		Ease of the returning process	W2
6	Service (Ardhy & Dahda [18], Mahendra [22], Kurniawan et al. [25])	Ease of communication	S1
		Ability to update information	S2

S-Cell which sometimes supply the unavailable material. All these suppliers were located in the same city.

The results of the calculation and analysis in the AHP method are presented in the result section. The last phase will be the Test which is the stage to evaluate the proposed solution in terms of selected supplier and risk management.

3 Result and Discussion

The eigenvalues of each criterion were 6.301, 6.428, 6.422, 6.312, 6.397, 6.324, subsequently. Based on the highest eigenvalue, the Consistency Index (CI) and Consistency Ratio (CR) can be calculated as follows:

$$\text{CI} = \frac{6.428 - 6}{6 - 1} = 0.086 \quad (1)$$

$$\text{CR} = \frac{0.086}{1.24} = 0.069 \quad (2)$$

The CR is 6.91% (less than 10%), so it can be stated that the assessment made at the criterion level is consistent. From the complete consistency ratio calculation, it is known that all appraisements on the criteria, sub-criteria, and alternatives are stated to be consistent as all the consistency ratios are less than 0.1 or 10%. The result of local priority calculation for each criterion and sub-criteria is presented in Table 2. Then, it is used for calculating the global weight. Finally, the overall priority for each alternative on each criterion is presented in Table 3. Hence, A-Cell is the supplier that ranks first, followed by B-Cell and then S-Cell.

In addition to prioritizing and selecting supplier, an additional solution was proposed in the form of a supplier risk management plan. Risk management for suppliers is carried out by conducting risk analysis, risk evaluation, as well as plans or efforts to handle these risks, so that the impacts that may occur from these risks can be minimized. The detail of risk events and risk agents are presented in Tables 4 and 5.

The final house of risk is then presented in Table 6. From the analysis using House of Risk three recommendations were proposed, those are (1) conduct periodic performance evaluations with suppliers, (2) re-confirm the order, and (3) make a legal contract with the supplier.

Table 2 Summary of criteria and sub-criteria weight

Number	Criteria	Value	Sub-criteria	Value	Global Weight
1	Price	0.287	P1	0.795	0.228
			P2	0.205	0.059
2	Availability	0.322	A1	0.12	0.039
			A2	0.13	0.042
			A3	0.75	0.241
3	Quality	0.181	Q1	0.855	0.155
			Q2	0.145	0.026
4	Delivery	0.061	D1	0.855	0.052
			D2	0.145	0.009
5	Warranty & Claim Policy	0.083	W1	0.750	0.062
			W2	0.250	0.021
6	Service	0.065	S1	0.436	0.028
			S2	0.564	0.037

Table 3 Overall priority of the alternatives

B-Cell		A-Cell		S-Cell	
Price	0.097	Price	0.160	Price	0.029
Availability	0.039	Availability	0.117	Availability	0.166
Quality	0.095	Quality	0.023	Quality	0.064
Delivery	0.017	Delivery	0.032	Delivery	0.012
Warranty & Claim Policy	0.019	Warranty & Claim Policy	0.047	Warranty & Claim Policy	0.017
Service	0.043	Service	0.016	Service	0.006
Total	0.310	Total	0.395	Total	0.295

Table 4 Risk events identification

	Risk event (Ei)	Severity of risk (Si)
E1	Raw materials are not available	8
E2	Late delivery of raw materials	7
E3	The price of raw materials is higher than other suppliers	6
E4	Incompatibility of the amount/type of raw materials that arrive	5
E5	Damages to incoming raw materials	5

Table 5 Risk agents identification

	Risk agent (Ai)	Occurrence of agent (Oi)
A1	Suppliers hold supply goods	5
A2	Not building supplier relationship management	7
A3	Do not have a backup supplier to supply raw materials	7
A4	Long mileage	6
A5	Disturbances/obstacles in the delivery process (traffic constraints, weather, etc.)	7
A6	Lack of coordination with suppliers	5
A7	Not negotiating with suppliers	5
A8	Damage during the shipping process	4
A9	Not being careful in storing/moving things	5

Table 6 House of risk presenting the relationship matrix and ARP calculation

Risk event (Ei)	Risk agent (Aj)									Severity of risk (Si)
	A1	A2	A3	A4	A5	A6	A7	A8	A9	
E1	9	9	9							8
E2	1			9	9	9				7
E3	3	3		1			9			6
E4						9				5
E5				1				9	9	5
Occ. of agent (Oi)	5	7	7	6	7	5	5	4	5	
Aggregate risk potential (ARP)	485	630	504	444	441	540	270	180	225	
Priority	4	1	3	5	6	2	7	9	8	

4 Conclusion

The Design Thinking Process is followed for solving the supply problem of Company X. This is part of learning process for students in solving engineering problems. Design thinking process is an iterative and solution-based problem solving approach by considering the needs of the stakeholders and possible alternative solutions.

In the Ideate phase, two alternative solutions and three alternative methods were considered. AHP method is selected and based on the analysis, A-Cell becomes the new supplier for the company. Then, to maintain the supplier performance, three recommendations were proposed considering the risk of supply: (1) conducting joint performance evaluations with suppliers as an action with priority, (2) reconfirming orders to suppliers, and (3) building cooperation contracts with suppliers.

Further research can be focused on other supplier relationship management such as information sharing.

References

1. Paparoidamis NG, Katsikeas CS, Chunpitaz R (2019) The role of supplier performance in building customer trust and loyalty: a cross-country examination. Ind Mark Manage 78:183–197
2. Fagundes MVC, Hellingrath B, Freires FGM (2021) Supplier relection risk: a new computer-based decision-making system with fuzzy extended AHP. Logistics 5(1):13
3. Li F, Wu CH, Zhou L, Xu G, Liu Y, Tsai SB (2021) A model integrating environmental concerns and supply risks for dynamic sustainable supplier selection and order allocation. Soft Comput 25:535–549
4. Rahmawati DU, Salimi N (2021) Sustainable and resilient supplier selection: the case of an Indonesian coffee supply chain. J Supply Chain Manag Sci 3(1–2):16–36
5. Sipos C, Szucs EG (2022) Supplier risk assessment strategy. Int Rev Appl Sci Eng 13(2):208–215
6. Pusca D, Northwood DO (2018) Design thinking and its application to problem solving. Glob J Eng Educ 20(1):48–53
7. Lin KY, Wu YT, Hsu YT, Williams PJ (2021) Effects of infusing the engineering design process into STEM project-based learning to develop preservice technology teachers' engineering deign thinking. Int J STEM Educ 8(1)
8. Foster MK (2021) Design thinking: a creative approach to problem solving. Manag Teach Rev 6(2):123–140
9. Alikhani R, Torabi SA, Altay N (2019) Strategic supplier selection under sustainability and risk criteria. Int J Prod Econ 208:69–82
10. Aouadni S, Aouadni I, Rebai A (2019) A systematic review on supplier selection and order allocation problems. J Ind Eng Int 15:267–289
11. Tam MCY, Tummala VMR (2001) An application of the AHP in vendor selection of a telecommunications system. Omega 29(2):171–182
12. Thor J, Ding SH, Kamaruddin S (2013) Comparison of multi criteria decision making methods from the maintenance alternative selection perspective. Int J Eng Sci 2(6):27–34
13. Singh PK, Sharma SK, Samuel C, Verma S (2017) Supplier relationship management and selection strategies—a literature review. In: Proceeding of the 4th international conference on industrial engineering, Surat, India
14. Paragony PP, Sifutra M, Wijayanto D (2020) Analisis implementasi supplier relationship management pada UMKM dan perannya terhadap kualitas pemasok dan kinerja UMKM di Indonesia. J TIN Universitas Tanjungpura 4(2):37–45
15. Tavana M, Shaabani A, Mohammadabadi SM, Varzgani N (2021) An integrated fuzzy AHP-fuzzy MULTIMOORA model for supply chain risk-benefit assessment and supplier selection. Int J Syst Sci: Oper Logistics 8(3):238–261
16. Izzuddin IA, Ernawati D, Rahmawati N (2020) Analisa dan mitigasi risiko pada proses supply chain dengan pendekatan house of risk di PT. XYZ. Juminten: J Manajemen Ind dan Teknologi 1(3):129–140
17. Magdalena R, Vannie V (2019) Analisis risiko supply chain dengan model house of risk (HOR) pada PT Tatalogam Lestari. J@ti Undip: J Teknik Ind 14(2):53–62
18. Ardhy MA, Dahda SS (2022) Pemilihan supplier buah kelapa dengan metode AHP dan TOPSIS di PT XYZ. Serambi Eng 7(2):3181–3190
19. Astanti RD, Mbolla SE, Ai TJ (2020) Raw material supplier selection in a glove manufacturing: application of AHP and fuzzy AHP. Decis Sci Lett 9:291–312

20. Lukmandono L, Basuki M, Hidayat MJ, Setyawan V (2019) Pemilihan supplier industri manufaktur dengan pendekatan AHP dan TOPSIS. J OPSI 12(2):83–88
21. Afandi A (2018) Penerapan AHP (analytical hierarchy process) terhadap pemilihan supplier di UD. Nagawangi Alam Sejahtera Malang. J Vatech 1(1):119–124
22. Mahendra TS (2019) Pemilihan supplier kayu mebel menggunakan metode AHP (analytical hierarchy process) di UD. Riyan Pasuruan. J Valtech 2(1):104–109
23. Taufik R, Sumantri Y, Tantrika FMT (2014) Penerapan pemilihan supplier bahan baku ready mix berdasarkan integrasi metode AHP dan TOPSIS. J Rekayasa dan Manajemen Sistem Ind 2(5):1067–1076
24. Pratama WD, Sianturi YR, Sitorus IH, Siboro BAH (2021) Pemilihan supplier lampu UV dalam pembuatan box sterilization menggunakan metode Analytical Hierachy Process. J Ilmiah Teknik Industri 9(2):158–166
25. Kurniawan C, Sudarwati W, Dewiyani L (2019) Pemilihan supplier part cover transmision case menggunakan Metode Analytical Hierarcy Process di PT XHI. Prosiding Seminar Nasional Sains dan Teknologi 1–12

Decision Support System for Tuition Subsidy in XYZ Senior High School Student with Families Affected by Pandemic

Irsyad Thalib Abdillah, Mia Amelia, Dana Azizah, Nurul Fitri Dwi Andini, Wali Hakim Pradana, Amelia Kurniawati, and Afrin Fauzya Rizana

Abstract According to UNICEF 2020, in the percentage of children aged 7–18 years according to their school status, 938 Indonesian children have dropped out of school due to the Covid-19 pandemic. Children from families whose parents have been laid off and need tuition. Regarding this issue, the authorities anticipate it with several policies; one of the policies that can be carried out is the tuition subsidy program with selected students considering the impact of Covid-19. To get optimal results, a decision support system (DSS) is needed to support policies until decisions are taken. Simple Additive Weighting (SAW) is used in designing the system design. In designing the assessment criteria system, it is based on previous literature studies. There are four selected assessment criteria in a row, namely KIP card recipients, Parental Income, number of Parent Dependents, and arrears in tuitions. In addition, system development is carried out using the Rapid Application Development (RAD) method, which has several phases: Requirements Planning, User Design, Construction, and Cutover. This research aims to assist XYZ High School in selecting prospective students who receive the tuition subsidy program by predetermined criteria and conditions. This research resulted in an application for selecting students who receive subsidies for tuition.

Keywords Covid-19 · Decision support system · Tuition subsidy · Simple additive weighting (SAW)

I. T. Abdillah (✉) · M. Amelia · D. Azizah · N. F. D. Andini · W. H. Pradana · A. Kurniawati · A. F. Rizana
Telkom University, Jl. Telekomunikasi No. 1, Bandung 40257, Indonesia
e-mail: Irsyad.thalib10@gmail.com

C. N. Rosyidi et al. (eds.), *Proceedings of the 6th Asia Pacific Conference on Manufacturing Systems and 4th International Manufacturing Engineering Conference*, Lecture Notes in Mechanical Engineering,
https://doi.org/10.1007/978-981-99-1245-2_42

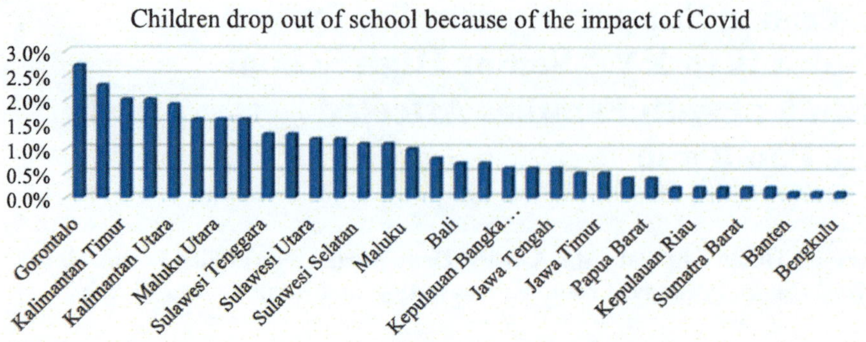

Fig. 1 Percentage of children dropping out of school due to the impact of the pandemic. *Source* [2]

1 Introduction

A United Nations International Children's Emergency Fund (UNICEF) survey noted nearly 60 million students are out of school. In addition, most children reported dropping out of school for economic reasons [1]. UNICEF's findings are limited to low-income families receiving the Family Hope Program (PKH) and Village Fund Direct Cash Assistance (BLT) with children aged 4–18. Figure 1 shows the percentage distribution of children dropping out of school due to the impact of the Cocid-19 pandemic [2].

In handling the Covid-19 pandemic, the Ministry of Education, Culture, Research, and Technology in 2020 and 2021 has distributed assistance of IDR 13.2 trillion and in handling the Covid-19 pandemic [3]. One of the policies that can be done is the tuition subsidy program with selected students considering the impact of Covid-19. To get optimal results, a decision support system is needed to support policies until decisions are taken. In this system, it is expected to be able to resolve decision-making issues to produce the best decisions by considering mathematical criteria [4]. A decision support system is very much needed and has been tried to be designed at the University level [5] and High School level [6], which is a scholarship program from the Department of Education for underprivileged students.

2 Methodology

This system is designed using RAD. Rapid Application Development (RAD) is a life cycle strategy that aims to provide much faster development and higher quality results than the traditional cycle or waterfall cycle, which is a rigid and time-consuming process. Therefore, the waterfall is less effective and is often only used in large-scale system development. [7]. This method does not spend much time in the planning stage [8]. The RAD method consists of four phases.

1. Requirements Planning Phase

This stage is the initial phase of system development. In this phase, problems are identified, and user data is collected to determine the system's goal or purpose and the required information. Therefore, the involvement of all stakeholders is significant.

2. User Design Phase

The system design stage uses the SAW method. Simple Additive Weighting (SAW) is a method of Fuzzy Multiple Attribute Decision Making (FMADM) that can solve multiple attribute decision-making problems by weighting all criteria and alternatives that produce the correct reference value [7]. Meanwhile, SAW itself has also been used in several DSS for scholarships, such as the Sultan Agung Jombang Vocational High School Scholarship [9] and the Dharma Karya Vocational High School Scholarship [10].

3. Construction Phase

At this stage, the system design that has been made and agreed upon is converted into a beta version of the application to the final version. At this stage, researchers must continue developing and integrating with other sections while considering feedback from users or clients. If the process runs smoothly, it can continue to the next stage, while if the application developed has not answered the needs, the researcher will return to the system design stage.

4. Cutover Phase

In this stage, the researcher applies the design of a system approved in the previous step. Before the system is implemented, the program testing process is conducted first to detect errors in the developed system. At this stage, it is customary to provide feedback on the system that has been made and get approval for the system.

3 Results and Discussion

The development cycle using the RAD method has a process of approval and gathering suggestions from users. The stages of the process carried out are as follows:

1. Requirements Planning Phase

Some relevant stakeholders are schools or the committee for subsidized tuition, students, parents, and researchers. The data needed in the research include student smart Indonesian card (KIP), parental income, number of parents dependent, and arrears in tuition.

2. User Design Phase

The user design stage is done by making a system design that can be seen as the workflow design or an overview of the system used [11]. The system design is designed where the platform used is web-based using the XAMPP server. For the framework using Codeigniter and using the PHP programming language. In addition, the database code uses MySQL and can use Google Chrome, Mozilla Firefox, and Opera browsers. Finally, the data processing system will use the Simple Additive Weighting (SAW) method. The following is the process of the system design that has been designed.

Figure 2 explains that the program owner oversees preparing and providing direction to the committee before running the assistance program. In addition, the program owner will also make decisions regarding policies in the program. Therefore, the committee will run the program from start to finish based on the direction of the program owner. Meanwhile, students will fill in the data needed by the committee to be able to select the program held.

3. Construction Phase

The next stage is to design the system as an application. The result of this stage is an application of the system. The following is an image of the application system user interface that has been designed (Figs. 3, 4, 5, 6, 7 and 8).

The user interface image of the application system above is designed based on the order of the system processes described in Fig. 2.

4. Cutover Phase

The next stage is testing to determine the suitability of the system by verifying the calculation of the system that has been designed. There are four criteria used. Where these criteria are taken from previous research (Table 1).

The following is a table of alternative parameters or input data based on the selection results from XYZ Senior High School for each prospective subsidy (Table 2).

The calculation of the weights on each criterion using the analytical hierarchy process (AHP). Furthermore, there are details of the attributes of each criterion and their weights generated from the results of discussions with XYZ senior high school (Table 3).

Based on the steps taken in the calculation using the SAW method, the following results were obtained (Table 4).

The ranking results above show that the alternatives entitled to be a priority in receiving subsidies for XYZ High School students whose families are affected by the pandemic are Wali (A5) and Dana (A1), so the system is appropriate.

While the disadvantages of this system are that student data entered by students can only be entered once, and detailed information on the selection results does not

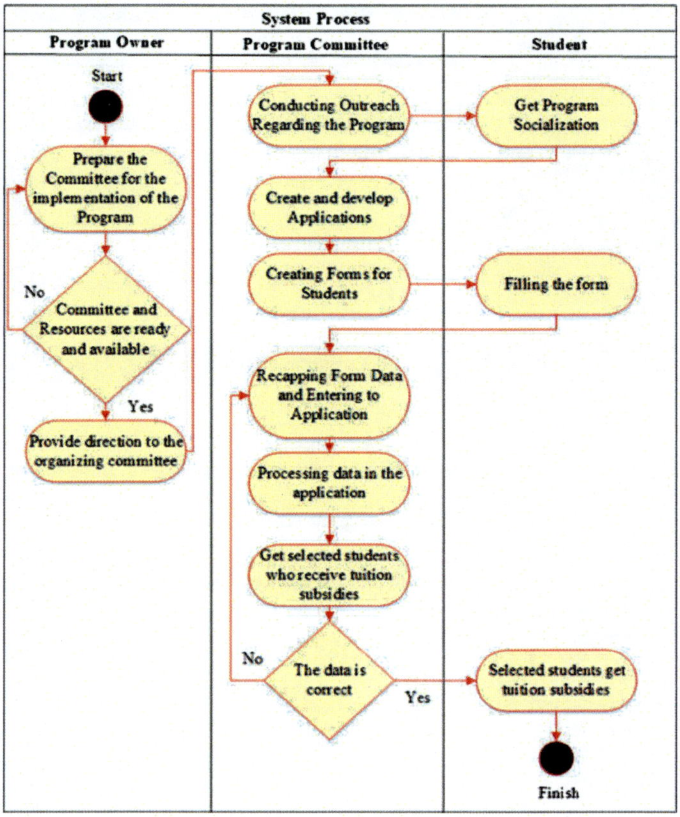

Fig. 2 System process

Fig. 3 Login page

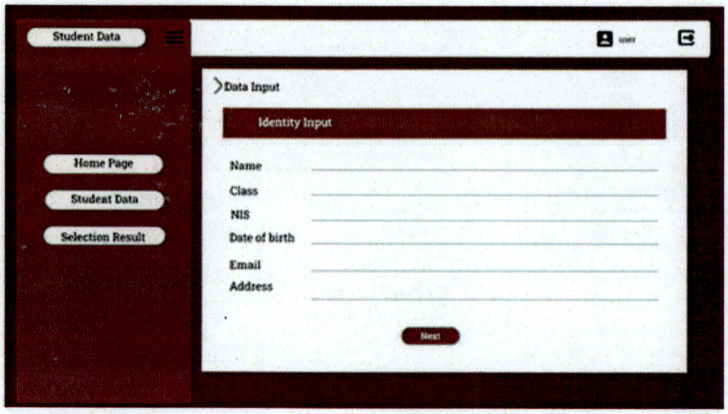

Fig. 4 Student data entry page

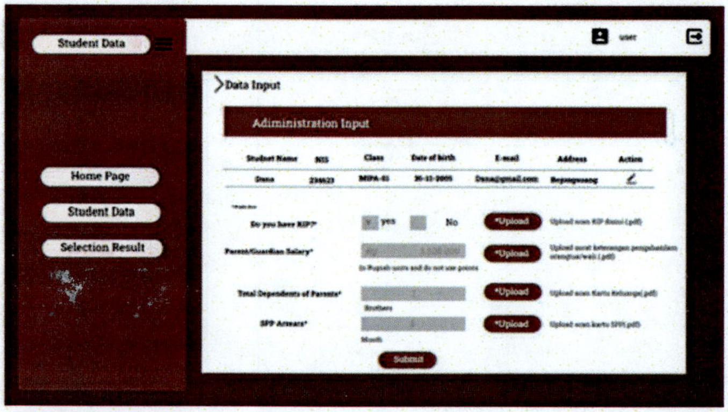

Fig. 5 Student criteria input page

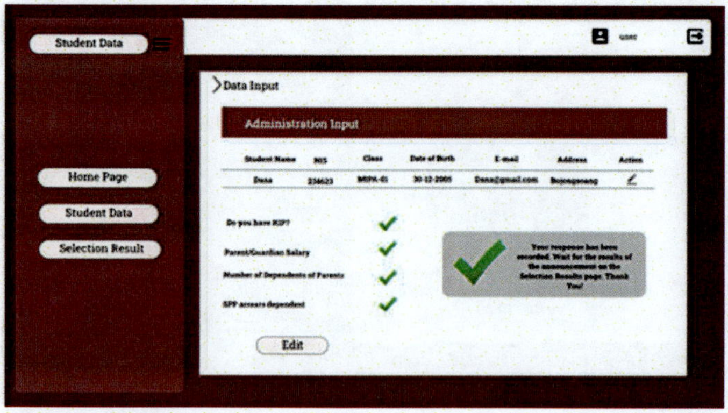

Fig. 6 Recorded data page

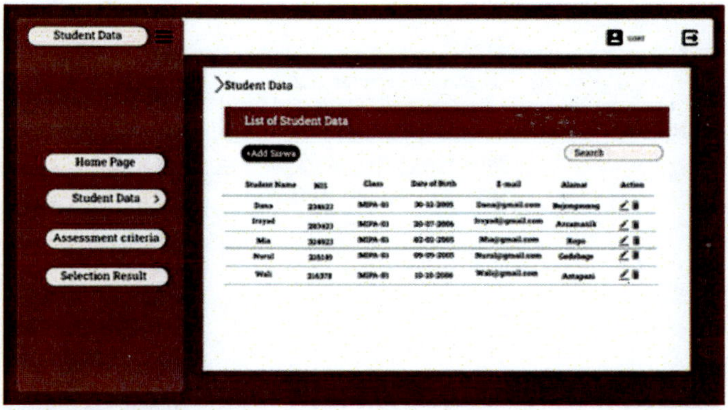

Fig. 7 Criteria page for committee

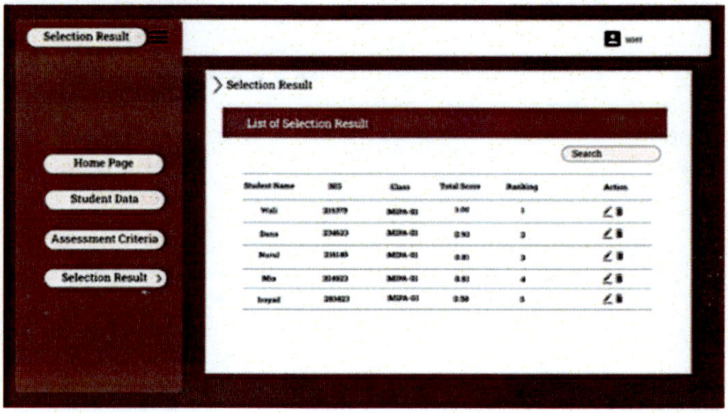

Fig. 8 Selection result page

Table 1 Selection criteria

Criteria name	Criteria code	Source
KIP card recipients	(C1)	[9] [10]
Parental income	(C2)	[7] [10] [9]
Number of parent dependents	(C3)	[10] [9]
Arrears in tuitions	(C4)	[7]

Table 2 Assessment based on criteria

Criteria	Attribute	Weight
KIP card recipients (C1)	Benefit	0.08
Parental income (C2)	Cost	0.24
Number of parent dependents (C3)	Benefit	0.09
Arrears in tuitions (C4)	Benefit	0.59

Table 3 Alternative calculation

Criteria	Attribute	Weight
KIP card recipients (C1)	Yes No	1 0
Parental income (C2)	> IDR 4,000,000 IDR 3,000,000—IDR 4,000,000 IDR 2,000,000—IDR 2,999,999 IDR 1,000.000—IDR 1,999,999 < IDR 1,000,000	0 0.25 0.5 0.75 1
Number of parent dependents (C3)	> 5 4–5 2–3 1	1 0.75 0.5 0.25
Arrears in tuitions (C4)	> 3 month 2–3 month 1 month 0 month	1 0.75 0.5 0.25

Table 4 Rank results

Alternative	Total	Rank
A1	0.93	2
A2	0.58	4
A3	0.61	3
A4	0.81	2
A5	1.00	1

display detailed calculation results. And the adjustments that need to be made by the school are socialization to students to participate in this program and evaluation and control during system implementation.

4 Conclusion

From the design of the decision support system application that has been completed, it can be concluded that the program for subsidizing the education costs of XYZ High School students with families affected by COVID will be easier to run using the application system developed by the researcher. Hence, the researcher recommends XYZ High School use this application. Furthermore, this application has also been verified based on manual data processing with Microsoft Excel and a Decision Support System (DSS) when selecting prospective students who receive tuition assistance. And it is hoped that further research can examine the features in the application that can further help facilitate the use of the application and the selection process.

References

1. United Nations International Children's Emergency Fund (UNICEF) (2020) COVID-19 and children in Indonesia
2. Ministry of Education, Culture, Research, and Technology of Indonesia (2021). Percentage of children dropping out of school due to the impact of the pandemic
3. Ministry of Education, Culture, Research, and Technology of Indonesia (2021) Provision of Internet data quota assistance and single tuition fee
4. Higher Education Service Institute of Indonesia (LLDIKTI) (2020) Scholarships and tuition fee assistance for students
5. Kustiyaningsih (2021) Decision tree C 4.5 algorithm for classification of poor family scholarship recipients. IOP Conf Ser: Mater Sci Eng
6. Huda M (2018) Decision support system of scholarship grantee selection using data mining. Int J Pure Appl Math 119(15)
7. McLeod Jr R, Schell GP (2008) Sistem Informasi Manajemen Edisi 10. Salemba Empat, Jakarta
8. Kendall E (2008) Rapid application development (RAD). PT Indeks. Klaten
9. Indriyanti AD (2019) Simple additive weighting algorithm to aid administrator decision-making of the underprivileged scholarship. J Phys 1402(6)
10. Oktaviani (2020) Comparison of weighted product method and simple additive weighting in scholarship recipient selection. IOS Sci: Manag, Inf Commun Technol 3(4)
11. Kurniawan R, Kurniawati A, Rizana AF (2020) Mapping of Telkom University Roadshow Location Using Geographic Information System. Article No. 50, pp 1–6. ICONETSI 2020

Cost Analysis for Warranted Products Protected by Lemon Laws Considering Failure Interaction

Fakhri I. Alifin, Andi Cakravastia, and Bermawi P. Iskandar

Abstract This study develops warranty cost models for a warranted product under the lemon laws protection. The product considered in this study is a multi-component system consisting of critical and non-critical components. The Failure of either component can cause the product to fail. The product is declared a lemon if recurrent failures of the critical component occur—i.e., the number of the failures reaches the threshold value, k (e.g., four failures). Furthermore, this study assumes that a failure of the critical component can occur naturally or induced by a failure of the non-critical component. Hence, two types of failures are considered—i.e., a natural failure and an induced failure. Whenever a lemon is declared, the customers can opt for either (i) a refund or (ii) a replacement warranty claim. Hence, this paper provides the warranty cost models for the two cases—(i) refund and (ii) replacement then encounters the optimal lemon law period that minimizes the expected warranty cost rate. The numerical examples are provided to illustrate the expected warranty cost and the optimal lemon law period.

Keywords Warranty cost · Lemon laws · Failure interaction · Refund · Replacement

F. I. Alifin (✉)
Graduate Program in Industrial Engineering and Management, Institut Teknologi Bandung, Bandung, Indonesia
e-mail: fakhri.ikhwanul@ft.unsika.ac.id

Department of Industrial Engineering, Universitas Singaperbangsa Karawang, Karawang, Indonesia

A. Cakravastia · B. P. Iskandar
Department of Industrial Engineering, Institut Teknologi Bandung, Bandung, Indonesia

C. N. Rosyidi et al. (eds.), *Proceedings of the 6th Asia Pacific Conference on Manufacturing Systems and 4th International Manufacturing Engineering Conference*, Lecture Notes in Mechanical Engineering,
https://doi.org/10.1007/978-981-99-1245-2_43

1 Introduction

In some marketplaces (e.g. USA, Singapore, South Korea, Australia), the warranted products are protected by lemon laws. In the automotive industry, a warranted car is also protected by the lemon law during the warranty period. The lemon laws aim to protect customers from recurring product failures caused by a defective product that does not meet the quality standards. The state of Connecticut was the first state in the USA to enact the lemon laws in 1982. Before the establishment of the lemon laws, customers frequently experienced frustration and excess expenditure to rectify subsequent product failures [1].

In general, a product is considered a lemon if (i) recurrent failures have occurred—(i.e., the number of failures exceeds the predefined threshold), or (ii) the product has been out of service for more than 30 days because the failed product was not satisfactorily repaired by the dealer/manufacturer [2, 3]. As a result, the manufacturer is more likely to receive more claims (and hence a higher warranty cost incurred) if the lemon law is enforceable during the warranty period. Consequently, the manufacturer should establish a better warranty strategy and product reliability to cope with this situation.

The study of warranty cost analysis for a warranted product with lemon laws has not yet received much attention. Only Husniah et al. [2, 3] consider cost analysis for the lemon law warranty among the research related to the warranty cost models and analysis. Reference [2] views a product as a single-component system and uses a black-box approach to modeling product failure, or the inner structure of the system is unconsidered. Whilst [3] considers a product as a multi-component system as well as the internal structure is considered in modeling the product failure. In Husniah et al. [3], the failure of each component occurs independently. However, in many cases, failures in one component can induce failures of other components (i.e., failure interaction) [4]. In automotive products, for example, a piston ring failure may generate an engine failure, and a bearing failure in a gearbox may accelerate the deterioration of gears and shafts [5, 6]. Some works in failure modeling that consider the failure interaction can be found in [5–7]. Yet, the failure interaction is often neglected by researchers in modeling failure and examining a warranty servicing strategy due to the complexity involved. They assume that the failure of a component occurs independently and does not influence or weaken the condition of other components (no failure interaction) [8–10]. Only references [11] and [12] examined warranty cost for a multi-component system with failure interaction. However, these works do not consider lemon laws enforceable during the warranty period. Thus, this paper studies warranty costs for a product protected by the lemon laws. The product is viewed as a multi-component system consisting of critical (C) and non-critical (NC) components as in [3], and failure of the C component may influence a failure of the NC component, or there is a one-way failure interaction between C and NC components. As a result, failure of the C component can be caused by internal factors (referred to as a natural failure) or external factors (i.e., induced by a failure of the NC component). The product is declared a lemon whenever the number of critical (C) component failures reaches k as this component failure may cause safety problems. Hence, this paper can be viewed as an extension of the work by Husniah et al. [3]. Two major contributions of this paper are—(i) developing a model of lemon law warranty occurrences considering a failure interaction between the C and NC components and (ii) carrying out warranty cost analysis based on (i) for refund and replacement cases.

The following is how this paper is organized. The notations used in the model formulation and the model assumptions are presented in Sect. 2. Then, Sect. 3 describes the model formulation consisting of the expected total warranty cost, expected warranty length, and expected warranty cost rate. Finally, the numerical examples, managerial implications, conclusions, and a brief description of future research topics are provided in Sect. 4.

2 Notations and Model Assumptions

2.1 Notations

Table 1 shows the notations used in this work.

2.2 Model Assumptions

The model assumptions are as follows:

1. The product is comprised of critical (C) and non-critical (NC) components
2. The time to the first failure distributions for C and NC components follow a Weibull distribution
3. All failures before reaching the kth failure are repaired minimally
4. The lemon condition is only triggered by the C component failure
5. The C component failure can be natural or induced by the NC component.

3 Model Formulation

3.1 Modeling Failure

Suppose that a warranted product has a warranty period (W) and the lemon period (W_L), where $W_L \leq W$. In general, W_L is equal to W, but in some cases, it is less than W. For instance, in Singapore, the lemon law period W_L is six months for a product being warranted with 12 months period. The product is declared lemon when the number of the C component failures has reached k in $(0, W_L]$. If the product is declared to be a lemon, the manufacturer must (i) refund the sales price

Table 1 Notations in this paper

Symbol	Description
W	Warranty period
W_L	Lemon laws period
C_p	Product sales price
C_m	Manufacturing cost of the product
C_{rc}	Repair cost of C component
C_{rn}	Repair cost of NC component
$C_1(W_L, W; k)$	Total warranty cost for refund scheme
$C_2(W_L, W; k)$	Total warranty cost for replacement scheme
k	Failure number thresholds
L_w	Warranty length
$U_n[V_n]$	n^{th} time to failure of C[NC] component
$N_1(t)[N_2(t)]$	Failure number of C[NC] component
$G_n(t)[H_n(t)]$	Cumulative distribution function (CDF) of C[NC] component
$g_n(t)[h_n(t)]$	Probability density function (PDF) of C[NC] component
$\overline{G}_n(t)\left[\overline{H}_n(t)\right]$	Survival function of C[NC] component
$\lambda_1(t)[\lambda_2(t)]$	Weibull intensity function of C[NC] component
$\Lambda_1(t)[\Lambda_2(t)]$	Cumulative hazard function of C[NC]
$\alpha_c[\alpha_{nc}]$	Weibull scale parameter of C[NC]
β	Weibull shape parameter

or (ii) replace the failed product with the new product. The warranted product is considered as a repairable system, and the failures before the kth value occurring under warranty are repaired minimally. Let $N_1(t)[N_2(t)]$ is the number of failures for C[NC] components over the $[0, t)$. As all failures before the kth failure are fixed by minimal repair and the repair times are very small relative to the mean time between failures, then $N_1(t)[N_2(t)]$ follows a Non-Homogeneous Poisson Process (NHPP) with the intensity function $\lambda_1(t)[\lambda_2(t)]$.

The two widely used intensity functions of NHPP are (i) a power-law function (often called a Weibull intensity function) and (ii) a log-linear intensity function. But in many practical cases, a Weibull intensity function is often used [13, 14]. Let $U_n[V_n]$ be the time of the nth failures of C[NC] component. It is assumed that the time to the first failure $U_1[V_1]$ follows the Weibull distribution function. Since all failures before the threshold (k) are repaired minimally, then the subsequent failures ($U_n[V_n]$) follow an NHPP with intensity function given by the Weibull intensity function. The Cumulative Distribution Function (CDF) of $U_n[V_n]$ is given by $G_n(t)[H_n(t)]$. Then, the probability of n successive failures for C [NC] components over (0, t] is given by,

$$P\{N_1(t) = n\} = G_n(t) - G_{n+1}(t)\,[P\{N_2(t) = n\} = H_n(t) - H_{n+1}(t)] \quad (1)$$

where,

$$P(U_n \le t) = G_n(t) = 1 - \sum_{i=0}^{n=1} e^{-\Lambda_1(t)} \frac{\Lambda_1(t)^i}{i!};$$

$$P(V_n \le t) = H_n(t) = 1 - \sum_{i=0}^{n=1} e^{-\Lambda_2(t)} \frac{\Lambda_2(t)^i}{i!}$$

While $\lambda_1(t)[\lambda_2(t)]$ and $\Lambda_1(t)[\Lambda_2(t)]$ are given by,

$$\lambda_1(t) = \frac{\beta}{\alpha_c}\left(\frac{t}{\alpha_c}\right)^{\beta-1}\left[\lambda_2(t) = \frac{\beta}{\alpha_{nc}}\left(\frac{t}{\alpha_{nc}}\right)^{\beta-1}\right];$$

$$\Lambda_1(t) = \int_o^t \lambda_1(\tau)\mathrm{d}\tau\left[\Lambda_2(t) = \int_o^t \lambda_2(\tau)\mathrm{d}\tau\right]$$

As the lemon condition is only invoked by the recurrent failures of the C component, the NC component failures are all-natural, but a single failure of the NC component can induce the C component to fail (i.e., one-way failure interaction). Hence, for the C component, there are two types of failures—'natural' and 'induced'. The failure of the C components is described as follows: if an NC component fails, it can induce a failure of a C component with probability p_2 or have no effect on a C component with probability $q_2 = 1 - p_2$. This is called as a type I failure interaction in [7]. Let $\tilde{N}_2(t)$ be the number of induced failures of the C component due to failures of NC components over $[0, t)$. Then the total number of C component failures over $[0, t)$ is $N_1(t) + \tilde{N}_2(t)$, which is a sum of two stochastic processes with intensity function given by $\lambda_1(t) + p_2\lambda_2(t)$ (see Fig. 1). The ways in which U_k can occur in the interval $[t, t + \delta t)$ must be formulated in the model. If U_k is the time of the kth failures of C component (or a lemon declared), then there are two ways that U_k can occur in the interval $[t, t + \delta t)$ as follows (See also Fig. 1).

(1) $N_1(t) = n_1 \le k - 1, \tilde{N}_2(t) = \tilde{n}_2 = k - 1 - n_1\left[N_1(t) + \tilde{N}_2(t) = k - 1\right]$ and a C component fails in $[t, t + \delta t)$
(2) $N_1(t) = n_1 \le k - 1, \tilde{N}_2(t) = \tilde{n}_2 = k - 1 - n_1\left[N_1(t) + \tilde{N}_2(t) = k - 1\right]$ and an NC component fails in $[t, t + \delta t)$ inducing a failure in a C component.

Using a conditional approach, the expression for $g_k(t)$ is given by,

$$g_k(t|N_1(t) = n_1, N_2(t) = n_2) = \binom{n_2}{k - 1 - n_1} p_2^{k-1-n_1}(1 - p_2)^{n_2-(k-1-n_1)}$$
$$[\lambda_1(t) + p_2\lambda_2(t)] \quad (2)$$

Note that the distributions for $N_1(t)$ and $N_2(t)$ are both Poisson and the distribution for $\tilde{N}_2(t) = k - 1 - n_1$ given $N_2(t)$ is binomial. The CDF and the probability of k successive failures for C component over (0,t] is given by,

$$G_k(t) = 1 - P\{N_1(t) + \tilde{N}_2(t) \le k - 1\}$$
$$= 1 - \sum_{i=0}^{k-1} \frac{e^{-[\Lambda_1(t)+p_2\Lambda_2(t)]}[\Lambda_1(t) + p_2\Lambda_2(t)]^i}{i!} \quad (3)$$

where,

$$\Pr\left\{N_1(t) + \tilde{N}_2(t) = n_1\right\} = \frac{e^{-[\Lambda_1(t)+p_2\Lambda_2(t)]}[\Lambda_1(t) + p_2\Lambda_2(t)]^{n_1}}{n_1!} \quad (4)$$

$\lambda_1(t) + p_2\lambda_2(t)$ and $\Lambda_1(t) + p_2\Lambda_2(t)$ are the intensity function and cumulative intensity function of the C component respectively.

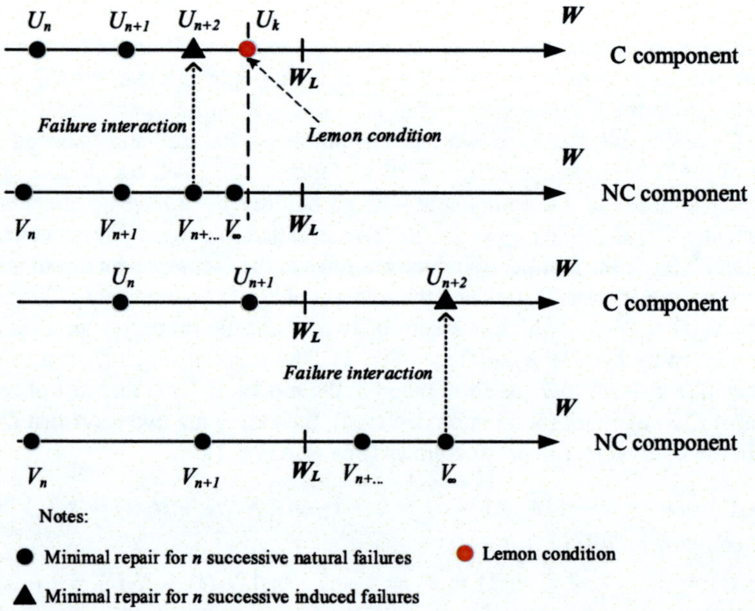

Fig. 1 Possible lemon conditions

3.2 Refund Warranty Scheme

In this scheme, the manufacturer will refund (i.e., equal to the sales price) if a product is returned due to a lemon claim. Consequently, the warranty will be terminated immediately at the time U_k (see Fig. 1). Note that $W_L \leq W$ (i.e., a lemon law period is less than or equal to the warranty period). The expected total warranty cost is obtained by using a conditional approach (i.e., if a lemon or no-lemon is declared). Two conditions are considered in this scheme,

(i) refund—i.e., $U_k \leq W_L, (N_1(W_L) + \tilde{N}_2(W_L) = k, N_2(W_L) > 0)$, and
(ii) no-refund—i.e., $U_k > W_L, (N_1(W_L) + \tilde{N}_2(W_L) \leq k - 1, N_2(W_L) > 0)$.

Hence,

$$\text{EWC} = \text{E[Refund]} + \text{E[No} - \text{refund]} \tag{5}$$

Using the conditional approach, the expected total warranty cost conditional on U_k is given by,

$$C_1(W_L, W; k) = \begin{cases} A, \text{ if } U_k \leq W_L \\ B, \text{ if } U_k > W_L \end{cases} \tag{6}$$

where,

$$A = [c_{rc}(k-1) + c_{mc} + c_{rn}E[N_2(W_L) = n_2]]$$

$$B = \left[c_{rc}E[N_1(W_L) + \tilde{N}_2(W_L) = n_1 | n_1 \leq k-1] + c_{rn}E[N_2(W_L) = n_2]\right] \\ + c_{rc}E[N_1(W_L, W) + \tilde{N}_2(W_L, W) = n_1] + c_{rn}E[N_2(W_L, W) = n_2]$$

Then, removing the condition yields,

$$E[C_1(W_L, W; k)] = c_{rc}\sum_{n_1=1}^{k-1} n_1 G_{n_1}(W_L) + (c_{mc})G_k(W_L) + c_{rn}\int_0^{W_L} \lambda_2(t)dt \\ + \left[c_{rc}\int_{W_L}^{W} \lambda_1(t) + p_2\lambda_2(t)dt + c_{rn}\int_{W_L}^{W} \lambda_2(t)dt\right]\overline{G}_k(W) \tag{7}$$

3.3 Replacement Warranty Scheme

Here, if the lemon is declared or $U_k \leq W_L$, the manufacturer will replace the failed product and issue a new warranty. Consequently, the warranty period will be renewed. Then, the expected total warranty cost (EWC) is given by,

$$\text{EWC} = \text{E}\big[\text{Replacement}\big] + \text{E}\big[\text{No} - \text{replacement}\big] \tag{8}$$

Using a similar approach as in the refund case, there are two conditions for this scheme, (i) Replacement ($U_k \leq W_L$) and (ii) No-replacement ($U_k > W_L$). Hence, the total warranty cost conditional on U_k is given by,

$$C_2(W_L, W; k) = \begin{cases} A, \text{ if } U_k \leq W_L \\ B, \text{ if } U_k > W_L \end{cases} \tag{9}$$

where,

$$\begin{aligned} A &= [c_{rc}(k-1) + c_m + c_{rn}E[N_2(W_L) = n_2]] \\ B &= \Big[c_{rc}E[N_1(W_L) + \tilde{N}_2(W_L) = n_1 | n_1 \leq k-1] + c_{rn}E[N_2(W_L) = n_2]\Big] \\ &\quad + c_{rc}E[N_1(W_L, W) + \tilde{N}_2(W_L, W) = n_1] + c_{rn}E[N_2(W_L, W) = n_2] \end{aligned}$$

Removing the condition yields,

$$\begin{aligned} E[C_2(W_L, W; k)] =& c_{rc} \sum_{n_1=1}^{k-1} n_1 G_{n_1}(W_L) + (c_m)G_k(W_L) \\ &+ E[C_2(W_L, W; k)] + c_{rn} \int_0^{W_L} \lambda_2(t)dt \\ &+ \left[c_{rc} \int_{W_L}^{W} \lambda_1(t) + p_2\lambda_2(t)dt + c_{rn} \int_{W_L}^{W} \lambda_2(t)dt \right] \bar{G}_k(W) \end{aligned} \tag{10}$$

Then, considering the renewal process yields,

$$E[C_2(W_L, W; k)] = \frac{\begin{array}{c} \left[c_{rc} \sum_{n_1=1}^{k-1} n_1 G_{n_1}(W_L) + (c_m)G_k(W_L) + \left[c_{rn} \int_0^{W_L} \lambda_2(t)dt \right] \right] \\ + \left[c_{rc} \int_{W_L}^{W} \lambda_1(t) + p_2\lambda_2(t)dt + c_{rn} \int_{W_L}^{W} \lambda_2(t)dt \right] \overline{G}_k(W) \end{array}}{\int_0^W \overline{G}_k(t)dt} \tag{11}$$

3.4 Optimization

We obtain the optimization function which will be used to find the optimal W_L that minimizes the expected warranty cost rate (EWR) given in (12).

$$\text{EWR} = \text{EWC}/L_W \tag{12}$$

The expected total warranty cost (EWC) is given in (8) and the expected warranty length obtained as follows. The warranty period will end either at U_k if $U_k \le W_L$, or at W_L if $U_k > W_L$. Thus, the warranty length conditional on U_k (L_W) is given by,

$$L_W = \begin{cases} U_k, \text{ if } U_k \le W_L \\ W, \text{ if } U_k > W_L \end{cases} \tag{13}$$

Then, removing the condition yields

$$E[L_W] = \int_0^{W_L} sf(s)ds + W\overline{G}_k(W_L)\overline{H}_k(W_L) \tag{14}$$

where,

$$F(s) = P(U_k < s) = \int_0^s \overline{G}_k(x)h_k(x)dx + \int_0^s g_k(x)h_k(x)dx,\ andf(s) = \frac{\partial F(s)}{\partial s}$$

Hence, the expected warranty cost rate for both refund and replacement scheme respectively is given by,

$$EWR[C_1(W_L, W; k)] = \frac{E[C_1(W_L, W; k)]}{E[L_W]};\ EWR[C_2(W_L, W; k)] = \frac{E[C_2(W_L, W; k)]}{E[L_W]} \tag{15}$$

Since the length of the warranty period (L_W) is dependent on the occurrence of a lemon—(i.e., can either end immediately in the refund scheme or be greater than W in the replacement scheme), the manufacturer should establish an effective lemon law period (W_L^*) that minimizes warranty cost. Hence, the optimization problem is given by,

$$\underset{W_L}{Min}\{EWR[C_1(W_L, W; k)|k]\}\ ;\ \underset{W_L}{Min}\{EWR[C_2(W_L, W; k)|k]\} \tag{16}$$

4 Numerical Example

The parameter values used in the numerical examples are presented in Table 2. Due to the complexity involved in EWR, the numerical method is utilized with uniform searching to obtain the sub-optimal solution.

Failure distributions for the C and NC components are Weibull distribution where $\alpha_c[\alpha_{nc}]$ is the scale parameter of the distribution for the C[NC] component. Two scenarios are considered in processing the numerical example, namely Scenario 1 and Scenario 2. In Scenario 1, the NC component has better reliability representing with $\alpha_c < \alpha_{nc}, \alpha_c = 0.4,\ \alpha_{nc} = 0.7$, while in Scenario 2 the C component reliability is better representing with $\alpha_c > \alpha_{nc}, \alpha_c = 0.7,\ \alpha_{nc} = 0.4$. Furthermore, the numerical method is utilized to evaluate the expected warranty cost rate and obtain the optimal lemon law period (applying a uniform search and coding by Matlab). The results are described in the following Subsection.

4.1 Refund Warranty Scheme

Tables 3 and 4 show the refund scheme numerical results for Scenario 1 and Scenario 2, respectively.

As seen in Tables 3 and 4 (See also Figs. 2 and 3), the expected total warranty cost ($E[C_1(W_L, W; k)]$) tends to decrease when the failure number threshold (k) increases. This is as expected since a greater k will lessen the possibility of lemon

Table 2 Parameters value used in this work

C_p	C_{rc}	C_{rn}	C_m	β	W	k
100	$0.05C_p$	$0.025C_p$	$0.7C_p$	2	1	3, 4, 5

Table 3 Numerical example of refund scheme (scenario 1)

k	$E[C_1(W_L, W; k)]$	$E[L_w]$	$EWR[C_1(W_L, W; k)]$	W_L^*
3	16.7260	0.3869	43.2294	0.3616
4	14.5042	0.4801	30.2102	0.4487
5	12.4696	0.5505	22.6508	0.5145

Table 4 Numerical example of refund scheme (scenario 2)

k	$E[C_1(W_L, W; k)]$	$E[L_w]$	$EWR[C_1(W_L, W; k)]$	W_L^*
3	15.2338	0.4572	33.3189	0.4273
4	13.6514	0.5433	25.1247	0.5078
5	11.9307	0.6178	19.3110	0.5774

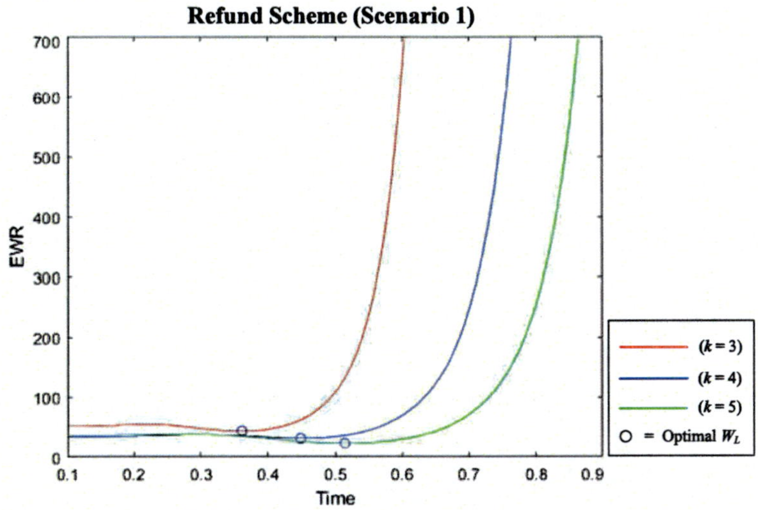

Fig. 2 Graphical illustration for the optimal lemon period on refund scheme (Scenario 1)

occurring, and this, in turn, decreases the expected total warranty cost on one side and increases the warranty length ($E[L_w]$) and the optimal lemon period (W_L^*) on the other side. Whilst, scenario 2 has a lower expected warranty cost for a given k as the C component has higher reliability and this results in a lower warranty cost. Moreover, the effectiveness of the warranty strategy can be evaluated by comparing the optimal lemon period (W_L^*) obtained with the regulated lemon law period, such as in Singapore (i.e., $\tilde{W}_L = 6$ months $= 0.5$). For instance, in scenario 1, the warranty strategy related to $k = 5$ has $W_L^* > \tilde{W}_L$, which indicates the sufficiency of the warranty strategy to meet the regulation (i.e., a lower possibility for a lemon claim occurring over $(0, \tilde{W}_L]$). The manufacturer should improve the product's reliability, especially on the component with higher repair costs (i.e., C component). This can be proved by the results in Scenario 2 for k = 4 and k = 5 with $W_L^* > \tilde{W}_L$.

4.2 *Replacement Warranty Scheme*

The results for the replacement scheme are shown in Table 5 (Scenario 1) and Table 6 (Scenario 2).

In this case, a similar pattern holds for the expected total warranty cost ($E[C_2(W_L, W; k)]$) when k increases—i.e., the expected total warranty cost decreases, whilst the warranty length ($E[L_w]$) and the optimal lemon period (W_L^*) increase (see also Fig. 3). Correspondingly, Scenario 2 results in the lower warranty cost. However, the expected total warranty cost of the replacement scheme is greater than the corresponding refund scheme. This pattern is due to in the replacement

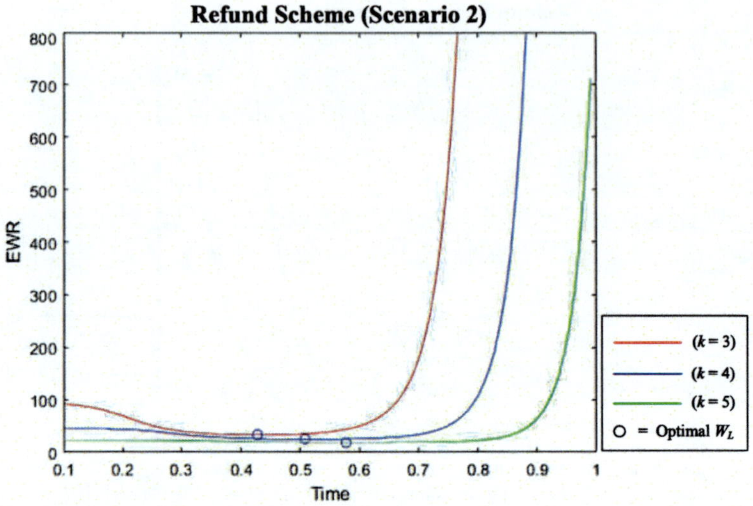

Fig. 3 Graphical illustration for the optimal lemon period on refund scheme (Scenario 2)

Table 5 Numerical example of replacement scheme (scenario 1)

k	$E[C_2(W_L, W; k)]$	$E[L_w]$	$EWR[C_2(W_L, W; k)]$	W_L^*
3	20.7097	1.0247	20.2106	0.4510
4	18.0324	1.0289	17.5263	0.5455
5	16.3912	1.1191	14.6469	0.6564

Table 6 Numerical example of replacement scheme (scenario 2)

k	$E[C_2(W_L, W; k)]$	$E[L_w]$	$EWR[C_2(W_L, W; k)]$	W_L^*
3	18.9311	1.0912	17.3485	0.4613
4	16.4982	1.1670	14.1375	0.5641
5	15.0682	1.2514	12.0412	0.6690

scheme, the number of lemon claims can be more than one per unit product, whilst under the refund scheme, the number of lemon claims is at most one. This issue, in turn, causes W_L^* for the replacement scheme is slightly greater compared to that of the refund case due to $E[L_w]$ as a function of W_L^* is larger such that to minimize the expected warranty cost rate. Figures 4 and 5 shows graphical illustration for the optimal lemon period on replacement scheme on scenario 1 and 2.

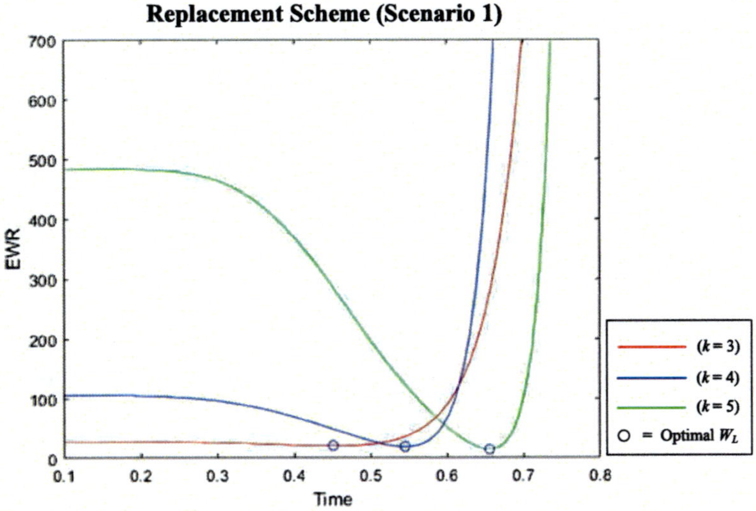

Fig. 4 Graphical illustration for the optimal lemon period on replacement scheme (Scenario 1)

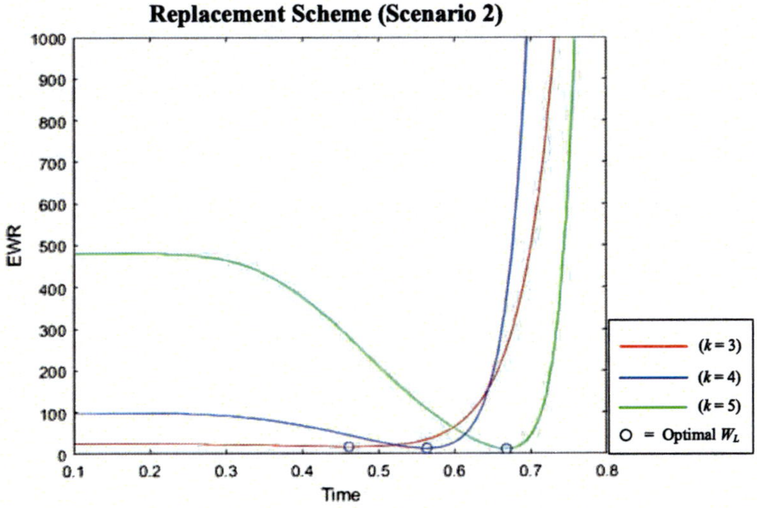

Fig. 5 Graphical illustration for the optimal lemon period on replacement scheme (Scenario 2)

4.3 Sensitivity Analysis

The sensitivity analysis aims to observe the effect of the change in reliability representing by the scale parameter values (α_c,α_{nc}) for both C and NC components to the

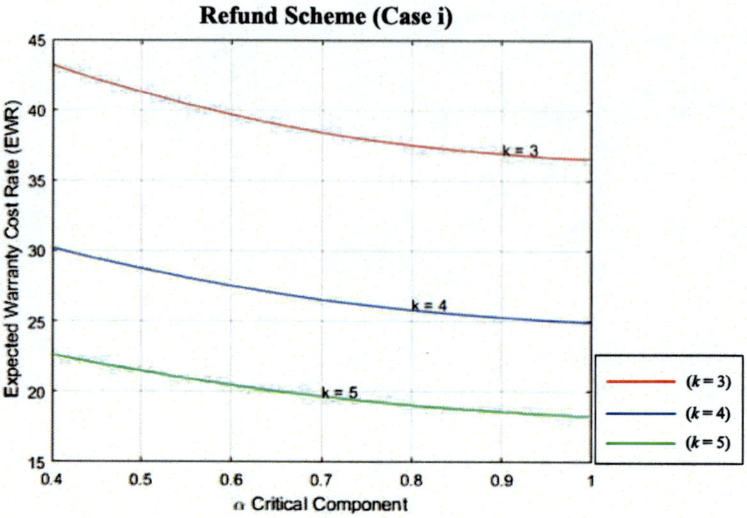

Fig. 6 Sensitivity analysis for refund scheme (Case i)

expected warranty cost rate (Note that the higher α_c and α_{nc} indicates better reliability of the components). Two cases are considered according to the Scenario 1 of numerical example (the NC component reliability is better)—i.e., Case (i): change the value of α_c (α_c = 0.4, …, 1.0) whilst α_{nc} is fixed (α_{nc} = 0.7), and Case (ii): change the value of α_{nc} (α_{nc} = 0.7, …, 1.3) whilst α_c is fixed (α_c = 0.4). The results are shown in Figs. 6 and 7.

As seen in Figs. 6 and 7, the EWR in refund schemes decreases if either the reliability of the C or NC component increases. Moreover, the EWR of Case (i) has a steeper slope than the EWR of Case (ii), and this pattern holds not only for refund scheme, but also replacement scheme. This pattern is due to the C component has higher repair and manufacturing costs. Note that the EWR for replacement scheme shows a similar pattern as that for refund scheme. As a result, a reliability improvement for the component with a higher repair cost will reduce the EWR significantly. In addition, the optimal lemon law period (W_L^*) will become longer as the reliability gets improved.

4.4 Managerial Implications

The lemon law regulation varies depending on the country or region. For example, in the USA, the lemon law period is 12 months, while in Singapore, the period is 6 months with a 12-month base warranty period. As the manufacturer must obey the standard regulation regarding the lemon law without any exception, the manufacturer should evaluate the optimal lemon period (W_L^*) for every product sold with

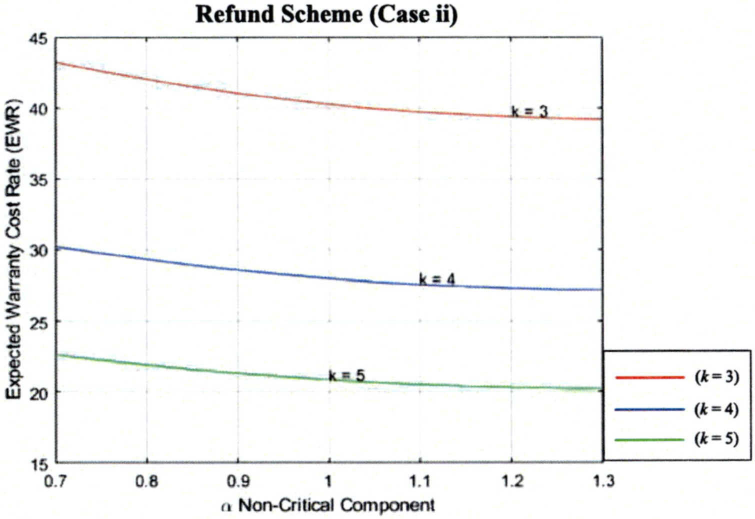

Fig. 7 Sensitivity analysis for refund scheme (Case ii)

a warranty. If W_L^* is greater than the lemon period stated in the regulation ($\tilde{W}_L$), then the reliability of the product is sufficient to meet the regulation and control the impact of the lemon law (i.e., keeping the warranty cost as low as possible). However, if it is not the case (i.e., W_L^* is less than $\tilde{W}_L$), then the warranty cost can be much larger as more lemon claims likely to happen. This issue, in turn, will erode the profit. Thus, the manufacturer must improve product reliability to cope with this situation. Additionally, for the policy allowing the customers to choose either a refund or replacement scheme, the manufacturer should determine a proper warranty cost model incorporating the customer's preference for the warranty service scheme offered. This topic becomes a potential topic for further research.

5 Conclusion

This study has developed two warranty cost models for a multi-component product (consisting of C and NC components) protected by the Lemon Laws. Two cases are considered—i.e., refund and replacement cases. Failure of the C (or NC) components are assumed to be dependent, or in other words, there is a failure interaction between the two types of components. This study would help the manufacturer to evaluate the warranty cost when the lemon law is enforceable during the warranty. Furthermore, to make this work more relevant to the actual condition, one can extend it to the case where the failure interaction happens in two ways instead of the one-way failure interaction assumed in this work. This model extension will allow the C component failure to induce a failure in the NC component. The other challenging research topics are (i) the lemon law warranty considering the customer's preference to choose a refund or replacement scheme, and (ii) the lemon condition considering the out-of-service time (downtime) of a product. The works on these topics are on the ways.

Acknowledgements This research is funded by the Ministry of Higher Education and Research Technology, Indonesia, with the contract number B/112/E3/RA.00/2021. The authors of this paper also would like to express our gratitude to those who have been involved in this work.

References

1. Kegley MB, Hiller JS (1986) 'Emerging' lemon car laws. Am Bus Law J 24(1):87–103. https://doi.org/10.1111/j.1744-1714.1986.tb00489.x
2. Husniah H, Hermawan K, Iskandar BP (2021) Cost analysis of warranty based on lemon law with multiple failures and total downtime. INOVASI (1):54–60. https://doi.org/10.29264/jinv.v17i1.9214.
3. Husniah H, Pasaribu US, Iskandar BP (2020) Warranty cost analysis for a multi-component product protected by lemon laws. IOP Conf Ser: Mater Sci Eng 1003(1). https://doi.org/10.1088/1757-899X/1003/1/012110
4. Liu B, Wu J, Xie M (2015) Cost analysis for multi-component system with failure interaction under renewing free-replacement warranty. Eur J Oper Res 243(3):874–882. https://doi.org/10.1016/j.ejor.2015.01.030
5. Sun Y, Ma L, Mathew J, Zhang S (2006) An analytical model for interactive failures. Reliab Eng Syst Saf 91(5):495–504. https://doi.org/10.1016/j.ress.2005.03.014
6. Murthy DNP, Nguyen DG (1985) Study of two-component system with failure interaction. Nav Res Logist Q 32(2):239–247. https://doi.org/10.1002/nav.3800320205
7. Murthy DNP, Nguyen DG (1985) Study of a multi-component system with a failure interaction. Eur J Oper Res 21:330–338
8. Iskandar BP, Murthy DNP, Jack N (2005) A new repair-replace strategy for items sold with a two-dimensional warranty. Comput Oper Res 32(3):669–682. https://doi.org/10.1016/j.cor.2003.08.011
9. Chien YH, Chen JA (2007) Optimal age-replacement policy for repairable products under renewing free-replacement warranty. Int J Syst Sci 38(9):759–769. https://doi.org/10.1080/00207720701599494
10. Iskandar BP, Jack N, Murthy DNP (2012) Two new servicing strategies for products sold with warranty. Asia-Pacific J Oper Res 29(3). https://doi.org/10.1142/S0217595912400222
11. Zhang N, Fouladirad M, Barros A (2017) Warranty cost analysis of a two-component system, with stochastic dependence. Math Methods Reliab [Online]. Available: https://hal.archives-ouvertes.fr/hal-01959831
12. Luo M, Wu S (2019) A comprehensive analysis of warranty claims and optimal policies. Eur J Oper Res 276(1):144–159. https://doi.org/10.1016/j.ejor.2018.12.034
13. Barlow RE, Proschan F (1996) Mathematical theory of reliability. Society of Industrial and Applied Mathematics, Philadelphia
14. Iskandar BP, Husniah H (2017) Optimal preventive maintenance for a two dimensional lease contract. Comput Ind Eng 113:693–703. https://doi.org/10.1016/j.cie.2017.09.028

Literature Review of Lean Manufacturing, Six Sigma and Lean Six Sigma Benefit

M. A. K. Tengtarto, M. L. Singgih, N. Siswanto, and P. Suwignjo

Abstract Researchers conducted a literature review on lean manufacturing, six sigma and lean six sigma to increase profits and productivity. The papers reviewed were papers published with a time frame of 1990 to 2022, selected from Emerald, ProQuest, and Wiley. The amount of publications search resulting about 1937 articles which were then filtered into 66 papers is not very extensive and it is somewhat dispersed. Thus, the main objective of this article is to collect and perform an analysis of the most recent literature about lean six sigma. Process improvement, defects & costs reduction have been the focus in several studies. On the other hand, quality improvement is not being researched quite intensively. The implementation of lean six sigma on food and construction sectors have been carried out, however there are still few on textiles and then automotive. Several studies show that there is an increase in waste management such as reducing defects, reducing machine breakdowns and eliminating root causes of problems. The success factors of lean six sigma are influenced by leadership, education, human resource management, and also supplier partnerships. However, the government does not have a big influence on the success of the implementation of lean six sigma.

Keywords Lean manufacturing · Six sigma · Lean six sigma · Profit improvement

1 Introduction

Manufacturing strategy plays an essential role in business success [1]. It is used as a competitive tool in volatile markets [2]. A manufacturing strategy is a set of coordinated action plans and policies to secure the desired long-term results [3]. The strategy is in line with the overall strategic direction of the business and provides a

M. A. K. Tengtarto (✉) · M. L. Singgih · N. Siswanto · P. Suwignjo
Department of Industrial and Systems Engineering, Institut Teknologi Sepuluh Nopember, Keputih, Kec. Sukolilo, Kota Surabaya, Jawa Timur 60111, Indonesia
e-mail: maxkredo@gmail.com

C. N. Rosyidi et al. (eds.), *Proceedings of the 6th Asia Pacific Conference on Manufacturing Systems and 4th International Manufacturing Engineering Conference*, Lecture Notes in Mechanical Engineering,
https://doi.org/10.1007/978-981-99-1245-2_44

competitive advantage [2]. Manufacturing strategies drive continuous improvement and enable organizations to meet various business, market, and social needs [4]. Thus, selecting the right manufacturing strategy at the right time and in the right direction has a vital role in business success [5]. One such strategy is "Lean Six Sigma". Lean six sigma is one of the most well-known methodologies to improve process efficiency and reduce waste [6]. Lean six sigma is an integration of 2 principles, namely lean manufacturing, and six sigma. An integrated lean six sigma approach can help improve process efficiency, customer satisfaction, resource optimization, increase profit and cost reduction [6].

By using literature survey, the purpose of this research is to identify related clauses or items and various drivers to have a successful implementation of lean six sigma. Following that, individual or reciprocal interactions of the driver can achieve desired results.. Therefore, management, organizations, stakeholders and governments can influence the implementation of any manufacturing policy and strategy by altering the driving forces. It is also crucial for companies to have self-assessment of current condition, such as strengths and weakness to formulate a successful plan for strategic approach in implementing six sigma. Goals and priorities must be set, and strategic planning, practices, enabling factors and proactive actions must be taken to meet the driving requirements, and consequently achieving the desired performance results and competitive advantage. The researcher conducted a literature review to systematically collect and critically analyze research in lean six sigma.

2 Contextualization of This Review

2.1 Lean Manufacturing

Lean manufacturing, firstly introduced by a Japanese automotive company Toyota Production Systems (TPS), aims to improve quality and reduce costs by eliminating non-value-adding activities and waste. What is considered waste is any activities besides the minimum number of materials, equipments, parts, processes, space, time and skills, to add value to the product [7]. Value flow mapping is used to identify waste. The success factors for implementing lean manufacturing are management style, availability of funds, leadership, organizational skills, organizational cultures and expertise [8]. It is important for an organization to develop the critical success factors mentioned to implement lean successfully [9]. Inappropriate lean implementation methods can disrupt the process and organization may not achieve the desired improvements [10].

2.2 Six Sigma

Six sigma is a data-based methodology to eliminate defects by controlling the process output at six-sigma quality level, which is towards six standard deviations between the mean and the specification limit [11]. Six sigma are used to develop world class quality products [12]. DMAIC (Define, Measure, Analyze, Improve, Control) and DMADV (Define, Measure, Analyze, Design and Verify) approaches are tools for six sigma quality levels [13]. DMAIC process is used for problem solving in a structured and systematic approach manner. The main goal of Six Sigma is to control process variation and reduce defects [14] as process variations have negative impacts on environmental performance.

DMAIC improves existing processes, whereas DMADV designs new six sigma processes [15]. DMAIC is a data-driven, consistency-based method for enhancing processes. DMAIC is typically used to guide Six Sigma implementation and attain business objectives. From a practitioner's point of view, problem-solving projects are carried out using quality tools and statistics derived from Total Quality Management (TQM) [16].

2.3 Lean Six Sigma

According to Byrne [17], the term "Lean Six Sigma" was used in the late 1990s and early 2000s to describe the combination of Lean Manufacturing and Six Sigma philosophies. This integration aims to overcome the shortcomings of both. By combining the two methodologies and utilizing the benefits, organizations may increase their improvement potential even further [18]. Lean Six Sigma is defined as business strategies and methodologies that improve process performance and develop customer satisfaction, leadership, and results by increasing quality, speed and cost [19]. Tools and techniques of both lean manufacturing and Six Sigma are utilized. The success of Lean Six Sigma has led organizations worldwide to adopt it to manage their operations and become more competitive, and thus making Lean Six Sigma as a popular hybrid continuous improvement methodologies.

Lean Six Sigma (LSS) combines two vital approaches: Lean and Six Sigma. Lean manufacturing validates minimizing waste by eliminating non-value-added activities, but the individual approach does not solve the problem statistically [20]. As Lean Manufacturing minimize all kinds of waste, e.g., useless movement and different processes inventory [21]. Six Sigma makes it possible to reduce defects and provide quality products and services [22]. It is imperative to assess and select projects suitable for continuous improvement [23]. Six Sigma reduces the variation of the process and consists of tools such as DMAIC (Define, Measure, Analyze, Improve, and Control). Six Sigma also consists of Statistic Process Control, DOE experimental histogram design, and Pareto graph [24]. Lean Six Sigma is helpful to set up proper and structured improvement procedures with operational capacities and approaches to reduce variation in organizational results [19].

3 Method

The systematic literature review method describes the Lean six sigma profit enhancement concept. The review process is done by identifying several selected articles. Identification of papers and selection process in this literature study, using the keywords "Lean manufacturing", "Six sigma", "Lean six sigma", "Productivity", "Profit Improvement", and "Profit enhancement" in 3 journal databases: Emerald, Wiley, ProQuest. The selected papers are from the period 1990 to 2022. Figure 1 shows an overview of the screening process for the journal.

At this stage, the 1937 papers were obtained. The next stage is filtering the selection of the journals by filtering papers without the words "Lean manufacturing", "Six sigma", "Lean six sigma", "Productivity", "Profit Improvement", and "Profit enhancement" in the title and abstract and eliminating the same papers. The remaining articles are 66 papers. Then it will be carried out on the selected paper content by identifying the definition of lean six sigma, the concept of lean six sigma, the benefits

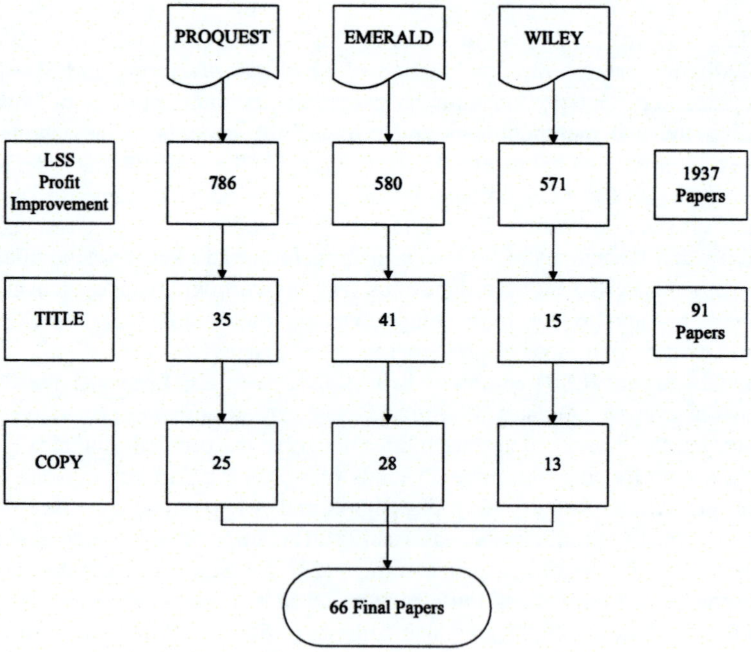

Fig. 1 Paper selection stages

of lean six sigma, and then the development that will be carried out. The results of the review will be analyzed and mapped.

4 Results and Discussion

4.1 Lean Six Sigma Benefits

Lean Six Sigma is also a business strategy and operations that can benefit the company. Table 1 show the benefit of using Lean Six Sigma. Various studies have shown that implementing Lean/Six Sigma may effectively have a positive effect on firm performance [25]. One of the benefits of Lean Six Sigma is increased productivity as a result of reduced costs, is a result of reduced lead times, and a consequence of reducing waste. Another practical benefit is the increased machine availability and processability, which is also subject to breakdown time and defect reduction. This benefit cannot be the primary driver because the measure of Overall Equipment Efficiency (OEE) can only be felt when the initiative is implemented [26]. Upadhye et al. [27] also presents detailed analysis of how strategic choices for implemented initiatives yield multiple benefits.

Lean six sigma offers significant profit potential with relatively low investment [41]. Based on the literature review conducted, most common studies have focused on process improvement, reducing defects and costs [25], but not many studies have focused on quality improvement.

Table 1 Lean six sigma benefit

Benefit LSS	References
Waste reduction	Bergmiller and McCright [25], Dakov and Novkov [28], Hajmohammad et al. [29], Moreira et al. [30]
Cost reduction	Bergmiller and McCright [25], Mollenkopf et al. [31], Tice et al. [32], Zhu and Sarkis [33], Browning and Health [34]
Quality improvement	Dakov and Novkov [28], Ng et al. [35], Tice et al. [32]
Productivity improvement	Bergmiller and McCright [25], Dakov and Novkov [28], Langenwalter [36], Ng et al. [35], Improta et al. [37], Bajaj et al. [38], Gholami et al. [39]
Customer satisfaction	Kidwell [40], Langenwalter [36], Ng et al. [35], Tice et al. [32]

Table 2 Lean six sigma scope

Scope	References
Food	Masooud et al. [42], Costa et al. [43], Dora et al. [44], Bigliardi et al. [45]
Textile	Maia et al. [46]
Construction	Banawi et al. [47], Li et al. [48], Sarhan et al. [49]
Automotive	Dayna [50], Azevedo et al. [51]

4.2 Lean Six Sigma Scope

Lean Six Sigma can be implemented in many sectors. Table 2 shows the Lean Six Sigma research field that implemented.

From the literature review conducted, several studies in the food and construction sectors have been carried out and there are still few on textiles and automotive.

4.3 Lean Six Sigma Tools

According to Costa et al. [52], the most widely used tools and methods are lean tools, such as Value Stream Mapping, 5S, Kanban and Kaizen, and they are considered universally applicable. This review is explained because the tool is simple and does not contain statistical equations or formulas [41]. Table 3 shows lean six sigma tools.

Lean Six Sigma tools offer benefits in improving process, better equipment efficiency and waste reduction. Several studies using Lean Six Sigma tools show that there is an increase in waste management such as reducing defects, reducing machine breakdowns and eliminating root causes of problems.

4.4 Critical Success Factors Lean Six Sigma

The success of Lean Six Sigma is not entirely tied to applying the right tools and methods but on knowing the critical success factors and obstacles that must be overcome. Critical success factors are the elements required for an organization or project to achieve its mission [61]. In addition, as changes will create forces or barriers of resistance, therefore in order for the organization to be sustainable, those barriers need to be identified and understood [62]. This knowledge is fundamental when sectors with lower levels of lean six sigma implementation are analyzed. Table 4 shows about critical success factors to implement lean six sigma.

According to the studies reviewed, the success factors of Lean Six Sigma are influenced by leadership, education, human resource management, and also supplier partnerships. However, the government does not have a big influence on the success of the implementation of Lean Six Sigma.

Table 3 Lean six sigma tools

Tools	Benefit	References
Value stream mapping	• Used as a life cycle analysis • Improve ergonomics, health, and safety of workers	Langenwalter [36], Park and Linich [53], Gholami et al. [39], Vinodh et al. [54], Chiarini [16], Ng et al. [35]
5s	• The work environment is clean and organized • Improve waste management	Langenwalter [36]; Vinodh et al. [54]; Chiarini [16]
Kaizen	• Develop a culture of problem-solving • Aid in the reduction of defective materials and pollution	Maxwell et al. [55], Vais et al. [56], Zhang and Awasthi [57], Vinodh et al. [54]
Kanban	• Minimize potential waste from damaged, defective, or deteriorated products • Maximize workshop space utilization	Kováčová [58], Ng et al. [35], Sobral et al. [59], Vinodh et al. [54]
SMED	• Reduced equipment setup time • Reduce some of the machine's environmental impact	Chiarini [16], Kováčová [58], Moreira et al. [30], Ng et al. [35]
DMAIC	• Elimination of the problem's root cause • Assist in the integration of human and process factors	Bajaj et al. [38], Vinodh et al. [54], Albliwi et al. [41], Tjahjono et al. [60], Gholami et al. [39]
TPM	• Increases equipment life • Reduce the rate of damage, production loss, and material consumption	Chiarini [16], Sobral et al. [59], Vais et al. [56], Vinodh et al. [54]

Table 4 CSF to implement lean six sigma

Critical success factors	References
Government	Park and Linich [53]
Leadership	Vinodh et al. [54], Bergmiller and McCright [25], Park and Linich [53]
Human resource management	Langenwalter [36], Zhang [63], Wong and Wong [64]
Supplier partnership	Galeazzo et al. [65], Dayna and Damien [50], Vachon and Klassen [66]
Education	Vinodh et al. [54], Sobral et al. [59], Park and Linich [53]

5 Conclusion and Future Research

In this study, researchers examined the existing literature on lean, six sigma, and lean six sigma to develop an integrated model. From 1990 to 2022, 66 publications were identified by researchers. This study also seeks to promote research in lean, six sigma, and lean six sigma, as well as its future course. Researchers identify the advantages of Lean Six Sigma tools and research that employs them. Researchers also identified the scope and critical success factor implementation of Lean Six Sigma. However, there are few papers for Lean Six Sigma with textile scope.

Based on literature review, Lean Six Sigma promotes benefits in waste reduction, cost reduction, quality improvement, productivity improvement, and customer satisfaction. Lean Six Sigma can be implemented sequentially or concurrently. Several studies have focused on process improvement and also reducing defects and costs, but not many studies have focused on quality improvement. From the literature review conducted, several studies in the food and construction sectors have been carried out but there are still few studies on textiles and automotive sectors. Several studies show that there is an increase in waste management such as reducing defects, reducing machine breakdowns and eliminating root causes of problems. The success factors of lean six sigma are influenced by leadership, education, human resource management, and also supplier partnerships. however, the government does not have a big influence on the success of the implementation of Lean Six Sigma. Future research could use this paper as insights for case study and survey to explore the field. More investigation about Lean Six Sigma practices implementation characteristics and barriers needed to provide practitioners with considering the sector context.

References

1. Chikán A, Czakó E, Kiss-Dobronyi B, Losonci D (2022) Firm competitiveness: a general model and a manufacturing application. Int J Prod Econ 243
2. Wang J, Cao DB (2008) Relationships between two approaches for planning manufacturing strategy: a strategic approach and a paradigmatic approach. Int J Prod Econ 115(2):349–361
3. Papke-Shields KE, Malhotra MK (2001) Assessing the impact of the manufacturing executive's role on business performance through strategic alignment. J Oper Manag 19(1):5–22
4. Saptioratri Budiono HD, Nurcahyo R, Habiburrahman M (2021) Relationship between manufacturing complexity, strategy, and performance of manufacturing industries in Indonesia. Heliyon 7(6):e07225
5. Sun H, Hong C (2002) The alignment between manufacturing and business strategies: its influence on business performance. Technovation 22(11):699–705
6. Cherrafi A, Elfezazi S, Chiarini A, Mokhlis A, Benhida K (2016) The integration of lean manufacturing, six sigma and sustainability: a literature review and future research directions for developing a specific model. J Clean Prod 139:828–846
7. Roberta R, Taylor B (2000) Operations management. Terc. New Jersey, Estados Unidos América
8. Achanga P, Shehab E, Roy R, Nelder G (2006) Critical success factors for lean implementation within SMEs. J Manuf Technol Manag 17(4):460–471
9. Al-Najem M, Dhakal H, Bennett N (2012) The role of culture and leadership in lean transformation: a review and assessment model. Int J Lean Think 3(1):119–138
10. Karim A, Arif-Uz-Zaman K (2013) A methodology for effective implementation of lean strategies and its performance evaluation in manufacturing organizations. Bus Process Manag J 19(1):169–196
11. Linderman K, Schroeder RG, Zaheer S, Choo AS (2003) Six sigma: a goal-theoretic perspective. J Oper Manag 21(2):193–203

12. Hariyani D, Mishra S (2022) Drivers for the adoption of integrated sustainable green lean six sigma agile manufacturing system (ISGLSAMS) and research directions. Clean Eng Technol 7
13. Breyfogle FW III (2003) Implementing six sigma: smarter solutions using statistical methods. Wiley
14. Pyzdek T, Keller P (2014) Six sigma handbook. McGraw-Hill Education
15. de Mast J, Lokkerbol J (2012) An analysis of the six sigma DMAIC method from the perspective of problem solving. Int J Prod Econ 139(2):604–614
16. Chiarini A (2014) Sustainable manufacturing-greening processes using specific lean production tools: an empirical observation from European motorcycle component manufacturers. J Clean Prod 85:226–233
17. Byrne G, Lubowe D, Blitz A (2007) Using a lean six sigma approach to drive innovation. Strateg Leadersh 35(2):5–10
18. Bhuiyan N, Baghel A (2005) An overview of continuous improvement: from the past to the present. Manag Decis 43(5):761–771
19. Snee RD (2010) Lean six sigma—getting better all the time. Int J Lean Six Sigma 1(1):9–29
20. Saieg P, Sotelino ED, Nascimento D, Caiado RGG (2018) Interactions of building information modeling, lean and sustainability on the architectural, engineering and construction industry: a systematic review. J Clean Prod 174:788–806
21. Ansah RH, Sorooshian S (2017) Effect of lean tools to control external environment risks of construction projects. Sustain Cities Soc 32:348–356
22. Schroeder RG, Linderman K, Liedtke C, Choo AS (2008) Six sigma: definition and underlying theory⋆. J Oper Manag 26(4):536–554
23. Zu X, Fredendall LD, Douglas TJ (2008) The evolving theory of quality management: the role of six sigma. J Oper Manag 26(5):630–650
24. Magnusson K, Kroslid D, Bergman B, Häyhänen P, Mills D (2003) Six sigma: the pragmatic approach. Studentlitteratur
25. Bergmiller GG, McCright PR (2009) Are lean and green programs synergistic. In: Proceedings of the 2009 industrial engineering research conference, 2009, pp 1155–1160
26. Dudbridge M (2011) Handbook of lean manufacturing in the food industry. Wiley
27. Upadhye N, Deshmukh SG, Garg S (2010) Lean manufacturing in biscuit manufacturing plant: a case. Int J Adv Oper Manag 2(1/2):108
28. Dakov I, Novkov S (2007) Assessment of the lean production effect on the sustainable industrial enterprise development. Verslas Teor ir Prakt 8(4):183–188
29. Hajmohammad S, Vachon S, Klassen RD, Gavronski I (2013) Lean management and supply management: their role in green practices and performance. J Clean Prod 39:312–320
30. Moreira F, Alves AC, Sousa RM (2010) Towards eco-efficient lean production systems. In: International conference on information technology for balanced automation systems, 2010, pp 100–108
31. Mollenkopf D, Stolze H, Tate WL, Ueltschy M (2010) Green, lean, and global supply chains. Int J Phys Distrib Logist Manag 40(1/2):14–41
32. Tice J, Ahouse L, Larson T (2005) Lean production and EMSs: aligning environmental management with business priorities. Environ Qual Manag 15(2):1–12
33. Zhu Q, Sarkis J (2004) Relationships between operational practices and performance among early adopters of green supply chain management practices in Chinese manufacturing enterprises. J Oper Manag 22(3):265–289
34. Browning TR, Heath RD (2009) Reconceptualizing the effects of lean on production costs with evidence from the F-22 program. J Oper Manag 27(1):23–44
35. Ng R, Low JSC, Song B (2015) Integrating and implementing lean and green practices based on proposition of carbon-value efficiency metric. J Clean Prod 95:242–255
36. Langenwalter G (2006) "Life" is our ultimate customer: from lean to sustainability. Target 22(1):5–15
37. Improta G, Cesarelli M, Montuori P, Santillo LC, Triassi M (2018) Reducing the risk of healthcare-associated infections through lean six sigma: the case of the medicine areas at the Federico II University Hospital in Naples (Italy). J Eval Clin Pract 24(2):338–346

38. Bajaj P et al (2021) Coupled effect of electronic medical record modifications and lean six sigma methodology on Rheumatoid Arthritis disease activity measurement and treat-to-target outcomes. ACR Open Rheumatol 3(3):164–172
39. Gholami H, Jamil N, Mat Saman MZ, Streimikiene D, Sharif S, Zakuan N (2021) The application of green lean six sigma. Bus Strateg Environ 30(4):1913–1931
40. Kidwell M (2006) Lean manufacturing and the environment. Target 22(6):13–18
41. Albliwi SA, Antony J, Halim Lim SA (2015) A systematic review of lean six sigma for the manufacturing industry. Bus Process Manag J 21(3):665–691
42. Massoud MA, Fayad R, El-Fadel M, Kamleh R (2010) Drivers, barriers and incentives to implementing environmental management systems in the food industry: a case of Lebanon. J Clean Prod 18(3):200–209
43. Costa LBM, Godinho Filho M, Fredendall LD, Devós Ganga GM (2021) Lean six sigma in the food industry: construct development and measurement validation. Int J Prod Econ 231
44. Dora M, Kumar M, Van Goubergen D, Molnar A, Gellynck X (2013) Operational performance and critical success factors of lean manufacturing in European food processing SMEs. Trends Food Sci Technol 31(2):156–164
45. Bigliardi B, Bottani E, Montanari R, Vignali G (2011) Successful new product development in the food packaging industry: evidence from a case study. Int J Eng Sci Technol 2(9):13–24
46. Maia L, Alves A, Leao C (2013) Sustainable work environment with lean production in textile and clothing industry. Int J Ind Eng Manag 4:183–190
47. Banawi A, Bilec MM (2014) A framework to improve construction processes: integrating lean, green and six sigma. Int J Constr Manag 14(1):45–55
48. Li X, Shen GQ, Wu P, Fan H, Wu H, Teng Y (2018) RBL-PHP: simulation of lean construction and information technologies for prefabrication housing production. J Manag Eng 34(2):4017053
49. Sarhan J, Xia B, Fawzia S, Karim A, Olanipekun A (2018) Barriers to implementing lean construction practices in the Kingdom of Saudi Arabia (KSA) construction industry. Constr Innov 18(2):246–272
50. Simpson DF, Power DJ (2005) Use the supply relationship to develop lean and green suppliers. Supply Chain Manag An Int J
51. Azevedo SG, Govindan K, Carvalho H, Cruz-Machado V (2012) An integrated model to assess the leanness and agility of the automotive industry. Resour Conserv Recycl 66:85–94
52. Costa LBM, Godinho Filho M, Fredendall LD, Gómez Paredes FJ (2018) Lean, six sigma and lean six sigma in the food industry: a systematic literature review. Trends Food Sci Technol 82:122–133
53. Park C, Linich D (2008) Green lean six sigma: using lean to help drive results in the wholly sustainable enterprise. Deloitte Consult, New York
54. Vinodh S, Arvind KR, Somanaathan M (2011) Tools and techniques for enabling sustainability through lean initiatives. Clean Technol Environ Policy 13(3):469–479
55. Maxwell J, Briscoe F, Schenk B, Rothenberg S (1998) Case study: Honda of America Manufacturing Inc.: can lean production practices increase environmental performance? Environ Qual Manag 8(1):53–61
56. Vais A, Miron V, Pedersen M, Folke J (2006) 'Lean and green' at a Romanian secondary tissue paper and board mill—putting theory into practice. Resour Conserv Recycl 46(1):44–74
57. Zhang M, Awasthi A (2014) Using six sigma to achieve sustainable manufacturing. In: Proceedings of the 2014 international conference on innovative design and manufacturing (ICIDM), 2014, pp 311–317
58. Kováčová Ľ (2013) The integration of lean management and sustainability. Transf inovácií 26:195–199
59. Sobral MC, de Sousa Jabbour ABL, Chiappetta Jabbour CJ (2013) Green benefits from adopting lean manufacturing: a case study from the automotive sector. Environ Qual Manag 22(3):65–72
60. Tjahjono B et al (2010) Six sigma: a literature review. Int J Lean Six Sigma 1(3):216–233
61. Ribeiro ATVB, Uechi JN, Plonski GA (2018) Building builders: entrepreneurship education from an ecosystem perspective at MIT. Triple Helix 5(1)

62. Jadhav JR, Mantha SS, Rane SB (2014) Exploring barriers in lean implementation. Int J Lean Six Sigma 5(2):122–148
63. Zhang M (2014) Using six sigma to achieve sustainable manufacturing: a case study in Aviation Company. Concordia University
64. Wong WP, Wong KY (2014) Synergizing an ecosphere of lean for sustainable operations. J Clean Prod 85:51–66
65. Galeazzo A, Furlan A, Vinelli A (2014) Lean and green in action: interdependencies and performance of pollution prevention projects. J Clean Prod 85:191–200
66. Vachon S, Klassen RD (2006) Extending green practices across the supply chain. Int J Oper Prod Manag 26(7):795–821

Design Process of Whippletree-Like Mechanism for Versatile Gripping in Arm Prosthesis

Wildan Trusaji, Dradjad Irianto, Ardianto Satriawan, Casi Setianingsih, Erza Nurtriandari, and Rachmat Zulkarnain Goesasi

Abstract One of the biggest challenges in arm prosthesis design is to resemble normal hand functionality.Some arm prostheses try to mimic all hand movements, while others try to produce several main hand grip patterns. We design a Versatile Gripping Technology (VGT) from a basic whippletree mechanism to produce the main hand grip pattern in an arm prosthesis. VGT development is the result of abstraction, negation, and systemic variation processes of existing solutions. To validate VGT, we test it to produce six patterns of hand grip from the Southampton Hand Assessment Procedure. VGT is capable to produce lateral, power, tip, extension, and spherical grip without changing the setting or mode of operation. In addition, VGT includes the thumb movement and it only uses one simple-shaped moving part. Thus, VGT is effective and efficient. With this result, we argue that the abstraction, negation, and systemic variation process of existing solutions can create a novel solution and it applies to another problem or domain.

Keywords Arm prosthesis · Versatile gripping · Design process

---

W. Trusaji (✉) · D. Irianto
Faculty of Industrial Technology, Institut Teknologi Bandung, Bandung, Indonesia
e-mail: trusaji@itb.ac.id

A. Satriawan
School of Electrical Engineering and Informatics, Institut Teknologi Bandung, Bandung, Indonesia

C. Setianingsih
School of Electrical Engineering, Telkom University, Bandung, Indonesia

E. Nurtriandari · R. Z. Goesasi
Department of Physical Medicine and Rehabilitation, Faculty of Medicine, Universitas Padjadjaran-Dr. Hasan Sadikin General Hospital, Bandung, Indonesia

C. N. Rosyidi et al. (eds.), *Proceedings of the 6th Asia Pacific Conference on Manufacturing Systems and 4th International Manufacturing Engineering Conference*, Lecture Notes in Mechanical Engineering,
https://doi.org/10.1007/978-981-99-1245-2_45

1 Introduction

The hand mechanism is intricate and complex [1]. A normal hand and wrist can make seventeen types of movement and at least six patterns of basic grip [2, 3]. Thus, the challenge of arm prosthesis design is not only substituting the structural arrangement but also designing an intuitive control to achieve functionality.

There are two main approaches to how to design functional arm prostheses. The first is to mimic all hand movements, and the second is to emulate a variety of hand grips. Arm prostheses using the first approach usually is electronically controlled. It has multiple actuators and sensors. Hence, it is complex and unaffordable. Some design uses the second approach, and their main goal is to help users manipulate daily objects by producing several basic gripping patterns. Most prostheses embracing this approach are mechanically controlled by using a whippletree mechanism.

This paper explains the design process of a novel whippletree mechanism that includes the thumb, has a minimum part, and is capable of versatile gripping. We called it the Versatile Gripping Technology (VGT). This paper emphasizes the design process since it is where the insight and the tacit knowledge reside [4]. This paper consists of 5 sections: (1) introduction, (2) state of the art, (3) design process, and (4) preliminary result & (5) conclusion.

2 State of the Art

2.1 Whippletree Mechanism in Everyday Life

Whippletree is a mechanism to distribute force equally through linkage. The main element of the whippletree mechanism is a pivotable bar at the center, where the force is applied from one direction at the center and the opposite direction at the tip. Figure 1 illustrates the basic element of Whippletree mechanisms.

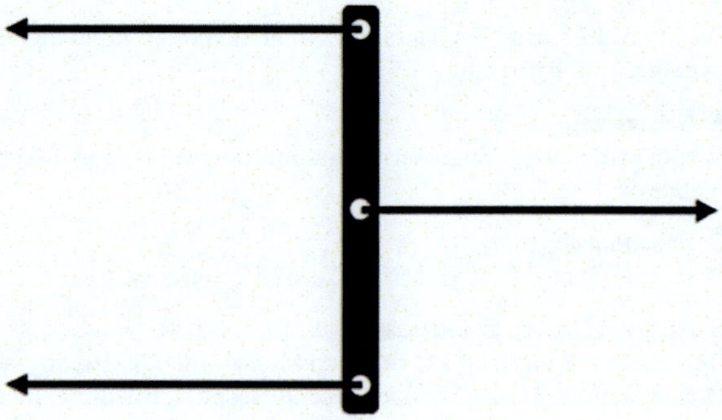

Fig. 1 Basic elements of the whippletree mechanism

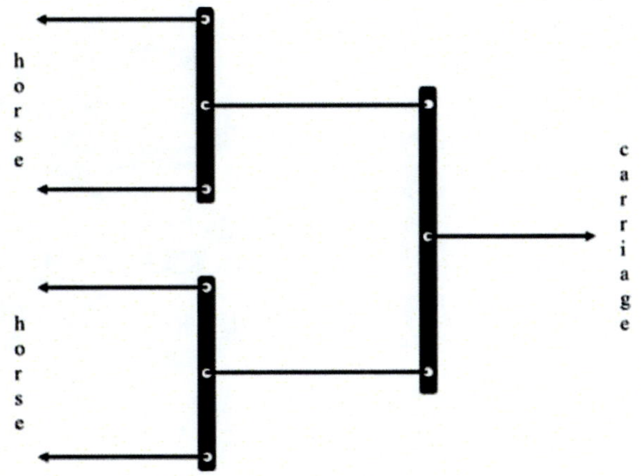

Fig. 2 Doubletree arrangement in a two-horse team

Several whippletree mechanisms can be used simultaneously by arranging bars in series. The most common of this whippletree mechanism example is a two-horse team for pulling a carriage. In the two-horses team, 3 three bars are used to distribute a force between two horses so they can share the work evenly. This arrangement is sometimes known as doubletree. Figure 2 illustrates the schematic of the doubletree arrangement.

2.2 *Whippletree Mechanism in Arm Prosthesis*

Fine-tuning every finger movement in an arm prosthesis to be capable of gripping varieties of an object is a tremendous task. Whippletree mechanism may become advantageous to solve this challenge. Instead of controlling every finger with multiple actuators, the whippletree mechanism distributes force evenly from one actuator to all fingers. Thus, the movement of fingers is mechanically controlled according to the shape of the object. This feature is achieved by regulating the force transmission into each finger based on the amount of resistance given by the object to each finger.

Groenewegen's Hand Prosthesis [5] uses a whippletree mechanism to achieve an adaptive gripping function. Instead of using bar and line, Groenewegen uses seven bars, three triangle-shaped bars, and nine pins. Those bars and pins are arranged like a double-tree arrangement. The design schematic is illustrated in Fig. 3.

Makerhand's arm prosthesis has a unique approach. It has only four fingers including the thumb [6]. The four fingers configuration is chosen since it is the most basic whippletree arrangement, which consists of only one bar and four lines including the actuator line. The schematic of Makerhand's whippletree mechanism is illustrated in Fig. 4. It is worth noting that Makerhand arm prosthesis is the winner of the CYBATHLON arm prosthesis race in 2020.

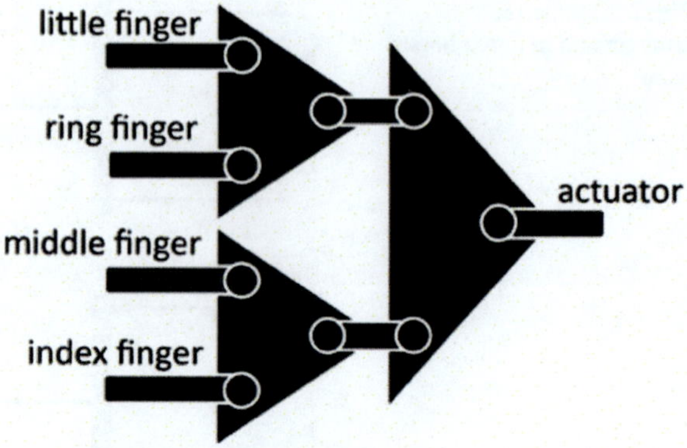

Fig. 3 Groenewegen's whippletree

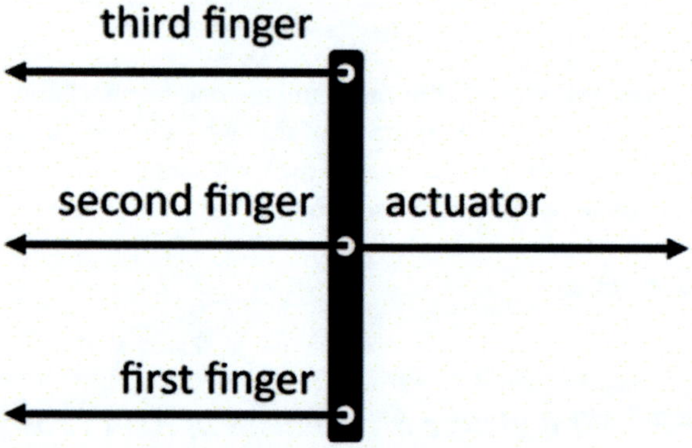

Fig. 4 Makerhand's whippletree

3 Design Process

The challenge is how to include the thumb in the mechanism since the thumb play important role in grasping [7]. To achieve that, we use three generally applicable methods from the fundamental of the Systemic Approach [8]. Those are abstraction, method of negation, and systemic variation. This section explains how we use it.

Abstraction is a cognitive process to find higher interrelationships to find an essential characteristic of a system/object [8]. In this case, we search for an underlying principle in varieties of whippletree mechanisms. We concluded that the main characteristic of the mechanism is multiple force transmissions coupled to each other by moving/pivoting objects. By this abstraction, the mechanism's main part is coupled force transmission and a moving object as a link between force transmission.

In the method of negation, a known solution is questioned, negated, or even inversed [8]. This process help designer finds alternative solutions by avoiding design fixation. Groenewegen negates the shape of the basic bar and comes with a triangle-shaped bar. Makerhand negates the amount of coupled force transmission in one

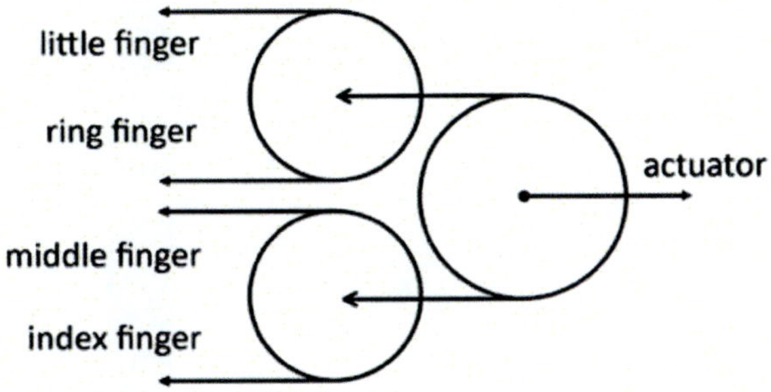

Fig. 5 Whippletree with multiple pulleys

bar. The bar in the basic whippletree mechanism is only coupled into two force transmissions while the bar in Makerhand is coupled into three force transmissions.

We negate several principles to develop our whippletree mechanism. In the basic whippletree mechanism, the line is relatively static, and the bar is relatively dynamic. What if we swap it so the line now is moving while the bar is standing still? Based on this negation we came up with including pulleys into the mechanism. The result is illustrated in Fig. 5. We feel another improvement can be made since introducing the pulley make the mechanism has more part. Thus, we change the pulley into a pin. This pin function is similar to the pulley, which is to guide the line movement. In another word, the pin acts like a tube. Figure 6 illustrates the pin base whippletree mechanism for arm prosthesis. The pin-based whippletree has fewer parts than Goenewagen and Makerhand whippletree since it only needs one bar to distribute force into four fingers.

Pin-based whippletree arrangement still not met the inclusion of thumb. To solve it, we do the method of negation again. We realize all the existing arrangements only used 2-dimensional space for their movement. The question is what if we exploit 3-dimensional space. Now we have forward, backward, yaw, and pitch movement. The thumb can be included in pin-base whippletree without adding a new bar and its movement is controlled by the pitch movement of the bar. This design schematic is in Fig. 7.

In the systemic variation, we create a general classification of solutions and generate possible solutions for each classification. We find that there are five general classifications in the whippletree mechanism; the shape of the balancer, means of force transmission, movement of linkage, dimensional space, and arrangement. All five general classifications and all possible solutions are listed in Table 1.

The possible solution in each general classification can be combined into a complete mechanism. We choose a doubletree arrangement with a triangle-shaped bar, line as force transmission, pivoting bar in the first level and pin as a guide in the second level, and 3D movement. The design result is in Fig. 8 and the detail of the triangle-shaped balancer is in Fig. 9. We call this novel mechanism Versatile Gripping Technology. This technology is patent pending in Indonesian Paten Office [9].

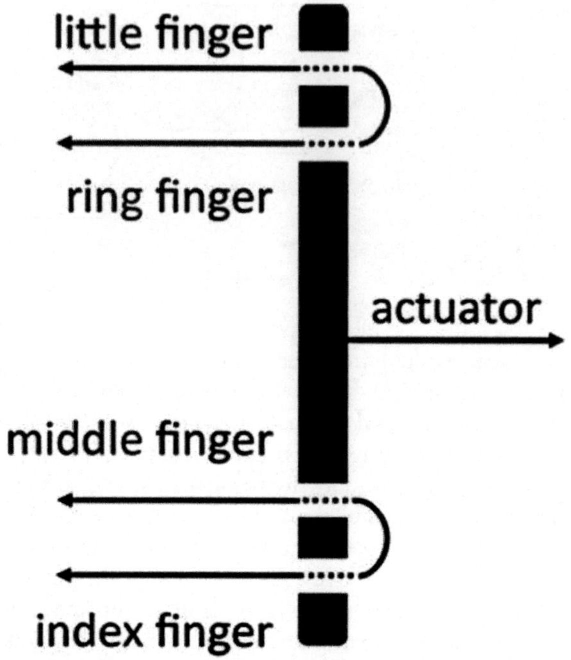

Fig. 6 Whippletree with pin

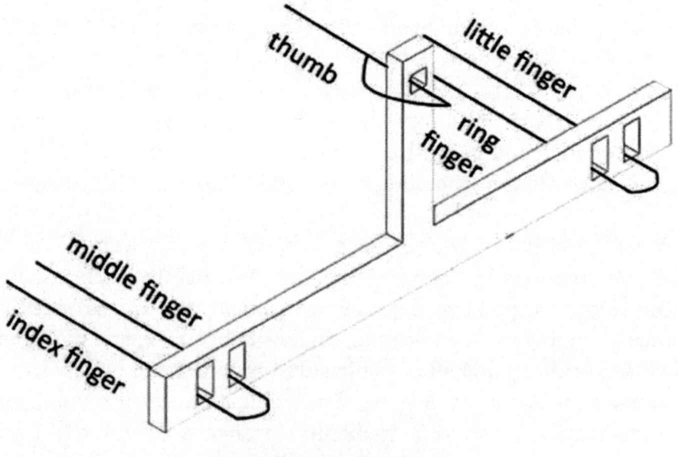

Fig. 7 Schematic of 3D pin-based whippletree mechanism

Table 1 Result of systemic variation

General classification	Possible solution
Shape of the balancer	Bar, triangle
Mean of force transmission	Line, bar and pin
Movement of linkage	Pivoting bar, pulley, pin as guide
Dimensional space	2D, 3D
Arrangement	Singletree, doubletree

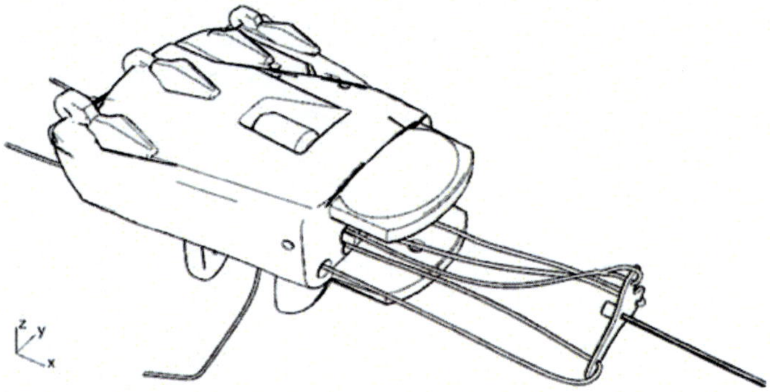

Fig. 8 Versatile gripping technology [9]

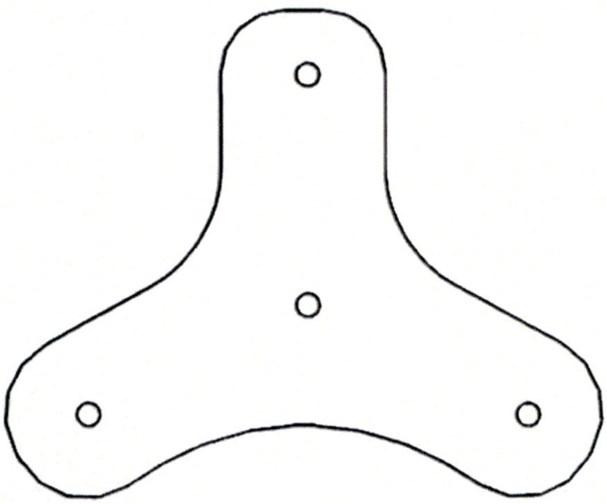

Fig. 9 The triangle shape balancer [9]

4 Preliminary Result

To validate the Versatile Gripping Technology (VGT) we adapt the Southampton Hand Assessment Procedure (SHAP) [3]. We only test the main six grips of SHAP consisting of lateral, power, tripod, tip, extension, and spherical grip with an everyday object. The VGT can produce 5 grips, but VGT cannot produce a tripod grip. However, we argue that the tripod grip can be substituted by the tip grip. The result for lateral, power, tip, extension, and spherical grip are illustrated in Figs. 10, 11, 12, 13 and 14, respectively. In addition to that, we compete in the CYBATHLON challenge in May 2022. In this competition, our pilot with the VGT-based arm prosthesis needs to transfer 8 small objects from one desk to another desk. Objects are marble, lego block, credit card, key ring, keys, poker chip, pen, and USB-C. He accomplished the task in under 3 min.

Fig. 10 Lateral grip

Fig. 11 Power grip

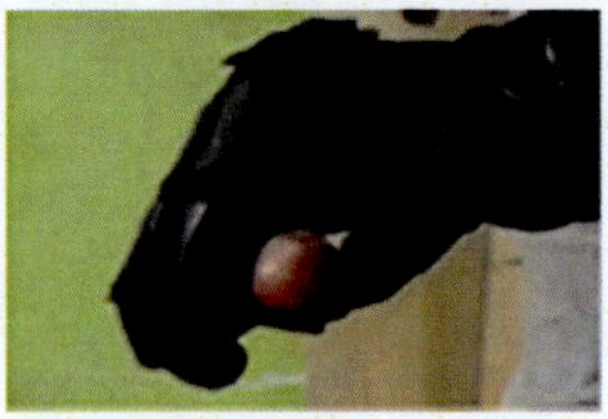

Fig. 12 Tip grip

Fig. 13 Extension grip

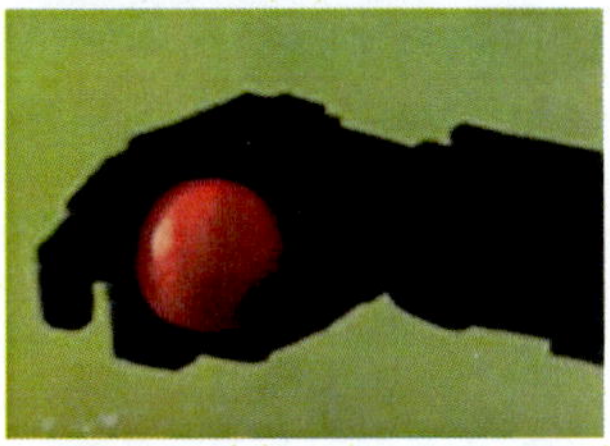

Fig. 14 Spherical grip

5 Disscusion and Conclusion

The VGT is developed from a basic whippletree mechanism through abstraction, method of negation, and systemic variation process. In abstraction, we find the main characteristics of the whippletree mechanism. In the method of negation, we explore the alternate solution and design space. In the systemic variation, we manifest all possible solutions and select the most suitable one. This process is similar to the analysis-synthesis bridge model [10] and TRIZ abstraction process [11]. In addition, the combination of a possible solution in the general classification is similar to the morphological method [12]. We argue that the process of abstraction, negation, and systemic variation can be a powerful methodology to generate a novel solution from an existing solution. We believe that this process can be used elsewhere.

The developed VGT is capable to produce lateral, power, tip, extension, and spherical grip without changing the setting or mode of operation. It also includes the thumb movement and uses only one simple-shaped moving part. Thus, VGT is effective and efficient to achieve hand functionality, especially in gripping. However further study is needed, especially in measuring the technical performance of VGT with a standardized test and evaluation for further medical trial or case report.

References

1. Taylor CL, Schwarz RJ (1955) The anatomy and mechanics of the human hand. Artif Limbs 2:22–35. https://doi.org/10.1016/B0-08-043749-4/09065-0
2. Ullah A, Ali S, Khan I, Khan MA, Faizullah S (2021) Effect of analysis window and feature selection on classification of hand movements using EMG signal. Adv Intell Syst Comput 1252:400–415. https://doi.org/10.1007/978-3-030-55190-2_30
3. Light CM, Chappell PH, Kyberd PJ (2002) Establishing a standardized clinical assessment tool of pathologic and prosthetic hand function: normative data, reliability, and validity. Arch Phys Med Rehabil 83(6):776–783. https://doi.org/10.1053/apmr.2002.32737
4. Stappers PJ, Giaccardi E (2014) Research through design. In: The Encyclopedia of human-computer interaction, Interaction Design Foundation
5. Groenewegen M (2014) Design of a compliant, multi-phalanx underactuated prosthetic finger. Delft University of Technology
6. Dukij A (2022) Makerhand. https://www.makerhand.com/. Accessed Sept 10 2022
7. Nanayakkara VK, Cotugno G, Vitzilaios N, Venetsanos D, Nanayakkara T, Sahinkaya MN (2017) The role of morphology of the thumb in anthropomorphic grasping: a review. Front Mech Eng 3. https://doi.org/10.3389/fmech.2017.00005
8. Pahl G, Beitz W (1996) Engineering design: a systematic approach. Springer
9. Trusaji W, Hammam S (2021) Tangan Prostetik Yang Memiliki Mekanisme Penggerak Adaptif Untuk Kelima Jari Serta Memiliki Metakarpus Keempat dan Kelima. P00202101141
10. Dubberly H, Evenson S, Robinson R (2008) The analysis-synthesis bridge model. Interactions 15(2):57–61. https://doi.org/10.1145/1340961.1340976
11. Hmina K, Sallaou M, Arbaoui A, Lasri L (2017) Clarification of a problem abstraction process for TRIZ technical contradiction model in preliminary design. Proc Int Conf Ind Eng Oper Manag no. Sallaou 2008, pp 1014–1023
12. Zwicky F (1948) The morphological method of analysis and construction. Interscience Publisher, New York

Factors that Affect the Job Performance of the Health Workers in Indonesia: A Configurational Analysis

Dicka Korintus Kurnianto, Amanda Syifa Ariqoh, Ferdinanda Pascha Hasian, Imanuella Indah Nugroho, Josyika Nisfullaili, and Retno Wulan Damayanti

Abstract Healthcare professionals are severely impacted by the Covid-19 epidemic. Due of the growing workload associated with managing Covid-19, many healthcare staff experienced a poor work environment and a tough night shift. Additionally, moderate to severe burnout syndrome, job dissatisfaction, and decreased emotional intelligence were common among healthcare workers. Based on these circumstances, researchers are looking for the ideal configuration model for the work environment, burnout, job satisfaction, emotional intelligence, and night shift to enhance the performance of healthcare workers. Two configurations produced high job performance in this study using fsQCA, while three configurations produced low job performance. To improve the analysis in future studies and perform comparative research when the results are circumstantial, it is advised to do some interviews or cross-checks with the stakeholders.

Keywords Covid-19 · Healthcare · Job performance · fsQCA

1 Introduction

Indonesia is still experiencing the Covid-19 pandemic. According to Covid19.go.id [1], the official website for Indonesian case development data, there were 5186 active cases nationwide as of the previous day, an increase of 32 cases. In the meantime, 208,266,720 people have received vaccinations. The increase in recovery is estimated to be 190 persons per day based on the cure rate. As many as 96.5% of those who caught the Covid-19 disease in Indonesia now recovered (4,109,865 persons). The healthcare system in Indonesia does not have the resources to be able to counter the COVID-19 pandemic.

D. K. Kurnianto (✉) · A. S. Ariqoh · F. P. Hasian · I. I. Nugroho · J. Nisfullaili · R. W. Damayanti
Industrial Engineering Department, Universitas Sebelas Maret, 57126 Surakarta, Indonesia
e-mail: dicka@student.uns.ac.id

C. N. Rosyidi et al. (eds.), *Proceedings of the 6th Asia Pacific Conference on Manufacturing Systems and 4th International Manufacturing Engineering Conference*, Lecture Notes in Mechanical Engineering,
https://doi.org/10.1007/978-981-99-1245-2_46

Several efforts have been made to deal with Covid-19, such as strengthening the health system to ensure that hospitals have good capabilities in handling patients, utilizing online treatment networks, utilizing online treatment systems, and preparing health sector emergency funds to minimize health financing [2]. To continue to increase the Covid-19 recovery rate in Indonesia, the performance of health workers is an important aspect. According to research conducted by the Faculty of Medicine University of Indonesia [3], it was recorded that more than 100 doctors and hundreds of medical personnel had died due to Covid-19 while carrying out health service duties. The results showed that 83% of health workers experienced moderate and severe burnout syndrome, 41% experienced moderate and severe emotional exhaustion, 22% experienced moderate and severe loss of empathy, and 52% experienced moderate and severe lack of confidence. Based on these conditions, researchers are interested in finding out the right configuration model to improve the work performance of health services so that the Covid-19 phenomenon can be handled properly.

This research uses the fuzzy-set qualitative comparative analysis (fsQCA) to investigates the variables (causal conditions) affecting healthcare workers performance during the COVID-19 period in Indonesia. The research's variables include job satisfaction and the work environment. Furthermore, this research adds burnout, emotional intelligence, and night shifts that may influence how well healthcare workers perform to produce a unique and specific configuration of factors that determine healthcare workers performance in Indonesia. Regarding that, these variables were combined using fsQCA method to achieve a high level of performance from healthcare workers in Indonesia during the COVID-19 period.

2 Related Study

2.1 Factors Affecting Work Performance

Several studies on the factors that affect work performance have been carried out previously. In a study conducted by Al-Omari and Haneen [4] as many as 85 respondents were collected in the data acquisition stage to prove the existence of a relationship between the work environment and employee work performance. The study was conducted on employees at an engineering company in Jordan. The results showed that the work environment really affects the motivation and ability of employees. The working environment factors in question are noise, ventilation, office facilities and lighting.

Subsequent research was conducted by Leskovic et al. [5] was carried out on healthcare workers during the pandemic in Slovenia. The results showed that the state of burnout affects the work performance of healthcare workers. As a result, healthcare workers' satisfaction with their work decreases, resulting in decreased work performance as well.

Alonazi [6] conducted a study to find out the effect of emotional intelligence on healthcare workers on their work performance in the midst of the covid-19 pandemic. Data from 340 healthcare nurses was collected to determine the current state of emotional intelligence of nurses and how they deal with their patients in the midst of a pandemic. The results showed that the emotional intelligence of nurses in Saudi Arabia belonged to the upper middle level. In addition, based on this study, it was also found that emotional intelligence had a significant effect on the work performance of healthcare workers.

Alsharari [7] examined the effect of the night shift on the performance of nurses in Saudi Arabia, the results showed that most (85.7%) nurses who worked the night shift experienced patient safety issues with physiological consequences of 93.6%. Thus, it can be concluded that the night shift indirectly affects the performance of healthcare nurses. Recent research by Kitsios and Kamariotou [8] found that job satisfaction is also a factor that affects the performance of healthcare nurses.

2.2 *Qualitative Comparative Methods (QCA)*

American sociologist Charles Ragin (1987) introduced qualitative comparative method (QCA) some 30 years ago as a third approach between case-oriented qualitative research strategies and variable-oriented quantitative research strategies [9]. Ragin [10] explains that case-oriented strategies have the advantage that they allow researchers to emphasize "complexity, variety, and uniqueness" and that they provide "a solid basis for case-historical interpretation". According to Ragin, QCA is a set theory method specially designed to explore case-oriented phenomena in social sciences, prove complex causal relationships such as feature configurations in minority cases, the body between independent and dependent conditions, and so on.

According to Mattke et al. [11], there are some terminologies in QCA that is different from the other common statistic methods. Some of the main terminologies are causal condition, outcome, and configuration. Causal condition is the membership set in a certain variable that explain the outcome while outcome is the membership set on a variable that is being explained by the causal condition. Lastly a configuration is a set of specific conditions where this set corresponds to the occurrence of a certain outcome. QCA analyzes the causal complexity in how the causal conditions are combined and how those conditions affect to the outcome [12]. Although QCA focuses on cases, QCA also can give the generalization pattern of all cases using Boolean algebra.

All causal conditions in the configuration could be labeled as "presence" (the condition is present in the configuration), "absence" (the condition is absent in the configuration), or "don't care" (whether the condition is presence or absent, it does not affect to the occurrence of a certain outcome). There are three types of QCA: crisp-set QCA (csQCA) using only conventional Boolean number, multi-value QCA (mvQCA) using multi value numbers, and fuzzy-set QCA (fsQCA) using fuzzy value

[13]. This research uses fsQCA because fsQCA is the newest type of QCA that can accommodate the weakness of csQCA and mvQCA in the condition variety.

3 Methods

3.1 Data Collection

The data collection method in this study used a questionnaire distributed to respondents. The criteria for respondents in this study are healthcare workers who directly or indirectly treat patients during the Covid-19 pandemic. The questionnaire media used is a google form which can be accessed via the following link bit.ly/SurveiKinerjaNakes.

This study used a 7 points Likert scale questionnaire. Lewis [14] concluded that the 7-point scale had a stronger correlation with the observed significance level than the 5-point scale. In addition, Finstad [15] shows that the 7-point scale tends to reflect respondents' actual subjective evaluations of the usability of the questionnaire items rather than the 5-point scale. Furthermore, Johns [16] concluded that data from Likert items and those with similar rating scales became significantly less accurate when the number of scale points fell below 5 or above 7 therefore Taherdoost [17] suggested the use of a 7-point scale because it was considered more effective by looking at specificity, reliability, and validity aspects. This study uses the fsQCA method to process the data that has been obtained, data processing is assisted using fsQCA 3.0 software. The causal condition indicator is shown in Table 1.

3.2 fsQCA Steps

Below are the steps in conducting fsQCA according to Pappas and Woodside [13] that is implemented in this research.

Data Entry. In the first stage, the data were input to the table where the statement was grouped according to the variable to be tested on each respondent. The value obtained is the average value of each statement variable. There are five input variables and one output variable.

Fuzzy Calibration. In the second stage, the data that were already grouped before were transformed into fuzzy values. Determination of each fuzzy value threshold is determined by the minimum value, maximum value, and average of each category. The thresholds of each category are shown in Table 2.

Truth Table. In the third stage, a truth table is created. The truth table is used to see the truth value of a statement that contains combinations of input variables that produce logical outputs. In this research the input variables consist of the work environment, burnout, job satisfaction, emotional intelligence, night shift and the

Table 1 Causal condition items

Variable	Indicator items	References
Workplace Environment (WE)	1. Environmental 2. Relationship with supervisors 3. Quality of work and equipment standards 4. Communication system	[18]
Burnout Absenteeism (BA)	1. Less emotional exhaustion 2. The absence of depersonalization 3. Personal accomplishment	[19]
Job Satisfaction (JS)	1. Workload 2. Salary 3. Relationships with colleagues 4. Work achievement	[8, 20]
Emotional Intelligence (EI)	1. Emotion management and impulse control 2. Social awareness 3. Optimism 4. Self-esteem	[21]
Night Shift (NS)	1. Number of nursing staff at night shift 2. Interruption at night shift work 3. Patient's cooperation 4. Nursing staff qualification	[7]
Job Performance (JP)	1. Contribution to the organization 2. Well in performing job 3. Good employee 4. Performance compared to other colleagues	[22]

Table 2 Thresholds

	WE	BA	JS	EI	NS	JP
Min	3.5	4	2.5	4	3	4
Mean	5.61	5.8	5.36	5.76	4.89	5.55
Max	7	7	7	7	7	7

output variable is job performance. The truth table is made based on the results of the fuzzy calibration values that have been carried out previously. Tables 3 and 4 shows the truth table that was carried out after processing the fuzzy calibration.

Delete and Code. In the fourth stage, after the truth table is created, delete and code operations are performed on the number of cases that appear. It is determined that the cases to be analyzed are cases with the number cases with several occurrences less than two and the consistency value is below 0.80. Cases with more numbers indicate the possibility of these cases to occur also more so that the analysis can be focused on that direction. Delete and code results are shown in Tables 5 and 6.

Table 3 Truth table for high job performance (fJP)

fEW	fBA	fJS	fEI	fNS	fJP	Num of cases	Raw coverage	PRI consistency	SYM consistency
0	1	1	1	1	1	3	0.971	0.863	0.886
0	1	0	1	1	1	1	0.968	0.825	0.825
1	1	1	1	1	1	7	0.961	0.904	0.922
1	0	1	1	1	1	1	0.924	0.669	0.672
1	0	0	0	1	1	1	0.897	0.214	0.214
0	1	0	0	1	1	2	0.889	0.110	0.110
1	0	1	0	0	1	2	0.857	0.110	0.110
0	0	1	0	0	1	1	0.850	0.116	0.116
0	0	0	0	1	1	2	0.841	0.240	0.240
1	0	0	0	0	1	2	0.811	0.067	0.068
0	1	0	0	0	0	3	0.775	0.071	0.076
0	0	0	0	0	0	8	0.651	0.095	0.108

Table 4 Truth table for low job performance (~fJP)

fWE	fBA	fJS	fEI	fNS	fJP	Num of cases	Raw coverage	PRI consistency	SYM consistency
0	1	0	0	1	1	2	0.986	0.890	0.890
1	0	0	0	0	1	2	0.984	0.921	0.932
1	0	1	0	0	1	2	0.982	0.890	0.890
0	0	1	0	0	1	1	0.980	0.884	0.884
1	0	0	0	1	1	1	0.972	0.786	0.786
0	1	0	0	0	1	3	0.967	0.863	0.924
0	0	0	0	1	1	2	0.950	0.760	0.760
0	0	0	0	0	1	8	0.916	0.783	0.892
0	1	0	1	1	1	1	0.849	0.175	0.175
1	0	1	1	1	1	1	0.845	0.326	0.328
0	1	1	1	1	1	3	0.812	0.111	0.114
1	1	1	1	1	0	7	0.623	0.077	0.078

4 Results and Discussion

This research found three configurations that produce in high job performance and three configurations resulted in low job performance that can be seen on Table 7.

Table 5 Delete and code results for high job performance (fJP)

fWE	fBA	fJS	fEI	fNS	fJP	Num of cases	Raw coverage	PRI consistency	SYM consistency
0	1	1	1	1	1	3	0.971	0.863	0.886
1	1	1	1	1	1	7	0.961	0.904	0.922
0	1	0	0	1	1	2	0.889	0.110	0.110
1	0	1	0	0	1	2	0.857	0.110	0.110
0	0	0	0	1	1	2	0.841	0.240	0.240
1	0	0	0	0	1	2	0.811	0.067	0.068
0	1	0	0	0	0	3	0.775	0.071	0.076
0	0	0	0	0	0	8	0.651	0.095	0.108

Table 6 Delete and code results for low job performance (~fJP)

fWE	fBA	fJS	fEI	fNS	fJP	Num of cases	Raw coverage	PRI consistency	SYM consistency
0	1	0	0	1	1	2	0.986	0.890	0.890
1	0	0	0	0	1	2	0.984	0.921	0.932
1	0	1	0	0	1	2	0.982	0.890	0.890
0	1	0	0	0	1	3	0.967	0.863	0.924
0	0	0	0	1	1	2	0.950	0.760	0.760
0	0	0	0	0	1	8	0.916	0.783	0.892
0	1	1	1	1	1	3	0.812	0.111	0.114
1	1	1	1	1	0	7	0.623	0.077	0.078

Table 7 Job performance configurations

	Configuration	Raw coverage	Unique coverage	Consistency
fJP	fWE*~fBA*~fEI*~fNS	0.377	0.037	0.775
	~fWE*~fJS*~fEI*fNS	0.411	0.061	0.812
	fBA*fJS*fEI*fNS	0.593	0.291	0.943
~fJP	~fWE*~fJS*~fEI	0.703	0.329	0.900
	fWE*~fBA*~fEI*~fNS	0.428	0.080	0.984
	~fWE*fBA*fJS*fEI*fNS	0.293	0.030	0.812

4.1 High Job Performance (fJP) Analysis

fWE*~fBA*~fEI*~fNS. Healthcare workers have a high degree of job performance due to a good work environment, high burnout rates, poor emotional intelligence, and unfavorable night shift conditions.

First, a supportive work environment is observed to result in high job performance in this configuration. This finding is consistent with Edem [18], who found that a positive work environment can lead to improved job performance. Four indicators are used to assess this: the communication system, room facilities, work equipment, and relationships with supervisors.

Second, burnout has a smaller impact on high job performance since, in this study, even though the burnout level was fairly high, job performance was still high. This is not the same as the research by Brady [19] that stated the high burnout results in low job performance.

Third, the emotional intelligence factor has a less impact on job performance since, even if emotional intelligence is low, job performance is still high. This finding contrasts with Alonazi [6] which claims that high emotional intelligence correlates with great job success.

Fourth, despite the lack of assistance for the night shift workplace, job performance remains high. This differs from the findings of Alsharari et al. [7], who discovered that night shifts have an impact on healthcare employees' job performance, especially on indicators of worker availability on night shifts.

~fWE*~fJS*~fEI*fNS. Even though the work environment was poor, job satisfaction was low, and emotional intelligence was low, with supportive night work conditions resulted in a high degree of job performance for healthcare employees in this configuration.

This is in line with the findings of Alsharari et al. [7], who discovered that the night shift had an impact on healthcare employees' job performance, particularly the indicator of workers' availability on the night shift. The job performance would suffer if the staff on the night shift are insufficient.

Then, healthcare personnel may experience poor job performance as a result of poor work environment conditions. This differs from the findings of Edem et al. [18], who claim that poor work environment planning can diminish workers' performance by lowering morale and decreasing work motivation, resulting in a lack of job satisfaction.

For healthcare employees, low job satisfaction factors might nonetheless lead to high job performance. This contrasts with the findings of Kitsios et al. [8], who found that high job satisfaction can lead to high job performance among healthcare personnel. This is also supported by Yáñez et al. [23]'s research, which claims that job satisfaction has little impact on worker performance and may be substituted by good leadership and staff engagement.

Also, the emotional intelligence factor has a less impact on job performance since, even if emotional intelligence is low, job performance is still high. This is in contrast to Alonazi [6] research, which claims that high emotional intelligence correlates with great job performance.

fBA*fJS*fEI*fNS. Low burnout rates, high job satisfaction, strong emotional intelligence, and supportive night shift conditions can all contribute to high job performance for healthcare personnel in this situation.

First, for healthcare personnel, a low burnout factor can result in good job performance. This is supported by research from Brady et al. [19], which claims that high burnout rates lead to poor job performance.

Second, high job satisfaction factors can lead to high job performance. Job satisfaction, according to Kitsios et al. [8], is a factor that motivates healthcare employees to work. Job satisfaction is influenced by several factors, according to his research, including the amount of compensation and the relationship between workers.

Third, a high level of emotional intelligence can result in great job performance.

According to Alonazi [6], research shows that having high emotional intelligence can lead to great job success.

Fourth, factors that encourage healthcare employees to work night shifts can result in great job performance. This is corroborated by the findings of Alsharari et al. [7], who found that the good conditions of the night shift affect healthcare workers' performance, especially on the indicator of the availability of workers on the night shift.

4.2 Low Job Performance (~fJP) Analysis

~fWE*~fJS*~fEI. A poor work environment, low job satisfaction, and a lack of adequate emotional intelligence can all lead to low job performance for healthcare workers in this condition.

First, healthcare workers may have low job performance as a result of poor work environment conditions. This is confirmed by research from Edem et al. [18], which claims that poor work environment planning can reduce workers' performance by lowering their morale and decreasing their enthusiasm to work, resulting in a lack of job satisfaction. This is a difficult task for management because the interaction between healthcare workers, their work, and the workplace environment is essential and fundamental to their jobs.

Second, low job satisfaction can lead to poor performance. This is supported by studies by Kitsios et al. [8], who found that high job satisfaction leads to high job performance among healthcare workers, and that poor job satisfaction leads to low job performance. Furthermore, this is supported by Yáñez et al. [23]'s research, which claims that job satisfaction has little impact on worker performance and may be replaced by good leadership and staff engagement in order to attain high job performance. As a result, low job satisfaction leads to low job performance.

Third, healthcare personnel, a high level of emotional intelligence can result in great job performance. According to Alonazi [6], research shows that having high emotional intelligence can lead to high job performance. As a result, low emotional intelligence, according to this research, can lead to low job performance.

fWE*~fBA*~fEI*~fNS. A good work environment, a high burnout rate, poor emotional intelligence, and an unsupportive night work environment result in a low degree of healthcare job performance in other circumstances.

First, the result of this configuration is that, while the work environment is supportive, job performance is low. This differs with the findings of Edem et al. [18] who found that a good work atmosphere can result in high job performance.

Second, it was discovered that a low level of burnout rate resulted in poor job performance. This is in line with Brady et al. [19], where high burnout results in low job performance.

Third, a high level of emotional intelligence can result in high job performance. According to Alonazi [6], research shows that having high emotional intelligence can lead to high job performance. As a result, low emotional intelligence, according to this study, can lead to poor job performance.

Also, for the night shift variable, low job performance was discovered because of the poor night shift conditions. This aligns with the findings of Alsharari et al. [7], who found that unfavorable working circumstances have an impact on healthcare workers' job performance.

~fWE*fBA*fJS*fEI*fNS. In some situations, a poor work environment combined with a low burnout rate, high job satisfaction, good emotional intelligence, and a supportive night work environment results in poor healthcare worker performance.

First, the unfavorable work environment has a significant impact on the low job performance in this case. This is in line with the findings of Edem et al. [18], who found that an unfavorable work environment had a significant effect on job performance.

Second, according to Brady et al. [19], low burnout factors can lead to high job performance among healthcare employees. This differs from the findings of this study because, while burnout is low, job performance stays low.

Third, high job satisfaction factors can lead to high job performance. Job satisfaction, according to Kitsios et al. [8], is a factor that motivates healthcare workers to work.

Fourth, good emotional intelligence factors can result in high job performance for healthcare workers, according to Alonazi's research [6]. This differs from the results of this study because, while the emotional intelligence obtained was high, the job performance produced was still low in this study.

Fifth, the night shift variable comes next. Unfavorable night shift conditions were discovered to be the cause of low job performance. This is in line with the findings of Alsharari et al. [7], who found that poor working conditions can affect healthcare workers' job performance.

4.3 Necessary Condition Analysis

Necessary conditions analysis was conducted to find which conditions that are considered as "necessary", meaning that those conditions must be present in order to make certain outcomes occur [24]. Necessary condition analysis can be seen on Table 8.

Table 8 Necessary condition analysis

High job performance (fJP)			Low job performance (~fJP)	
	Consistency	Coverage	Consistency	Coverage
fWE	0.722	0.762	0.799	0.763
~fWE	0.679	0.580	0.560	0.662
fBA	0.732	0.774	0.502	0.595
~fBA	0.617	0.526	0.810	0.772
fEI	0.839	0.873	0.499	0.582
~fEI	0.598	0.516	0.891	0.861
fNS	0.841	0.774	0.587	0.605
~fNS	0.571	0.552	0.781	0.846
fJS	0.728	0.786	0.513	0.620
~fJS	0.648	0.543	0.823	0.772

Based on the results, there are no necessary conditions in both high job performance (fJP) and low job performance (~fJP) because there are no consistency values in both outcomes that reach the threshold value of 0.90 [25]. This concluded that there are no certain conditions that highly guarantee high or low outcomes if those conditions present in the configuration.

However, there are some conditions that almost reach the threshold value in their consistency value such as fEI (0.839) and fNS (0.841) in fJP outcome. These can be concluded that the good healthcare worker's emotional intelligence and good night shift scheduling play a big role in the occurrence of high job performance. However, it is not sure if in the configuration fEI and fNS are present, the configuration will produce high job performance. While in ~fJP outcome there are ~fBA (0.810), ~fEI (0.891), and ~fJS (0.823) that almost reach the threshold value. These can be interpreted that the presence of burnout in high rate, the lack of emotional intelligence, and the low job satisfaction of the healthcare workers contribute big portion to the occurrence of low healthcare workers performance. Same as before, it cannot be sure if those causal condition presence in the configuration, the outcome will be low job performance.

4.4 The Best Configurations

Based on the results, there are three configurations in both high and low job performance outcomes. However, first configuration of high job performance and second configuration of low job performance have both the same configuration which is fWE*~fBA*~fEI*~fNS although they result on the different level of outcome. This situation is called "prime implicants" and researcher must exclude one configuration that less resembling the real situation.

Table 9 The best configurations

Outcome	Configuration	fEW	fBA	fJS	fEI	fNS
fJP	1	⊗		⊗	⊗	•
	2		•	•	•	•
~fJP	3	⊗		⊗	⊗	
	4	•	⊗		⊗	⊗
	5	⊗	•	•	•	•

• = Presence (high degree)
⊗ = Absence (low degree) Blank space = don't care

The exclusion is done by comparing the value of consistency. The consistency value of the first configuration of high job performance is 0.775 and the value of the second low job performance configuration consistency is 0.984. So, the second configuration of low job performance was chosen because it has a greater consistency value. The best configuration for job performance is shown on the Table 9.

From the high job performance outcomes (fJP), we can conclude that night shift scheduling on the healthcare service facilities is giving the biggest contribution in the configuration to the occurrence of high job performance because although the other causal conditions are on the low degree (fWE, fJS, and fEI) or on the don't care condition (fBA), if the night shift is on the high degree, all variables still result a high job performance in the outcomes (see configuration 1). However, it does not mean that the other conditions are "not important". If fBA, fJS, fEI, and fNS are on high degree, together with fWE in don't care condition, the configuration can result high job performance (see configuration 2). For the low job performance outcome, work environment and emotional intelligence give great contribution to low job performance outcome because both conditions are not in the don't care condition, means that both conditions are either on high degree or low degree (see configuration 3, 4, and 5). Emotional intelligence become condition that gives influence on all five configurations both high and low job performance because fEI appears in all configurations are either on high degree or low degree.

5 Conclusion

The results show two configurations that contribute to the high job performance of healthcare workers. The healthcare stakeholder management could try to create an environment based on the conditions to achieve a high degree of healthcare worker's job performance and avoid the occurrence of low degree healthcare workers job performance.

For future research, conducting some interviews or cross checks with the stakeholder could be implemented so they can strengthen the arguments in the analysis. This research focuses in Covid-19 pandemic and shows that the results highly affect the job performance of healthcare workers due to higher load and pressure than the times before pandemic. However, when the condition is back to normal, the load and pressure probably go lower than in the pandemic. So future research can be conducted in the normal era when the pandemic is over. The last thing is although QCA does not prohibit the usage of small samples, we recommend taking a bigger number of samples to avoid bias sampling.

References

1. Covid19.go.id. Angka Kesembuhan Covid-19 Meningkat Mencapai 4109865 Orang. Available: https://covid19.go.id/berita/angka-kesembuhan-covid-19-meningkat-mencapai-4109865-orang. Last accessed 2021/12/15
2. Putri RN (2020) Indonesia dalam menghadapi pandemi Covid-19. J Ilmiah Univ Batanghari Jambi 20(2):705–709
3. Fk.ui.ac.id (2020) 83% Tenaga Kesehatan Indonesia Mengalami Burnout Syndrome Derajat Sedang dan Berat Selama Masa Pandemic Covid-19, Available: https://fk.ui.ac.id/berita/83-tenaga-kesehatan-indonesia-mengalami-burnout-syndrome-derajat-sedang-dan-berat-selama-masa-pandemi-covid-19.html. Last accessed 2021/12/10
4. Al-Omari K, Okasheh H (Yong) The influence of work environment on job performance: a case study of engineering company in Jordan. Int J Appl Eng Res 12(24):15544–15550
5. Leskovic L, Erjavec K, Leskovar R, Vukovic G (2020) Burnout and job satisfaction of healthcare workers in Slovenian nursing homes in rural areas during the COVID- 19 pandemic. Ann Agric Environ Med 664–671
6. Alonazi WB (2020) The impact of emotional intelligence on job performance during COVID-19 crisis: a cross-sectional analysis. Psychol Res Behav Manag 13:749
7. Alsharari AF, Abuadas FH, Hakami MN, Darraj AA, Hakami MW (2021) Impact of night shift rotations on nursing performance and patient safety: a cross- sectional study. Nurs Open 8(3):1479–1488
8. Kitsios F, Kamariotou M (2021) Job satisfaction behind motivation: an empirical study in public health workers. Heliyon 7(4):e06857
9. Rihoux B (2013) Qualitative comparative analysis (QCA)
10. Ragin CC (1987) The comparative method: moving beyond qualitative and quantitative strategies. University of California Press, Berkeley
11. Mattke J, Maier C, Weitzel T, Gerow JE, Thatcher JB (2022) Qualitative comparative analysis (QCA) in information systems research: status quo, guidelines, and future directions. Commun Assoc Inf Syst 50(1):8
12. Yong T, Park C (2017) A qualitative comparative analysis on factors affecting the deployment of electric vehicles. Energy Procedia 128:497–503. https://doi.org/10.1016/j.egypro.2017.09.066
13. Pappas IO, Woodside AG (2021) Fuzzy-set qualitative comparative analysis (fsQCA): guidelines for research practice in information systems and marketing. Int J Inf Manag 58:102310.e
14. Lewis JR (1993) Multipoint scales: mean and median differences and observed significance levels. Int J Hum-Comput Interact 5(4):383–392
15. Finstad K (2010) Response interpolation and scale sensitivity: evidence against 5-point scales. J Usability Stud 5(3):104–110
16. Johns R (2010) Likert items and scales survey question bank: methods fact sheet, 1. University of Strathclyde. Available at: https://dam.ukdataservice.ac.uk/media/262829/discover_likertfactsheet.pdf
17. Taherdoost H (2017) What is the best response scale for survey and questionnaire design; review of different lengths of rating scale/attitude scale/Likert scale. Hamed Taherdoost 1–10

18. Edem MJ, Akpan EU, Pepple NM (2017) Impact of workplace environment on health workers. Occup Med Health Aff 5(2):1–5
19. Brady KJ, Ni P, Sheldrick RC, Trockel MT, Shanafelt TD, Rowe SG, Kazis LE (2020) Describing the emotional exhaustion, depersonalization, and low personal accomplishment symptoms associated with Maslach Burnout Inventory subscale scores in US physicians: an item response theory analysis. J Patient-Rep Outcomes 4(1):1–14
20. Shoja E, Aghamohammadi V, Bazyar H, Moghaddam HR, Nasiri K, Dashti M, Asgari A (2020) Covid-19 effects on the workload of Iranian healthcare workers. BMC Public Health 20(1):1–7
21. Treglown L, Furnham A (2020) Birds of a feather work together: the role of emotional intelligence and cognitive ability in workplace interaction and advice networks. Pers Individ Differ 158:109833
22. Vandenabeele W (2009) The mediating effect of job satisfaction and organizational commitment on self-reported performance: more robust evidence of the PSM—performance relationship. Int Rev Adm Sci 75(1):11–34
23. Yáñez-Araque B, Gómez-Cantarino S, Gutiérrez-Broncano S, López-Ruiz VR (2021) Examining the determinants of healthcare workers' performance: a configurational analysis during COVID-19 times. Int J Environ Res Public Health 18(11):5671
24. Schneider CQ, Wagemann C (2006) Reducing complexity in qualitative comparative analysis (QCA): remote and proximate factors and the consolidation of democracy. Eur J Polit Res 45(5):751–786
25. Schneider MR, Schulze-Bentrop C, Paunescu M (2010) Mapping the institutional capital of high-tech firms: a fuzzy-set analysis of capitalist variety and export performance. J Int Bus Stud 41(2):246–266

Best-Fit Heuristic Algorithm for Multi-stack Rectangular Packing Problem of NC Router Machine

Irfansyah Ali, Wisnu Aribowo, and Sukoyo

Abstract The research is motivated by a nesting problem in a CNC router machine in a metalworking company, where efficient cutting patterns are desired in order to minimize metal sheets. In the routing process under study, the metal sheets vary in size, the material leftovers after cutting may be used again, and the machine can cut the sheets in stacks simultaneously. This paper reports the development of an efficient algorithm to solve problems with such combined characteristics. It is based on the Best Fit algorithm with a multi-start strategy. A numerical example is provided to demonstrate the algorithm. It is found that the algorithm can find a good solution with 10.06% difference from the optimum in a quick time despite the relatively simple procedure.

Keywords Nesting problem · Bin-packing · Multi-stack

1 Introduction

The object of this research is the nesting process for a CNC router machine in an aerospace manufacturer. Nesting is one of the important processes in the whole production process. It can minimize so much cost if it is done appropriately and efficiently. In this research case, there are several features and characteristics that have not been considered and aligned in the existing nesting process with the real condition. First, the size of metal sheets used in the shop varies, which means the width and length of each sheet may differ. This condition has not yet been considered in the existing process yet, which often makes the sheet selection inefficient. The cutting patterns formed would not be compact and leave so much unused space. Second, the CNC router machine can cut several stacked sheets at a time on the

I. Ali · W. Aribowo (✉) · Sukoyo
Department of Industrial Engineering, Faculty of Industrial Technology, Institut Teknologi Bandung, Jl. Ganesa 10, Bandung, Indonesia
e-mail: wisnu@itb.ac.id

C. N. Rosyidi et al. (eds.), *Proceedings of the 6th Asia Pacific Conference on Manufacturing Systems and 4th International Manufacturing Engineering Conference*, Lecture Notes in Mechanical Engineering,
https://doi.org/10.1007/978-981-99-1245-2_47

condition that the sheets have the exact same cutting patterns. This feature has not been considered and applied in the existing shop. Applying this feature would save much setup and cutting time. Third, the cutting patterns formed by the current process tend to be scattered and not compact. While on the contrary, the ideal condition is that the parts allocation or cutting patterns must be designed compactly to minimize the unused space material that will be scrapped, if there are any. The unused space can be used again for the next cutting process if the leftover size is large enough, at least as big as the smallest length or width of the parts [1–6]. If those three conditions are being considered and applied in the shop, the company will have so much benefit from it.

Those issues mentioned before are developed as specific additional characteristics to solve in this research. They are called variable-bin sizes (VS), usable leftover (UL), and stacked sheets (SS) in this research. These characteristics had also been formulated in a mathematical model in a previous research article [2]. Considering that the optimal model is not suitable for large-scale data, this current research develops a heuristic algorithm to solve the problem with the objective of obtaining the best solutions within a reasonable period of time.

2 Literature Review

The nesting problem is similar to the two-dimensional rectangular packing problem (2DRP), which belongs to a subset of cutting and packing problems that are considered to be NP-Hard. Many instances and many research of the cutting and packing problems arise from practical issues in the industry.

In this research case, the shop uses a CNC router machine that has the capability to cut sheets in a non-guillotine manner that has similar manner with [1]. Andrade et al. [1] developed a MIP model for the two-dimensional non-guillotine cutting problems with usable leftovers consideration. Andrade et al. [1] also used a single bin size and variable bin size with space for usable leftovers consideration in the numerical example. Similar characteristics to [1], Summethapiwat et al. [7] proposed a column generation on a two-dimensional cutting stock problem with fixed usable leftover and multiple stock sizes. Without usable leftover consideration, Liu et al. [5] proposed a dynamic programming-based heuristic for the variable-sized two-dimensional bin packing problem. Several articles [3, 4] proposed an heuristic approach for two-dimensional packing problems. Except for [2], the stacked sheet aspect has not been investigated, especially using the heuristic approach. With this stacked sheets consideration, we conclude that it would make the routing process more efficient.

In this research context, exact algorithms can only solve small-sized problems and are not capable of solving larger practical problems because it requires an expensive computational time to obtain desired solutions. Most researchers have devoted their efforts to fast approximate heuristics and meta-heuristics to identify practical and highly effective solutions within a reasonable period of time [3–5]. Therefore, in this research, we also develop an heuristic algorithm for the two-dimensional nesting problem with variable-bin size, usable leftover, and stacked sheets consideration.

3 Problem Formulation

This research adopts the performance criteria used in the optimal model [2], which is to minimize the usage of materials [material costs (CM)] and the number of cutting batches [processing costs (CP)]. In order to achieve minimum materials usage, it is necessary to select a sheet that can pack as many parts as compactly possible in one sheet so that the remaining space can be allocated for usable leftover and all parts can be packed by using less number of sheets. With the tendency to use large size sheets, that condition can be achieved [7]. And with multi-stack materials consideration, more material can be processed in one batch so that the processing cost can be reduced. Each characteristic mentioned above becomes the fundamentals in developing the proposed algorithm in this research.

4 Proposed Algorithm

The proposed algorithm, called best-fit heuristic algorithm with variable-size sheets, usable leftover, and stacked sheets consideration (BFHA-VS-UL-SS), developed in this research is based on the BFHA in Chen et al. [3] for the heuristic packing strategy structure, which is then adapted to the specific characteristics as inspired by Liu et al. [5], Andrade et al. [1], Aribowo et al. [2], and Sumetthapiwat et al. [7]. In the proposed algorithm, adjustments and modifications are carried out so that each characteristic can be applied with the BFHA as a synergized algorithm. Figure 1 shows the outline of the proposed algorithm.

The algorithm process starts from the identification of the input (type of parts, sheets, and demand). Then, the initiation process starts (includes determining the size of the space for usable leftover (SUL), observing priority parts (usually the largest and specific parts that can only be packed in a specific sheet and in a specific orientation) potential, and grouping-sorting the data (type of parts, sheets, demand, area, n and j in the list in a descending manner). The index j represents the packing order of the parts and n represents the number of part types in the list. The number of solution alternatives that will be found is basically $2n$ because each part generates two possible alternatives: the original orientation and 90-degree rotated orientation. But if there are square parts or priority parts, the formed alternatives will be less than $2n$ because a square part or a priority part cannot be rotated and thus will generate only one initial layout and solution.

The packing iterations are carried out as many as the number of part types. In each iteration, a selected part type will be packed first. It needs to be noted that priority parts, if any, are always the first parts to pack in every iteration, even preceding the

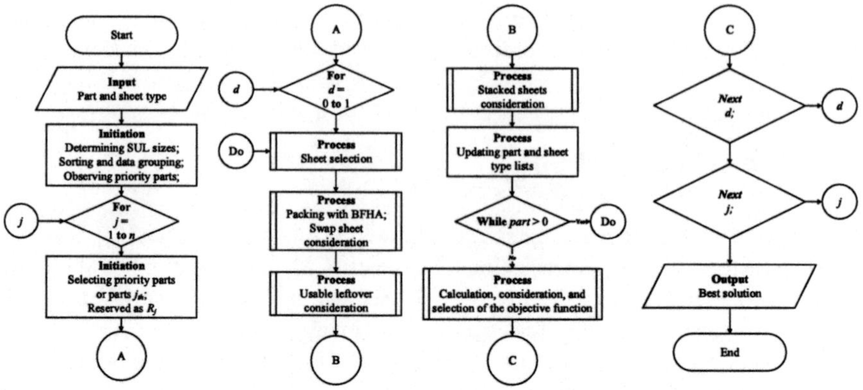

Fig. 1 The outline flow diagram of the proposed algorithm (BFHA-VS-UL-SS)

selected R_j parts. After that, the looping function for orientation (d) is executed. The value of d determines the orientation for R_j parts: $d = 0$ means unrotated, and $d = 1$ means rotated. The algorithm also will recheck the R_j parts shapes; if it is square or it is cannot be rotated, then the looping process will only be done once, for $\text{d} = 0$ or 1.

The sheets selection process is accomplished based on three conditions. The first condition is when there are priority parts in the list. The algorithm will select the largest size sheets or the one that can pack the priority parts or the selected R_j *parts.* The second condition is when there are no priority parts. In general, the algorithm will select the sheets based on descending order of available sheets size that can pack the R_j *parts.* The third condition is when the remaining number of parts that need to be packed is less than or equal to 20%. The 20% parameters value was decided based on trial and error using an hypotethical data set and in purpose to generate more compact cutting patterns using smaller sheets. The parameters value is adjustable, it can be modified and adjusted to the most fit number as required. Then, the algorithm will select the sheets based on ascending order of available sheets size that can pack the R_j *parts.* After that, the algorithm will pack the priority parts or R_j *parts* into the selected sheets and form an initial layout.

Then the packing process continues in accordance with BFHA procedures and stages. From the initial layouts formed, the algorithm calculates every space in the sheet and produces an AOP or angle that can be used in packing parts. Each AOP is tentatively calculated and compared by packing the parts in the list sequentially with different orientations. Then BFHA will select the parts that provide the highest fit degree (number of angles occupied by the part) then pack it to the chosen angle. The packing process continues until the sheet is full and no more parts remain to be packed. If the packing process is complete and the algorithm finds that there are much unused spaces in a sheet, the algorithm will consider swapping the used sheets with a smaller sheets size available. After that, the algorithm will observe and calculate the remaining space of the used sheets as usable leftover if possible. The initial usable leftover size is determined based on the smallest width or length of parts in the list. The algorithm will incrementally add the size up until it fills up the remaining space without overlapping any packed parts. And the last added up size becomes usable leftover size (ρ_k). Then the stacked sheets consideration step starts. The stack consideration is done only if the number of each packed parts in a sheet has not fulfilled the demand and only if there are still sheets available. The consideration is done by multiplying the number of each packed parts with a multiply factor. The multiply factor value is incrementally added until each or one of the packed parts meet, not exceeding its demand number and not exceeding the maximum number of stacks ($m^{\max}$) in a batch (δ_k). If the number of stacks formed is greater than the maximum number, then it will affect the processing cost and the number of batches. Then, the multiply factor value will be applied as the number of sheets stacked (m_k).

The procedure above is done repeatedly, forming diverse solutions until the value of j is equal to n (the *multi-start strategy*). Lastly, the algorithm will calculate and select one with the best objective function as the best solution from the solution alternatives formed.

5 Numerical Example

In order to illustrate the packing process flow and to verify the algorithm, we use the same case, same objective, and same small hypothetical data that are also used in the optimal model of previous research [2]. The hypothetical data sets are used only to demonstrate the usage of our proposed algorithm and does not represent any measurement units. So unit measurements in this numerical example data sets is not required and not used (see Table 1).

However, it should be noted that there are two differences between the proposed algorithm and the optimal model. The proposed algorithm considers (90-degree) rotation, while the optimal model does not. The proposed algorithm takes into account usable leftover that is formed on the top side as well as on the right side of sheets, while the optimal model only allows usable leftover on the right side. Consequently, for equivalent comparison and analysis purposes, we make a temporary adjustment in the proposed algorithm by limiting the allocation of usable leftover on the right side only and by avoiding rotation. The objective function for this numerical example is as follows:

Table 1 Sorted data for the numerical example

Parts						Sheets			
j	Type	Width	Height	Area	Demand	Type	Width	Height	Area
1	3	4	9	36	3	1	6	10	60
2	2	7	2	14	2	2	7	5	35
3	1	3	3	9	14				

$$\min F = \left(CM \cdot \sum_{k \in K} m_k \cdot L_{s(k)} \cdot \left(W_{s(k)} - SV \cdot \rho_k\right)\right) + \left(CP \cdot \sum_{k \in K} \delta_k\right) \quad (1)$$

where the objective function (F) in Eq. (1) consists of material cost and processing cost. The parameters CM and CP are material cost and processing cost factors, respectively. The parameters value used are as follows: processing cost factor (CP) = 50; material cost factor (CM) = 1; salvage value (SV) = 0.8; minimum usable leftover size ($\rho^{\min}$) = 2; maximum number of stacks ($m^{\max}$) = 3. Sheet s is provided in a limited number C_s for routing with a certain width W_s and length L_s, while sheet item k is the sheet entity used for packing parts in a sheet (forming patterns). The number of sheets stacked is denoted as m_k. We use $s(k)$ to denote the sheet s of sheet item k. Assuming that data in Table 1 has gone through the process of determining the size of the SUL, through grouping-sorting process, through observing-selecting of priority parts process, and through sheets selection process, then the packing process of one of the iterations is done (see Fig. 2).

After the algorithm has finished, it is known that the proposed algorithm obtained a solution with an objective function value of 522 (see Fig. 3).

6 Analysis and Discussion

It is found that, using the same case, same data, and same objective as [2], the proposed algorithm can yield a good solution with an objective function value of 522, which is relatively close to the optimal solution as obtained by the optimal model with objective function value of 472. The difference is 50 points or 10.06% from optimum. The optimal solutions are difficult to find for deterministic heuristic algorithms that work in a greedy manner like the proposed algorithm in this research. The proposed algorithm is indeed a simple constructive procedure without further improvement iterations, which limits the searching capability. But with the multi-start strategy, the algorithm can obtain diverse solution alternatives, as many as twice the number of part types. However, compared to the optimal model, the proposed algorithm has a clear advantage in the very fast computation time. The proposed algorithm is a simple one-pass algorithm or heuristic search algorithm that used to generate a limited number of solutions in a multi-start manner which can be done instantaneously as opposed to optimal model that is an exact methods or exhaustive search algorithm which will enumerate all possible solutions within the constraints. In that regard, the proposed algorithm has the potential to be used to quickly generate initial solutions for various meta-heuristic algorithms.

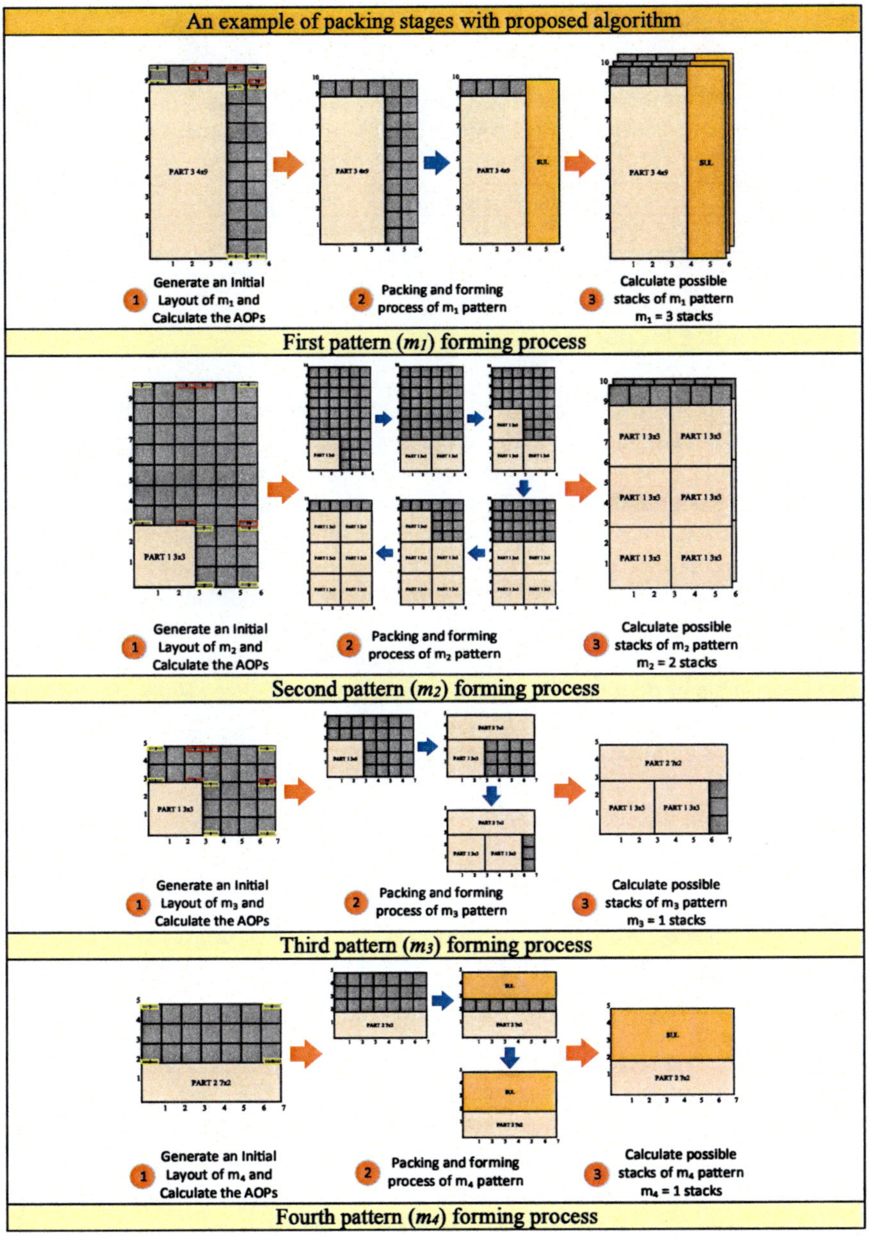

Fig. 2 An illustrative example of packing stages

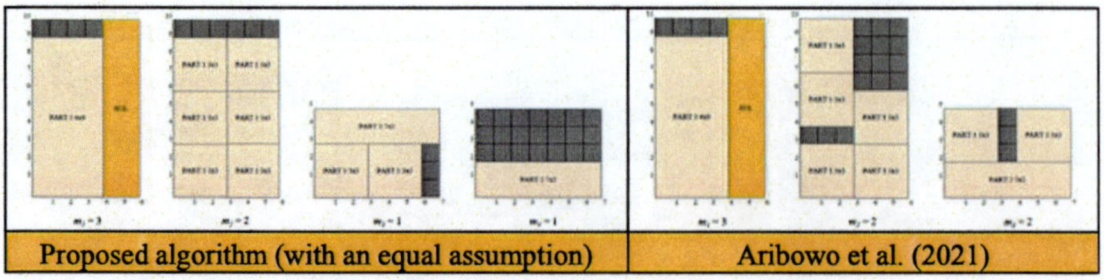

Fig. 3 Comparison of the proposed algorithm solution with the solution of Aribowo et al. [2]

7 Concluding Remarks

The proposed algorithm has succeeded in solving the multi-stack rectangular packing problem. The problem class investigated in this paper has a few particular characteristics, i.e. variabel sheet size, multi-stack cutting, and usable material leftovers. The numerical example shows that the algorithm can generate a pretty good solution with 10.06% difference compared to the optimal solution. For this stage of research, we deem the result good enough as our early intention is to have a quick algorithm for solving large practical problems. From the results, we could also identify some weak points in the algorithm to be improved, one of which is that the algorithm needs to be able to take parameter values into consideration, e.g. unit material cost and salvage value of material leftovers. For further research, we also plan to further develop the proposed algorithm to be able to solve various part shapes and develop a meta-heuristics approach to improve the algorithm's searching ability.

References

1. Andrade R, Birgin EG, Morabito R, dan Ronconi DP (2014) MIP models for two-dimensional non-guillotine cutting problems with usable leftovers. J Oper Res Soc 65(11):1649–1663
2. Aribowo W, dan Ali I (2021) An optimization model of variable size two-dimensional nesting with stacked sheets and usable leftovers for the CNC routing process. J Phys: Conf Ser 1858(1):012043
3. Chen M, Wu C, Tang X, Peng X, Zeng Z, dan Liu S (2019) An efficient deterministic heuristic algorithm for the rectangular packing problem. Comput Ind Eng 137:106097
4. Leung SC, dan Zhang D (2011) A fast layer-based heuristic for non-guillotine strip packing. Expert Syst with Appl 38(10):13032–13042
5. Liu Y, Chu C, dan Wang K (2011) A dynamic programming-based heuristic for the variable sized two-dimensional bin packing problem. Int J Prod Res 49(13):3815–3831
6. Resende MG, Martí R, dan Panos P (2017) Handbook of heuristics
7. Sumetthapiwat S, Intiyot B, dan Jeenanunta C (2020) A column generation on two-dimensional cutting stock problem with fixed-size usable leftover and multiple stock sizes. Int J Logist Syst Manag 35(2):273–288

Cost Minimization Model for Raw Materials Procurement Planning at an Indonesian Aluminium Producer

Audira Hutami, Praditya Ajidarma, and Dradjad Irianto

Abstract Aluminum production requires raw materials with fluctuating prices. The procurement of raw materials must be executed with the least possible cost, subject to demand fulfillment. Currently, the Indonesian aluminium producer aims to minimize procurement costs for the three primary raw materials: alumina, calcified petroleum cokes (CPC), and coal tar pitch (CTP) which are used to produce aluminum. This study conducted procurement planning by minimizing the purchasing value of the future price of raw materials, as well as ensuring the adequacy of raw materials to meet the needs of aluminium production. The product demand and raw materials price is obtained by time-series forecasting using the ARIMA method with predictors. The final results show the monthly procurement quantity for each raw material subject to the predicted demand.

Keywords Aluminium raw material procurement · ARIMA · Cost minimization · Linear programming

1 Introduction

A primary aluminum producer in Indonesia annually produces 250,000 metric tons of aluminum products, consisting of three types of products: ingots, billets, and alloys. As a commodity product, aluminum products have a fluctuating selling price. Events such as COVID-19 in 2020 could lower aluminum prices further than previous price

A. Hutami (✉) · P. Ajidarma · D. Irianto
Industrial Engineering Program Study, Bandung Institute of Technology, Jl. Ganesha 10, Bandung, West Java 40132, Indonesia
e-mail: pundiaudira@gmail.com

P. Ajidarma
e-mail: ajidarmap@itb.ac.id

D. Irianto
e-mail: dradjad@mail.ti.itb.ac.id

C. N. Rosyidi et al. (eds.), *Proceedings of the 6th Asia Pacific Conference on Manufacturing Systems and 4th International Manufacturing Engineering Conference*, Lecture Notes in Mechanical Engineering,
https://doi.org/10.1007/978-981-99-1245-2_48

fluctuations. The limited production capacity indicates that even when produced at the maximum value, the fluctuating price of the product often introduces uncertainty and an uncontrollable impact on the income earned by the company. As price is determined entirely by the market, the only way for the company to optimize its profit is to minimize the incurred operating cost. The cost of goods sold (COGS) consists of the four main raw material costs and other supporting raw materials. To ensure consistent profit attainment, the company must conduct procurement planning optimally by managing the stock of the primary raw materials. This study proposes a model to optimally procure three primary raw materials of alumina, calcified petroleum cokes (CPC), and coal tar pitches (CTP). The optimization model is expected to produce an improved value by reducing procurement costs incurred by the company.

2 Literature Study

2.1 *Aluminium Industry*

The aluminum industry consists of primary aluminum production, the use of aluminum raw materials by manufacturers and users, and secondary (recycled) aluminum production [4]. Aluminum is widely used as an essential material in the manufacturing industry. Primary aluminum production involves using raw materials such as bauxite alumina, coal tar pitch, calcified petroleum cokes, aluminum fluoride (AlF_3), and others [1]. The general aluminum production process is smelting alumina with the help of carbon blocks produced from the combination of calcified petroleum cokes and coal tar pitch raw materials. Most aluminum smelters do not produce the required raw material in-house, instead, they followed a five-step procurement process: assessment, selection and negotiation, design collaboration, procurement, and analysis [3].

2.2 *Time Series Forecasting*

According to Kengpol et al. [8], the value of time series prices must be determined first before planning. Actual data from time series data may contain missing data or missing data, so preprocessing is required before forecasting, in which missing data is imputed with an estimated value using the appropriate imputation method [10]. Forecasting for time series data types can be accomplished using a variety of methods, including simple exponential smoothing, Holt's double exponential smoothing, Holt-Winter's seasonal exponential smoothing, and autoregressive integrated moving averages (ARIMA) [12]. The ARIMA method combines an autoregression (AR) model with a moving average (MA) model. Forecasting a commodity value using the ARIMA method can also be done by referring to a predictor value in the form of related commodity stocks, but only aggregate stock prices or related commodity index prices can be used [11]. At the same time, the SP500 index has a negative relationship with several metal commodities (aluminum, zinc, gold) and agricultural commodities, so it cannot be used to predict the price of aluminum and its raw materials [2]. Prospective predictors are chosen under the condition that predictors that can be used in time series regression can be seen from the frequency of similarity [7].

2.3 Mathematical Model Reference

The OR problems researched in this study are classified as linear programming, and Hillier and Lieberman [6] claim that they can be solved using the simplex method or with software assistance. This optimization concept is based on standard linear programming, which includes objective functions, limiting functions, parameter values, and decision variables. This study refers to previous research using linear programming conducted by Hariga [5] and Kengpol et al. [8].

Hariga [5] studied "A Production Planning Model for an Aluminum Company." This time, the reference model generates a cost-minimization model that can help management make decisions. This design of the research's model is a system in aluminum companies with multiple products, stages, and periods. This reference model performs cost optimization modeling by taking into account the movement of materials from the entry of raw materials to the fulfillment of final product demand within the organization, which has a broad scope.

Kengpol et al. [8] published "A Procurement Planning Improvement Using Linear Programming and Forecasting Models." Kengpol et al. [8] conducted research with the goal of improving the procurement system carried out on research objects in the paper industry by using linear programming and forecasting models, where forecasting is dependent on each procurement condition (e.g., type of material or source of producer).

3 Methodology

3.1 Mathematical Model Formulation

It is possible to obtain previous scientific works that have solved similar problems to those that will be solved by conducting a literature study. Section 2 explains that the reference model used in this study is based on previous research by Hariga [5] and Kengpol et al. [8]. The reference model is then adjusted to the actual conditions of the research object, the Indonesian aluminium producer, which differs from the object studied by previous researchers. This time, the cost minimization problem model can be seen in Eqs. (1)–(18).

The objective function reveals the mathematically written purpose of the problem. The objective function in Eq. (1) minimizes the total cost of purchasing and storing alumina, CPC, and CTP raw materials over a 12-month period beginning with period 1. CAlumina, CCPC, and CCTP values are derived from raw material purchase costs, while HAlumina, HCPC, and HCTP are derived from raw material storage costs. XAlumina, XCPC, and XCTP variables as decision variables for purchasing raw materials in month t, and finally, the function objective of IAlumina, ICPC, and ICTP as raw material stock units in the company's warehouse at the end of t month.

$$Min\,Z = \sum_{t=1}^{12}[(CAlumina_t XAlumina_t + CCPC_t XCPC_t + CCTP_t XCTP_t) + HAlumina_t IAlumina_t + HCPC_t ICPC_t + HCTP_t ICTP_t)] \quad (1)$$

The limiting function equations for this problem begin with Eq. (2). Before entering the limiting function equation, some values that will be used as constraints are calculated. The demand for alumina, CPC, and CTP raw materials is depicted in Eqs. (2)–(4), respectively. The value of dAlumina, dCPC, and dCTP, or the ratio of raw material demand, is multiplied by the demand for the final product in month t to get the value of raw material demand in month t.

$$dAlumina_t = 1.904 \times demand_t \quad \forall t = 1, \ldots, 12 \quad (2)$$

$$dCPC_t = 0.38 \times demand_t \quad \forall t = 1, \ldots, 12 \quad (3)$$

$$dCTP_t = 0.066 \times demand_t \quad \forall t = 1, \ldots, 12 \quad (4)$$

Furthermore, Eqs. (5)–(7) show the TAlumina, TCPC, and TCTP requirements, or the three raw material requirements needed to meet production in month t. To calculate the total production needs of alumina, CPC, and CTP in month t, multiply the value of the raw material need ratio by the weight of the monthly production plan in month t. The value of planned production is the estimated value of aluminum smelting capacity each month.

$$TAlumina_t = 1.904 \times produksi_t \quad \forall t = 1, \ldots, 12 \quad (5)$$

$$TCPC_t = 0.38 \times produksi_t \quad \forall t = 1, \ldots, 12 \quad (6)$$

$$TCTP_t = 0.066 \times produksi_t \quad \forall t = 1, \ldots, 12 \quad (7)$$

The containing function begins with the stock balance equation for each raw material, as shown in Eqs. (8)–(10). According to the three equations, the number of raw materials purchased in month t plus the stock from the previous period (t − 1) minus the stock for the current period (t) must equal the number of raw materials required for production in month t. The constrain for alumina, CPC, and CTP raw materials are shown in Eqs. (8)–(10), respectively.

$$XAlumina_t + IAlumina_{t\%1} - IAlumina_t = TAlumina_t \quad \forall t = 2, \ldots, 12 \tag{8}$$

$$XCPC_t + ICPC_{t\%1} - ICPC_t = TCPC_t \quad \forall t = 2, \ldots, 12 \tag{9}$$

$$XCTP_t + ICTP_{t\%1} - ICTP_t = TCTP_t \quad \forall t = 2, \ldots, 12 \tag{10}$$

The constraint functions (11)–(13) limit the stock of alumina, CPC, and CTP raw materials, respectively. The raw material stock in month t must be less than or equal to the raw material storage capacity minus the amount of alumina required for production in month t.

$$IAlumina_t \leq KapasitasAlumina - TAlumina_t \quad \forall t = 1, \ldots, 12 \tag{11}$$

$$ICPC_t \leq KapasitasCPC - TCPC_t \quad \forall t = 1, \ldots, 12 \tag{12}$$

$$ICTP_t \leq KapasitasCTP - TCT \quad \forall t = 1, \ldots, 12 \tag{13}$$

Furthermore, the constraint function (14) demonstrates that the quantity of alumina raw material demand in month t must be at least equal to, greater than, or equal to the amount of alumina raw material purchased in month t. This constraint is used to ensure that the maximum stock of alumina raw materials is available at the start of the period [5]. Similar to the constraint function (14), these two equations are used to ensure the maximum stock of each raw material at the start of the period for CPC and CTP raw materials written in Eqs. (15) and (16).

$$XAlumina_t \leq dAlumina_t \quad \forall t = 1, \ldots, 12 \tag{14}$$

$$XCPC_t + dCPC_t \quad \forall t = 1, \ldots, 12 \tag{15}$$

$$XCTP_t + dCT \quad \forall t = 1, \ldots, 12 \tag{16}$$

According to the limiting function (17), the stock of alumina raw materials must always be greater than or equal to 6000 metric tons, which is the dead stock value or the minimum storage capacity of alumina raw materials. The minimum storage limit for CPC and CTP raw materials is 0, so it is not recorded.

$$IAlumina_t \geq 6000 \quad \forall t = 1, \ldots, 12 \tag{17}$$

Lastly, the constraint function (18) is a non-negative function that requires all decision variable results to be greater than or equal to 0.

$$XAlumina_t, XCPC_t, XCTP_t, IAlumina_t, ICPC_t, ICTP_t \geq 0 \ \forall t = 1, \ldots, 12 \tag{18}$$

3.2 Price and Demand Forecasting

The main component of the objective function is the purchase price of the main raw materials, where the actual value varies and trends differently for each raw material. Companies that plan for the future will require a good value estimate so that the planning results are as close to the actual value as possible when the period is over. Forecasting in this study was done in three stages: preprocessing data for the value of the price of the main raw materials, selecting predictors for forecasting the primary raw materials, and predicting the value of the price of raw materials and demand for the final product.

At the data preprocessing stage, missing price data for alumina, CPC, and CTP raw materials are imputed so that the forecasting process can begin. The steps taken are to complete the preprocessing stage. At this point, the data is aggregated monthly, and data imputation is used to fill in the gaps. RStudio software is used to perform the imputation process on data within a monthly period.

In this study, predictors are used to forecasting trends from the price data to be forecasted using the auto ARIMA method. The predictor value is matched with the raw material price to be forecasted, and the combination with the uncorrelated residual value is used to become a forecasting predictor.

Finally, price and demand forecasting is performed using the exponential smoothing method described by Kengpol et al. [8], as well as the ARIMA method, whether predictors are used or not. Forecasting is done for the price of alumina, CPC, and CTP raw materials, as well as the demand value of the finished product.

4 Results and Discussion

4.1 Forecasting Results

Forecasting is done on the final product demand as well as the prices of the three main raw materials, which are alumina, calcified petroleum cokes, and coal tar pitch. The three prices are paired with the appropriate predictor data, and forecasting is performed using a variety of methods, with the best forecasting method for each price chosen based on the smallest error value. After comparing the Holt-Winter Additive and Multiplicative Seasonal Exponential Smoothing methods, the demand value is predicted using the auto ARIMA method. The comparison results show that the auto ARIMA method has the smallest RMSE error for forecasting trend values with a stationary shape, moving horizontally, and having significant value fluctuations.

Auto-ARIMA forecasting results conclude the tuned ARIMA(0,0,0) to forecast the average price and demand for the next 12 months. Figure 1 depicts the demand forecast results in graphical form. The final product demand forecast results are then converted into the ratio of the main raw material requirements, which are called dAlumina, dCPC, and dCTP in the modeling. The following figures also show the forecasting results for the three raw material demands as a time series.

Forecasting the price of the primary raw materials is accomplished by ensuring that the best method is to use the fitted auto ARIMA method or the auto ARIMA method using predictor values as a reference for forecasting values. The fitted forecasting method is chosen by first comparing whether the fitted method has a lower error value when the three raw material prices are forecasted using the Simple Exponential Smoothing (SES), Holt's Double Exponential Smoothing, Holt-Additive Winter's, and Multiplicative Smoothing methods, and auto ARIMA without predictor value.

To forecast using the fitted auto ARIMA method, first, choose a candidate for the predictor value that is correlated with the three prices to be forecasted. Manufacturing IDX, Aluminum Futures, IDX Mining, Infrastructure IDX, China Shenhua Energy, and Altium Australia are the predictor values that deserve to be used as prospective

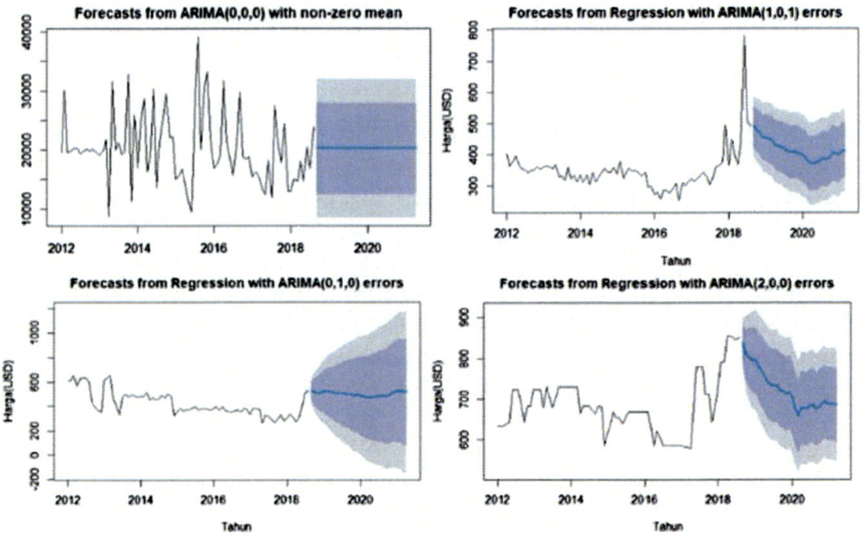

Fig. 1 ARIMA forecasts for aluminium price and the demand for raw materials

Table 1 Procurement planning of raw material purchase in metric tons

Period	t	IAlumina	ICPC	ICTP
April 2020	1	0.00	0.00	0.00
May 2020	2	13,818.71	2146.59	0.00
June 2020	3	38,492.79	7682.38	1249.24
July 2020	4	38,492.79	7682.38	1334.31
August 2020	5	37,981.37	7580.32	1316.58
September 2020	6	38,492.79	7682.38	1334.31
October 2020	7	38,224.70	7628.88	1325.02
November 2020	8	38,492.79	7682.38	1334.31
December 2020	9	38,492.79	7682.38	1334.31
January 2021	10	38,492.79	7682.38	1334.31
February 2021	11	38,492.79	7682.38	1334.31
March 2021	12	38,492.79	7682.38	1334.31

predictors in this study. There are 18 forecasting results because the three main raw material prices are forecasted with each of the six predictor values.

The RMSE and MAPE values show that the predictor that produces the most appropriate forecast results with the original value is if the Alumina price is forecasted with the Aluminum Futures predictor, the CPC price is forecasted with the IDX Mining predictor, and the CTP price is forecasted with the Manufacturing IDX predictor. The fitted auto ARIMA method produces the smallest error value when compared to the other five forecasting methods, so the forecasting results using the fitted auto ARIMA method are chosen to be the values that will be used in modeling.

4.2 *Procurement Planning Results*

The results of the objective function can be obtained by mathematical modeling the results of adjusting the reference model to the actual conditions of the research object. The decision variables whose values will be generated by the modeling are the amount of alumina, CPC, and CTP raw materials that must be purchased each month, as well as the value of the needs for the production of the three raw materials. The results of the decision variable values obtained through modeling are consistent with those obtained when computing with LINGO software [9]. The modeling results for the primary raw material purchases of XAlumina, XCPC, and XCTP are as follows (see Table 1).

Furthermore, another decision variable whose results are obtained through modeling is the amount of raw material stock. Table 2 shows the total stock values for IAlumina, ICPC, and ICTP.

Table 2 Monthly aluminium raw material levels in metric tons

Period	t	IAlumina	ICPC	ICTP
April 2020	1	33,000.00	6000.00	1500.00
May 2020	2	6444.39	88.69	100.47
June 2020	3	7601.65	319.66	55.52
July 2020	4	6000.00	0.00	0.00
August 2020	5	7462.65	291.92	50.70
September 2020	6	6000.00	0.00	0.00
October 2020	7	6481.71	96.14	16.70
November 2020	8	8314.88	462.00	80.24
December 2020	9	6551.40	110.05	19.11
January 2021	10	6434.87	86.79	15.07
February 2021	11	7632.11	325.74	56.58
March 2021	12	6000.00	0.00	0.00

5 Conclusion

This research yields the optimal quantity of the primary raw materials that must be purchased, as well as the raw materials that will be used in the monthly production of liquid aluminum for the research object company at the lowest possible cost. The results of this modeling show a cost reduction of 4.994% of the total cost for raw material procurement. The use of predictor values to aid in the forecasting of raw material prices for modeling has a positive impact on forecasting results, with the error value decreasing. The low error value indicates that the forecasting results are more consistent with the shape and actual value of the forecasted price during the same period.

Table 1 depicts the purchase price of the forecasted results of the main raw materials. Forecasting is performed for 12 periods, each lasting one month, from April 2020 to March 2021. Table 2 shows the value of the number of main raw materials that must be purchased as well as the raw materials that must be produced. The value of raw material obtained is also for 12 periods, with alumina, CPC, and CTP, respectively, having 13,818.71 metric tons, 2146.59 metric tons, and no purchases in the May 2020 period. The study's results may be improved if a forecasting predictor that is more in line with the predicted price value is discovered, and if the duration of the period is reduced, such as from one month (4 weeks) to two weeks.

References

1. Brough D, Jouhara H (2020) The aluminium industry: a review on state-of-the-art technologies, environmental impacts and possibilities for waste heat recovery. Int J Thermofluids 1–2
2. Charfeddine L, Benlagha N (2016) A time-varying copula approach for modelling dependency: new evidence from commodity and stock markets. J Multinatl Financ Manag 37:168–189
3. Chopra S, Meindl P (2007) Supply chain management: strategy, planning, and operation. 3rd penyunt. Prentice Hall, Upper Saddle River, NJ
4. Haraldsson J, Johansson MT (2018) Review of measures for improved energy efficiency in production-related processes in the aluminium industry—from electrolysis to recycling. Renew Sustain Energy Rev 93:525–548
5. Hariga MA (1994) A production planning model for an aluminum company. J King Saud Univ—Eng Sci 6(2):241–263
6. Hillier FS, Lieberman GJ (2010) Introduction to operations research, 9th edn. McGraw Hill, New York
7. Hyndman RJ, Athanasopoulos G (2018) Forecasting: principles and practice. 2nd penyunt. OTexts, Melbourne
8. Kengpol A, Kaoien P, Tuominen M (2007) A procurement planning improvement by linear programming and forecasting models. IEEE, Portland
9. LINDO Systems, Inc. (2015) Optimization modeling with LINGO. 6th penyunt. LINDO Systems, Inc., Chicago, IL
10. Moritz S, Bartz-Beielstein T (2017) imputeTS: time series missing value imputation in R. R J 9(1):207–218
11. Ntantamis C, Zhou J (2015) Bull and bear markets in commodity prices and commodity stocks: is there a relation? Resour Policy 43(C):61–81
12. Parmezan ARS, Souza VMA, Batista GEAPA (2019) Evaluation of statistical and machine learning models for time series. Inf Sci 484:302–337

Measuring Customer Satisfaction on Educational Management Services in University

Cindy Saraswati, Hari Prasetyo, and Gusti Fauza

Abstract Higher education institutions in the territory of Indonesia should provide high quality education in order to produce qualified young generations who are creative, innovative, disciplined, responsible, and able to compete in the millennial era. The Government of Indonesia through the National Accreditation Board of Higher Education (BAN-PT) establishes a quality standard of education which are summarized in nine accreditation criteria. The accreditation requires the university to evaluate the customer satisfaction of those nine criteria. Limited universities have instruments for measuring the customer satisfaction based on the BAN PT accreditation. Therefore, this research is aimed to create an integrated survey system to measure the customer satisfaction specifically on learning management. The Servqual method is applied where the questions are produced to measure five service quality dimensions e.g. tangibility, reliability, responsiveness, assurance, and empathy. The study was conducted in Universitas Muhammadiyah Surakarta (UMS) involving 894 respondents. The results show that the university has met customer expectations of 84.44% with an average gap between perception and expectation value equal to − 0.423, meaning that they are satisfied with the existing services, but it is still necessary to continuously improve the quality of these services.

Keywords Servqual · Higher education · Customer satisfaction

C. Saraswati · G. Fauza (✉)
Universitas Sebelas Maret, Surakarta, Indonesia
e-mail: gustifauza@staff.uns.ac.id

C. Saraswati
e-mail: cindysaraswati007@student.uns.ac.id

C. Saraswati · H. Prasetyo (✉)
Universitas Muhammadiyah Surakarta, Kartasura, Indonesia
e-mail: hari.prasetyo@ums.ac.id

C. N. Rosyidi et al. (eds.), *Proceedings of the 6th Asia Pacific Conference on Manufacturing Systems and 4th International Manufacturing Engineering Conference*, Lecture Notes in Mechanical Engineering,
https://doi.org/10.1007/978-981-99-1245-2_49

1 Introduction

Higher education institutions in the territory of Indonesia should provide high quality education in order to produce qualified young generations who are creative, innovative, disciplined, responsible, and able to compete in the millennial era. The higher education system is a fundamental player in the development and growth of society and the economy [1]. The organizers of higher education are universities. According to Indonesian Law on Higher Education [2], universities have a function in educating the nation's life through providing a good learning management, conducting excellent research, and doing community service (called higher education 'tridharma') in order to produce creative and innovative graduates, who are competitive, and able to develop science and technology without disregarding the human values [2]. Recently, the number of universities either public or private has increased, requiring each university to continuously make improvements, especially in terms of quality. Improving the quality of higher education must be carried out through a continuous process. As the world becomes borderless, the threat not only comes from the national competitors but also from the internationals. Every year new universities are established offering various services, e.g., competitive university tuition fee, excellent facility, outperformed collaboration, in order to produce distinguished graduates that enable to be pioneers and successors of science and technology. in the future [3, 4]. Therefore, in order to be able to compete regionally and globally quality improvement must be continuously carried out on an ongoing basis [5].

In competitive business, examining whether a product provided by an organization meets or exceeds customer expectations is a key factor, hence measuring customer satisfaction is critical in organizations [6, 7] including higher education institutions [8]. However, there is no consensus has been achieved regarding what values that drive the customer satisfaction [9]. Majority research on customer satisfaction in higher education such as in [7–11] did not highlight satisfaction in the educational management system.

National board accreditation (BAN-PT) develops an accreditation system (called nine accreditation criteria) to evaluate and grade universities in Indonesia. The accreditation system requires universities to measure the user/customer satisfaction of all criteria except vision and mission. Learning/educational management system is one of the criteria that need to be evaluated i.e., how satisfy the customers to the university's services regarding this criterion. Still very few universities have a system for measuring customer satisfaction. Therefore, in this research, an integrated survey system is designed where the instruments source from BAN PT guidelines, specifically Criterion No. 6 i.e., assessing the service quality of educational management systems.

This study could be used to develop the existing education management system by evaluating the customer satisfaction of users or stakeholders, e.g. students, lecturers, deans and head of study program. The Service Quality (SERVQUAL) method is applied to analyze five dimensions of service quality i.e. tangible dimensions (physical evidence), reliability, responsiveness, assurance, and empathy [12]. The result of this study would be beneficial for universities not only to meet the BAN PT accreditation requirement but also to improve the quality of their educational management services.

2 Method

The instrument is constructed based on university business processes related to educational management service. Firstly, the design process was started by creating a framework that contains the service activities provided by this system. The results of this framework (as seen in Fig. 1) are used as a reference for constructing the educational questionnaire statements [13]. The study is conducted in Universitas Muhammadiyah Surakarta (UMS). The steps are (i) identifying the business process of UMS educational management service, (ii) generating the related attributes based on literatures including BAN-PT guidelines, (iii) validating the instruments using expert overview (head of division that related to educational management service activities), (iv) revised the instrument based on experts' suggestion, (v) test the validity and reliability of the instrument, (vi) applying the valid instrument into UMS survey information system (vii) collecting and analyze the data. The questionnaire on the educational management system consists of two questions, namely closed questions, which means that respondents can only choose the answers given by the researcher, and open questions containing suggestions and criticisms of respondents on the management of education [14]. In closed questions, there are two kinds of answers, namely the level of importance answers on a scale of 1–5 (very unimportant, not important, quite important, important, very important) and the satisfaction level answers which also consist of a scale of 1–5 (very dissatisfied, dissatisfied, moderately satisfied, satisfied, and very satisfied).

Content validation is a review process by people who are experts or experienced in their fields [15]. The expert who became the reviewer in this research was the Head of the Learning Innovation Division of the UMS Quality Assurance Institute. In content validation, reductions, editorial improvements, and additions of statements to the questionnaire were carried out. The final result of content validation is a statement that has been declared valid by the expert and is ready to be used as a data retrieval tool.

Validity testing is testing the level of reliability and validity of measuring instruments, especially questionnaires [16]. The instrument is valid if the value of Rcount is greater than the value of Rtable. This study applies the Pearson validity test (product moment) [17]. The results of the validity test show that the entire statement has a value of Rcount greater than Rtabel which is worth 0.361, meaning that all statements are valid and can be used to retrieve user satisfaction data [15, 18, 19].

Reliability testing aims to determine the consistency of the use of the instrument measurement, or if the measurement instrument is reused at a different time it will obtain a fixed result [20]. Cronbach's alpha validity test is used for testing educational instruments and if Cronbach's alpha value > 0.6 means that the instrument is reliable.

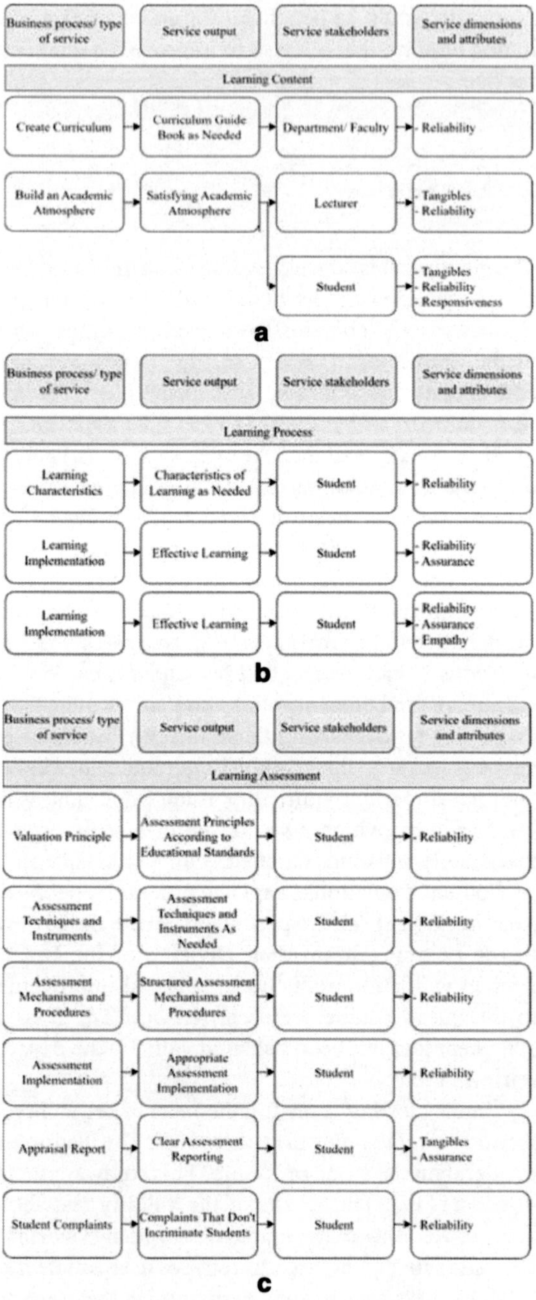

Fig. 1 Research framework **a** Learning content, **b** learning process, **c** learning assessment

The results of the reliability test showed that the value of Cronbach's alpha was 0.768, so the instrument was reliable to be used as a tool to retrieve data. The valid instrument can be seen at Table 1 in Appendix 1 (original version) and at Table 2 in Appendix 2 (English translation). The data collection process is carried out through a survey information system owned by UMS. Through this system, the resulting data can be directly recorded in the form of an excel file and the data retrieval process can be done centrally.

3 Result and Discussion

The respondent profile shows the number of respondents who filled out the survey based on certain characteristics. There are 894 respondents involved, 821 students, 66 lecturers, 4 heads of study programs and 3 deans as can be seen in Fig. 2. The respondents are from several faculties e.g., Faculty of Law 444, Faculty of Health Science 276, Faculty of Engineering 104, Faculty of Psychology 26 and Faculty of Medical 11 as seen in Fig. 3.

The SERVQUAL gap in accordance with the educational management system is the difference between the services received (expected service) and the services desired (perceived service) by customers [6]. The value of actual servqual score (ASC) is calculated to measure customer satisfaction level. The value of ASC is a ratio between the value of customer perception and the value of expectation as seen in Table 3. As presented by Table 3, overall the UMS education services has met the customer expectations since the minimum ASC score is 78.18% and the average of those scores is equal to 84.439%. These values have exceeded the minimum value of good quality that has been determined by the UMS quality assurance office, which is 75%. The ASC value is influenced by the importance and satisfaction values. The higher the satisfaction value, the higher the ASC value obtained. On the other hand, if the value of interest is greater, the ASC value will be smaller. From Table 3, some attributes still have a high gap, e.g. Q1 and Q3 The gaps value is higher than 0.8 showing that the attributes are important but the satisfaction value is quietly lower than expectation. To improve the customer satisfaction, UMS need to pay more attention to these 2 attributes.

The five quality attributes, namely, student assessment of the sufficiency, accessibility, quality of facilities and infrastructure for learning (Q1) have the smallest actual service quality scores. For the attribute with the lowest ASC, Q1, even though its ASC value is above 75%, indicating that the quality of the service is good, the satisfaction rate of 3.5 indicates that users are quite satisfied with the services provided but customer expectation on this quality attribute is quite high. It means that students expect more of the quality of facilities and infrastructure for learning. This value relates to the result in the open ending question that majority complaints come from facilities, such Wi-Fi, classroom facilities, e.t.c. These complaints can be reduced by conducting regular monitoring of the facilities, especially in the classrooms. Such

Table 1 Original version of the valid instrument (statements are in Bahasa Indonesia)

Kode	Pernyataan	Keterangan
Tangible		
Qi	Penilaian mahasiswa terhadap kecukupan, aksesibilitas, kualitas sarana dan prasarana untuk pembelajaran	Mahasiswa
Q2	Sistem Informasi penilaian pembelajaran (STAR UMS) informatif dan mudah dioperasikan	Mahasiswa
Q16	UMS menyediakan fasilitas pembelaj aran yang memadai guna menunjang sarana akademik yang kondusif	Dosen
Reliability		
Q3	Kemampuan dosen, tenaga kependidikan dan pengelola dalam memberikan pelayanan kepada mahasiswa seperti menjadi fasilitator dalam kegiatan yang melibatkan mahasiswa	Mahasiswa
Q4	Mutu dan kuantitas interaksi dosen dan mahasiswa dalam bentuk seminar, kuliah, penelitian pengabdian masyarakat dan kegiatan UKM telah sesuai untuk mewujudkan suasana akademik yang kondusif	Mahasiswa
Q5	Proses pembelajaran mata kuliah yang dilakukan oleh dosen bersifat interaktif. mengutamakan kreativitas, serta sesuai tuntutan bidang keilmuan atau jurusan yang diambil	Mahasiswa
Q6	Dosen melaksanakan pembelajaran dengan menggunakan metode dan ilustrasi yang mudah dipahami dengan memanfaatkan lingkungan sekitar, serta materi dan pustaka pendukung yang digunakan mudah diakses mahasiswa	Mahasiswa
Q7	Dosen melakukan pembelajaran sesuai dengan Rancangan Pembelajaran Semester (RPS) dan kesepakatan dengan mahasiswa	Mahasiswa
Q8	Dosen menggunakan hasil PKM dalam pembelajaran mata kuliah	Mahasiswa
Q9	Bentuk pembelajaran yang digunakan dosen sesuai dengan konteks mata kuliah dan membantu meningkatkan pemahaman mahasiswa terhadap materi	Mahasiswa
Q10	Dosen melakukan penilaian pembelajaran secara objektif transparan, sesuai kriteria dan prosedur, sesuai dengan rencana pembelajaran yang disepakati di awal perkuliahan, serta memberikan hak banding bagi mahasiswa yang kurang puas terhadap nilai yang didapatkan	Mahasiswa
Q11	Syarat kelulusan mahasiswa telah disampaikan di awal perkuliahan dan sesuai dengan Standar Nasional Pendidikan Tinggi (minimal sks yang ditempuh 144 sks, dan lain – lain)	Mahasiswa
Q17	UMS menyediakan pelatihan untuk peningkatan kualitas pembelajaran untuk dosen dengan mutu yang sesuai guna menunjang suasana akademik yang kondusif	Dosen

(continued)

Table 1 (continued)

Kode	Pernyataan	Keterangan
Q15	UMS menyediakan panduan kurikulum yang mudah diimplementasikan dan selalu melakukan pendampingan dan peninjauan ke Prodi dalam menyusun kurikulum	Prodi/Fakultas
Responsiveness		
Q12	Kemauan dosen, tenaga kependidikan, dan pengelola dalam membantu mahasiswa memberikan pelayanan dengan cepat	Mahasiswa
Assurance		
Q13	Kemampuan dosen, tenaga kependidikan dan pengelola untuk memberi keyakinan kepada mahasiswa bahwa pelayanan yang diberikan telah sesuai dengan ketentuan	Mahasiswa
Empathy		
Q14	Kesediaan/kepedulian dosen, tenaga kependidikan, dan pengelola untuk memberi perhatian kepada mahasiswa, seperti menyediakan waktu untuk berkonsultasi di luar jam perkuliahan	Mahasiswa

monitoring can be carried out by special officers or security officers or cleaners who clean the classroom every day.

The second statement that has small ASC value is the statement about The ability of lecturers, education staff, and managers in providing services to students such as being a facilitator in activities that involve students (Q3). For the attribute with the low ASC, Q1, even though its ASC value is above 75%, indicating that the quality of the service is good, the satisfaction rate of 3.58 indicates that users are quite satisfied with the services provided but customer expectation on this quality attribute is quite high. This value relates to the result in the open ending question that majority complaints come from The ability of lecturers, education staff, and manager in providing services to students such as being a facilitator in activities that involve students. These complaints can be reduced by optimization of organizations in the campus to collect student aspirations which can then be submitted to the university for follow-up.

Table 4 related to the calculation of ASC for each dimension, shows that the dimensions with the lowest ASC values are responsiveness. Thus, it can be concluded that the university excels in delivering the promised services. The responsiveness dimension relates to the willingness of the organizers to provide services quickly. Complaints felt by users regarding these dimensions include student aspirations that are not accommodated and not responded to. This is based on the results of open questions expressed by 29 respondents. These complaints can be reduced by maximizing the function of existing organizations in the campus environment as student representatives to filter the aspirations of incoming students and then submit them to the university.

Table 2 English version of the valid instrument (statements are translated into English)

Code	Statement	Description
Tangible		
Q1	Student assessment of the sufficiency, accessibility, quality of facilities and infrastructure for learning	Student
Q2	The learning assessment information system (STAR UMS) is informative and easy to operate	Student
Q16	UMS provides satisfy learning facilities to support a conducive academic atmosphere	Lecturer
Reliability		
Q3	The ability of lecturers, education staff, and manager in providing services to students such as being a facilitator in activities that involve students	Student
Q4	The quality and quantity of interaction between lecturers and students in courses, seminars, teaching and learning activities, research, and community service are appropriate to create a conducive academic atmosphere	Student
Q5	The learning process for courses carried out by lecturers is interactive, prioritizing creativity, and according to the demands of the scientific field or majors	Student
Q6	Lecturers carried out learning using methods and illustrations that are easy to understand by utilizing the surrounding environment, as well as supporting materials and libraries that are easily accessible to students	Student
Q7	Lecturers doing the coursework according to the Semester Learning Plan (RPS) and an agreement with students	Student
Q8	Lecturers use PKM results in learning courses	Student
Q9	The learning method used by the lecturer is in accordance with the context of the course and helps improve student's understanding of the material	Student
Q10	Lecturers conduct learning assessments objectively, transparently, according to criteria and procedures, in accordance with the agreed learning plan at the beginning of the lecture, and provide appeal rights for students who are not satisfied with the scores obtained	Student
Q11	Student's graduation requirements have been submitted at the beginning of the lecture and are in accordance with the National Higher Education Standards (minimum credits taken are 144 credits, and others)	Student
Q17	UMS provides training to improve the quality of learning for lecturers with appropriate quality to support a conducive academic atmosphere	Lecturer

(continued)

Table 2 (continued)

Code	Statement	Description
Q15	UMS provides curriculum guidebooks that are easy to implement and always provide Department/assistance and review to study programs in preparing curriculum	Department/faculty
Responsiveness		
Q12	Willingness of lecturers, education staff and managers in helping students provide services quickly	Student
Assurance		
Q13	The ability of lecturers, education staff, and manager to give confidence to students that the services provided are in accordance with the provisions	Student
Empathy		
Q14	Willingness/concern of lecturers, education staff and managers to pay attention to students, such as providing time for consultation outside lecture hours	Student

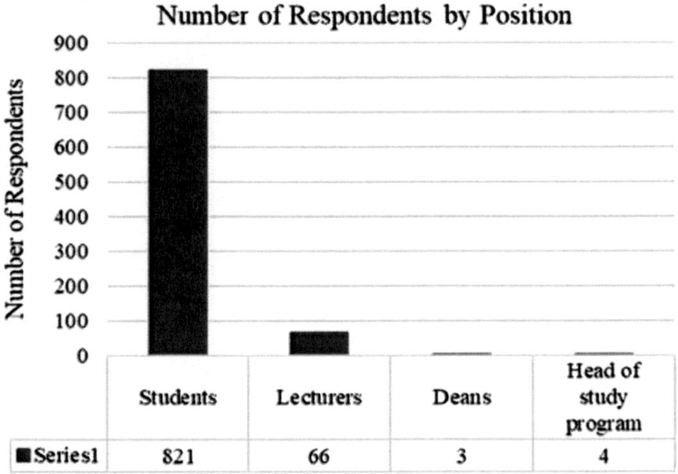

Fig. 2 Number of respondent by position

4 Conclusions

In this study, the instruments to assess the service quality of educational management have been established and implemented into the UMS information system. The data has been collected and analyzed to measure the customer satisfaction involving several stakeholders, e.g., students, lecturers, head of study programs and deans of faculties at UMS. Based on the measurement results, it can be seen that generally, users are satisfied with the educational management services at UMS. However,

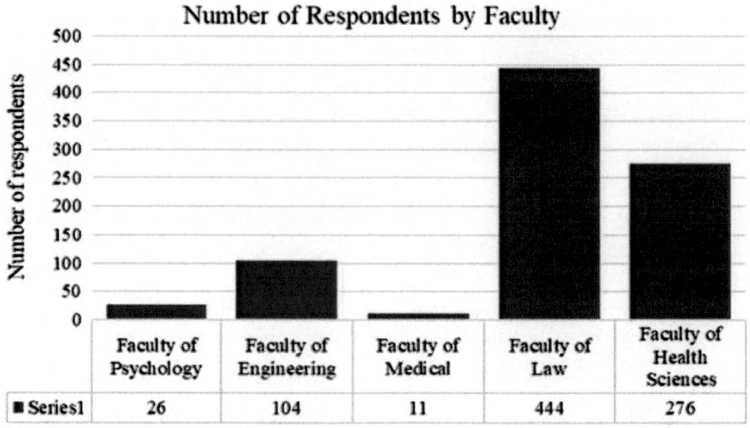

Fig. 3 Number of respondents by faculty

Table 3 ASC calculation recapitulation

Code	P	H/E	GAP	ASC (%)	Code	P	H/E	GAP	ASC (%)
Q1	3.5	4.48	− 0.98	78.18	Q10	3.4	3.91	− 0.48	87.72
Q2	3.8	4.52	− 0.75	83.36	Q11	3.8	4.51	− 0.69	84.67
Q3	3.6	4.52	− 0.94	79.22	Q12	3.7	4.54	− 0.85	81.23
Q4	3.6	4.41	− 0.77	82.65	Q13	4	4.56	− 0.56	87.82
Q5	3.6	4.47	− 0.83	81.47	Q14	4.1	4.55	− 0.42	90.72
Q6	3.7	4.4	− 0.66	85.10	Q15	4.3	4.86	− 0.57	88.24
Q7	3.7	4.45	− 0.71	84.05	Q16	4	4.7	− 0.74	84.19
Q8	3.7	4.51	− 0.85	81.24	Q17	4.2	4.79	− 0.59	87.66
Q9	3.9	4.45	− 0.54	87.98					

P—perception, H/E—expectation, ASC—actual servqual score

Table 4 ASC Results for each Dimension

No.	Dimension	P	(H/E)	GAP	ASC (%)
1	Reliability	3.81	4.46	− 0.65	85.355
2	Tangible	3.94	4.65	− 0.71	84.687
3	Assurance	3.64	4.41	− 0.77	82.648
4	Empathy	3.64	4.47	− 0.83	81.465
5	Responsiveness	3.58	4.52	− 0.94	79.215
	Average	3.72	4.5	− 0.78	82.674

P—perception, H/E—expectation, ASC—actual servqual score

based on the open-ended question results of, there are several things that need to be reviewed or improved, including inadequate learning facilities, inappropriate learning methods or systems, transparency of assessments and responses to students regarding value comparisons, dissemination of information related to learning, and Regular Academic Integrated System (STAR) UMS which often an error occurs when accessed simultaneously. The results of this study could help UMS to increase its quality services to achieve a higher customer satisfaction. However, this study only focuses on assessing the customer satisfaction of educational management with only one section of nine accreditation criteria. Therefore, developing instruments to measure the customer satisfaction of stakeholders on other criteria would be necessary for the forthcoming research. Future research in the field of service quality perceptions should consider using different data gathering methods in an attempt to achieve a degree of data triangulation that further supports the validity of the instrument.

Appendix 1

See Table 1.

Appendix 2

See Table 2.

References

1. Rozak LA et al (2022) Empirical evaluation of educational service quality in the current higher education system. Emerg Sci J 6(Special Issue):55–77. https://doi.org/10.28991/ESJ-2022-SIED-05
2. Undang-Undang Republik Indonesia Nomor 12 Tahun 2012 Tentang Pendidikan Tinggi. Pemerintah Republik Indonesia, Indonesia, 2012, p 32. Accessed Sept 15 2022. [Online]. Available: https://lldikti8.ristekdikti.go.id/2019/02/05/undang-undang-republik-indonesia-nomor-12-tahun-2012-tentang-pendidikan-tinggi/
3. Kurniasih N, Hasyim C, Wulandari A, Setiawan MI, Ahmar AS (2018) Comparative case studies on Indonesian Higher Education Rankings. J Phys: Conf Ser 954(1). https://doi.org/10.1088/1742-6596/954/1/012021
4. Chankseliani M, Qoraboyev I, Gimranova D (2021) Higher education contributing to local, national, and global development: new empirical and conceptual insights. High Educ 81(1):109–127. https://doi.org/10.1007/s10734-020-00565-8
5. Delgado BC, Farrapo AC, Mergulhão RC, Silva DAL (2022) Assessment of the quality of digital services provided by an e-learning platform focused on sustainability issues. Production 32. https://doi.org/10.1590/0103-6513.20210110
6. Jones JL, Shandiz M (2015) Service quality expectations: exploring the importance of SERVQUAL dimensions from different nonprofit constituent groups. J Nonprofit Public Sect Mark 27(1):48–69. https://doi.org/10.1080/10495142.2014.925762

7. Tarí JJ, Dick G (2016) Trends in quality management research in higher education institutions. J Serv Theory Pract 26(3):273–296. https://doi.org/10.1108/JSTP-10-2014-0230
8. Munteanu C, Ceobanu C, Bobâlcă C, Anton O (2010) An analysis of customer satisfaction in a higher education context. Int J Public Sect Manag 23(2):124–140. https://doi.org/10.1108/09513551011022483
9. Khoo S, Ha H, McGregor SLT (2017) Service quality and student/customer satisfaction in the private tertiary education sector in Singapore. Int J Educ Manag 31(4):430–444. https://doi.org/10.1108/IJEM-09-2015-0121
10. Teeroovengadum V, Kamalanabhan TJ, Seebaluck AK (2015) Quality assurance in education article information. Qual Assur Educ 24(2)
11. Annamdevula S, Bellamkonda RS (2016) The effects of service quality on student loyalty: the mediating role of student satisfaction. J Model Manag 11(2):446–462. https://doi.org/10.1108/JM2-04-2014-0031
12. Parasuraman A, Zeithaml VA, Berry LL (1988) Servqual: a multiple-item scale for measuring consumer perceptions of service quality. J Retail 64(1):12–40. https://doi.org/10.1002/9781118785317.weom090654
13. Aboubakr RM, Bayoumy HMM (2022) Evaluating educational service quality among dentistry and nursing students with the SERVQUAL model: a cross-sectional study. J Taibah Univ Med Sci 17(4):648–657. https://doi.org/10.1016/j.jtumed.2022.01.009
14. Gilavand A, Torabipour A (2022) The quality of services of Iran University Hospitals based on SERVQUAL's evaluation model: a systematic review and meta-analysis. Front Public Health 10. https://doi.org/10.3389/fpubh.2022.838359
15. Ihsan H (2015) Validitas Isi Alat Uukur Penelitian: Konsep Dan Panduan Penilaiannya. Pedagog J Ilmu Pendidik 13(3):173. https://doi.org/10.17509/pedagogia.v13i3.6004
16. Sugiyono (2004) Statistik untuk Penelitian. CV Alfabeta, Bandung
17. Budiastuti D, Bandur A (2018) Validitas dan Reliabilitas Penelitian [Online]. Available: www.mitrawacanamedia.com
18. Rahayu AP, Santoso HB, Rahayuningsih S (2019) Analisa Kepuasan Sistem Informasi Akademik Mahasiswa Menggunakan E-Servqual. JATI UNIK J Ilm Tek dan Manaj Ind 2(1):55. https://doi.org/10.30737/jatiunik.v2i1.387
19. Yusup F (2018) Uji Validitas dan Reliabilitas Instrumen Penelitian Kuantitatif. J Tarb J Ilm Kependidikan 7(1):17–23. https://doi.org/10.18592/tarbiyah.v7i1.2100
20. Latif KF, Latif I, Farooq Sahibzada U, Ullah M (2019) In search of quality: measuring higher education service quality (HiEduQual). Total Qual Manag Bus Excell 30(7–8):768–791. https://doi.org/10.1080/14783363.2017.1338133

Perception of Employees on Lean Six Sigma Practices and Corporate Sustainability in Automotive Manufacturing Industry

Nur Syahirah Mohamad Azlan and Noraini Mohd Razali

Abstract This study aims to examine the perception of employees in the Malaysian automotive industry towards their company regarding the level of awareness in the implementation of Lean Six Sigma (LSS) and corporate sustainability. Two sets of questionnaires each containing two sections were developed in this survey. Surveys were given to Top Management and Executive positions, and Shop Floor employees. The evaluation method is generally in the form of a Five-Point Likert Scale Questionnaire. The targeted respondents were employees from three different automotive manufacturing companies. The data obtained was analyzed using Minitab. The results show a good perception of the implementation of LSS and awareness of corporate sustainability among employees in the automotive sector. However, the study revealed that no correlation between LSS practices and the sustainability awareness in companies where high LSS implementation will not affect corporate sustainability awareness. Findings from this study also revealed that the environmental aspect is the most impacted by LSS practices. There is also a correlation between the positions and the level of awareness in LSS implementation and sustainability performance in which the upper management shows a better awareness towards the practices compared to Shop Floor employees. In general, employees in automotive sector do have good perception on LSS implementation and corporate sustainability awareness. This will reinforce the culture of sustainability awareness among the engineers and encourage them to think on sustainability ideas during the concept and design stages of product development.

Keywords Lean · Six sigma · Sustainability · Automotive

N. S. M. Azlan · N. M. Razali (✉)
Universiti Malaysia Pahang, Pekan, Malaysia
e-mail: norainimbr@ump.edu.my

C. N. Rosyidi et al. (eds.), *Proceedings of the 6th Asia Pacific Conference on Manufacturing Systems and 4th International Manufacturing Engineering Conference*, Lecture Notes in Mechanical Engineering,
https://doi.org/10.1007/978-981-99-1245-2_50

1 Introduction

Manufacturing businesses utilize a lot of energy and generate a lot of waste all over the world. Mallak et al. [1] stated that companies have become more conscious of the strategic position of sustainable performance for competitive advantage as customer pressure for businesses to be economically, socially, and financially responsible has grown. According to Deloitte [2], Lean Six Sigma (LSS) can be an effective way for generating a green road map. Those combined strategies can also help businesses improve their environmental performance while moving toward sustainability and economic growth.

The term "Lean Six Sigma" was used in the late 1990s and early 2000s to describe the combination of lean and Six Sigma philosophies [3]. LSS is defined as a business strategy and methodology that increases process performance, and develops customer satisfaction, leadership, and bottom line results by improving quality, speed and costs [4]. On the other hand, sustainability is based on the three pillars; environmental, economic and social. If any of one pillar is weak then the system as a whole is unsustainable. According to Gladwin [5], the three dimensions must be satisfied at the same time to achieve a better position in the market.

Sustainable manufacturing is a sort of industrial development that fulfils current requirements without compromising future generation's ability to meet their own. The United States Department of Commerce coined the term "sustainable production" to describe the production of manufactured goods that minimize negative environmental impacts, conserve energy and natural resources, and are safe and economically sound for workers, communities, and consumers [6].

Nowadays, the pressure of competition from multinational corporations has intensified, and the automotive manufacturing industry is one of them. It has an impact as the level of competition rises when the manufacturing of automobiles moves from a national to a global level. Understanding the application of LSS ideas is very useful for Malaysia to be a good competitor in Asian as part of the competition [7]. Moreover, incorporating sustainability into all aspects of business and industry is a strategic need for company competitiveness as it can support business continuity and long-term value for stakeholders and society at large. However, the impact of LSS on sustainability performance is still unclear today because some organizations just apply the principle of LSS without evaluating the sustainability performance.

Previously the LSS is successfully implemented in manufacturing of automotive components [8–10]. In the context of automotive industry in Malaysia, these are the following questions that this study aims to answer:

(a) If the level of LSS implementation in the company is high, will the sustainability awareness also be high?
(b) Do the employees from such companies have good perceptions on the company's sustainable business practices according to the LSS theory?

(c) If the employee is from the upper management, does that mean he has a high level of understanding in sustainability?

Therefore, this study aimed to identify the current level of awareness on LSS implementation and the sustainability aspects in the automotive company by analyzing the perceptions from the employees themselves since they are directly involved in the activities executed based on the company's corporate sustainability policy.

2 Methodology

The study begins by a thorough review of the literature on the subject as comparable studies have been conducted in various businesses around the world. Figure 1 shows the flowchart of methodology for the study.

Three chosen automotive manufacturing company (i.e. Company A, B, and C) that apply LSS techniques make up the study's sampling frame. Company A is a private limited company which founded in 1993 with the mission of providing Malaysians with total mobility through useful, high-quality products and services that are given with the utmost care. The company's portfolio includes pre-owned cars as well as body and paint repair, sales, after-sales service, and car manufacturing. The company has two manufacturing plants which are located in Sungai Choh, Rawang. The company sold nearly 4 million vehicles as of the end of 2021,

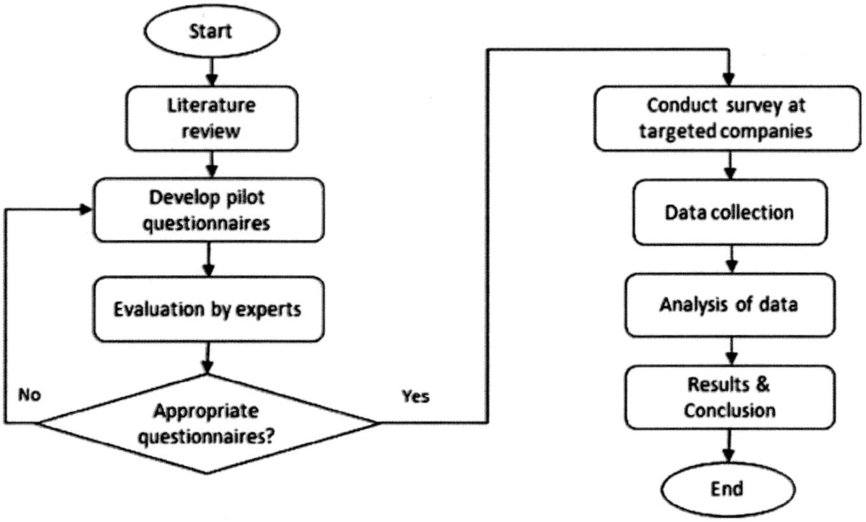

Fig. 1 Flowchart of methodology for the study

assisting numerous Malaysian suppliers, dealers, and business partners. Both manufacturing plants can produce up to 320,000 vehicles a year. On the other hand, Company B is a Malaysian automobile manufacturer company with its headquarters in Tanjung Malim, Malaysia. It is a part of the Miscellaneous Durable Goods Wholesalers industry. The company was officially founded on May 6, 2003 where the business creates, develops, and provide sports, luxury, and commercial vehicles. In 2020, the company's net sales revenue increase by 79.77%. The third company which is Company C produces and assembles automobiles with its headquarters in Pekan, Malaysia. The company is the biggest conglomerate in Malaysia founded in 1992. Currently, the company assembles Mercedes-Benz and Volkswagen vehicles for the Malaysian market. With a goal of becoming the preferred assembler in the automotive industry, the company has built about 500,000 vehicles of various worldwide marques throughout the course of its 30-year history. In 2020 to 2021, the company's net sales revenue increase by 33.03%.

Top management (Manager, Head of Department), executive (Engineer, Supervisor), and shop floor (Operator, Technician) workers are most likely to participate in the survey. The questionnaire's essential characteristics were derived and constructed from reviewed the existing literature, LSS Success Factors, human and process aspect of improvements, most relevant impacts caused by LSS based on most citation, and sustainability indicators.

The survey study consists of two sets of questionnaires to avoid bias in the survey data as the level of awareness in LSS practices and sustainability awareness in the organization may be different among the respondents. Different sets of questions were constructed based on their understanding and work experiences in the company. Set 1 of Section A contain questions pertaining to the current level of awareness on LSS practices from Top Management and Executive perception while Section B contain questions about the sustainability awareness in term of three aspects (Social, Environment and Economic) resulting from the implemented of LSS practices from Top Management and Executive perception. Set 2 of Section C contain questions pertaining to the current level of awareness on LSS practices from Shop Floor employees' perception while Section D contain questions about the sustainability awareness in term of three aspects (Social, Environment and Economic) resulting from the implemented of LSS practices from Shop Floor employees' perception. The ratings are made using Five-point Likert scales with possible responses from 'strongly disagree' to 'strongly agree'.

The data is analyzed using the Cronbach's Alpha, Levene's test, ANOVA, Two Sample t-Test, and Fisher Least Significant Difference (LSD). Depending on the Levene's Test result, the one-way ANOVA test with the same variance or the Two Sample t-Test with different variances will be used. The hypothesis of each test is described in Table 1. To evaluate the sustainability performance of the companies, the rating in each question will be used to calculate the mean score for each aspect as shown in Table 2.

Table 1 Hypothesis of each test

Levene's test	Ho: There is no difference in variance between the populations
	Ha: There is a difference in variance between the populations
One-Way ANOVA	Ho: There is no difference between the groups and equality between means
	Ha: There is a difference between the means and groups
Two-Sample T-test	Ho: $\mu A = \mu B$
	Ha: $\mu A \neq \mu B$

Table 2 Overall sustainability performance

Legend	Rating
Negative impact high	1
Negative impact medium	2
Neutral	3
Positive impact medium	4
Positive impact high	5

3 Results and Discussion

3.1 The Awareness of LSS Practices in the Organization Between Top Management, Executives, and Shop Floor of the Companies

In total, there are 81 respondents involved and all respondents are direct employees of the automotive sector. As the majority of the respondents (61.11%) are from the Top Management and Executive level, therefore high expectation is expected from them in this study as we assumed they were involved in company's strategic plan and directives in automotive industry. As for Shop Floor employees (28.89%), the data is considered reliable based on the assumption that they are more knowledgeable and familiar with the LSS practices in the production floor of automotive industry and relative performance of their organization. Moreover, they ought to have easier access to important and reliable data from their organization to complete the survey.

Figure 2 shows the Levene's Test on the awareness of LSS practices between Top Management, Executives, and Shop Floor employees in Company A, B, and C. The *P*-value of Levene's Test is 0.739 which is greater than $\alpha = 0.05$ indicates that there is no difference between the variances. Thus, the null hypothesis, H_0 is accepted. Therefore, we have insufficient evidence to reject the null hypothesis and ANOVA test is carried out to obtain the detailed result for each question. From the data, it can be said that the Top Management and Executive have better understanding towards LSS implementation in the companies compared to Shop Floor employees.

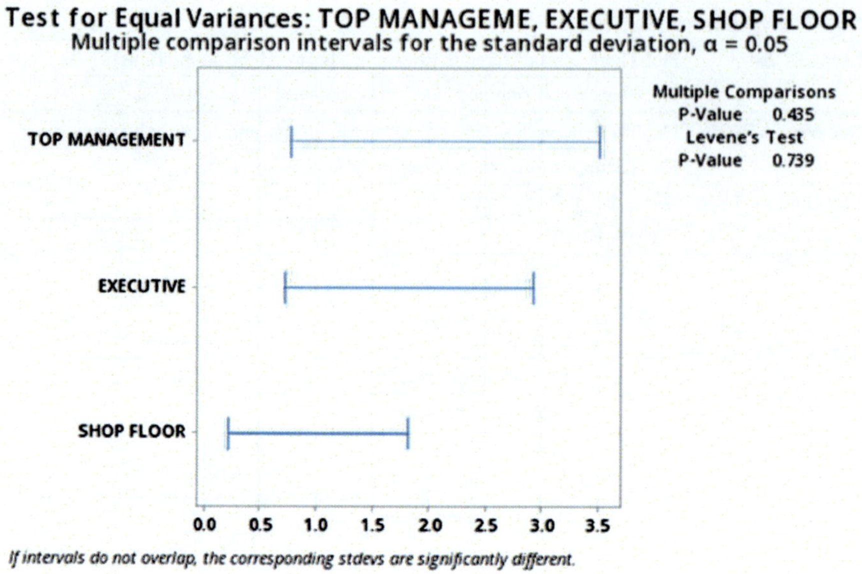

Fig. 2 Test for equal variances for LSS implementation between positions

3.2 *The Relationship Between Level of Implementation of LSS Practices and Corporate Sustainability Performance of Company A, B, and C (Top Management and Executives)*

The result in Fig. 3 shows the level of LSS implementation is higher compared to the sustainability awareness showing that in Malaysia, the employees in automotive sector are aware of LSS implementation but they have less understanding on sustainability. However, both top management and executives believes that the higher the understanding in LSS tools and practices, the higher the positive impact it brought towards the sustainability performance. It can be shown through the result of Company A with the highest average mean score for sustainability performance awareness which is 4.22. Apart from that, the result for the most impacted sustainability aspect from the LSS implementation is environmental with value of mean score 4.54.

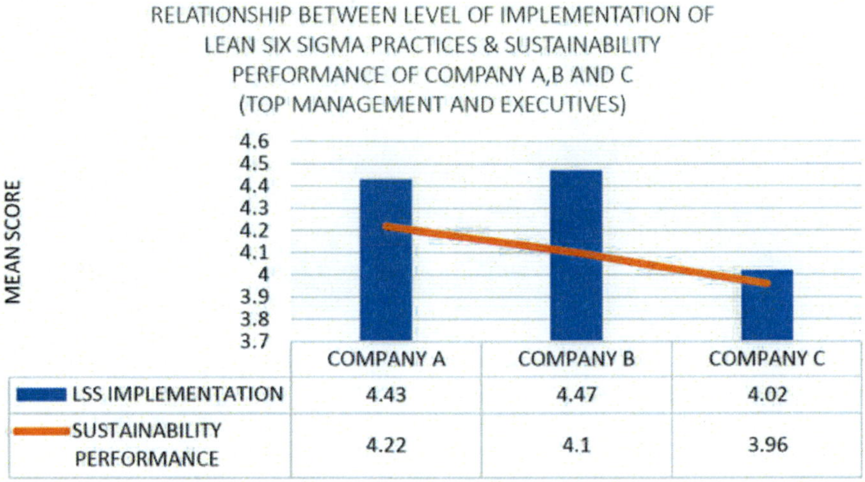

Fig. 3 The Relationship between level of LSS implementation and sustainability awareness from the top management and executives perspective

3.3 The Relationship Between Level of Implementation of LSS Practices and Corporate Sustainability Performance of Company A, B, and C (Shop Floor)

The result in Fig. 4 shows no correlation between the level of implementation of LSS practices and sustainability awareness among the Shop Floor employees. From the data, it can be shown that the sustainability awareness among the shop floor employees is higher compare to the understanding of the LSS implementation. This result may be influence by the technical tasks performed in their daily basis. Company C shows the highest sustainability performance awareness with the value of 4.42 indicates high positive impact of corporate sustainability in the company. However, the result for the LSS implementation awareness is the lowest among the companies. Therefore, it can be assume that they were exposed more to the practical knowledge which is the LSS tools and techniques on the production floor compare to the concept and theories about LSS itself. As for the sustainability performance, the result from the detailed analysis shows that the environmental aspect is the most impacted with the LSS implementation and the social aspect score the least.

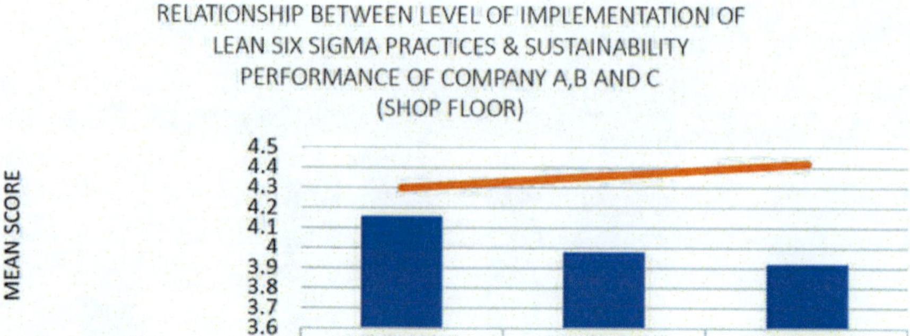

	COMPANY A	COMPANY B	COMPANY C
LSS IMPLEMENTATION	4.16	3.98	3.92
SUSTAINABILITY PERFORMANCE	4.3	4.36	4.42

Fig. 4 The relationship between level of LSS implementation and sustainability awareness from the shop floor employees perspective

4 Conclusion

The three companies have the same company background which is car manufacturers. Based on the overall result, employees in automotive sector do have good perception on LSS implementation and corporate sustainability awareness. This is due to the maturity of lean use in automotive sector is high compare to another sector. Moreover, the companies are manufacturing plant in which the production process is based on an ideology of maximizing productivity while simultaneously minimizing waste within a manufacturing operation and the LSS practices such as 5S, Kaizen, FMEA, and leadership are visible in the management especially on the production floor. In terms of sustainability, the results obtained are vary for different level of employees in the automotive sector which in the form of understanding about the sustainability concept and the technical job that gives benefits towards the sustainability. Since the companies are manufacturing plant, the result for the sustainability awareness is high among the shop floor employees which may be due to their understanding about the technical job that give benefit towards the sustainability.

In summary, few conclusions are observed from this study. The employees in automotive sector do notice that elements of LSS initiatives are important to drive the sustainability in the business performance of an organization. This could serve as the motivation for organization to incorporate sustainability as a core component of their strategic plan. This can be proven through the achievement of Company A that has high level of awareness in both LSS implementation and sustainability. In general, employees in automotive sector do have good perception on LSS implementation and corporate sustainability awareness. Additionally, this will reinforce the culture of sustainability awareness among the engineers and encourage them to think about sustainability ideas during the concept and design stages of product development.

Furthermore, there is a correlation between the positions and the level of awareness in LSS implementation and sustainability in which the upper management shows a better awareness towards the practices compared to Shop Floor employees.

Future research can expand the study by adding open-ended interviews since the Likert Scale Questionnaire limits the feedback and responses from the respondents. It is also recommended that the number of respondents should same for each of the companies or sectors preferred in order to get more precise data and analysis.

Acknowledgements This research is fully funded by the Faculty of Manufacturing & Mechatronics Engineering Technology, Universiti Malaysia Pahang under the Grant number RDU210370.

References

1. Mallak SK, Ishak MB, Mohamed AF, Iranmanesh M (2018) Toward sustainable solid waste minimization by manufacturing firms in Malaysia: strengths and weaknesses. Environ Monit Assess 190(10)
2. Deloitte Development LLC (2008) Green lean six sigma: using lean to help drive results in the wholly sustainable enterprise
3. Byrne G, Lubowe D, Blitz A (2000) Using a lean six sigma approach to drive innovation. Strat Leadersh 35(2):5–10 (Sheridan JH)
4. Snee RD (2010) Lean six sigmae getting better all the time. Int J Lean Six Sigma 1(1):9–29
5. Gladwin TN, Kennelly JJ, Krause TS (1995) Shifting paradigms for sustainable development: implications for management theory and research. Acad Manag Rev 20(4):874–907
6. Martin Steel (2016) What is sustainable manufacturing? https://www.marlinwire.com/blog/what-is-sustainable-manufacturing-2
7. Zamri FIM, Hibadullah SN, Fuzi NM, Desa AFNC, Habidin NF (2013) Green lean six sigma and financial performance in Malaysian automotive industry. Bus Manage Strategy 4(1):97
8. Gupta V, Jain R, Meena ML, Dangayach GS (2018) Six-sigma application in tire manufacturing company: a case study. J Ind Eng Int 14(3):511–520
9. Kowang TO, Yong TS, Rasli A, Long CS (2016) Lean six sigma sustainability framework: a case study on an automotive company. Asian J Sci Res 9(5):279–283
10. Ruben R, Ben Vinodh, S, Asokan P (2017) Implementation of lean six sigma framework with environmental considerations in an Indian automotive component manufacturing firm: a case study. Prod Plann Control 28(15):1193–1211

Cognitive-Affective Model of Perceived User Satisfaction (CAMPUS) on the Design of E-Learning Interface: A Literature Review

Melati Salma and Ari Widyanti

Abstract Quality is seen as a tool to achieve competitive advantage, including in design of learning-interface. Purpose of this study is to review literature of Cognitive-Affective Model of Perceived User Satisfaction (CAMPUS) on the design of e-learning interface in increasing competitive advantage. To achieve the objective of this study, the method used are a literature review from past papers from Google Scholar. Result shows that there are some constructs to be considered in e–learning design that are aesthetic, efficiency, effectiveness, and playfulness. A model is proposed based on this construct. Implications of the result is discussed.

Keywords E-Learning · Cognitive-Affective model · Satisfaction · Motivation

1 Introduction

The use of information technology as well as internet is growing rapidly in Indonesia. It was reported that there are 204.7 million internet users in Indonesia per January 2022. Start-up companies that are using internet are starting to be active in offering online services through applications and software that support today's human needs [1]. The use of the internet is currently developing not only for access to entertainment and business, but also contributes to the world of education to support learning both as a tool to access wider information as well as supporting teaching and learning activities [2, 3]. According to Lestari [2], there are several forms of information and communication technology that can be implemented in the world of education, such as learning media, learning resources, and educational administrative tools.

According to Al-Sofi [4], the sudden spread of the Covid-19 pandemic prompted institutions to think of innovative alternatives for the smooth running of teaching and learning activities. Therefore, since the Covid-19 pandemic emerged, teaching and

M. Salma (✉) · A. Widyanti
Bandung Institute of Technology, Jawa Barat, Bandung 40116, Indonesia
e-mail: 23421013@mahasiswa.itb.ac.id

C. N. Rosyidi et al. (eds.), *Proceedings of the 6th Asia Pacific Conference on Manufacturing Systems and 4th International Manufacturing Engineering Conference*, Lecture Notes in Mechanical Engineering,
https://doi.org/10.1007/978-981-99-1245-2_51

learning activities have shifted from face-to-face activities at schools and campuses to online learning activities (e-learning). Schools and companies finally see this problem as an opportunity to develop and utilize e-learning technology even further.

The success of implementing e-learning is so great that the use of technology for education is not only considered as an emergency measure in the context of the Covid-19 pandemic, but also as a means for educational institutions in the context of new teaching methods (remote mode) to be adapted in their various classes [5]. Maloney and Kim [6] also say that institutions will harness the full potential of e-learning while maintaining the fundamental core of the offline learning experience.

Moreover, according to the report published by Research and Market [7], the market analysis of global e-learning industry trends will increase at a CAGR of 16.6% over the next decade and is expected to reach a profit of US $2424.7 billion. This increase is said to grow due to the high rate of learning through games (gamification), security through information technology, cloud-based solutions, rapid growth in digitization, and innovation from existing technologies that make learning switch to cloud-based systems.

Currently in Indonesia, there are many start-up companies that are developing educational service businesses as online learning media. Among them are RuangGuru, Zenius, Quipper, IndonesiaX, and others. The article from Bisnis.com quoted Rohan Monga as CEO of Zenius, that currently Zenius has more than 20 million users as of 2020 [8]. Not only Zenius, RuangGuru also has more than 22 million users as of the end of 2020 [9]. This shows that about 7% of Indonesia's total population of 273 million people [10] use e-learning-based learning applications and websites.

Quality is seen as a tool to achieve competitive advantage, because quality is one of the main factors that determine the selection of products and services for consumers. Customer satisfaction will be achieved if the quality of the products and services provided is in accordance with their needs [11]. Customer satisfaction is the level of one's feelings from the results of comparing perceived product performance in relation to one's expectations [12] and is an evaluation process between what is received and what is expected [13]. Feeling satisfied means feeling happy when customers get what they want [14].

According to Coursaris [15], the aspect that affects customer satisfaction in using information systems, including in this case e-learning, is the cognitive-affective aspect. The satisfaction that a person feels is the result of the perception process on the cognitive aspect and the process of the emergence of emotions from the affective aspect. Cognition is an activity or process of acquiring knowledge, including awareness, feelings, and so on, or an attempt to recognize something through one's own experience [16]. Meanwhile, according to Krathwohl, the affective aspect (affection) is a process that burdens feelings, emotions, or the degree of rejection or acceptance of an object. These processes are concerned with the interaction between customers or users with the services or products they use.

Coursaris [15] said that the visual design of an information system (IS) has a strong potential to assess the emotions of users and have an impact on user behavior and satisfaction. Previous studies have found that aspects such as color, image, shape, and design composition affect a person's perception of a brand, including their perception of utility [16]. However, not only the physical aspect of visual design affects the emotion of user, but studies show that collective culture influenced the impact to both an individual and organization [17]. Therefore, cultural aspects are considered in this study.

At the other hand, the use of e-learning also has direct and significant relationship for the student's motivation. Harandi [18] stated that motivation emerges from e-learning application in class which led to student's engagement. When engagement done successfully, the student will be more likely to achieve the learning objectives which should led to the effectiveness indicator of usability [19]. Therefore, we would like to know if there any significant effect from student's satisfaction to their learning motivation.

This study will discuss the development of cognitive-affective models that affect customer satisfaction and learning motivation of e-learning applications by considering the aspects and dimensions of the website design process (interaction design). The CAMPUS model is used as the basis for evaluating e-learning system satisfaction and learning motivation because the usability and experience variable of the website is part of human computer interaction (HCI), where human behavior can be evaluated [20].

2 Methods

To achieve the objective of this study, the method used are a literature review from past papers from Google Scholar with keywords of "e-learning", "satisfaction", "motivation", "cognitive", "affective", and "higher education" from year 2012 to 2022. This study is divided into three categories. First, concluding the state of the are from the previous research to explain the factors that influence satisfaction. The second step is to position the research between the state of the art. The third is to explain the constructs that will be used in further research and its relevance to the study.

3 Result and Discussion

There are many cognitive-affective aspects that affect customer satisfaction. The following is a state-of-the-art table that shows the comparison of research variables conducted by previous researchers and the position of this study. Mapping of searching result can be seen in Table 1.

Based on the search result, some construct that must be considered in framework development of CAMPUS are as follows:

Table 1 Existing research dimensions

	Sarsam [21]	Coursaris [15]	Parsazadeh [22]	Klett and Pharow [16]	Lindgaard [23]	Suryani [24]	Prastawa [25]	Research Position
Usability								
Efficiency		x		x	x	x	x	x
Effectivity		x		x	x		x	x
User Interface	x	x	x	x				x
Interactivity		x	x	x			x	x
Aesthetic		x			x			x
Color		x					x	x
Brightness							x	x
Harmony		x					x	x
Salience							x	x
Comfort							x	x
Attractiveness							x	x
Clarity		x						x
Symmetry		x						x
Contrast		x						x
Originality		x						x
Creativity		x						x
Involvement					x			
Attitude					x			x
Motivation					x			x

3.1 Cognition

Cognition is an activity or process of acquiring knowledge, including awareness, feelings, etc., or an attempt to recognize something through one's own experience [26]. Cognition is all processes that enter the human sensor and are processed either transformed, reduced, elaborated, stored, and used [20]. It can be concluded that cognition is a process that occurs in the human brain regarding the process of processing incoming data through sensors. According to Wickens [27] the process of cognition occurs through several stages, namely the process of receiving sensors from stimuli, perception, attention, selection of responses, and implementation of responses. These processes occur in the brain which consists of memories.

3.2 *Affection*

According to Krathwohl [28], the affective aspect (affection) is a process that burdens feelings, emotions, or the degree of rejection or acceptance of an object. Based on Schiffman [29], affection is a direct experience and information obtained from various sources in the form of knowledge and perceptions which generally take the form of beliefs [30]. The affective component or affection is an emotional aspect that describes a person's feelings towards an object. According to [31], the constructs forming the affective aspect are emotion [31], motivation [32], and attitude [33]. Lindgaard [23] said that from previous neurophysiological studies, researchers found that emotional responses occur quickly, about 3–4 ms since the stimulus is shown.

User satisfaction is an indication of system success [34]. According to ISO 9241-11, user satisfaction is a contribution forming usability together with effectiveness and efficiency [35]. While Lindgaard [23] cites Hassenzahl [36] who thinks that feelings of pleasure or satisfaction with a system (user satisfaction) arise from good usability. The satisfaction model referenced for the study is CAMPUS Model of Coursaris [15] which contains five variables of effectiveness, efficiency, classical aesthetic, expressive aesthetic, and playfulness as stated in the CAMPUS model.

3.3 *CAMPUS Model*

The CAMPUS model that is proposed can be seen in Fig. 1.

Utility: Cognition Aspect

In this study, usability group refers to how the cognitive flow is being processed objectively to assess the instrumental value that people derive from interacting with a technology.

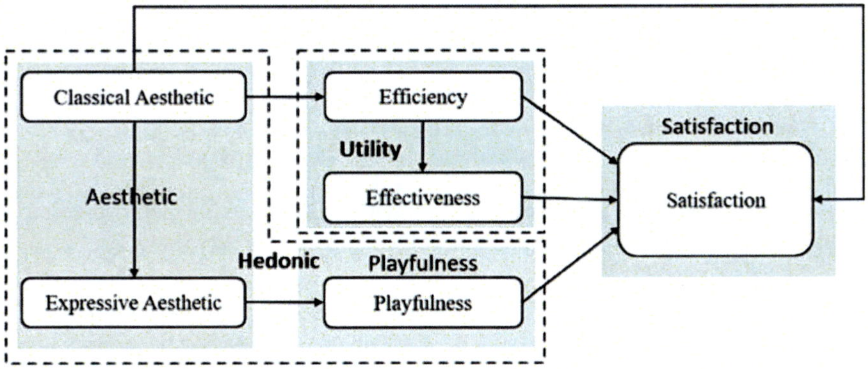

Fig. 1 CAMPUS model of information system satisfaction

Coursaris [15] stated that satisfaction aspects emerge from two usability aspects that is effectiveness and efficiency. Effectiveness is often defined as the quality of the product or desired result. According to Rubin [37], effectiveness in the context of usability refers to how a product is produced in accordance with user expectations and how easy and appropriate something is to the user's wishes. Effectiveness, as one aspect of usability, has an important role in assessing user satisfaction. In a cognitive context, effectiveness relates to accuracy and completeness in completing a goal [15].

Efficiency is the speed of a person (user) in achieving the desired goal or goal accurately, completely, and according to a predetermined time period [37]. In a cognitive context, efficiency focuses on how resources are used and utilized to achieve a goal effectively. In this case, efficiency provides a measurement of how the performance of the time and effort made by the user in using a technology [15, 38, 39]. Cognitive efficiency (CE) or mental efficiency can be interpreted by three things, which is neurological efficiency, where this is evidenced by the location and level of brain activity during the learning process; as instructional efficiency, which is the relative effectiveness of a method in studying a thing or teaching material as measured by learning outcomes. or as the efficiency of performance and the learning process, which is determined by regulating an effort in the learning process [40].

Hedonic: Affection Aspect

While the utility describe satisfaction driven by the objective values, hedonic aspects emphasize the satisfaction by subjective and self-fulfilling values that derive from interacting with technology. Coursaris [15] analyzed two dimensions of hedonic aspect of technology that is aesthetic and playfulness.

Classical aesthetic as stated emphasize clear and orderly design and associated with the design rules by usability experts which consist of attributes of clarity, cleanliness, symmetry, contrast, and others. Meanwhile expressive aesthetic addresses the originality, sophistication, and creativity in terms of graphics, layout, and typography. These dimensions evaluated as there are studies explains the relation between aesthetic and usability that focus on cognitive load, increased motivation, focused attention, and felt involvement where when the high consideration of aesthetic decreases cognitive load and increases user satisfaction [15].

3.4 Proposed Construct for Further Research

The following subchapter lists the factors that will be inspect as the factors predicted to influence satisfaction and its relevance to the further study (see Table 2).

Table 2 Proposed construct and its relevance

Construct	Description	Relevance to this study
Classical aesthetic	Classical aesthetic as stated emphasize clear and orderly design and associated with the design rules by usability experts which consist of attributes of clarity, cleanliness, contrast, symmetry, and others. [15]	According to Coursaris [15], classical aesthetic in few studies have pointed positive effects on perception of efficiency and satisfaction regarding cognitive load, increased motivation, and involvement
Expressive aesthetic	Expressive aesthetic addresses the originality, sophistication, and creativity in terms of graphics, layout, and typography [15]	According to Coursaris [15] expressive aesthetic indirectly affect satisfaction through playfulness
Perceived effectiveness	Effectiveness in the context of usability refers to how a product is produced in accordance with user expectations and how easy and appropriate something is to the user's wishes [37]	Users are more likely to be satisfied if they can perform a task effectively
Perceived Efficiency	Efficiency is the speed of a person (user) in achieving the desired goal or goal ac-curately, completely, and according to a predetermined time period [37]	According to Coursaris [15], perceived efficiency has a role in the user acceptance of technologies by assisting and supporting perceived effectiveness
Perceived playfulness	Playfulness refers to subjective characteristic of experience [15]	It has been reviewed that playful state emerges from the perceived aesthetic of the system which would affect the satisfaction
Motivation	Motivation serves as the impetus behind direction, regulation, and steadfastness in human conduct. [41]	Motivation will likely occur when the user felt any satisfaction regarding the interaction they made with an object. Therefore, motivation can be considered as the parameter to assess the performance of the satisfaction
Demography	Demographics encompass a range of population attributes, including age, gender, marital status, race, ethnicity, income, occupation, education, and additional factors. [42]	The human or user aspect of the HCI system is an aspect that needs to be reviewed. This is because user satisfaction is an affective aspect that arises from human perception when using a system
Attitude	Attitude is defined as the student's impression of the e-learning program [43]	According to Law [44] attitude, expectations, goals, pressure from the environment, and competition are factors that affect the motivation of the learning process. The factors that influence this motivation, then affect the perceived user satisfaction [45]

4 Conclusion

Purpose of this study is to review literature on factors that influence competitive advantage of e-learning interface. Result shows that there are some constructs that are important for—learning design that are aesthetic, efficiency, effectiveness, and playfulness. Further model will be proposed based on the constructs reviewed.

References

1. Heldal F, Sjøvold E, Heldal A (2004) Success on the Internet—optimizing relationships through the corporate site. Int J Inf Manage 24(2):115–129. https://doi.org/10.1016/j.ijinfomgt.2003.12.010
2. Lestari S (2018) Peran Teknologi dalam Pendidikan di Era Globalisasi Edureligia. Jurnal Pendidikan Agama Islam 2(2):94–100. https://doi.org/10.33650/edureligia.v2i2.459
3. Tondeur J, van Braak J, Sang G, Voogt J, Fisser P (2012) Ottenbreit-Leftwich a: preparing pre-service teachers to integrate technology in education: a synthesis of qualitative evidence. Comput Educ 59:134–144. https://doi.org/10.1016/j.compedu.2011.10.009
4. Al-Sofi B (2021) Student satisfaction with e-learning using blackboard LMS during the Covid-19 circumstances: realities, expectations, and future prospects. Pegem J Educ Instr. https://doi.org/10.47750/pegegog.11.04.26
5. Costa L (2021) The status of e-learning after the corona pandemic. Brazil
6. Maloney EJ, Kim J (2021) Learning From COVID-19. https://www.insidehighered.com/blogs/learning-innovation/learning-covid-19
7. Research and Market (2020) E-Learning—market analysis, trends, and forecasts. A report by Global Industry Analysts, Inc. Online at: https://www.researchandmarkets.com/reports/2832322
8. Sugandi A (2022) Kinerja Edtech Moncer, Ini Capaian Zenius Sepanjang 2021 | Teknologi—Bisnis.com, from https://teknologi.bisnis.com/read/20211209/266/1475742/kinerja-edtech-moncer-ini-capaian-zenius-sepanjang-2021. Retrieved 11 Apr 2022
9. Ruangguru (2022) Ruangguru Tutup Tahun 2020 dengan Melayani Lebih dari 22 Juta Pengguna di Indonesia. From https://www.ruangguru.com/blog/ruangguru-tutup-tahun-2020-dengan-melayani-lebih-dari-22-juta-pengguna-di-indonesia. Retrieved 20 Apr 2022
10. Sekertariat Dukcapil: Dukcapil Kemendagri Rilis Data Penduduk Semester I Tahun 2022, Naik 0, 54% Dalam Waktu 6 Bulan. In: Dukcapil.kemendagri.go.id. https://dukcapil.kemendagri.go.id/berita/baca/1396/dukcapil-kemendagri-rilis-data-penduduk-semester-i-tahun-2022-naik-054-dalam-waktu-6-bulan#:~:text=Jakarta%20%2D%20Ditjen%20Dukcapil%20Kementerian%20Dalam,tercatat%20sebanyak%20275.361.267%20jiwa. Retrieved 23 Sept 2022
11. Kohler T (2022) Autonomy, relatedness, and competence in UX design, from https://www.nngroup.com/articles/autonomy-relatedness-competence/, Retrieved 10 Apr 2022
12. Kotler P (1994) Reconceptualizing marketing: an interview with Philip Kotler. Eur Manage J 12(4):353–361. https://doi.org/10.1016/0263-2373(94)90021-3
13. Edward LD, Ryan RM (2012) The Oxford handbook of human motivation. In: Khadka K, Maharjan S (eds) Customer satisfaction and customer loyalty, 1st ed. Centria University of Applied Sciences. Oxford University Press, Oxford, UK (2017)
14. Khadka K, Maharjan S (2017) Customer satisfaction and customer loyalty. Centria Univ Appl Sci
15. Coursaris C, van Osch W (2016) A cognitive-affective model of perceived user satisfaction (CAMPUS): the complementary effects and interdependence of usability and aesthetics in IS design. Inf Manage 53(2):252–264. https://doi.org/10.1016/j.im.2015.10.003

16. Klett F, Pharow P (2006) How to achieve user satisfaction in complex e-learning environments. Spec Session Future Technol
17. Aparicio M, Bacao F, Oliveira T (2016) Cultural impacts on e-learning systems' success. Internet Higher Educ 31:58–70. https://doi.org/10.1016/j.iheduc.2016.06.003
18. Harandi S (2015) Effects of e-learning on students' motivation. Procedia Soc Behav Sci. 181:423–430. https://doi.org/10.1016/j.sbspro.2015.04.905
19. Kyoung-Jee K, Theodore WF (2011) Changes in student motivation during online learning. J Educ Comput Res 44:1–23
20. Neisser U (2014) Cognitive psychology. In: Machmud, R (ed) Kepuasan Penggunaan Sistem Informasi (p. 9). Psychology Press, Ideas Publishing, Gorontalo
21. Sarsam SM, Al-Samarraie H (2018) A first look at the effectiveness of personality dimensions in promoting users' satisfaction with the system. SAGE Open 8(2):215824401876912. https://doi.org/10.1177/2158244018769125
22. Parsazadeh N, Ali R, Zainuddin N (2014) Technological aspects of E-learning system in Malaysian context. In: International conference on teaching and learning in computing and engineering. https://doi.org/10.1109/LaTiCE.2014.21
23. Lindgaard G, Dudek C (2002) User satisfaction, aesthetics and usability: beyond reductionism. In: IFIP World computer congress. Kluwer Academic Publisher, Ontario
24. Suryani NK, Sugianingrat I (2021) Student e-Learning satisfaction during the covid-19 pandemic in Bali, Indonesia. J Econ 17.1. https://journal.uny.ac.id/index.php/economia
25. Prastawa H, Ciptomulyono U, Singgih ML, Hartono M (2019) The effect of cognitive and affective aspects on usability. Theoret Issues Ergon Sci 20(4):507–531. https://doi.org/10.1080/1463922X.2018.1547458
26. Kognisi (2016) At KBBI daring. Retrieved 05 Sept 2020, from https://kbbi.kemdikbud.go.id/entri/kognisi
27. Wickens C, Helton W, Hollands J, Banbury S (2013) Engineering psychology and human performance
28. Krathwohl DR, Bloom BS, Masia BB (1964) Taxonomy of educational objectives, the classification of educational goals. In: Handbook II: affective domain. David McKay Co., Inc., New York
29. Schiffman LG, Kanuk LL, Wisenblit J (2010) Consumer behavior, 10th edn. Pearson Education, Inc., New Jersey
30. Tjandra E, Tjandra S (2013) Hubungan antara komponen kognitif, komponen afektif dan komponen perilaku terhadap sikap konsumen memanfaatkan teknologi internet. J Manajemen XVII:42–52
31. Arroyo I, Shanabrook D, Burleson W, Woolf B (2012) Analyzing affective constructs: emotions 'n attitudes. Intell Tutoring Syst 714–715
32. Mueller CM, Dweck CS (1998) Praise for intelligence can undermine children's and performance. J Pers Soc Psychol 75(1):33–52
33. Eccles J, Wigfield A, Harold RD, Blumenfeld P (1993) Age and gender differences in children's self and task perceptions during elementary school. Child Dev 64:830–847
34. Sharples M (1996) An introduction to human-computer interaction. In: Boden M (ed) Artificial intelligence. Academic Press, pp 293–323
35. Lestariningsih T (2017) Analisis Sitem Informasi Dalam Perspektif Human Computer Interaction. Jurnal AKSI (Akuntansi Dan Sistem Informasi), 1(1):5–9. https://doi.org/10.32486/aksi.v1i1.98
36. Hassenzahl M, Tractinsky N (2006) User experience—a research agenda. Behav Inf Technol 25(2):91–97. https://doi.org/10.1080/01449290500330331
37. Rubin J, Chisnell D (2007) Handbook of usability testing, 2nd edn. Wiley Pub, Indianapolis, IN
38. Widyanti A, Hanna, Muslim K, Sutalaksana IZ (2017) The sensitivity of galvanic skin response for assessing mental workload in Indonesia. Work 56(1):111–117
39. Widyanti A, Sofiani NF, Soetisna HR, Muslim K (2017) Eye blink rate as a measure of mental workload in a driving task: convergent or divergent with other measures? Int J Technol 8(2):283–291

40. Hoffman B (2012) Cognitive efficiency: a conceptual and methodological comparison. Learn Instr 22(2):133–144. https://doi.org/10.1016/j.learninstruc.2011.09.001
41. Tohidi H, Jabbari M (2012) The effects of motivation in education. Retrieved 29 Sept 2022, from https://doi.org/10.1016/j.sbspro.2011.12.148
42. Conrick K (2020) Different ways to identify user segments — roles, demographics, need states and personas. Retrieved 10 Apr 2022, from https://uxdesign.cc/different-ways-to-identify-user-segments-roles-demographics-need-states-and-personas-405e60d6c161
43. Suryani N, Sugianingrat I (2021) Student e-learning satisfaction during the covid-19 Pandemic in Bali Indonesia. J Economia 17(1):141–151. https://doi.org/10.21831/economia.v17i1.33196
44. Law K, Lee V, Yu Y (2010) Learning motivation in e-learning facilitated computer programming courses. Retrieved 29 Sept 2022, from https://doi.org/10.1016/j.compedu.2010.01.007
45. Todorova M, Karamanska D (2015) A study of motivation and satisfaction of students in e-learning environment. Appl Technol Innovations 11(2):82–89. https://doi.org/10.15208/ati.2015.09

PSO-LFDE Algorithm on Constrained Real-Parameter Optimisation Test Functions

Nafrizuan Mat Yahya and Nur Iffah Mohamed Azmi

Abstract This paper introduces a new version of the particle swarm optimisation (PSO) algorithm particle swarm optimisation with Lévy Flight and the Doppler Effect (PSO-LFDE), maintaining an optimal balance between the exploration and exploitation phases of the optimisation process. The proposed algorithm will hold a better exploration–exploitation equilibrium if the contributions of convergence and diversity and global and individual bests in attracting particles are maintained in balance. The proposed PSO-LFDE algorithm is compared with the PSO algorithm by Gaing on single-objective constrained real-parameter optimisation test functions. The results indicated that the PSO-LFDE has achieved competitive results on the single-objective constrained real-parameter optimisation test functions as compared to PSO algorithm by Gaing. Thus, the PSO-LFDE is validated as a stable, well-designed algorithm and can be a functional alternative approach to deal with various single-objective constrained real-parameter optimisation problems.

Keywords PSO · Optimisation · Single objective function

1 Introduction

Currently, the PSO algorithm attracts researcher attention and is wide implements in various fields [1–4]. Particle swarm optimisation algorithm (PSO) is an Evolutionary Computation technique that is less expensive computer-based and can converge faster than other approaches. The terms swarm are kept in the PSO algorithm by Kennedy and Eberhart under the swarm intelligence principles of accessibility, reliability, variety of response, stability, and flexibility [5–9].

N. M. Yahya (✉) · N. I. M. Azmi
Faculty of Manufacturing and Mechatronic Engineering Technology, Universiti Malaysia Pahang, 26600 Pekan, Pahang, Malaysia
e-mail: nafrizuanmy@ump.edu.my

C. N. Rosyidi et al. (eds.), *Proceedings of the 6th Asia Pacific Conference on Manufacturing Systems and 4th International Manufacturing Engineering Conference*, Lecture Notes in Mechanical Engineering,
https://doi.org/10.1007/978-981-99-1245-2_52

Maintaining a proper equilibrium between the exploration and exploitation stages of the optimisation method is the main limitation of PSO and every other meta-heuristic algorithm. Some algorithms, such as PSO, have a good capability for exploiting a high convergence rate to encouraging areas in the search space encountered [8]. PSO has more appealing characteristics than the standard evolutionary estimate. It also includes keeping the memory of the PSO algorithm, building cooperation and exchanging knowledge between particles with the others particle [10].

Additionally, to a brief theoretical background and happy programming and coding of the machine, the PSO procedure can produce high-quality solutions in a undersized amount of time [11]. In PSO, the proximity principle is represented as a multi-dimensional scheming that is performed in each repetition. Swarms of particles react to quality variables in both the immediate environment and the individual's optimum location simultaneously [6, 12].

In PSO algorithms, the clauses of the interactions among the personal best position and the best position of the neighbours well constitute the standard of equality and par- ticipation response [6]. The ability of swarms to adjust the environment only when the position of the individual or collective best changes is another aspect of the stability principle. The flock will pursue whether the personal best position or the overall best position changes till the PSO algorithm adaptation principle is complied [5, 6, 13, 14].

This paper presents a new swarm intelligence algorithm entitled particle swarm optimisation algorithm with Lévy Flight and the Doppler Effect (PSO-LFDE), an progressed form of the novel PSO [15–17]. This algorithm proposed using Lévy flights for local and global searching as a random factor to update each particle's end position. Using this theory, the particle moves in each iteration with slight jumps combined with longer jumps or "flights". As a result, this particle's movement will be triggering the variance of the spreading to diverge [18].

The Doppler Effect's purpose is to substitute the inertia weight by the use of a wave concept to control the influence of the prior velocity on the present velocity as exag-generated by the mutual fitness functions of *pbest* and *gbest* [12]. In order to overcome single-objective optimisation issues, the PSO-LFDE has been implemented.

This paper distributed into several topics. Section 2 discussed the new proposed algorithm, a particle swarm optimisation algorithm with Lévy Flight and the Doppler Effect (PSO-LFDE). While in Sect. 3 discusses results on simulation, and in conclusion are concluded in Sect. 4.

1.1 Particle Swarm Optimisation

A group of birds or a swarm of fish display the swarming social behaviour that serves as the basis for the typical particle swam optimization [15] proposed the standard PSO algorithm that is easy to apply and has the stable convergence element with good com- putation efficiency.

A PSO algorithm development scheme is created, primarily using Algorithm 1's PSO continuous function optimization.

Algorithm 1 Particle swarm optimisation (PSO) [10, 19–22]
Begin
for each particle in the swarm
Initialise its position & velocity randomly
end for
while iter < max_iter **do**
for each particle, j **do**
update velocity ( ) ( )

$$j, g v_{k+1} = v_{j,g^k} + c_1 r_1 \text{ pbest }_{j \cdot g} - s_{j,g^k} - c_2 r_2 \text{ gbest }_{j \cdot g} - s_{j,g^k}$$

check the velocity boundaries
update position $x_{i,g^{k+1}} = x_{j,g^k} + v_{j,g^{k+1}}$
calculate fitness value f
if
the fitness value is better than the best fitness value. *pbest* in the past set the current value as the new *pbest*
end if
end for
from all the particles or neighbourhood, choose the particle with the best fitness value as the *gbest*
end while

1.2 Lévy Flight

Lévy Flight (LF) is a type of unsystematic walk in which the step-size probability density function is heavy-tailed, which indicates that a particle moving according to LF performs infrequent "big" steps interspersed with numerous frequent "small" ones [23, 24]. A typical motion pattern is as follows: the particle moves locally at first, taking a series of slight movements, then takes a significant step, and then moves locally again. The LF can be mathematically defined as follows:

$$\sigma_u = \{\Gamma(1+\beta)\sin(\pi\beta|2)\beta\Gamma[(1+\beta)22^{(\beta-1)f}\}^{1\beta} \quad (1)$$

$\Gamma\Gamma$ represents the gamma function. The swarm's behaviour should be unpredictable. The LF equation is required to create a chaotic swarm movement. The field of quest should explore rather than exploit. It does not mean that the solution leads to local optima.

1.3 The Doppler Effect

According to the definition of the Doppler Effect (DE), a modification in frequency or wavelength results from a shift in the wave source of the observers. When related to the transmitted frequency, the obtained frequency is greater throughout the method, identical during the passing, and lower during the recession [25].

These phenomena are mathematically represented as follows:

$$f = (c \pm v_r)/(c \mp v_s) f_0 \tag{2}$$

The receiving and emission frequencies are denoted by the characters f and f_0. While v_r denotes the receiver's speed, v_s denotes the source's speed, and c denotes the speed of the wave in the air, which is 340 m/s.

The most familiar to us is DE based on a sound wave. For example, when emergency vehicles passed us on the highway with their sirens blaring, the pitch of the siren sound was loud at first, then quickly dropped as the vehicle passed by. The DE defined by an apparent change in frequency for a sound wave caused by a moving source.

These days, the DE used to tackle a wide range of optimization issues in a variety of areas. Doppler Effect, for example, used to track higher-speed targets in acoustic wire- less sensor networks, as well as in ultrasonic broadband transmissions that used an indoor acoustic simulator to apply DE. All of these versions have shown to be quite efficient in their own circumstances, but as we know, DE may improve the response of a varied range of applications.

2 Proposed PSO-LFDE Algorithm

PSO-LFDE is a new variation method in which a part of LF idea is used to enhance the random walk expression on the position particles update. Besides that, an element of DE theory is used to replace the inertia weight theory, which is used to manage the effect of the previous velocity on the present velocity, which is influenced by the integrated fitness functions of *pbest* and *gbest* [26].

Algorithm 2 Particle swarm optimisation with Lévy Flight and the Doppler Effect
Begin
for each particle in the swarm
Initialise its position & velocity randomly
end for
while iter < max_iter **do**
for each particle, j **do**
Update velocity with the Doppler Effect equation replaces the inertia weight equation.

$$v_{j,g^{k+1}} = dev_{j,g^k} + c_1 r\left(pbest_{j\cdot g} - s_{j,g^k}\right) - c_2 r_2\left(gbest_{j,g} - s_{j,g^k}\right)$$

Check the velocity boundaries
Update position $x_{j,g^{k+1}} = x_{j,g^k} + v_{j,g^{k+1}}$ with lévy flight equation
Calculate fitness value f
if
The fitness value is better than the best fitness value. *pbest* in the past set the current value as the new *pbest*.
end if
end for
From all the particles or neighbourhood, choose the particle with the best fitness value as the *gbest*.
end while

Next, there are eight steps in the PSO-LFDE algorithm—the following eight steps to find the best value of *pbest* and *gbest*.

1. Set the lower and upper bounds for the three controller parameters, and then randomly seed the particle populations with the searching points, velocities, *pbest*, and gbest [27, 28].
2. Setting iterations = 1, if ite = 1, the best *pbest* particle is then recorded in an external archive as the best-so-far solution obtained until the current iteration.
3. Using the evaluation function f, compute the evaluation value for each particle in the population [29].
4. Compare each particle's evaluation function to its *pbest*. *gbest* is the best evaluation value among the *pbest*.
5. Change the velocity of each particle's member, v.

$$v_{j,g^{k+1}} = \mathrm{dev}_{i,g^k} + c_1 r_1 (pbest\ j, g - s_{i,g^k}) - c_2 r_2 (gbest_{j\cdot g} - s_{i,g^k}) \quad (3)$$

6. Change the position of each particle's member, k.

$$x_{i,g^{k+1}} = x_{j,g^k} + v_{j,g^{k+1}} + Le'vy \quad (4)$$

   Such that $x_{g^{\min}} \leq x_{j,g^{k+1}} \leq x_{g^{\max}}$, where $x_{\min} x_g^{\max}$ and represent the lower and upper bounds, respectively, for the member g of the particles, [30].
7. Go to Step 7 if the maximum number of iterations has been reached. If not, proceed to Step 2.

An optimal controller parameter is the particles that generate the most recent gbest.

3 Results and Discussion

PSO-LFDE is validated on single-objective constrained real-parameter optimisation test functions. [31] The selected functions come from different types; quadratic, cubic, nonlinear, linear, and others. All test functions are to minimise values. Table 1 contains information about the selected test functions. Then, 30 runs were executed for each test function. The population size of all algorithms is set to 100 to be compared fairly. For all algorithms, a limit of 1000 iterations are considered.

As the main performance indicators, the algorithm's 30 runs' best optimal, mean, worst, and standard deviation are considered. Table 2 shows all the detailed results of the implementation of the algorithms expected. The PSO-LFDE method based on the result achieved, and some of them will achieve the desired global optimum values in 30 runs. When evaluated on F01, F02, F03, F04 and F05, the PSO-LFDE give a mini- mum value of the best optimum value than on the initial PSO algorithm. Furthermore, if the algorithm achieves or is less than the optimum global value of the functions, the algorithm is essentially providing the best performance on it.

Table 1 The single-objective constrained real-parameter optimisation $f(x)$ and the bound [6, 31–35]

Fun No.	Dim	Benchmark f(x)	Bounds
F01	13	$f(x) =$ $5\sum_{i4=1} x_i - 5\sum_{i4=12} x_i - \sum_{i135} x_i$	$0 \le xi \le 1; \quad i = 1, 2, 3, \ldots, 9$ $0 \le x_i \le 100; \quad i = 10, 11, 12$ $0 \le x_{13} \le 1$
F02	4	$f(x) =$ $3x_1 + 0.000001x_1^3 + 2x_2 + (0.0000023)x_2^3$	$0 \le x_1, x_2 \le 1200;$ $-0.55 \le x_3, x_4 \le 0.55;$
F03	2	$f(x) = x_1^2 + (x_2 - 1)^2$	$-1 \le x_1, x_2 \le 1;$
F04	5	$f(x) = e^{x_1x_2x_3x_4x_5}$	$-2.3 \le x_1, x2 \le 2.3;$ $-3.2 \le x_3, x_4, x_5 \le 3.2;$
F05	2	$f(x) = -x_1 - x_2$	$0 \le x_1 \le 3;$ and $0 \le x_2 \le 4;$
F06	5	$f(x) = 5.3578547x_3^2 + 0.8356891x_1x_5$ $+ 37.293239x_1 - 40792.141$	$78 \le x_1 \le 102; \quad 33 \le x_2 \le 45;$ $27 \le x_i \le 45; \quad i = 3, 4, 5$
F07	2	$f(x) = (x_1 - 10)^3 + (x_2 - 20)^3$	$13 \le x_1 \le 100;$ and $0 \le x_2 \le 100;$
F08	8	$f(x) = x_1 + x_2 + x_3$	$100 \le x1 \le 10000;$ $1000 \le x_2, x_3 \le 10000;$ $10 \le x_i \le 1000; \quad i = 4, \ldots, 8$

Table 2 Results of single-objective constrained real-parameter optimization tests with n = 30 for minimization

Fun No.	Algorithm	Optimum F(x)	Best	Mean	Worst	Standard deviation
F01	PSO	– 15.00	– 116.3307	– 116.3307	– 116.3307	5.778E–14
	PSO_LF_DE		– 127.8058	– 127.8058	– 127.8058	5.778E–14
F02	PSO	5126.49671	4539.1641	4539.1641	4539.1641	1.849E–12
	PSO_LF_DE		1674.5755	1674.5755	1674.5755	6.933E–13
F03	PSO	0.7499	0.0000	0.0000	0.0000	0.0000
	PSO_LF_DE		– 0.238719	– 0.238719	– 0.238719	8.464E–17
F04	PSO	0.053941514	0.035919	0.035919	0.035919	0.0000
	PSO_LF_DE		0.030158	0.030158	0.030158	1.058E–17
F05	PSO	– 5.508013272	– 6.43144	– 6.43144	– 6.43144	3.611E–15
	PSO_LF_DE		– 7.21243	– 7.21243	– 7.21243	5.417E–15
F06	PSO	– 30665.53867	– 32154.43	– 32154.43	– 32154.43	2.227E–11
	PSO_LF_DE		– 32154.43	– 32154.43	– 32154.43	2.227E–11
F07	PSO	– 6961.81388	– 7973	– 7973	– 7973	0.0000
	PSO_LF_DE		– 7973	– 7973	– 7973	0.0000
F08	PSO	7049.24802	2100	2100	2100	0.0000
	PSO_LF_DE		2100	2100	2100	0.0000

Additionally, the standard deviation will determine the algorithm's accuracy since the most negligible standard deviation effects prove that the algorithm has the highest precision. A low standard deviation suggests that the majority of the results are more accurate and closely resemble the average. On the other hand, a high standard deviation indicates that the measurements are inconsistent and less accurate. According to the results on the single-objective constrained real-parameter optimisation test functions in Table 2, the PSO-LFDE can also achieve a low standard deviation.

Figure 1 shows the average percentage deviation between PSO and PSO-LFDE on single-objective constrained real-parameter optimisation test functions with different dimensions in the form of a graph. In order to make the difference between the algorithms easier to understand, this graph shows the standard deviation of PSO and PSO- LFDE in percentage.

Table 3 compares the performance of PSO and PSO-LFDE on the mean value ∓ standard deviation of the optimal global value for single-objective constrained real-parameter optimisation test functions using one-way analysis of variance (ANOVA) [6]. Moreover, the ANOVA test will be used to determine whether the findings are significant.

The table illustrates that the PSO-LFDE findings on single-objective constrained real-parameter optimisation test functions are substantial and significant in indicating that this method can be accomplish a minimum global optimum value.

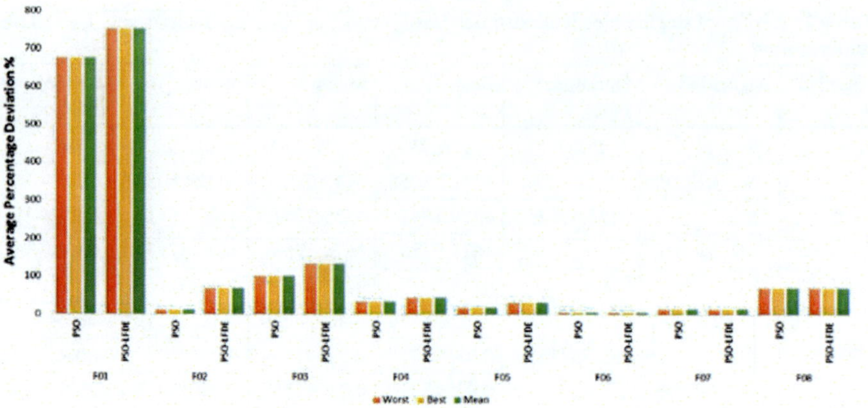

Fig. 1 Comparison average percentage deviation for the single-objective constrained real-parameter optimisation test functions over 30 independent runs

Table 3 Performance evaluation using one-way analysis of variance (ANOVA) between PSO and PSO-LFDE for single-objective constrained real-parameter optimisation test functions over 30 independent runs

Function num ber	PSO	PSO_LF_DE	Significantly
F01	− 116.3307	− 127.8058	Yes
F02	4539.1641	1674.5755	Yes
F03	0.0000	− 0.238719	Yes
F04	0.035919058	0.030157576	Yes
F05	− 6.431436578	− 7.212427034	Yes
F06	− 32,154.43104	− 32,154.43104	Yes
F07	− 7973.000	− 7973.000	Yes
F08	2100.000	2100.000	Yes

4 Conclusion

A PSO-LFDE algorithm has presented. By maintaining this balance between the roles played by global and personal bests in attracting particles, the new strategy has been added to maintain best equilibrium among the exploration and exploitation stages of the optimisation process. For all of the single-objective constrained real-parameter optimisation test functions that have considered, the PSO-LFDE has completed competitive outcomes on single-objective constrained real-parameter optimisation test functions at a comparatively better optimum function value.

The proposed algorithm is rated as a robust, well-designed algorithm capable of han- dling any optimisation problem based on these results. Furthermore, the virtual research demonstrates that the PSO-LFDE is reliable over a wide range of functions and dimen- sions. It also demonstrates that when objective function topography is unknown in an engineering situation, PSO-LFDE will offer a worthwhile solution in a slight number of objective function evaluations. We intend to further validate the PSO LFDE method in the future by analysis it on engineering applications as practical applications in the area of optimisation.

Acknowledgements The authors would like to thank Universiti Malaysia Pahang for providing financial support through the UMP Internal Grant (Grant No. RDU210370) and Faculty of Manufacturing and Mechatronics Engineering Technology, Universiti Malaysia Pahang for laboratory facilities.

References

1. GirirajKumar SM, Jayaraj D, Kishan AR (2010) PSO based tuning of a PID controller for a high performance drilling machine. Int J Comput Appl (0975–8887), 1(19):12–18
2. Liu Y, Zhang J, Wang S (2004) Optimization design based on PSO algorithm for PID controller. Proc World Congr Intell Control Autom 3:2419–2422
3. Chen D, Zhang G (2011) PSO algorithm based PID parameters optimization of hydraulic screwdown system of cold strip mill. Int Conf Fluid Power Mechatron 2011:113–116
4. Kennedy J, Eberhart R (1995) Particle swarm optimization. In: ICNN'95—international conference on neural networks, vol 4, pp 1942–1948
5. Engelbrecht AP (2005) Computational intelligence: an introduction. Wiley
6. Yahya NM (2016) Bats echolocation-inspired algorithms for global optimisation problems. University of Sheffield
7. Panigrahi BK, Suganthan PN, Das S, Satapathy SC (eds) Swarm, evolutionary, and memetic computing. Springer, Berlin, Heidelberg, vol 7077
8. Rezaei F, Safavi HR (2020) GuASPSO: a new approach to hold a better exploration–exploitation balance in PSO algorithm. Soft Comput 24(7):4855–4875
9. Radhika S, Chaparala A (2018) Optimization using evolutionary metaheuristic techniques: a brief review. Brazilian J Oper Prod Manag 15(1):44–53
10. Shi Y, Eberhart R (1998) A modified particle swarm optimizer. In: 1998 IEEE international conference on evolutionary computation proceedings. IEEE world congress on computational intelligence (Cat. No.98TH8360), pp 69–73
11. Kamesh, Sakthi Priya N (2012) A survey of cyber crimes yanping. Secur Commun Netw 5(June 2011):422–437
12. Azlan NA, Yahya NM (2019) Modified adaptive bats sonar algorithm with doppler effect mechanism for solving single objective unconstrained optimization problems. In: 2019 IEEE 15th international colloquium on signal processing and its applications, CSPA 2019, March, pp 27–30
13. Liu Y, Gu X (2007) Skeleton-network reconfiguration based on topological characteristics of scale-free networks and discrete particle swarm optimization. IEEE Trans Power Syst 22(3):1267–1274
14. Yang Q, Webb G (eds) (2006) PRICAI 2006: trends in artificial intelligence 4099. Springer, Berlin Heidelberg
15. Gaing Z (2014) A particle swarm optimization approach for optimum design of PID controller in AVR system. IEEE Trans Energy Convers 19(2):384–391
16. Yahya NM, Tokhi MO (2017) A modified bats echolocation-based algorithm for solving constrained optimisation problems. Int J Bio-Inspired Comput 10(1):12–23
17. Yahya NM, Tokhi MO, Kasdirin HA (2016) A new bats echolocation-based algorithm for single objective optimisation. Evol Intell 9(1–2):1–20

18. Kamaruzaman AF, Zain AM, Yusuf SM, Udin A (2013) Levy flight algorithm for optimization problems—a literature review. Appl Mech Mater 421(September):496–501
19. Jimenez-Nunez JJ, Cedeno-Maldonado JR (2005) A particle swarm optimization approach for reactive power dispatch. In: Proceedings of the 37th annual North American power symposium, pp 198–205
20. Eberhart R, Shi Y (2001) Particle swarm optimization: developments, applications and resources. In: Proceedings of the 2001 congress on evolutionary computation (IEEE Cat. No.01TH8546), pp 81–86
21. Kao C-C, Chuang C-W, Fung R-F (2006) The self-tuning PID control in a slider–crank mechanism system by applying particle swarm optimization approach. Mechatronics 16(8):513–522
22. Panigrahi BK, Suganthan PN, Das S, Dash SS (eds) (2013) Swarm, evolutionary, and memetic computing, vol 8297. Springer International Publishing, Cham
23. Iacca G, dos Santos Junior VC, Veloso de Melo V (2020) An improved jaya optimization algorithm with lévy flight. Expert Syst Appl, 16, 113902
24. Naresh T, Jaya Lakshmi A, Reddy VK 2019) Resource optimization using cloud scheduling. Int J Innov Technol Explor Eng 8(6 S2):184–189
25. Possel M (2017) Waves, motion and frequency: the Doppler Effect. In: Einstein Online, vol 5, 2011. Retrieved October 15 2018, from http://[Online], 2017. [Online]. Available: www.einstein-online.info/spotlights/doppler
26. Mafarja M, Jarrar R, Ahmad S, Abusnaina AA (2018) Feature selection using binary particle swarm optimization with time varying inertia weight strategies. ACM Int Conf Proceeding Ser
27. Khaehintung N, Kunakorn A, Sirisuk P (2010) A novel fuzzy logic control technique tuned by particle swarm optimization for maximum power point tracking for a photovoltaic system using a current-mode boost converter with bifurcation control. Int J Control Autom Syst 8(2):289–300
28. Kao CC, Chuang CW, Fung RF (2006) The self-tuning PID control in a slider-crank mechanism system by applying particle swarm optimization approach. Mechatronics 16(8):513–522
29. Malarvizhi K, Kiruba R (2015) Design and implementation of two tank conical interacting system using intelligent technique (PSO). ICACCS 2015—Proc. 2nd Int Conf Adv Comput Commun Syst 2:4–8
30. Aboura F (2019) Tuning PID controller using hybrid genetic algorithm particle swarm optimization method for AVR system. In: 2019 International aegean conference on electrical machines and power electronics (ACEMP) & 2019 international conference on optimization of electrical and electronic equipment (OPTIM), pp 570–574
31. Liang J et al (2006) Problem definitions and evaluation criteria for the CEC 2006 Special session on constrained real-parameter optimization. Nangyang Technol Univ Singapore Tech Rep 41
32. Barkat Ullah ASSM, Sarker R, Cornforth D, Lokan C (2009) AMA: a new approach for solving constrained real-valued optimization problems. Soft Comput 13:741–762
33. Mallipeddi R, Suganthan PN (2008) Evaluation of novel adaptive evolutionary programming on four constraint handling techniques. 2008 IEEE Congr Evol Comput IEEE World Congr Comput Intell:4045–4052
34. Kaveh A, Akbari H, Hosseini SM (2020) Plasma generation optimization: a new physically-based metaheuristic algorithm for solving constrained optimization problems. Eng Comput
35. Mahmoodabadi MJ, Bagheri A, Nariman-zadeh N, Jamali A (2012) A new optimization algorithm based on a combination of particle swarm optimization, convergence and divergence operators for single-objective and multi-objective problems. Eng Optim 44(10):1167–1186

A Condition-Based Maintenance Lease Contract with Continuously Monitored Equipment

H. Husniah, R. Wangsaputra, and B. P. Iskandar

Abstract Rapid development of sensors and maintenance technologies have supported a better implementation of Condition-Based Maintenance (CBM) policy. Meanwhile CBM with continuously monitored equipment during a lease contract (LC) period has limited attention. This equipment is employed in dynamic environments and can be either employed in normal stress (N) or severe stress (H) randomly and follows a two-state continuous time Markov chain. The failure is dependent on the frequency of use of the equipment. The frequency of usage is modeled as a continuous random variable. Since the equipment's failure and degradation is dependent on usage, this has an effect on the expected lease contract cost. Finally, we have the optimal preventive measure that maximizes the total LC profit through non-cooperative game theory.

Keywords Lease contract · Condition based maintenance · Non-Cooperative game theory

1 Introduction

Industrial equipment are subject to deterioration due to age and environmental influences. Without interfering the deterioration eventually leads to equipment failure that require corrective maintenance (CM). However in most cases, it is best to prevent failures through preventive maintenance (PM). PM interventions can be prescribed according to the age of the devices. However, the further development of sensor

H. Husniah (✉)
Department of Industrial Engineering, Langlangbuana University, Karapitan 116, Bandung, Indonesia
e-mail: hennie.husniah@gmail.com

R. Wangsaputra · B. P. Iskandar
Department of Industrial Engineering, Bandung Institute of Technologi, Ganesha 10, Bandung, Indonesia

C. N. Rosyidi et al. (eds.), *Proceedings of the 6th Asia Pacific Conference on Manufacturing Systems and 4th International Manufacturing Engineering Conference*, Lecture Notes in Mechanical Engineering,
https://doi.org/10.1007/978-981-99-1245-2_53

technology makes it possible to do real-time status monitoring of industrial equipment [1]. That resulting information flow can be used to carry out maintenance work depending on the actual condition of the equipment; this type of maintenance is called the Condition-Based Maintenance (CBM) policy, the condition of the system that is affected by deterioration can be monitored using several types of monitoring equipment, such as sensors. The rapid development of technology and sensors supports the implementation of better CBM policies [2]. Optimal implementation of CBM policies will be able to reduce or even eliminate system failures so as to reduce maintenance costs [2] and increase system equipment at the same time.

Many researches related to CBM policy have been done before. Such in [3] conducted research on CBM policies by considering the maintenance costs and equipment capacity of the system. The research was conducted on a single unit system and assumed non-periodic inspection scheduling, perfect types of maintenance action, and system deterioration influenced by the age of the system. The method of finding a solution in this research is to use an analytical approach. Then, [4] developed their research by changing the assumption of inspection scheduling to be periodic at certain time intervals and the type of maintenance to be imperfect. The deterioration of the system is still assumed to be influenced by the age of the system and the solution finding method is still using an analytical approach.

A research related to CBM policies on leased system which is influenced by satisfaction of the lessee had been conducted by [5]. This research uses an analytical approach to find the optimal solution and uses the assumption of the perfect type of maintenance action, periodic inspection scheduling, and the deterioration of the system is influenced by the age of the system. Just like [3] research, system failures in this study are assumed to only be detected on the inspection schedule. Also [6] conducted a study on lease contracts for remanufacturing systems using the CBM policy as a maintenance policy for the system. This research uses an analytical approach to find the optimal solution and uses the assumptions that [5] related to system failures that can only be detected on the inspection schedule. The development carried out in this research is that the deterioration of the system is not only influenced by age, but also by factors of use and operating condition. This deterioration modeling was conducted using the Accelerated Failure Time (AFT) formulation.

In this study, we consider CBM optimization for leased stand-alone assets, where maintenance interventions involve the entire system. To simulate degradation, we use a continuous scale approach. We assume that a sensor is installed that continuously monitors the degree of degradation. Furthermore, we assume that maintenance is done in certain period time. It is only limited number of existing studies considering continuous monitoring of gradually deteriorating equipment.

Many LC models studied assume continuous use of the equipment regardless of the external environment and the mode of operation of the equipment [7–11]. Equipment may have a different failure rate under normal conditions than under severe conditions. To assess LC costs from a real system, consideration of failure models and dynamic environment are needed. To make formulation cost model accurate, we focus on LC costs for equipment under dynamic environments over the life of the equipment. The equipment can be used in both normal and severe conditions, failure depends on the intensity of use of the equipment. Intensity of use varied across user groups and hence gives an impact on the expected LC cost. This usage intensity was modeled as a continuous random variable. The optimal decisions will be determined using Nash game theory.

This article is organized as follows. Sections 1 and 2 present background and example formulas for the leased maintenance contracts examined. Section 3 includes model analysis for optimal PM time and optimal LC benefits. Before conclusion and next research in Sect. 5, illustration of the model is explained.

2 Model Formulation

2.1 *Lease Contract*

Lease contract as part of servicing the lessor to the lessee is categorized in two categories of leasing as follows.

Leasing 1 (L1): The lessee performs the PM measures internally as a tenant, but the CM measures are outsourced to the lessor. With this option, if the equipment breaks down, the lessee can call the lessor to have the equipment repaired. The lessor charges the lessee a fixed cost C_S per repair.

Leasing 1 (L2): The lessor undertakes to perform a complete LC including PM and CM in a certain time with a price P_G. The lessor promises that the total downtime D in $(0, t]$ due to repaired and idle time per failure will be less than the contracted target E. Since the leased is fully covered, lessor will bear the cost of penalties if the downtime per-failure X_i is greater than the downtime target E, $D > E$, the lessor have to pay a penalty. The penalty cost C_P is considered compensation received by the lessee.

2.2 *Failure Modelling*

Modeling in this paper follows the work of Husniah et al. [8]. We model transitions by a time-continuous two-state Markov chain formulation $X(t)$. $X(t) = 1$ if the equipment is heavily loaded at time t, and $X(t) = 0$ if the equipment is normally loaded. Since the utilization rate U = u, the probabilities $\{X(t+\delta t) = j|X(t) = i\}, 0 \leq i, j \leq 1$ are given by the following matrix:

$$X(t) = \begin{matrix} 1 \\ 0 \end{matrix} \overset{X(t+\delta t)}{\begin{bmatrix} 1-\lambda_1\delta\tau & \lambda_1\delta t \\ \lambda_0\delta t & 1-\lambda_0\delta t \end{bmatrix}}$$

A two-dimensional lease policy has been proposed, and we model equipment failures using a one-dimensional approach. And also the Weibull AFT model is used to account for the effect of utilization in normal stresses (i = 0) and high stresses (i = 1) on equipment degradation. We assume that the failure rate when normal is always less than the failure rate when in high and it can be shown by

$$r_y(t) = r_y^1(t|X(t) = 1)p_1(t) + r_y^0(t|X(t) = 0)\ p_0(t) \tag{1}$$

with

$$r_y^i(t) = \frac{\beta}{\alpha_i (u/u_0)^{\eta_i}} t^{\beta-1}, i = 0, 1$$

where $p_i(t, u)$, $0 \le i \le 1$ is the probability the Markov-chain $X(t)$ is in state i at time t. By substituting $p_i(t, u)$ into Eq. (1) then we have

$$\begin{aligned} r_y(t) = r_y^1(t|X(t) = 1)\left[\frac{\lambda_0}{\lambda_0+\lambda_1} + \frac{\lambda_1}{\lambda_0+\lambda_1} e^{-(\lambda_0+\lambda_1)t}\right] \\ + r_y^0(t|X(t) = 0)\left[\frac{\lambda_0}{\lambda_0+\lambda_1} - \frac{\lambda_1}{\lambda_0+\lambda_1} e^{-(\lambda_0+\lambda_1)t}\right] \end{aligned} \tag{2}$$

To control equipment degradation, perform PM periodically at $k.\tau_y$ with continuously monitored throughout the lease period can be done in-house by the lessee or by lessor. In this model, PM performed by lessee is less effective than by lessor and it is given by

$$r_y(t) = \rho r_y(t)\ 0 \le t < \Gamma_y \tag{3}$$

where $\rho > 1$ and $\rho = 1$ for PM by lessee and lessor respectively.

Symbols

$\Omega_T = [0, \Gamma_0) \times [0, U_0)$	LC area, time and usage limits
D_y	PM stage
Y	Usage rate
C_m	Repair cost performed by lessor
C_s	Repair fee charged to the lessee for each failure
C_{pm}	PM cost per unit of time

(continued)

(continued)

C_P	Penalty cost per unit of time
C_b	The annual equipment fee over LC period
P_O	PM cost done in-house over the contract period
$F(t,\alpha y)$	Conditional failure distribution for a given usage rate y
$ry(t)$, $Ry(t)$	Hazard, and Cumulative hazard functions associated with $F(t, \alpha_y)$
k	Number of PM during LC

3 Model Analysis

3.1 Lessee Decision Problem

Leasing-1 (L1): For condition $y \le \gamma$, if K is the revenue ($/hour) received by the lessee, then the expected profit is given by,

$$\begin{aligned} E\left[\phi_y(L_1; C_s)\right] &= K\left\{\Gamma_0 - R_{1y}(\Gamma_0)E[X_{ji}]\right\} - C_s R_{1y}(\Gamma_0) - P_0 \\ P_0 &= kC_0 - \sum_{j=1}^{k_y+1}\left[r\left(j\tau_y\right) - r\left((j-1)\tau_y\right)\right]\left[C_m\left(\Gamma_0 - j\tau_y\right) - C_v\right] \end{aligned} \tag{4}$$

For case $y > \gamma$, the expected profit of the lessee is given by (4) replacing Γ_0 with Γ_y.

Leasing-2 (L2): For $y \le \gamma$, the lessee expected profit is

$$\begin{aligned} &E\left[\phi_y(L_2; P_G)\right] = K\left\{\Gamma_0 - R_{2y}(\Gamma_0)E[X_{ji}]\right\} + R_{2y}(\Gamma_0)EP(\Gamma_0) - P_G, \\ &EP(\Gamma_0) = C_p\overline{G}(D)N(k_y, \tau_y) \end{aligned} \tag{5}$$

$EP(\Gamma_0)$ is the expected penalty viewed as a compensation received by the lessee. For case $y > \gamma$, as in L1, the expected profit of the lessee is given by (5) replacing Γ_0 with Γ_y.

3.2 Lessor Decision Problem

Leasing-1: For $y \le \gamma$, the expected profit of lessor is given by

$$E\left[\pi_y(L_1; C_s)\right] = (C_s - C_m)R_{1y}(\Gamma_0) - C_b \tag{6}$$

For $y > \gamma$, the expected profit of lessor is given by (6) replacing Γ_0 with Γ_y.

Leasing-2: For $y \leq \gamma$, whilst the down time of the equipment is greater than the intended target, the lessor is charged a penalty value, repair cost and a PM cost. Whilst the revenues of the lessor will be the price of the contract. We calculate expected penalty cost, the expected repair cost, and expected PM cost in Γ_0 per equipment as follows.

Expected cost with PM and minimal repair,

$$C(k_y, \tau_y) = C_m R_y(\Gamma_0) + kC_0 - \sum_{j=1}^{k_y+1} \left[r(j\tau_y) - r((j-1)\tau_y)\right]\left[C_m(L - j\tau_y) - C_v\right] \tag{7}$$

Expected penalty cost

The lessor incurs penalty cost when the down time caused by a failure exceeds the predetermined target. Let D denote down time for each failure occuring during the contract and down time allowed. The expected penalty cost is given by $C_P\overline{G}(D)N(k_y, \tau_y)$ where C_P is the penalty cost and $N(k_y, \tau_y)$ denotes the expected number of failure in interval $(0, \Gamma_0]$.

As a result, the total expected profit of the lessor is

$$E[\pi(L_2)] = P_G - C(k_y, \tau_y) - C_b - C_P\overline{G}(D)N(k_y, \tau_y) \tag{8}$$

For $y > \gamma$, the expected profit of the lessor choosing L_2 is given by (8) replacing Γ_0 with Γ_y. We consider a situation where both the lessee and the lessor want to negotiate the terms and condition of the contract in order to achieve a mutually beneficial agreement. As a result, a Nash solution of the bargaining game is employed to obtain the optimal solution C_s^* and P_G^* as follows

$$C_s^* = \frac{\left[K\left\{\Gamma_0 - R_{1y}(\Gamma_0)E\left[X_{ji}\right]\right\} + C_m R_{1y}(\Gamma_0) - P_0 - C_b\right]}{2R_{1y}(\Gamma_0)} \tag{9}$$

$$P_G^* = \frac{1}{2}\left[K\left[\Gamma_0 - R_{2y}(\Gamma_0)E\left[X_{ji}\right]\right] + 2EP_y(\Gamma_0) + C(k_y, \tau_y) - C_b\right] \tag{10}$$

Then the optimal expected profit of each party (the lessee or the lessor) for L_1 and L_2 given by

$$E[\pi_y(L_1; C_s^*) = \frac{1}{2}\left[K\left\{\tau - R_{1y}(\Gamma_0)E\left[X_{ji}\right]\right\} - P_0 - C_b - C_m R_{1y}(\Gamma_0)\right] \tag{11}$$

$$E[\pi_y(L_2; P_G^*) = \frac{1}{2}\left[K\left\{\tau - R_{2y}(\Gamma_0)E\left[X_{ji}\right]\right\} - C(k_y, \tau_y) - C_b\right] \tag{12}$$

For $y > \gamma$, C_s^* and P_G^* and the optimal expected profit are given by (9)–(12) replacing Γ_0 with Γ_y.

4 Numerical Example

In this part, the failure distribution is given by the Weibull distribution, and the parameter values be as follows.

$\alpha_0 = 236.28$ (days), $\beta = 2.06$, $U = 2$ (3×10^4 km), ($\gamma = U/\Gamma_0 = 1$), $\rho = 2.5$, $\Gamma_0 = 3$(years), $C_0 = 0.7 \times 10^6$, $C_v = 4.3 \times 10^6$, $C_m = 6.8 \times 10^6$, $C_b = 269 \times 10^6$, and the transition matrix is $p_H = 0.474$, $(1 - p_H) = 0.526$, $p_N = 0.567$, $(1 - p_N) = 0.433$.

Figure 1 shows the optimal number of PM and the optimal expected cost in leasing-1 2 for the lessee and and leasing-2 for the lessor. With the number of PM $k^* = 3$ for the leasing-1 and leasing 2, which means the lessee and the lessor have to consider PM to gain maximum expected profit. This result also show that option 1 is preferable for both players. This is due to the increase of penalty cost in option 2 as the number of failure increase by usage and time (Fig. 1).

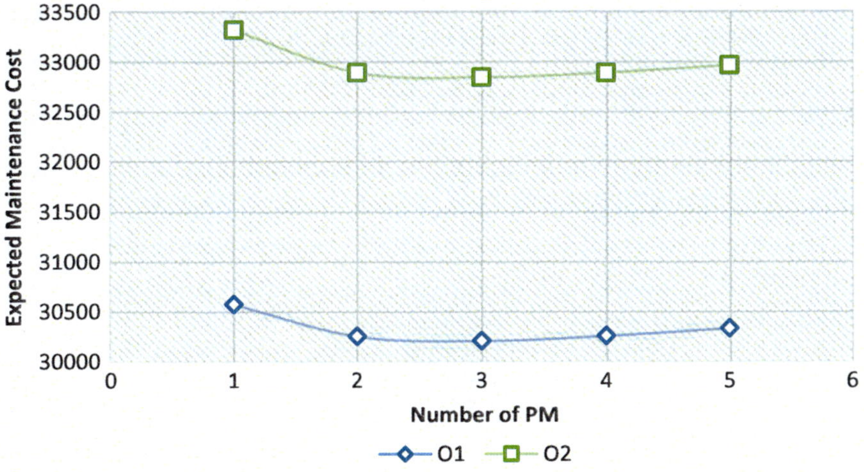

Fig. 1 Expected LC cost for leasing-1 and leasing-2

5 Conclusion

Through Condition Based Maintenance, we can continuously monitor the equipment's condition to determine the optimal maintenance cost. Additionally, dynamic operating conditions make this feasible. For both the manufacturer and the lessee, we can determine the best lease options and maintenance costs. However, we only consider equipment without warranty in this paper. This way, we can focus on equipment with two players: the lessor and lessee. Lease maintenance contract providers often sell a repairable piece of equipment with warranty. They also have multiple representatives from the maintenance contract provider who are lessor— partial, moderate, or full coverage. Currently, this area of research is being looked into further with additional topics.

Acknowledgements This work is funded by ITB research Grant 2022 with contract number 1103A/IT1.C07.2/TA.oo/2022.

References

1. Feng Q, Shanthikumar JG (2018) How research in equipmention and operations man agement may evolve in the era of big data. Prod Oper Manag 27:1670–1684
2. Bérenguer C, Grall A, Dieulle L, Roussignol M (2003) Maintenance policy for a continuously monitored deteriorating system. Probab Eng Inf Sci 17:235–250
3. Van DP, Berenguer C (2010) Condition based maintenance model for a equipmention deteriorating system. In: Conference on control and fault tolerant systems—IEEE, 2010
4. Van DP, Berenguer C (2011) Condition-Based maintenance under an imperfect repair policy for a deteriorating equipmention system. In: 7th International conference on modelling in industrial maintenance and reliability
5. Zhang Y et al (2019) Lessees' satisfaction and optimal condition-based maintenance policy for leased system. Realibility Eng Syst Saf 191:106532
6. Husniah H, Pasaribu US, Iskandar BP (2019) Optimal condition-based maintenance policy for leased remanufactured system. In: International conference
7. Ashgarizadeh E, Murthy DNP (2000) Maintenance contracts. Math Comput Model 31:11–20
8. Husniah H, Pasaribu US, Andi CV, Iskandar BP (2015) Miantenance contract with dynamic operating conditions. In: The 2015 international 9th international conference on mathematical methods in reliability (MMR 2015) Tokyo, Japan, June 1–4, 2015
9. Iskandar BP, Husniah H, Pasaribu US (2014) Maintenance service contract for equipments sold with two dimensional warranties. J Quant Qual Manage
10. Jackson C, Pascual R (2008) Optimal maintenance service contract negotiation with aging equipment. Eur J Oper Res 189:387–398
11. Mirzahosseinian and Piplani (2011) Compensation and incentive modeling inperformance-based contracts for after-market service. In: Proceedings of the 41st international conference on computing industrial engineering, pp 739–744. Singapore

Designing Drug Inventory Management System Design Using ABC-VED and Probabilistic Model to Minimize Total Inventory Cost in Public Health Service

Inaya Izzati, Ari Yanuar Ridwan, and Hardian Kokoh Pambudi

Abstract The Regional General Hospital (RSUD) of the Mentawai Islands Regency is located on North Sipora Island, precisely on Jl. Raya Tuapejat KM. 09 North Sipora District, Mentawai Islands Regency, West Sumatra Province. Hospitals must pay attention to how to manage drug supplies as well as possible so that there are no losses due to expired drugs caused by excess stock. There is a large gap between supply and demand, there is no inventory policy so that orders and inventories of medicines are stored in excess. Based on these problems, the design of an inventory policy using the probabilistic continuous review (r, Q) method of drugs is needed to reduce the number of expired drugs and losses incurred. In this final project, as many as 133 drugs of medical ingredients will be analyzed using the ABC-VED matrix then only category I (AV, AE, AD, BV, CV) will be counted. The classification process using the ABC-VED matrix resulted in as many as six types of medical materials belonging to the AV group, 16 types of medical materials from the AE group and one type of medical material from the AD group. Then the results of calculations using continuous review (r, Q) with fixed lifetime on the type of medical material category I resulted in 0% percent of expired drugs and Rp. 0 losses and the total cost of the proposed inventory of Rp. 329,835,062.48 or a savings of 28%.

Keywords Drugs · Matrix ABC-VED · Probabilistic continuous review (r Q) with fixed life time

I. Izzati (✉) · A. Y. Ridwan · H. K. Pambudi
Telkom University, Jl. Telekomunikasi No.1, Bandung, Indonesia
e-mail: Inayaizzati88@gmail.com

C. N. Rosyidi et al. (eds.), *Proceedings of the 6th Asia Pacific Conference on Manufacturing Systems and 4th International Manufacturing Engineering Conference*, Lecture Notes in Mechanical Engineering,
https://doi.org/10.1007/978-981-99-1245-2_54

1 Introduction

The Regional General Hospital (RSUD) of the Mentawai Islands Regency is located on North Sipora Island, precisely on Jl. Raya Tuapejat KM. 09 North Sipora District, Mentawai Islands Regency, West Sumatra Province. The operation of this hospital was inaugurated by the Governor of West Sumatra on March 20, 2006. As an organization engaged in hospital services, of course, it requires good management in managing all existing components, so that every existing business process can run well. Logistics is one of the managements that plays an important role in the service process both internally and externally.

Medicine is an essential component that must be available in health services, medicine is a link between patients and health service facilities, because the availability or not of medicines suggested that health services will have a positive or negative impact on the quality of service [1]. Therefore, it is necessary to have good and correct drug management to aim to ensure the continuity of the availability and affordability of drug services. Damaged or expired drugs are drug conditions whose concentration has been reduced by between 25 and 30% of the initial concentration as well as the physical form that has undergone changes, drugs whose form or condition can no longer be used. Damaged drugs can no longer be used because they have suffered damage accompanied by changes in shape, color, smell, taste, or consistency [2].

Pharmaceutical installations have medical materials with different types and uses, namely Tablet Drugs, Syrup Drugs, Ointments, Injections, Liquids, BMHP (Consumable Medical Materials), Laboratory Reagent Materials and Dentistry Materials. Pharmaceutical drug storage standards that have been set by SK. Mankes No. 1197/SK/X/2004, the percentage of expired drugs can still be accepted if the percentage value of the drug is below 1% of the total expired drugs [3]. If the percentage of drugs is above 1% this indicates ineffectiveness and efficiency in the management of drug supplies. Figure 1 shows the Percentage of Expired Drugs:

The planning of drug supplies at the Mentawai Islands Regional Hospital has not gone well because there are still 5 types of drugs whose percentage expires in 2020 is above the tolerance limit, namely the type of drug Tablets 9%, Syrup 34%, Injection 19%, Liquid 9%, Bmhp 15%. this happens because drug evaluation is still not optimal and drug stock data collection is not accurate. The occurrence of expired drug stocks can lead to the emergence of material losses. The biggest disadvantage of the five types of drugs whose percentage exceeds the tolerance limit is in the type of BMHP drugs which is Rp. 45,033,866.95, Consumable medical materials are medical devices intended for single use (PMK N0. 58 Year 2014) some examples of BMHP: syringes (syringes), urine collection devices (Urine bags), gauze, masks etc. The Presentation Value of BMHP expired drugs by 15% occurs because the total inventory exceeds the total usage. So, there was an overstock. There is a gap between supply and demand, which is 83% of the total average needs in 2020. The difference between total inventory and usage can cause damage to BMHP inventory because the item has an expiration time. The problem of the total cost of inventory exceeding

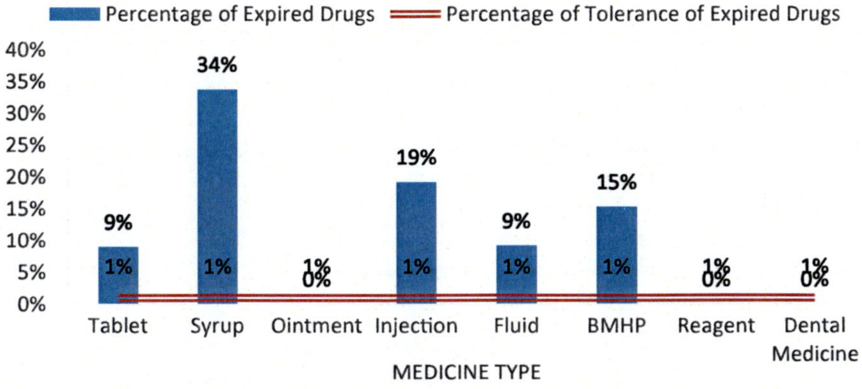

Fig. 1 Percentage of expired medicine in 2022

the budget occurs due to overstock of 83% of the total needs, the existence of this overstock is caused by the strength of the ordering the drug is purchased in excessive quantities. Hospitals have not implemented a willingness policy; drug evaluation is still not optimal and drug stock data collection is not accurate. Based on the previous exposure of the problem, it is necessary to design a drug supply policy by paying attention to how much the supply is to minimize the drug piling up which results in the large number of expired drugs as well as the total cost of excessive inventory using the continuous review method.

2 Theoretical Review

This study will use continuous inventory control where the control is carried out using a continuous review (r, Q) system model. The continuous review model is a control model for types of inventories whose demand is volatile, but demand data is normally distributed [4]. The major advantage of continuous review is that, to provide the same level of customer service, it requires less SS [5]. Inventory policy on perishable products that is carried out continuous review and ordering as much as Q When the inventory level reaches r [6]. In the continuous review policy (r, Q) that has been developed by Chiu (1995), inventory checks are carried out continuously [7]. If the inventory position is below the reorder point (r) at the time of checking, then an order of Q is made to return the inventory position to Q + r. This policy is evaluated by varying the product life span and lead time [8]. The notation used in the calculation of the probablistic method of the Continuous review model (r, Q), is as follows:

D Total product demand per year
S Standard deviation request

L	Lead time
A	Product ordering fee (Rp)
h	Product storage cost (Rp)
Cu	Cost of deficiency (Rp)
α	Possible shortcomings
$Z\alpha$	Normal deviation
$f(Z\alpha)$	Ordinate
$\Psi(Z\alpha)$	Partial expectations
N	The amount of inventory shortage each cycle
Ss	Safety stock
r	Reorder point
$q0n$	Ordering lot size
$\overline{X}$	Total average needs
σ	Standard deviation
Tc	Total cost
m	Product life span
E[O]	Expectation expires
E[I]	Annual inventory level of the unit of time
E[T]	Cycle length
E[S]	Expectations of shortcomings

Before carrying out calculations by the method of continuous review of medical materials classified first by ABC-VED analysis. ABC analysis is a classification of types of goods based on the level of annual investment absorbed in the supply of inventory for each type of goods [9] which is shown in Table 1. The purpose of ABC analysis is to distinguish all types of inventories in three groups, A, B and C. Then according to each group, the inventory level that will be controlled in general is determined [10].

VED analysis is a method used to control inventory, including the control of drugs and medical devices based on the criticality of inventory items [11]. Table 2 shows the VED Categories in this study.

Matrix ABC-VED is a combination of cross-tabulation results between ABC and VED analysis [11]. The matrix is shown in Table 3. The results of the analysis resulted in three categories, namely category I, category II, and category III. This Matrix is one of the important tools in pharmaceutical or drug inventory management which serves to simplify drug management and narrow down drug categories that require tight control. Category I (large investment value and high priority) category

Table 1 ABC Analysis

Group	Dolar usage (%)	Inventory item (%)	Quantitative control used
A	80	20	Yes
B	15	30	In some cases
C	5	50	No

Table 2 VED categories

VED categories	
V	Vital items without which the hospital cannot function
E	Important items without which the hospital can function, buying can affect the quality
D	Desired goods that if not available do not interfere with the function

Table 3 Matrix ABC-VED

Categories	Group				
I	AV	AE	AD	BV	CV
II	BE	CE	BD		
III	CD				

II (medium investment value/low price high priority) category III (low investment value and low priority).

3 Metodology

Figure 2 is the systematics of research. The explanation of the steps is as follows [12–14]:

1. Literature Review: collects information relevant to the research topic taken.
2. Observation: obtaining information or data by directly
3. Problem identification: this process identifies the problem, sets research goals and arrangements as well as the boundaries of the problem.
4. Data collection and processing This process collects all the data needed to conduct research. Conducting a data distribution test using the Kolsmogorovskmirnov test on demand data for drugs that have been classified using VED analysis, namely in categories I, AV, AE, AD, BV, CV. Perform calculations to determine inventory policy using continuous review (r, Q)
5. Data analysis this process analyzes data against the results of previous data processing
6. The conclusions and suggestions of this process conclude the results as well as provide the best advice.

Fig. 2 Systematic research

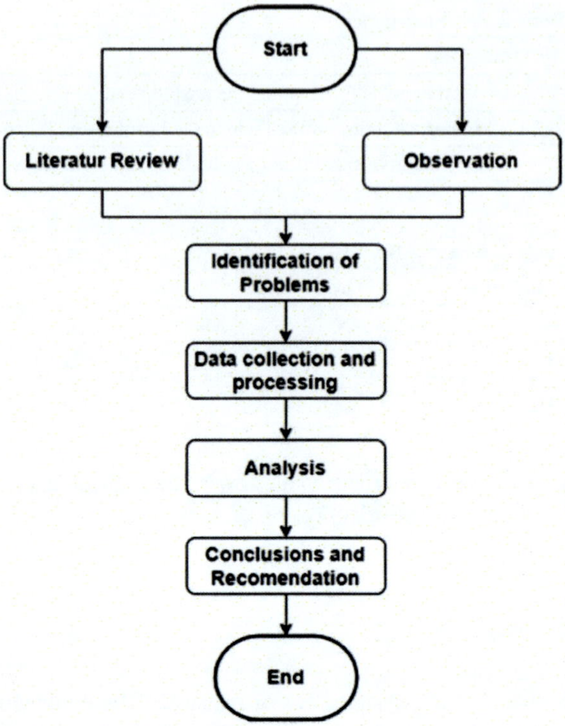

4 Results and Discussion

4.1 ABC Analysis

Figure 3 shows the classification of drugs based on groups A, B and C with a total of 133 types of medical materials, where in groups A and B there are 23 types of medical materials, and group C there are 87 types of medical materials.

Fig. 3 ABC analysis

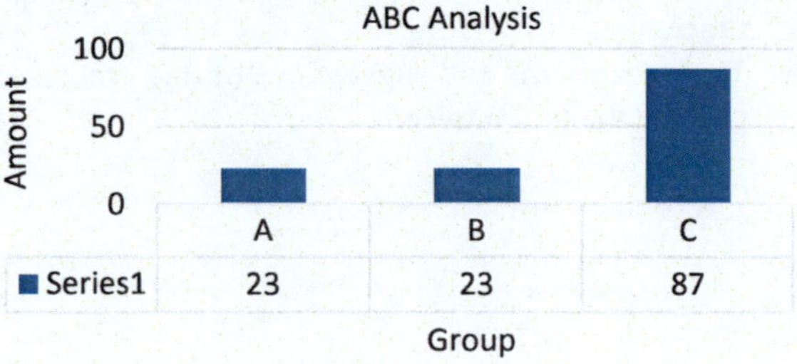

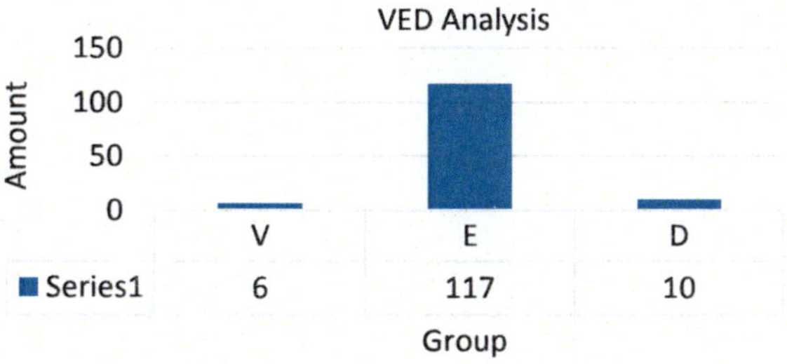

Fig. 4 VED Analysis

4.2 *VED Analysis*

Figure 4 shows the classification of drugs based on groups V, E and D with a total of 133 types of medical materials, where in groups V there are six types of medical materials and E there are 117 types of medical materials, and group D there are 10 types of medical materials.

4.3 *Matrix ABC-VED*

Medical materials that have been grouped using ABC-VED analysis as many as 133 types of medical consumables, will then be categorized into categories I, II, III. Table 4 is the number of medical materials based on the ABC-VED matrix.

Category I inventory requires strict inventory control management, Category I consumables management is monitored by management resulting in better control over annual expenses and at the same time providing vital consumables [4]. will use the continuous review method (r, Q). The use of the continuous review method (r, Q) determines the optimal order of Q when the inventory has reached the point r or below it. The Matrix ABC-VED Category 1 is shown in Fig. 5.

Table 4 Matrix ABC-VED result

Medicine category	A		B		C	
		Amount		Amount		Amount
V	AV	6	BV	0	CV	0
E	AE	16	BE	22	CE	79
D	AD	1	BD	4	CD	5
Total		23		26		84

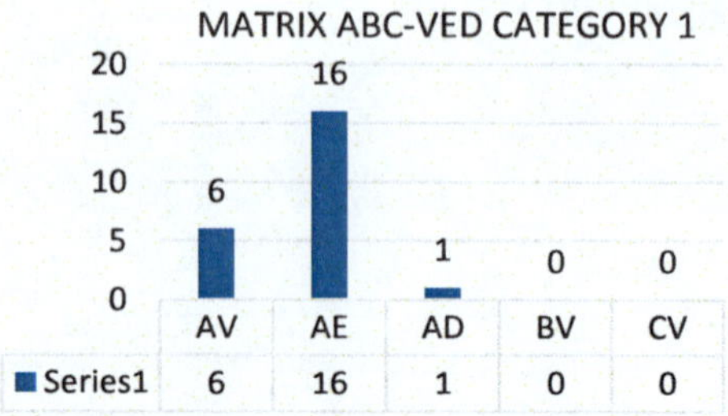

Fig. 5 Matrix ABC-VED Category 1

4.4 Continuous Review Formulation Model Hadley-Within

Calculation of inventory policy for types of consumable medical drugs with abc-ved classification category I use Continuous Review (r, Q) with fixed lifetime. This method performs inventory control using the parameters of the reorder point (r) and the optimum number of orders (Q), where a reorder is made as much as (Q) when the inventory level has reached (r). to perform calculations, namely using Wilson's formula and iterative algorithms that minimize total costs [6]. The following will be explained the calculation of Continuous Review (r, Q) on consumable medical materials, namely Handscoon No Sterile/2020 as an example.

Known:

Annual demand (D)	190
Mean demand (μ)	15.83
Stdev demand (σ)	6699
Lead time (LT)	30 days/365 = 0.082
Message charges (A)	IDR 20,000
Storage cost (h)	IDR 13,920
Cost of deficiency (Cu)	IDR 174,000
Unit price (C)	IDR 174,000
Cost of expired products (W)	IDR 174,000

Iteration 1

a. Look up the initial Q value (q01*) with Wilson's formula

$$q01 = \sqrt{\frac{(2AD)}{h}} \tag{1}$$

$$q01* = \sqrt{\frac{2.A.D}{h}} = \sqrt{\frac{2 \times Rp\,20.000 \times 190}{Rp\,13.920}} = 23$$

b. Looking for values to get possible inventory shortages based on the results of q01* calculations that have been previously obtained α

$$\alpha = \int_{r}^{\infty} f(x)dx = \frac{h.q01}{(Cu.D)} \tag{2}$$

$$\alpha = \int_{r}^{\infty} (x)dx = \frac{h.q01}{(Cu.D)} = \frac{Rp13.920 \times 23}{(Rp174.000 \times 190)} = 0.0098$$

Then get the value of Z $\alpha = 2.35\alpha = 2.25$

c. Looking up the value of r1*

$$r1* = DLT + Z\alpha.\sigma\sqrt{LT} \tag{3}$$

$$r1* = DLT + Z\alpha.\sigma\sqrt{LT} = ((190 \times 0.082)) + 2.35 \times 10\sqrt{0.082} = 22$$

d. Looking up the value of q02*

$$q02* = \sqrt{\frac{2.D[A + Cu \int_{r}^{\infty}(x - r1*)f(x)dz]}{h}} \tag{4}$$

where:

$$\int_{r}^{\infty} (x - r1*)f(x)dx = \sigma LT[f(Z\alpha) - Z\alpha.\Psi(Z\alpha)] = N$$

$$= 10 \times \sqrt{0.082}[0.0252 - (2.35 \times 0.0032)] = N = 0.1 \tag{5}$$

$$q02* = \sqrt{\frac{2 \times 190[Rp20.000 + Rp174.000 \times 0,0339]}{Rp13.920}} = 28$$

e. Looking for values to get possible inventory shortages based on previously obtained q02* calculation results α

$$\alpha = \int_{r}^{\infty} f(x)dx = \frac{h.q02}{(Cu.D)}, \alpha = \frac{Rp13.920 \times 25}{(Rp174.000 \times 190)} = 0.0111$$

Earned Z grade $\alpha = 2.25\alpha = 2.25$

f. Looking for the value of r2

$$r2* = DLT + Z\alpha.\sigma\sqrt{LT} = ((190 \times 0.082)) + 2,25 \times 10\sqrt{0.082} = 22$$

g. Compare r1* with r2* (22 with 2 2), then the values r1* and r2* are the same, the iteration is enough to get here.

The optimal inventory policy, namely:

$$\begin{aligned} q0* = q02* &= 28 \text{ types} \\ r* = r2* &= 22 \text{ types} \\ ss = Z\alpha S\sqrt{L} &= 2.25 \times 10\sqrt{0.082} = 7 \text{ types} \end{aligned}$$

Based on the calculations that have been done, the value of r1*r2* then the iteration is continued until r1* = r2*, so that the optimum Q value is obtained. In handscoon consumable medical materials No Sterile/2020, only the 1st iteration calculation is carried out so that the optimum Q value is obtained, namely q0* = 28 types and r0* = 22 types. Because the values of r and Q have been obtained, then determine the number of *safety stocks*, which are as many as 7 types.

4.5 Continuous Review Formulation (r, Q) with Fixed Lifetime

This model approach considers five cost parameters, namely purchase costs, ordering costs, holding costs, product use costs, and shortage cost [7]. The following is an example of a calculation on medical material handscoon no steril/2020:

1. Possible number of obsolete products (E[O])

$$\int_0^{r+Q} (r + Q - x) fm + LT(dm + LT)dx - \int_0^{r} (r - x) fm + LT(dm + LT)dx \tag{6}$$

where fm + LT(dm + LT) is a probability function of dm + LT, where m is a fixed lifetime which is 1 year or 365 days and d is demand.

$$f(x) = \frac{1}{\sqrt{2\pi\sigma^2}} e^{\left(-1/2\frac{(m+LT)-((m+LT)\times d))}{\sigma}\right)1^2}$$

$$f(x) = \frac{1}{\sqrt{2 \times 3,14 \times 10^2}} e^{\left(-1/2\left(\frac{((365+30)-(365+30)\times 15)}{10}\right)\right)^2} = 0$$

$$\int_0^{22+28} (22 + 28 - x)(0)dx - \int_0^{20} (22 - x)(0)dx = 0$$

$$\int_0^{22+28} ((25 \times 0) + (28 \times 0) - (x \times 0))dx - \int_0^{22} ((22 \times 0) - (x \times 0))dx$$

The following is the result of the integral above:

$$= \left((22 \times 0)50 + (28 \times 0)50 - \left(x^2 \times 0\right) - 0\right)$$

$$-\left((22\times 0)22-\left(x^{2}\times 0\right)-0\right)=0$$

2. Possible number of product shortages (E[S])

$$\int_{r1*}^{\infty}(x-r1*)f(x)dx=\sigma LT[f(Z\alpha)-Z\alpha.\Psi(Z\alpha)=N$$

$$=(10\times\sqrt{0.082}\times(0.0317-2.25\times 0.0042)=0.1\,jenis \quad (7)$$

3. Inventory level per unit time (E[I])

$$E[I]=r-DLTD+\frac{Q}{2}+DLTD\frac{E[s]-E[O]}{2(Q-E[O])} \quad (8)$$

$$E[I]=22-16+\frac{28}{2}+16\frac{0,1-0}{2(28-0)}=21$$

4. Inventory cycle length (E[T])

$$E[T]=\frac{Q-E[O]}{D} \quad (9)$$

$$E[T]=\frac{28-0}{190}=0.15\,tahun$$

5. Total inventory cost

$$TC=\frac{A+C.Q+W.E[O]+Cu.E[S]}{E[T]}+h.E[I] \quad (10)$$

$$TC=\frac{Rp20.000+(Rp174.000\times 28)+(Rp174.000\times 0)+(Rp174.000\times 0,1)}{0,15}$$

$$+Rp13.920\times 21=Rp35.838.032/tahun$$

4.6 Result (r, Q) Optimal

With a lead time of one month, the number of requests for medical consumables for Handscoon No Steril/2020 is 16 types and safety stock are 7 types. For the optimal inventory policy, the hospital will place an order of Q, which is 28 types, if the inventory has reached the point r, which is 22 types (Table 5).

Table 5 Result (r, Q) optimal

No.	Medical materials	R	Q	ss	SL (%)
1	Handscoon Steril 7.5	615	800	164	100
2	Handscoon No Steril/2020	22	28	7	100
3	Supercare Surgical Face Mask/Earloop 3 Ply/2020	48	62	15	100
4	Handscoon Steril 7.0/Tro - sensosurge, surgical glove steril 7.0/ 2020	345	591	123	100
5	BAJU APD TCC WR/Biru/120/2020	3	4	1	100

4.7 Dashboard Monitoring System

Furthermore, the results of the monitoring dashboard design from this final project aim to provide brief information and assist in making decisions. The image is a display of the monitoring dashboard on medical materials. The dashboard could display data visualization of several metrics includes medical material classification, purchase cost, demand [15]. Figure 6 shows the Dashboard Monitoring System.

Figure 7 is a monitoring dashboard display using Excel software, which displays information on demand data for medical materials for 12 months in 2020. If you want to know the demand data for one type of medical material, then click on the column for the preparation of medical materials that has displayed the name of the medical material then the graph will show the number of requests for 12 months.

Figure 8 is a monitoring dashboard display using Excel software, which displays the ABC-VED group total information. If you want to know a group of medical materials, then click on the Group ABC-VED column, for example, when you want to know the number of BE groups, the graph will show the number of BE groups as many as 22 types.

Figure 9 is a monitoring dashboard display using Excel software, which displays information on purchasing costs for a year. If you want to know one of the costs of

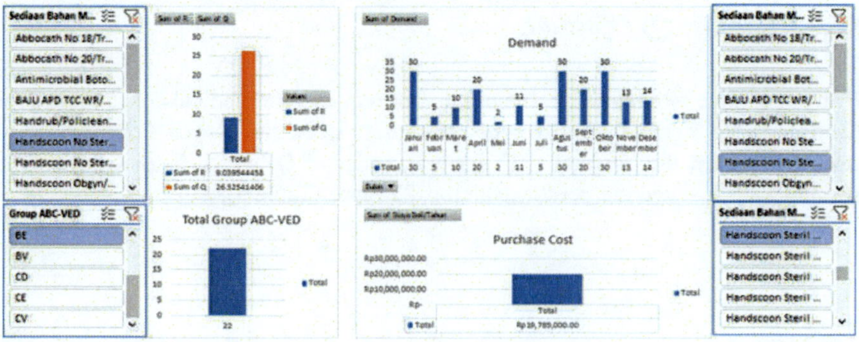

Fig. 6 Dashboard monitoring system

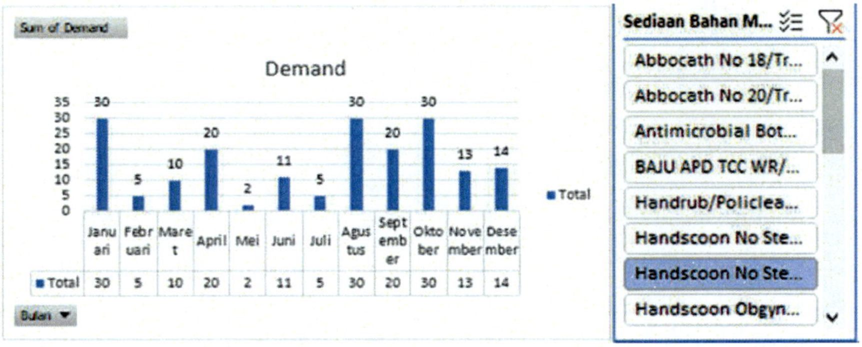

Fig. 7 Dashboard monitoring demand

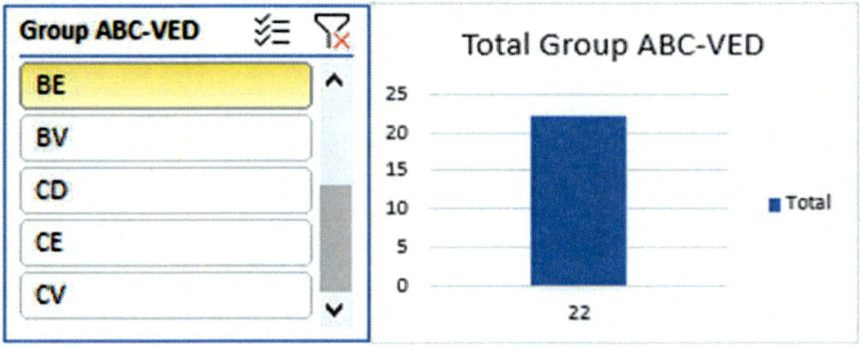

Fig. 8 Dashboard monitoring matrix ABC-VED

buying medical materials, then click on the medical material preparation column, for example, when you want to know the cost of buying a sterile handscoon in 2020, the graph will show the purchase cost of Rp. 19,785,000.00.

Figure 10 is a monitoring dashboard display using Excel software, which displays optimal order information (Q) and reorder point (R) for each medical material. If

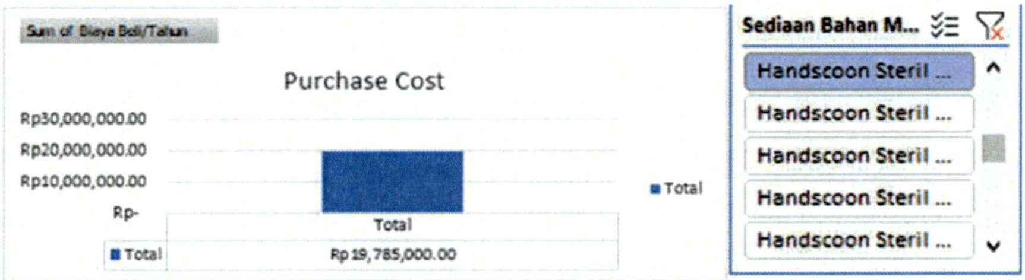

Fig. 9 Dashboard monitoring purchase cost

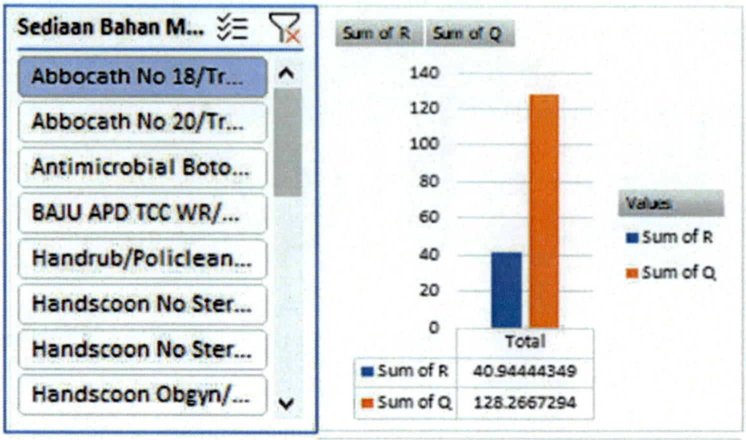

Fig. 10 Dashboard monitoring Q optimal

you want to know how much the optimal order is for one of the medical materials, then click on the column for the preparation of medical materials, for example, when you want to know the optimal purchase amount of Abbocath No. 18, the graph will show the number of (Q) 128 types and (R) 41 types.

4.8 *Expected Amount of Expired BMHP*

Table 6 shows that the number of expired medical consumables has decreased, where the Polifix 5 cm × 5 M/2020 preparations that have expired are 242 types after calculating the continuous review policy (r, Q) with fixed lifetime by considering expectations expired E[O] then obtained the number of expired as many as zero types. By obtaining the expected number of expired types of BMHP drugs, the proposed loss value is also Rp. 0.

Table 6 Expected amount of expired

No.	Medical material	Unit	Amount expired of existing	Expected amount expired
1	Handscoon No Steril/2020	Box/50 pairs	10	0
2	Spuite 3 cc	Pcs	145	0
3	Three Way/	Pcs	780	0
4	Handscoon Steril Panjang	PSG	6	0
5	Handscoon No Steril Powder/ Protos/2020	Box/50 pairs	15	0

Table 7 Total cost inventory

	Existing	Research proposal
Purchase cost	Rp 380,253,433.00	Rp 310,993,735.00
Holding cost	Rp 30,420,274.00	Rp 3,338,861.00
Order cost	Rp 460,000.00	Rp 1,969,962.00
Shortage cost	Rp –	Rp 13,527,588.00
Expiry cost	Rp 45,033,866.00	Rp –
TC	Rp 456,167,573.00	Rp 329,830,146.00

4.9 Comparative Analysis of Total Inventory Cost

Based on the results of calculations that have been carried out using the continuous review method (r, Q) with a fixed lifetime, the total cost of inventory has decreased by 28% from the total cost of the existing inventory. The total cost inventory is shown in Table 7.

5 Conclusion and Recommendation

5.1 Conclusion

Based on the drug classification of medical substances in the pharmaceutical installation of the Mentawai Islands Hospital using the ABC and VED analysis approach, six types of medical materials were obtained including the AV group, 16 types of medical materials in the AE group and one type of medical substance in the AD group, the BV group and zero CV. which is category I in the ABC-VED matrix.

Based on the results of data processing using the continuous review method (r, Q) with a fixed lifetime, it is obtained that the expectation of expired medical materials is zero, so the number of medical materials that initially expired was 15% with a loss of Rp.45,033,866.95, in the proposed condition it became 0% and the loss becomes Rp 0.

Based on the results of calculations that have been carried out using the continuous review method (r, Q) with a fixed lifetime, the total cost of inventory has decreased by 28% from the total cost of the existing inventory.

The dashboard monitoring system on the drug inventory helps provide information about the classification of drugs that have been carried out so that they can find out the priority of each category can find out the number of requests each month and the cost of buying per year.

5.2 Recommendation

Based on the results of the design that has been carried out, the hospital needs to implement an inventory policy in the pharmaceutical warehouse. Warehouse Officers can reduce the total cost of inventory by implementing an inventory policy, where it is necessary to conduct a continuous inventory review on drug groups that have absorption of funds is about 80% of the overall amount of the drug. Officers can find out which drugs absorbed as much as 80% of the funds, namely by conducting an ABC-VED analysis, after which they determined the calculation of inventory policies using the continuous review method (r, Q) with fixed lifetime so that warehouse officers can find out what is the optimal purchase amount (Q) that must be made to meet demand when the inventory reaches the reorder point or below.

References

1. Chaira S, Zaini E, Augia T (2016) Evaluasi Pengolahan Obat Pada Puskemas di Kota Pariaman
2. Kareri (2018) Pelaporan Obat Rusak dan Kadaluarsa di Seksi Kefarmasian Dinas Kesehatan Kabupaten Sumba Timur Tahun 2017–2019
3. Nugroho (2008) Keperawatan Gerontik, Jakarta
4. Bahagia SN (2006) Sistem Inventori
5. Silver EA, Pyke DF, Thomas DJ (2017) Inventory and production management in supply chain
6. Kouki C (2010) Perishable items inventory management and the use of time temperature integrators technology. Education
7. Chiu HN (1999) A good approximation of the inventory level in a (Q, r) perishable system
8. Madduri (2009) Inventory policies for perishable product with fixed shelf lives. Ind Manufact Eng
9. Trihudiyatmanto M (2018) Riset operasional. 1
10. Indartono S (2006) Operational reserach. 17
11. Gupta (2007) ABC and VED analysis in medical store inventory control. 326
12. Wahyuni NKC, Ridwan AY, Santosa B (2018) Penentuan Kebijakan Persediaan Auto Part menggunakan Metode Continuous Review (s,S) Untuk Meningkatkan Service Level dan Meminimasi Biaya Akibat Backorder Price Discount Pada PT.XYZ. J Ind Serv
13. Permatasari PM, Ridwan AY, Santosa B (2017) Inventory policy determination for raw material in ILY pharmaceutical using periodic review (R,s,s) and periodic review (R,S) method to minimize total inventory cost. In: 8th International conference on mechanical and manufacturing engineering 2017 (ICME'17)
14. Maulaya A, Ridwan AY, Santoso B (2019) Spare part inventory policy planning based on FRMIC (fuzzy-rule-based approach for multi-criteria inventory classification) using base-stock policy method. In: Proceedings of the 2019 1st international conference on engineering and management in industrial system (ICOEMIS 2019)
15. Fathoni FA, Ridwan AY, Santosa B (2018) Development of inventory control application for pharmaceutical product using ABC-VED cycle counting method to increase inventory record accuracy. In: Proceedings of the 2018 international conference on industrial enterprise and system engineering (IcoIESE 2018)
16. Shiddieq NF, Ridwan AY, Santosa B (2020) Antibiotic inventory policy design for minimizing total inventory cost in pharmacies based on ABC-fuzzy classification analysis approach using probabilistic continuous review method
17. Raihan R, Ridwan AY, Santosa B (2018) Development of e-Kanban application using stock-needs rule prioritizing policy to reduce 0-pick for pharmaceutical warehousing

Elaborate on Sustainable Production and Consumption in Mocaf Agro-Industry

Dessy Srikandi, Muhammad Hisjam, and Eko Pujiyanto

Abstract Sustainability supply chain management (SSCM) is related to supply chain practices that synergize with sustainable practices along the supply chain. This study tries to describe the scope of sustainable production and consumption, especially related to the use of raw materials in the mocaf agro-industry. Mocaf is a commodity of modified cassava flour. This paper only discusses environmental aspects and several ways related to the implementation of sustainability in the mocaf agro-industry that can be carried out by business actors. The data used are secondary data taken from the internet and previous research on mocaf and the related environment in one production cycle. The results are expected to be a consideration for mocaf agro-industry companies to decide to implement environmental sustainability as the first step in SSCM. Mocaf agro-industry can implement sustainability as follows, residues from the sieving process can be used as animal feed, and the indoor flouring process can minimize flour residue in the form of fine dust, switch to using environmentally friendly electricity-based equipment, minimize the use of fossil fuels.

Keywords Sustainability supply chain management (SSCM) · Sustainable production and consumption · Agro-industry · Mocaf

---

D. Srikandi (✉)
Master Program of Industrial Engineering Department, Faculty of Engineering, Universitas Sebelas Maret, Surakarta, Indonesia
e-mail: srikandi0401@student.uns.ac.id

M. Hisjam · E. Pujiyanto
Department of Industrial Engineering, Faculty of Engineering, Universitas Sebelas Maret, Surakarta, Indonesia

C. N. Rosyidi et al. (eds.), *Proceedings of the 6th Asia Pacific Conference on Manufacturing Systems and 4th International Manufacturing Engineering Conference*, Lecture Notes in Mechanical Engineering,
https://doi.org/10.1007/978-981-99-1245-2_55

1 Introduction

Environmental, social, and economic variables that contribute to sustainability include reducing carbon emissions and waste, improving water efficiency, boosting variety, prosperity, and equal opportunity, creating strength, and a stable, efficient, and equitable economy. Sustainability supply chain management (SSCM) is related to supply chain practices that synergize with sustainable practices along the supply chain.

Agroindustry becomes an interesting discussion regarding long supply chains, seasonality, and product categories that are not durable (perishable) and have more uncertainties compared to other manufacturers. The agro-industry is still a promising opportunity in the context of research and business. In terms of agro-industry research, strives for zero waste for every raw material used in the manufacture of its products. From a business perspective, it is a great opportunity because Indonesia is an agricultural country with various kinds of abundant agricultural products.

Mocaf agro-industry company is a company that produces cassava-based flour as a substitute for wheat flour. Mocaf is one of the modified starch products that has been widely used in various food products. According to [1], mocaf is cassava flour produced by modifying cassava cells by fermentation. These changes occur because, during fermentation, microbes work to produce enzymes and organic acids that change the structure of starch in such a way that it produces starch characters that are much better than without fermentation [2]. If in other flours the aroma of cassava is still attached, in mocaf flour the aroma is almost certainly not present anymore. Interestingly, the color of mocaf flour is whiter and its nutritional content is equivalent to rice flour and wheat flour. Mocaf flour is low in gluten so it is good for autistic children and diabetics to consume.

Mocaf agro-industry companies need to implement sustainable supply chain management to improve business performance, increase economic value, benefit the social environment and minimize environmental impacts so that companies have competitiveness, and maintain business continuity so that they can create value for customers. This study describes how the implementation of sustainable production and consumption can be carried out at the mocaf agro-industry focus with environmental dimensions, as consideration for company management in implementing SSCM.

2 Literature Review

SSCM is the creation of a coordinated supply chain through the voluntary integration of economic, environmental, and social considerations with inter-organizational key business systems intended to efficiently and effectively manage material, information, and capital flows associated with the procurement, production, and distribution of goods or services to meet stakeholder requirements and improve the organization's profitability, competitiveness, and resilience in the marketplace [3]. The approach in SSCM with a triple bottom line, the company's focus is not only on the economy but also on the environment and society [4].

Sustainable manufacturing is the ability to use natural resources for the manufacturing process in a way that is behaviorally coherent and meets economic, environmental, and social goals, achieving environmental sustainability while continuing to improve the quality of human life, this is accomplished through the use of new technologies, and behaviorally coherent regulatory and social measures [5]. At the heart of a civilized lifestyle, manufacturing is heavily influenced by sustainability issues and plays an important role in building sustainability, manufacturing-based technology is expected to interact with culture and business to provide tools and options to provide new solutions to sustainable manufacturing concepts [5]. Key elements of a sustainable manufacturing system include process planning, production planning, and forward and reverse supply chains [6].

Sustainable consumption meets basic human needs, and minimizes the use of natural resources, toxic substances, waste emissions, and pollutants throughout the life cycle, endangering the needs of future generations, however, it can be defined as the use of goods and services to create a better quality of life [7].

The Oslo Symposium in 1994 defined sustainable consumption and production (SCP) as the use of products and services that are used as a basic need for a better quality of life by minimizing the use of natural resources and toxic materials as well as waste and pollution emissions during the service or product life cycle so that does not interfere with the needs of future generations [8]. UNEP 2011 defines SCP as a holistic approach to reducing the negative environmental impacts of consumption and production systems while improving the quality of life for all [9].

3 Methods

This study discusses the sustainable production and consumption of the mocaf agro-industry by using secondary data in the form of previous research. The methodology in this research departs from a literature study on sustainable supply chain management and sustainable production and consumption. Data is sourced from Scopus, Google Scholar, and Google search engines. Secondary data is used for searching with keywords agro-industry. Then our data is narrowed by filtering agroindustry with

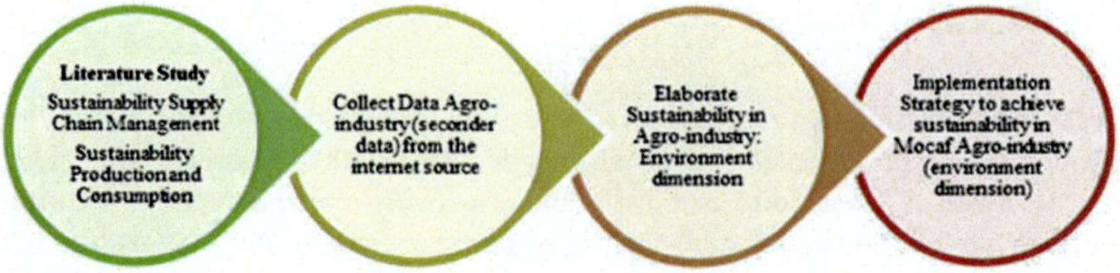

Fig. 1 Methods

environment dimension. The elaboration is carried out using an environment dimension approach and the last of the study is the implementation that can be applied to achieve sustainable production and consumption in the mocaf agro-industry. Figure 1 is an illustration of this research.

4 Result and Discussion

Mocaf agro-industry supply chain management involves a long supply chain. Cassava from farmers is the upstream raw material chain, and the home industry carries the semi-finished raw material processing stage. The result of the semi-finished raw material is cassava chips which are the main raw material for mocaf which is processed by the manufacturer. Mocaf is distributed to customers and the food industry sector. This is shown in Fig. 2 to illustrate the supply chain. The supply chain will be described in terms of sustainable production and consumption with an environmental dimension approach at the end of the discussion of this paper.

A simple cycle in the cassava planting process is land cultivation, planting cassava, plant processing time, harvesting, and distribution to industry. The process of planting to harvest by the farmer takes 8–10 months. The planting process of cassava can be seen in Fig. 3. Cassava is distributed to small industries to be processed into cassava chips. Figure 4 is the process of making cassava chips. The manufacture of cassava chips is carried out by home industries or can be done by farmers. However, most of the farmers sell it directly in the form of raw materials as. A great opportunity for home industries to process cassava into semi-finished raw materials in the form of cassava chips with a higher selling value than raw materials. The dried cassava chips are then distributed to the mocaf factory for the flouring and packaging process. An overview of the mocaf flour production process by the Mocaf Agro-industry,

Fig. 2 Supply chain Mocaf agro-industry

Fig. 3 The planting process of cassava by farmer

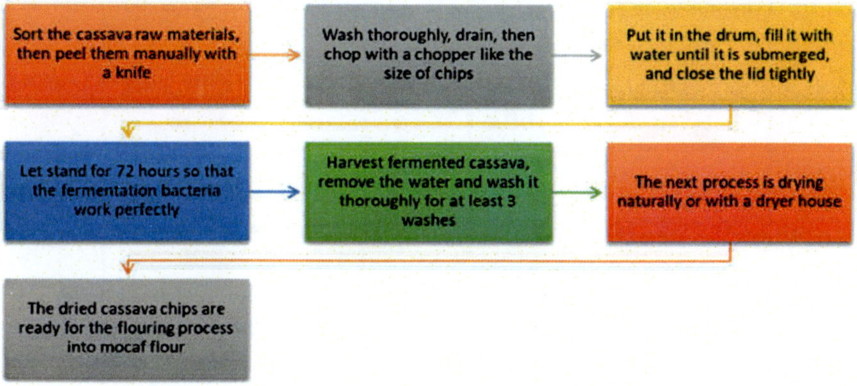

Fig. 4 The process of making cassava chips by the home industry

namely raw material selection (cassava chip), flouring process, shifting process, and packaging process.

Mocaf flour making process can be seen in Fig. 5. The dried cassava chips are sorted and then floured, after the flouring process is done sifting with a fineness of 80–100 mesh [10], the mocaf is packed (25 kg, 5 kg, 1 kg). The process of making mocaf uses assumptions based on the capacity of the flour machine 500 kg/h. The mocaf processing process follows the guidelines of the Badan Standarisasi Nasional (BSN) and can be adapted to local conditions, as long as it has the same goal, namely to produce product quality according to requirements [11]. Mocaf processing can also be carried out by farmer groups in the regions with guidance from the relevant agencies [12], with the process of making cassava raw materials into mocaf flour [13].

The influence diagram of sustainable consumption and production of Mocaf can be seen in Fig. 6. The picture shows a causal relationship that affects the consumption of Mocaf and its production. The amount of mocaf production is influenced by cassava production decisions. Mocaf consumption by industry is influenced by how much food production will be carried out. Mocaf consumption on a household scale is influenced by the large number of people in Indonesia who decide to buy or not the mocaf product. Influence diagrams are used to describe causality with a process approach with several symbols, namely clouds for uncontrollable input, rectangles for control inputs such as decision-making, ovals as outputs, circles representing system variables, and arrows representing influence relationships [14].

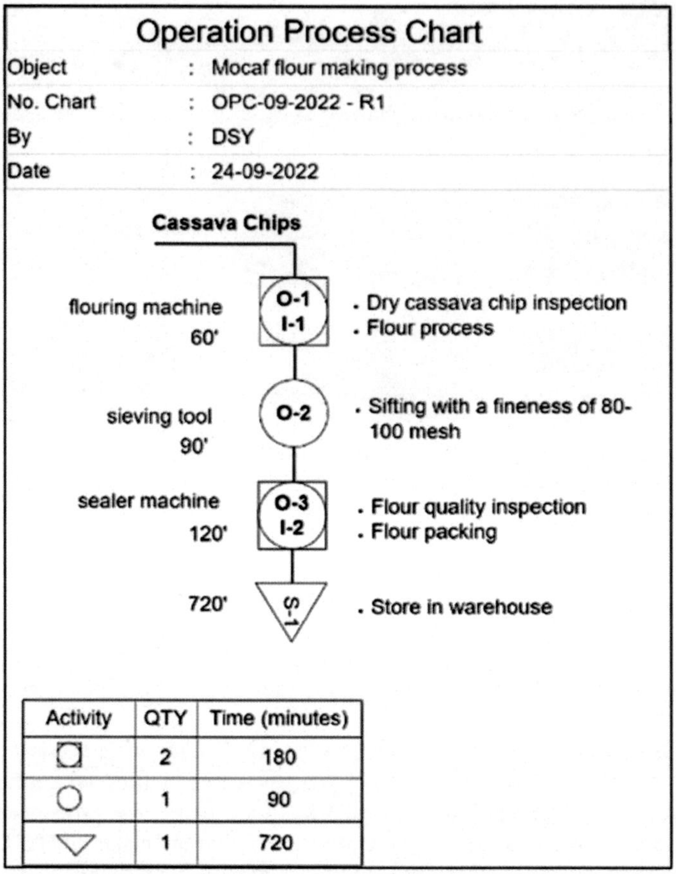

Fig. 5 Mocaf flour-making process by Mocaf agro-industry

Sustainable supply chain management can function as a tool to increase the value of a product which is automatically able to increase the value of the company. Creating value for the company means being able to create profit. So that the application of SSCM in the agro-industry is not something to be afraid of with the triple bottom line parameter. In practice, valid data and information can produce policies and strategies that are right on target. Sustainability consumption and production of mocaf agro-industry are explained by adopting from Lsaputri et al. [15] sustainability indicators that have three aspects: environmental, social, and economic. However, in this study, we take an environmental dimension approach with an external scope of discussion using secondary data. Lsaputri et al. [15] in their research explain that the part of environmental dimension is natural resources, solid waste, air emissions and water, transport modes and distances, global warming, biodiversity, acid deposition, landscapes, final waste disposal, product recyclability, forest, and ecosystem restoration,

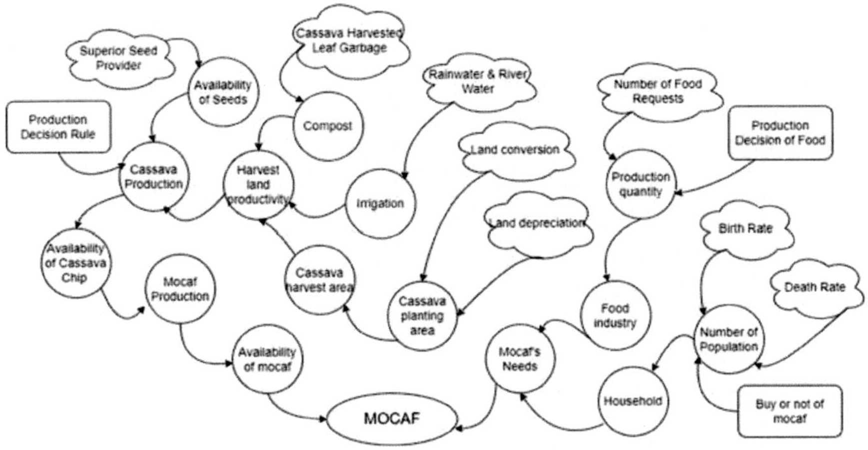

Fig. 6 Influence diagram of sustainable consumption and production Mocaf

deforestation, conservation of primary forests, use of GMOs used, use of chemicals, pest control, and fertilizer. This study will elaborate on its application in the mocaf agro-industry, namely, natural resources, solid waste, water and water emission, and acid deposition. Table 1 shows sustainability of consumption and production of raw material.

Table 1 Sustainability of consumption and production of raw material

The environment dimension	Previous research	Sustainability of consumption and production of raw material
Nature resources	Organic fertilizer is a form of a natural resource that is useful in restoring soil quality to increase productivity and agricultural sustainability [16]. The sustainable principle in making packaging must be bio-based and easily biodegradable by natural processes, bio-based fiber materials have the potential to meet these criteria [17]	• Standardize the use of raw materials to prevent waste • Implement fast cycles in the use of raw materials • Utilization of cassava stems for replanting • Making compost from cassava leaves • Processing of cassava peels for animal feed, as well as for making bioplastics • Using biodegradable packaging such as sacks made from textile fibers and bioplastics • Using sunlight in the drying process • Speed up the drying process with a renewable energy-based dryer

(continued)

Table 1 (continued)

The environment dimension	Previous research	Sustainability of consumption and production of raw material
Solid waste	The processing of solid waste using the composting technique has the advantages of being environmentally friendly, and financially profitable, and the composting results can be returned as nutrients for new plants [18]	• Use of compost in the planting process • The sifting process residue can be used as animal feed
Air emissions and water	Cassava starch sedimentation wastewater treatment can be used as biogas on a household scale [19]	• Utilization of soaking water as raw material for household-scale biogas • The indoor flouring process can minimize flour residue in the form of fine dust
Acid deposition	Acid deposition is the emission of NO_x and S02 released from the burning of fossil fuels [20]. Atmospheric acid deposition or acid rain is the result of burning fossil fuels and agricultural activities [21]. Acid deposition of NO_x and S02 produced from the burning process of fossil fuels is released into the air which causes damage to the human, plant, and animal ecosystems [22]	• Switch to eco-friendly electric-based devices • Minimize the use of fossil fuels

5 Conclusion

This study describes the implementation of sustainable production and consumption that can be carried out at the mocaf agro-industry focus with environmental dimensions, as consideration for company management in implementing SSCM. The implementation of sustainable consumption and production can be conducted by farmers, including the use of cassava stems for replanting, making compost from cassava leaves, and using compost in the planting process. Applications for the home industry are processing cassava peels for animal feed, drying cassava peels as a bioplastic-making material, using sunlight in the drying process, accelerating the drying process using environmentally friendly energy-based dryers, and utilizing soaking water as a raw material for biogas on a household scale. Mocaf agro-industry can implement sustainability as follows, residues from the sieving process can be used as animal feed, and the indoor flouring process can minimize flour residue in the form of fine dust, switch to using environmentally friendly electricity-based equipment, minimize the use of fossil fuels.

The limitation of this research is that the literature and data are obtained only from internet sources. Further research can be done using data on the actual condition of the company that makes more accurate data and analysis obtained.

References

1. Subagio A, W. SW, Y. Witono, and F. Fahmi, *Prosedur OPerasi Standar (POS) Produksi Mocal Berbasis Klaster*. 2008.
2. Putri NA, Herlina H, Subagio A (2018) Karakteristik Mocaf (modified cassava flour) berdasarkan Metode Penggilingan dan Lama Fermentasi. Jurnal Agroteknologi 12(01):79. https://doi.org/10.19184/j-agt.v12i1.8252
3. Ahi P, Searcy C (2013) A comparative literature analysis of definitions for green and sustainable supply chain management. J Clean Prod 52:329–341. https://doi.org/10.1016/j.jclepro.2013.02.018
4. Elkington J (2004) Chapter 1: enter the triple bottom line
5. Garetti M, Taisch M (2012) Production planning & control: the management of operations sustainable manufacturing: trends and research challenges. Taylor & Francis, pp 37–41
6. Haapala KR et al (2013) A review of engineering research in sustainable manufacturing. J Manuf Sci Eng Trans ASME 135(4). https://doi.org/10.1115/1.4024040
7. Hornibrook S, May C, Fearne A (2015) Sustainable development and the consumer: exploring the role of carbon labelling in retail supply chains. Bus Strateg Environ 24(4):266–276. https://doi.org/10.1002/bse.1823
8. SDG Homepage. Sustainable consumption and production. In: Sustainable development goals. https://sustainabledevelopment.un.org/topics/sustainableconsumptionandproduction/. Accessed 24 Sept 2022
9. Akenji L et al (2015) Sustainable consumption and production
10. Azra RA (2014) Tepung Mocaf. bbp2tp.litbang.pertanian.go.id. http://bbp2tp.litbang.pertanian.go.id/images/Download/Kementan-Rumah_Mocaf-Mocafine-Banjarnegara.pdf
11. Harahap YE (2014) SPO Mocaf—standar prosedur operasional modified cassava flour (MOCAF). Direktorat Pengolahan Hasil Pertanian Direktorat Jenderal Pengolahan Dan Pemasaran Hasil Pertanian Kementerian Pertanian 2014:1–44

12. Rosyidhana Z (2021) Mengenal Mocaf (modified cassava flour). DPKP DIY. https://dpkp.jogjaprov.go.id/baca/Mengenal+Mocaf+%28Modified+Cassava+Flour%29/161121/2fd4ffd3878ba7d31d6aec01c1c9dae55e4211336dc22c46e761e6827d31da89400. Accessed 26 Sept 2022
13. Helmi RL, Khasanah Y, Damayanti E, Kurniadi M, Mahelingga DE (2020) Modified cassava flour (Mocaf): Optimalisasi Proses dan Potensi Pengembangan Industri Berbasis UMKM. LIPI Press. https://doi.org/10.14203/press.43
14. Daellenbach HG, McNickle DC (2005) Management science: decision-making through systems thinking. Pelgrave Macmillan
15. Lsaputri VH, Hisjam M, Sutopo W (2020) A review on sustainable metrics for sustainability measurement in supply chain. IOP Conf Ser Mater Sci Eng 943(1). https://doi.org/10.1088/1757-899X/943/1/012056
16. Mrunalini K et al (2022) Nature-based solutions in soil restoration for improving agricultural productivity. Land Degrad Dev 33(8):1269–1289. https://doi.org/10.1002/ldr.4207
17. Srivastava V, Singh S, Das D (2022) Biodegradable fibre-based composites as alternative materials for sustainable packaging design. In: Smart innovation, systems and technologies, vol 262SIST, pp 87–98
18. Gaspar LMR, Inácio CT, Quintaes BR, Carvalho LSQ, Peres AAC (2020) Economic and financial analysis of the management of organic solid waste from a small-scale agro-processing industry. Engenharia Sanitaria e Ambiental 25(3):477–488
19. Colin X, Farinet JL, Rojas O, Alazard D (2007) Anaerobic treatment of cassava starch extraction wastewater using a horizontal flow filter with bamboo as support. Biores Technol 98(8):1602–1607. https://doi.org/10.1016/j.biortech.2006.06.020
20. Bricker OP, Rice KC (1993) Acid rain. Annu Rev Earth Planet Sci 21:151–174. https://doi.org/10.1146/annurev.ea.21.050193.001055
21. Driscoll CT, Fallon-Lambert K, Chen L (2005) Acidic deposition: sources and effects. Encycl Hydrol Sci. https://doi.org/10.1002/0470848944.hsa095
22. Kementerian Negara Lingkungan Hidup (2009) Deposisi Asam. Sekretariat East Asia Network on Acid Deposition (EANET). http://perpustakaan.menlhk.go.id/pustaka/images/docs/DEPOSISI_ASAM.pdf. Accessed 28 June 2022

Manufacturing Automation

Deep Learning Implementation for Rami Fiber Image Classification Using Convolutional Neural Network Algorithm

Anisa, Muhammad Nihri Jundullah, Latifu Itsnaini Khoirun Nafiah, Loviana Ajeng Sari, and Pringgo Widyo Laksono

Abstract As a process of categorizing and labeling groups of images, image classification can be applied in various work fields to make work easier. PT Retota Sakti Magelang is a company that processes rami fiber into sun shade curtains, carpets, upholstery basketry and wallcovering. Currently, in this company after the bleaching process, there is still no inspection process to ensure that all fibers are perfectly bleached so there are needs to be an inspection after bleaching process. To realize a fast and effective rami fiber inspection process, a deep learning implementation in the form of image classification is used. In this study, image classification was formed using a pre-trained Convolutional Neural Network (CNN) which is constructed using the GoogLeNet architecture. Two hundred and six pieces of augmented fiber images in JPEG format data were processed into two classifications to assist in the inspection and separation process of bleached and unbleached rami fiber. Rami fiber database was divided into two sets 70% data for training set and 30% for validation set, then processed using the algorithm. As a result from the validation accuracy, the value of the training and testing that has been carried out is 100% as well as the core of accuracy, precision, recall, specificity, and F1-score obtained from confusion matrix. After the training results are used to test other images besides the input data, it is found that the results of the training managed to classify most of the images correctly. Of 25 images, only 2 images of unbleached rami fiber being wrongly predicted as bleached, which means the accuracy in classifying rami fiber is 92%.

Keywords Convolutional neural network · Deep learning · Image classification · Rami fiber

Anisa (✉) · M. N. Jundullah · L. I. K. Nafiah · L. A. Sari · P. W. Laksono
Sebelas Maret University, Jl. Ir. Sutami No. 36, Surakarta 57126, Indonesia
e-mail: nisa.anisa3112@student.uns.ac.id

C. N. Rosyidi et al. (eds.), *Proceedings of the 6th Asia Pacific Conference on Manufacturing Systems and 4th International Manufacturing Engineering Conference*, Lecture Notes in Mechanical Engineering,
https://doi.org/10.1007/978-981-99-1245-2_56

1 Introduction

Technological developments in the fields of mechanics and electronics encourage rapid development in the field of robotics [1]. More and more jobs that were previously done by humans are now carried out by robots because of the sophistication it has, one of which is in the industrial sector. Through the use of robots in various industrial sectors, the consistency of product quality and quantity can be improved [2]. One of the operations carried out in the natural fiber weaving industry in this research is the sorting of objects, namely rami before bleaching and rami after bleaching. The process of sorting rami which was previously done manually is still time consuming and the use of human operators in its implementation.

Research in automatic sorting of objects is nothing new. One research that has been done [3] shows that object sorting can be done automatically using the parameters of color and shape with machine learning. Machine learning is assisted by USB cam hardware and servo motors. While in other research [4] segmentation of the image is carried out using the thresholding method. Thresholding method is considered to have a great efficiency in segmenting images with high contrast levels, but this method is less reliable for images with many variations. So in other research conducted performing image classification using CNN or Convolutional Neural Network [5].

In this study, a method was developed for the rami sorting operation using an image sensor. This method was optimized using random search (RS) and grid search (GS), where a convolutional neural network (CNN) was used to classify the rami fiber. CNN uses pattern recognition to detect essential features that help distinguish between different classes. In this study, using a Convolutional Neural Network (CNN), features will be learned automatically rather than by human manipulation. So the expected result is that rami fiber can be classified according to which category has been bleached and which has not been bleached (Fig. 1).

There are two neural networks algorithm that form the basis for most pre-trained models in deep learning other than Convolutional Neural Network which are Artificial Neural Network (ANN) and Recurrent Neural Networks (RNN). But according to

Fig. 1 Unbleached rami fiber (Left) and bleached (Right)

Table 1 Comparison of ANN, CNN, and RNN

	Artificial neural network (ANN)	Convolutional neural network (CNN)	Recurrent neural networks (RNN)
Data type	Tabular data, text data	Image data	Sequence data
Performance	ANN is considered less powerful than CNN, and RNN	CNN is considered to be more powerful than ANN, and RNN	Less feature compatibility compared to CNN
Application	Computer vision and face recognition	Text digitization, natural language processing and face recognition	Text to speech conversion
Advantages	Having fault tolerance, and ability to work with incomplete knowledge	High accuracy in image recognition, and weight sharing	Remembers each and every information, Time series prediction

a course [6], CNN is more powerful than ANN, and RNN. Moreover, CNN specialization is very good at image classification. The following is a summary of the comparison of the three networks (Table 1).

This study is an initial concept for simple applied application of quality control in small industry, with a simple method to solve the problem. Implementation of CNN in the Indonesian textile sector in the field of quality control can be carried out so that companies can do the material inspection process with fewer resources with a good level of accuracy.

2 Method

In a feature-based model, the classification performance of branch fibers is highly dependent on the features extracted from the dataset. In this study, a deep learning model, namely a Convolutional Neural Network (CNN), was used to classify rami fiber images. Deep learning models such as CNN exhibit better generalization and noise tolerance. Deep learning CNNs also do not require human intervention for feature extraction, making them more robust choice when using unbiased classifiers. CNN can extract unique features from input data using pooling and convolutional operations. The research aims at automatic classification of rami fiber using CNN.

The rami fiber dataset contains 206 augmented fiber images in JPEG format [7]. There are 103 images each with categories that have been bleached and have not been bleached. The dataset is processed simultaneously with augmentation and repositioned before CNN input to avoid overfitting. The rami fiber database was divided into two sets: About 70% (144 images) of the data is for the training set and 30% (62 images) is for the validation set. The training set consists of 72 images for each category that have been bleached and have not been bleached, while the validation set contains 31 images for each category that have been bleached and not bleached.

In a convolutional neural network, it is often used for the topology of 2D images and has been successfully applied to many problems in visual classification, pattern recognition, segmentation, and related tasks. With appropriate filters, spatial and temporal dependencies in images can be detected. The advantage of CNN is that it can reduce the image to a more tractable type while reducing the loss of important information.

In this study the GoogLeNet architecture is used. GoogLeNet is a 22-layer CNN, an inception network, and a deep convolutional neural network developed by Google researchers. The GoogLeNet architecture for rami fiber image classification is shown in Fig. 2. In addition, as a neural network, the CNN model has a few layers (convolution layers) to extract features. In this case, the features are mainly extracted from colors and textures at the beginning of the network. The first convolutional layer weight that has a role in feature extraction is given in Fig. 3.

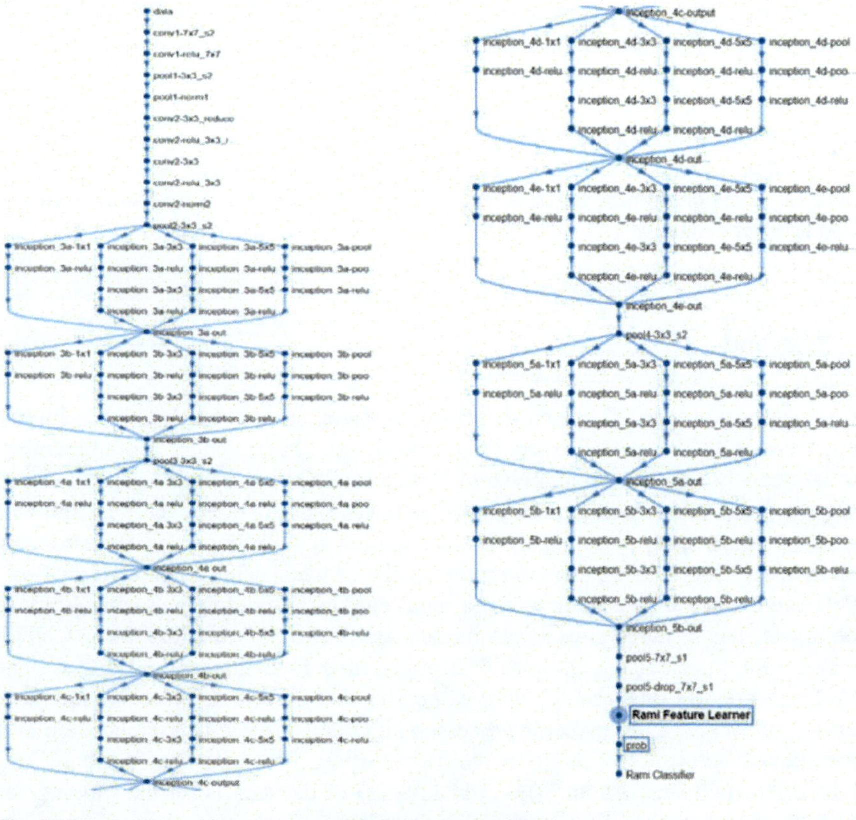

Fig. 2 GoogLeNet architecture on rami fiber image classification

Fig. 3 First convolutional layer

3 Results and Discussion

The CNN input data in this study is a collection of rami images that have two different specifications in color quality, namely rami that has gone through the bleaching process and rami that is still natural. The image data for each specification is 103 pieces and the image size is 3472 × 4624. The concept of this research is to distinguish and detect specifications between two different types of rami.

3.1 Accuracy and Plot Losses

Graphs of accuracy and loss are built during training and testing, depending on the epochs. The accuracy curve shows information related to the level of accuracy obtained during the training process. A significant gap in the value of the accuracy of training and testing is a sign of overfitting. Figure 4 shows the validation accuracy value of the training and testing that has been carried out at 100%. While the function of the loss plot is to measure network performance on the training set and provide a snapshot of the training process. During each epoch, the loss function is computed by evaluating the packet during the forward pass. The loss curve in Fig. 4 shows that the model performs well with each optimization iteration, as the loss decreases with each iteration. In general, there is always lose more in the training set than in the validation set for the first few epochs. Also, training loss continues to decrease as

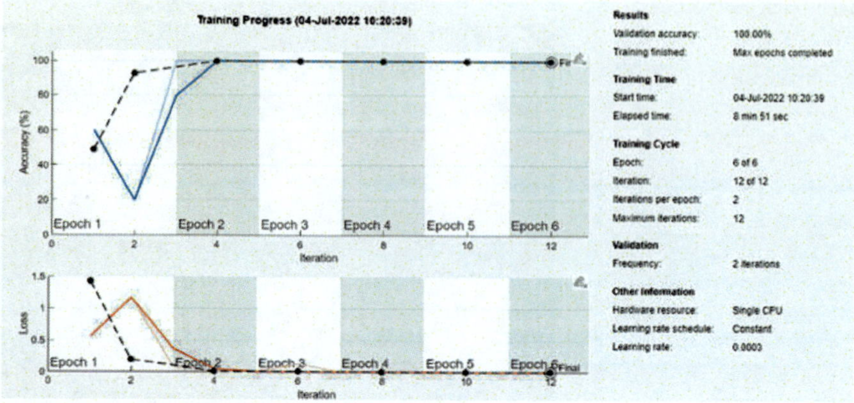

Fig. 4 Accuracy and loss test results

the number of epochs increases. As the number of epochs increases, the loss in the validation set increases, implying possible overfitting.

3.2 *Confusion Matrix*

The confusion matrix shown in Fig. 5 present how well CNN correctly identified rami fiber categories and relatively misclassified cells. This matrix reveals the confusion between rami types. Based on this confusion matrix, several other important performance measures can be calculated. The evaluation metrics are used to evaluate the diagnostic capabilities of the classifier. Metrics such as accuracy and F1 scores are calculated to determine how well a classifier can distinguish between different classes. From Fig. 5, one can determine the true positive (TP), true negative (TN), false positive (FP), and false negative (FN) result for each type of rami. In this study, TP occurs when the classifier correctly predicts a particular class. TN occurs when a classifier diagnoses that an image does not belong to a particular class. FP occurs when a classifier incorrectly predicts a negative sample as positive. FN occurs when a classifier incorrectly predicts a positive sample as negative. The confusion matrix in Fig. 5 shows, for example, that 0 predicted images were misclassified as bleached as not bleached, and 31 predicted images that were bleached were correctly classified as bleached.

Performance calculations can be calculated by:

$$Accuracy = \frac{(\mathrm{TP} + \mathrm{TN})}{(\mathrm{TP} + \mathrm{FP} + \mathrm{TN} + \mathrm{FN})} \tag{1}$$

$$Precision = \frac{\mathrm{TP}}{(\mathrm{TP} + \mathrm{FP})} \tag{2}$$

Fig. 5 Confusion matrix

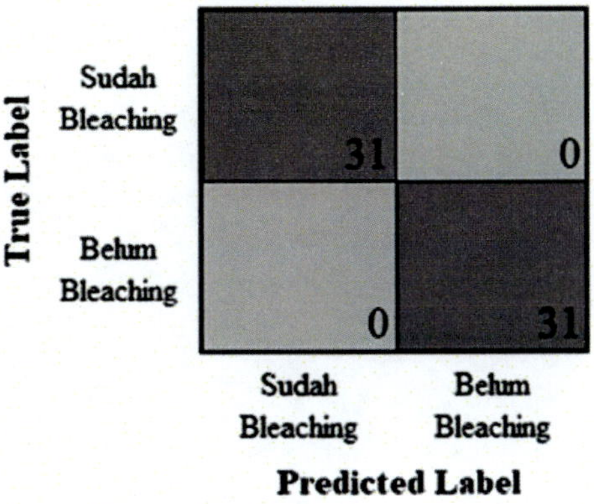

$$Sensitivity/recall = \frac{\text{TP}}{(\text{TP} + \text{FN})} \quad (3)$$

$$Specificity = \frac{\text{TN}}{(\text{FP} + \text{TN})} \quad (4)$$

$$F1 - score = \frac{2 \times \text{TP}}{(2 \times \text{TP} + \text{FP} + \text{FN})} \quad (5)$$

Since it is generally positive, all precision, precision, sensitivity or recall, specificity, and F1-score are all equal to 100%. The accuracy percentage indicates the proportion of images whose network generates the right class label for the test data. This information can be estimated by measuring the error rate, which is the proportion of an image whose model predicts the wrong class. The recall measures the TP sample from the entire set of positive samples that the model has identified. Precision indicates the level of relevant results, not irrelevant results. Recall and precision are competing metrics because attempts to maximize precision usually result in lower recall values, and vice versa. The F1 scoring measure, which takes into account the class imbalance in the dataset, is considered because it is a combined measure of the classifier's performance in terms of precision and recall.

3.3 Test Result

From the results of the training managed to classify most of the images correctly. Examples of correctly classified image are shown in Fig. 6 and examples of an incorrectly classified image are shown in Fig. 7. The image used to test is an image of rami fiber which can be downloaded freely from the internet. The number of new testing data is 25 images, with 2 images of unbleached rami fiber being wrongly predicted as bleached, so the new level of accuracy in classifying rami fiber is 92%.

Fig. 6 Examples of correct classification results unbleached and bleached

Fig. 7 Wrong classification results have not bleached as bleached

4 Conclusion

This study presents a comparison of the class classification of rami with before bleaching and rami after bleaching, some conclusions are obtained, namely the CNN model in this study uses an input shape measuring 3472×4624, learning rate value 0.003, filter size 3×3, Number of epochs 6, training data 144, and data testing 62.

Produces a 100% level of accuracy for training and testing in classifying rami fiber images. This study uses 25 new testing data to be tested into the model that has been made. The test results resulted in a new level of accuracy in classifying this rami fiber of 92%.

In the future research, it would be great to increase the number of datasets used. To improve data quality and overcome sample shortages, appropriate dataset sizes and other data augmentation approaches such as image inversion and intensity modification need to be considered. This exploration can be improved if the CNN architecture is applied to a larger dataset.

References

1. Sitompul E, Sodri (2020) Prototipe manipulator lengan robot berbasis Arduino dengan metoda kendali lead-through. ELEMENTER J 6(1):1–14
2. Perzylo A, Rickert M, Kahl B (2019) SMErobotics: smart robots for flexible manufacturing. IEEE Robot Autom Mag 26(1):78–90
3. Alfian F (2019) Rancang Bangun Robot Pemilah Sampah Organik Dan Non Organik. Thesis, Syarif Hidayatullah State Islamic University, Jakarta
4. Braiki M, Abdesslam B, Nasreddine K, Hymery N (2020) Automatic human dendritic cells segmentation using K-means clustering and Chan-Vese active contour model. Comput Methods Programs Biomed. https://doi.org/10.1016/j.cmpb.2020.105520
5. Hosseini M, Bani-Hani D, Lam SS (2022) Leukocytes image classification using optimized convolutional neural networks. Expert Syst Appl 205:1–12
6. Abhishek G. https://www.geeksforgeeks.org/difference-between-ann-cnn-and-rnn/. Last accessed 26 Sept 2022
7. Nihri J (2022) Personal documentation image dataset. Magelang

Control and Electrical System Design Consideration for a 3D Concrete Printer

Fadlika Septiansyah, Sonki Prasetya, and Djoko Nursanto

Abstract PT Solusi Bangun Indonesia Tbk through the Research and Development working with State Polytechnic of Jakarta want to make a 3D concrete printer machine with a larger dimension of 6 m^3. According to research that has been done on 3DCP 1 m^3, the problem of delay in the data transfer process from local control to the driving motor often occurs. The purpose of this research is to create an optimal electrical and control system for a 6 m^3 3D concrete printer. The method used in the control system and electrical 3D concrete printer machine must be systematic and scientific. Such as determining the driving motor, determining the controller, determining the data communication system, and designing the placement of cables in the mechanical circuit of a 3D concrete printer. After analyzing and testing the optimal data communication system is to use a cable. The amount of torque needed is 41.7 Nm so that the right driving motor is a stepper motor with a leadshine NEMA 51 13,045 HS with a holding torque of 45 Nm and a DM2282 motor driver. Transferring data 30 times with a cable length of 30 m using Arduino controller obtained a linear graph with an R square between RPM and delay microsecond of 0.9969.

Keywords 3D concrete printer · Data communication · Arduino · Stepper motor

1 Introduction

3D printer is a technology to produce a 3D object with various shapes and sizes using software computer. 3D printers are also known as additive manufacturing. Additive manufacturing is the process of combining materials to create an object from a 3D model using a layer by layer [1]. In 2015 a company from China succeeded in

F. Septiansyah · S. Prasetya (✉)
Mechanical Engineering, State Polytechnic of Jakarta, Jakarta, Indonesia
e-mail: sonki.prasetya@mesin.pnj.ac.id

F. Septiansyah · D. Nursanto
EVE Workshop, PT Solusi Bangun Indonesia Tbk, Bogor, Indonesia

C. N. Rosyidi et al. (eds.), *Proceedings of the 6th Asia Pacific Conference on Manufacturing Systems and 4th International Manufacturing Engineering Conference*, Lecture Notes in Mechanical Engineering,
https://doi.org/10.1007/978-981-99-1245-2_57

building a 5-story mansion-style villa and tower, using 3D Printing [2]. This building stands to be the first complete construction structure built using 3D Printing. The advantage of this construction 3D printer is its ability to create various objects and parts that are difficult to work with conventionally. Another advantage is that it can save labor and time for a construction [3]. To build a type 36 house, a 3d printer is required with dimensions of at least $6 \times 6\ m^2$. With these dimensions, the cable required will be longer. Problems that often arise in long cables are failures in the data transfer process. According to research that has been done on 3DCP 1 m3, the problem of delay in the data transfer process from local control to the driving motor often occurs [4]. Which resulted in the performance of the 3D printer is not optimal. In this project, there are two focus of discussion, namely the mechanical system and the electrical system. And in this paper the author only focuses on the discussion of the electrical and control system of this 3D concrete printer machine. With the aim of determining the most optimal data communication system and 3D concrete printer control operating system.

2 Theoretical Review

2.1 3D Concrete Printer

The application of 3D printer technology in the construction sector is a new thing in the industrial sector. 3D Concrete Printer (3DCP) is a technology where the product structure is arranged using concrete with a layer by layer system. With the application of 3D printer technology in the construction sector, the number of workers involved will be reduced, therefore labor costs will be reduced and the risk of work accidents will be minimized. And by using 3DCP the waste due to the construction process will be reduced by 30–60%. The most prominent advantage of the creation of this 3DCP machine is that it can save time from working on a construction which is usually done conventionally [5].

2.2 Data Communication

Data communication consists of two meanings of the words communication and data. Data refers to things like images, audio, video, text. Communication is the activity of sending or receiving data. Therefore the definition of data communication is the exchange of data between two or more devices through various transmission media such as cables. There are several components for data communication to occur. The first is sender, which is the component that sends data. Then there is the transmission medium, namely as a component of sending data. And is the receiver, i.e. the component that receives the data [6].

3 Research Methodology

At this stage, the authors design based on consumer needs, the results of the selection of concepts and literature studies. The author will determine the communication system that will be used, the controller, and the motor to drive the 3DCP machine. The design is done by determining the indicators as the basis for determining the object to be selected.

3.1 Data Communication System

To determine the data communication system that will be implemented in the 3DCP circuit. There are two types of data communication systems, namely wire and wireless. there are criteria as the basis for selection, including:

- The speed of data transfer

The speed of data transfer is a benchmark in the selection of a data communication system. This is in accordance with the objectives to be achieved, namely the accuracy of data transfer (real time).

- Ability to deal with distractions

Disturbances that often occur during the data transfer process become a problem that the data from the controller does not reach the motor. Therefore, we need a data communication system that is resistant to external interference.

- Prices and availability of goods

Prices and availability of goods are criteria to avoid budget shortages, so that the existing budget can be maximized optimally.

- Ease of fabrication

Ease of fabrication includes the criteria to be used, the easier it is to fabricate, the higher the value to be selected.

3.2 Define Drive Motor

In determining the driving motor that will be used to drive the 3DCP machine, there are the following criteria:

- High torque at low speed

With the large dimensions of the 3DCP to be made, a motor with high torque at low speed is needed.

- Accuracy of motion

To become the driving force of the 3d printer requires high accuracy of motion. Therefore, the accuracy of the motion is one of the criteria for the selection of the driving motor.

- Price

Price and become a criterion to avoid a shortage of budget costs, so that the existing budget can be maximized optimally.

- Mass of the motor

The mass of the motor will affect the total force that will work. The greater the mass of the motor, the greater the force that will work.

3.3 Data Transfer

There are two systems being tested at this stage, namely data transfer using a cable and data transfer using a wireless.

1. Transfer data by cable

Figure 1 is show a block diagram used for testing data transfer by cable. Using Arduino controller, NEMA 17 stepper motor, and TB6600 motor driver [7]. The purpose of this test is to identify when the data transfer process from Arduino to the stepper motor is there whether there is interference or delay.

For data transfer by cable, there are any a flowchart diagram carried out, as shown in Fig. 2.

2. Transfer data by wireless

For this trial using the NRF24L01 module as a data transfer media with wireless [8]. The purpose of this trial is to see the potential use of a wireless system in the application of data transfer on a 6 m^3 3DCP machine. Figure 3 is show a trial block diagram for a wireless system.

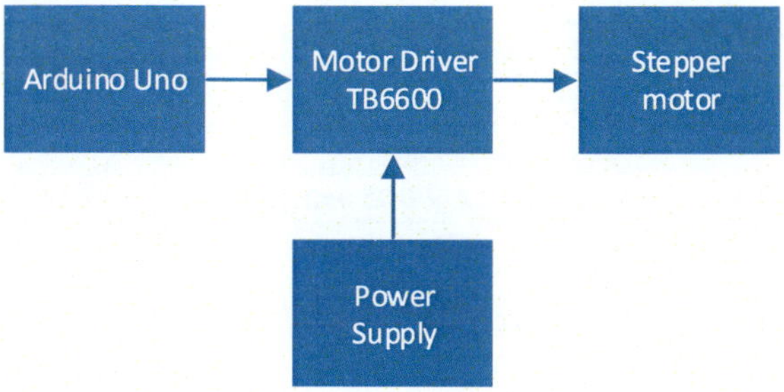

Fig. 1 Block diagram for cable test

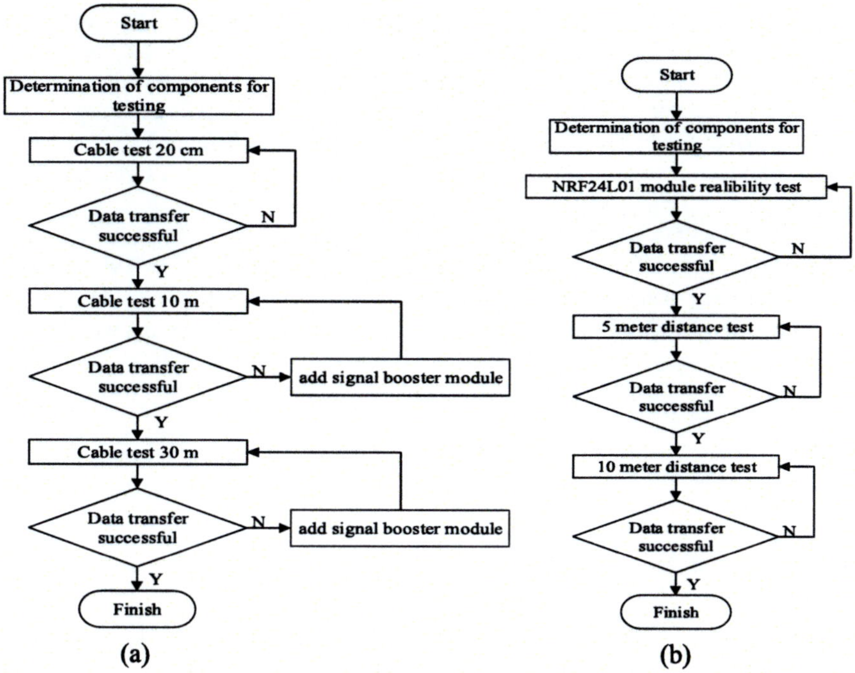

Fig. 2 Flowchart transfer data by cable. **a** Transfer data with different distance. **b** Motor realibility test

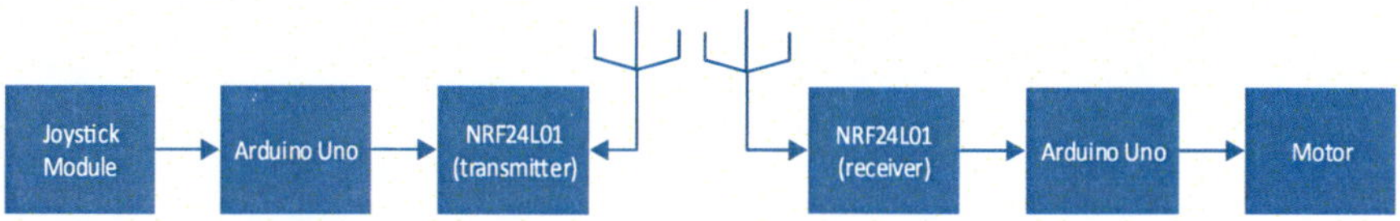

Fig. 3 Block diagram wireless test

Table 1 Results transfer data by cable

Cable type	Length (M)	Delay transfer data (S)
Jumper Arduino	0.2	3.5
NYAF 0.75 mm	10	3.56
YSLY-JZ 4 × 0.75 mm	30	3.58

4 Implementation and Results

4.1 *Transfer Data by Cable*

This test aims to find out during the data transfer process from Arduino to the motor driver whether there is interference or delay in the movement of the stepper motor.

From the data in Table 1, it is obtained from 3 stages of transfer testing where the *delay* is the time during the upload process from the Arduino IDE until the stepper motor moves. It can be seen that the change in time between different cable lengths is not too significant, so for a cable length of 30 m it does not require a signal amplifier module.

Figure 4 displays a graph of the stepper motor reliability test results. It can be seen that the shape of the graph tends to be linear and the R square is 0.9969, which means that changes in the microsecond delay affect the resulting speed value (RPM). And to find out the RPM value of the motor, you can use the following equation:

$$Y = -0.0375X + 130.4 \tag{1}$$

Information: Y = rotation per minute (RPM), X = delay microsecond.

4.2 *Transfer Data by Wireless*

The test is done by moving the servo motor with a joystick continuously. And at 56 s, as shown in Fig. 5, the servo motor turns off and does not move again. The use of the NRF24L01 module as a wireless system data transfer medium is not effective, This is due to risk of signal loss often occurs even at close distances.

4.3 *Data Communication System Selection*

Figure 6 shows the radar diagram of the results of the assessment of the type of data communication system. In the diagram above, wire is shown with a blue line, while wireless is shown with a red line. It can be seen that the widest area is indicated by the blue lined area which represents the wire. Thus, the right type of data communication

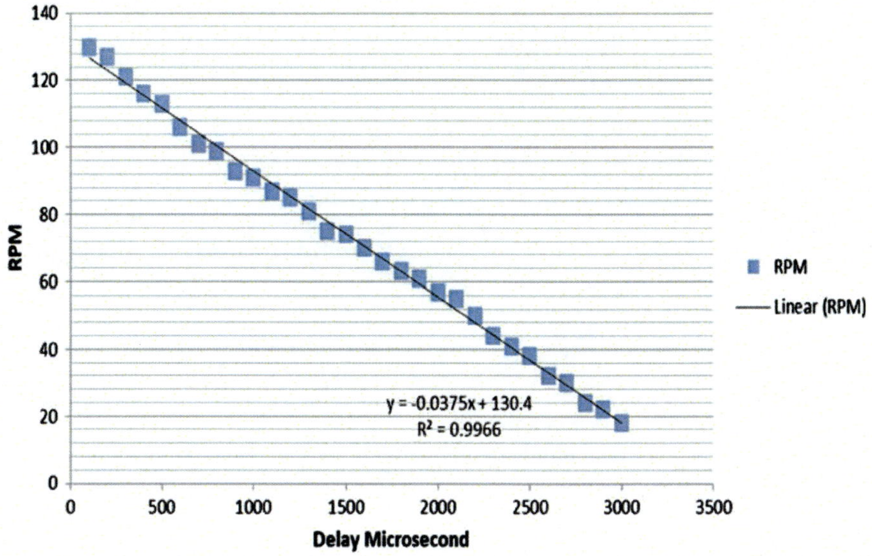

Fig. 4 Reliability test results graph

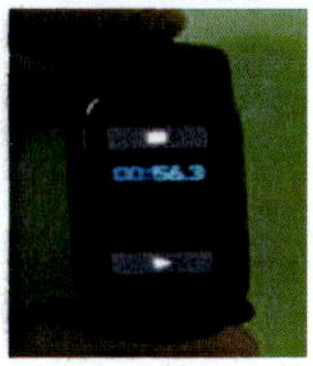

Fig. 5 Servo motor turn on long time

to be used in this research is wire. Therefore, several points connections for cables are required in order to connect different parts (plug-and-play) for the system.

4.4 Drive Motor Selection

To move the structure of each axis, it is necessary to calculate the amount of torque needed to move the structure [9]. Torque is obtained from the calculation of the force acting on each axis. The force acting on the X axis is calculated from the total weight of the extruder that has been filled with mortar and the weight of the transmission, the total weight is 200 kg. After that, the working force can be calculated.

$$
\begin{aligned}
F &= m \times g \\
F &= 200\ \text{kg} \times 9.81\ \text{m/s}^2 \\
F &= 1962\ \text{N}
\end{aligned} \tag{2}
$$

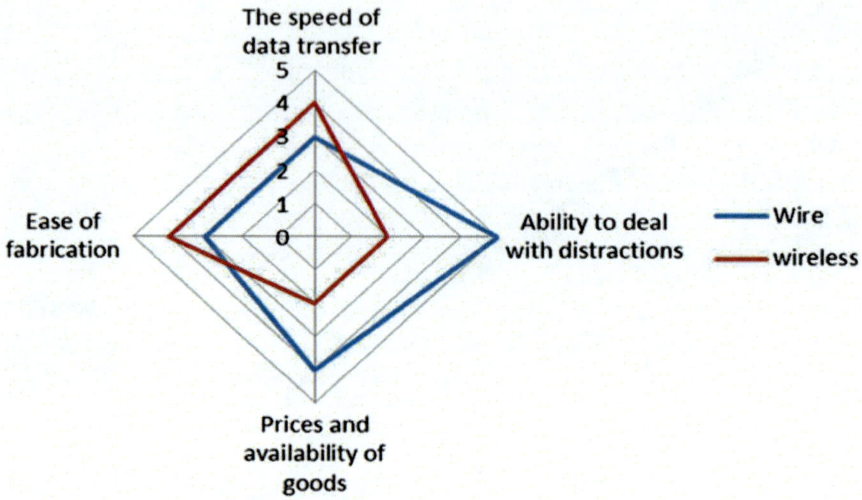

Fig. 6 Radar chart data communication selection

This calculation is represented on the X-axis motor, because the largest load is experienced on the X-axis motor 1962 N, here is the formula for calculating torque [10].

$$T = F \times l$$
$$T = 1962 \text{ N} \times 0.02125 \text{ m}$$
$$T = 41.7 \text{ Nm} \quad (3)$$

Information: T = torque (Nm), F = force (N), L = rack radius (M).

Figure 7 shows the radar diagram of the results of the motor selection assessment. In the diagram above, the stepper motor is shown with a blue line, while the servo motor is shown with a red line. It can be seen that the widest area is indicated by the blue lined area which represents the stepper motor. Thus, the right motor to be used in this study is a stepper motor.

After doing research on each type of stepper motor, the right type of stepper motor is the NEMA 51 130HS45 stepper motor. With the following specifications [11]:

- Step angle : 1.8°
- Holding torque : 45 Nm
- Current : 7 A
- Length : 283 mm
- Weight : 19 kg
- Match drives : DM2282/DM1182

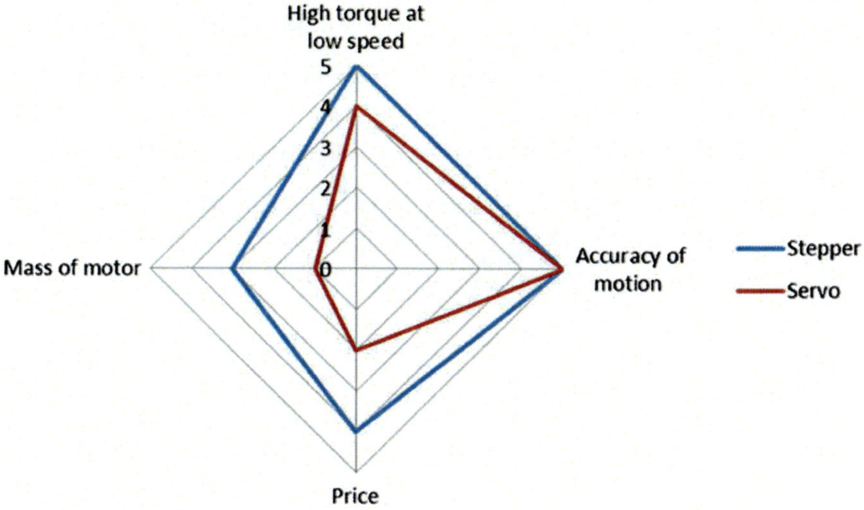

Fig. 7 Radar chart drive motor selection

5 Conclusion

After conducting analysis and testing, it was found that the optimal data communication system for use on 3DCP 6 m^3 is a wire system because it has higher data transfer stability compared to wireless systems. And also after testing the data transfer 30 times using an arduino controller with a cable length of 30 m, a linear graph is obtained with an R square value between RPM and microsecond delay of 0.9969. And finally from the calculation results it was found that the type of driving motor used for 3DCP 6 × 6 × 6 meters is a stepper motor with a leadshine type NEMA 51 130HS45 which has a holding torque of 45 Nm.

References

1. Rehman AU, Kim JH (2021) 3D concrete printing: a systematic review of rheology, mix designs, mechanical, microstructural, and durability characteristics 14(14)
2. Sevenson B (2015) Shanghai-based WinSun 3D prints 6-story apartment building and an incredible home. 3dprint.com. https://3dprint.com/38144/3d-printed-apartment-building/
3. Fouda K, Zayed N, Kamal M. 3D construction printer. [Online]. Available: https://www.researchgate.net/publication/353193531
4. Abdillah RF, Prasetya S, Nursanto D (2021) Electrical system design and control 3D printer mortar 1 × 1 × 1 m, pp 678–686
5. El-Sayegh S, Romdhane L, Manjikian S (2020) A critical review of 3D printing in construction: benefits, challenges, and risks. Arch Civ Mech Eng 20(2):1–25. https://doi.org/10.1007/s43452-020-00038-w

6. Stojcev M (2005) Data communication. Microelectron Reliab 45(2):403–404. https://doi.org/10.1016/j.microrel.2004.07.004
7. Swarnkar MV, Yamini K, Vani K (2022) Stepper motor controller using Arduino (ATMEGA 328) 10(5):32–34
8. Sidqi R, Rio Rynaldo B, Hadi Suroso S, Firmansyah R (2018) Arduino based weather monitoring telemetry system using NRF24L01+. IOP Conf Ser Mater Sci Eng 336(1). https://doi.org/10.1088/1757-899X/336/1/012024
9. Adam MT, Prasetya S, Nursanto D, Suhandi A (2022) Design and build mechanical system 3D printer mortar 1 × 1 × 1 m. J Mek Terap 3(1):34–43. https://doi.org/10.32722/jmt.v3i1.4547
10. Bagia IN, Parsa IM (2018) Electrical motor
11. Leadshine (2012) Motion control products 2012/2013

Comparative Study on the Automated Warehouse with Cartesian Robot Using PLC and Arduino

Thitisak Aussawarangkul, S. A. Che Ghani, Pakanun Wattanasinbumrung, and Worapong Sawangsri

Abstract Nowadays, automation system is necessary in every warehouse. It can assist operators in easily managing warehouses furthermore reduce unexpected errors from humans and surrounding environments. Nevertheless, a number of control methods have been available for the warehouse system. Each method has advantages and disadvantages considered regarding ease of use, stability as well as cost-effectiveness, etc. Therefore, in order for comparative performances of control algorithms between the PLC and the Arduino, the models of automatic warehouse with PLC and Arduino boards are implemented in this paper. The comparative results show that the average time difference between two methods is 7% with Arduino being faster and lower cost rather than PLC. This ability of using microprocessor as the Arduino board presents some advantages even lesser stability. Thus, alternative methods with current and higher technological devices are possibly selected instead of the old-fashion control algorithm in order to gain more benefits and advantages. Furthermore, integration of the ROS with an automation unit to upgrade an existing system have currently been conducted on the warehouse system to evaluate and compare the system's performances. The results will be presented in a separated paper.

Keywords Warehouse robot · Arduino processor robot · PLC processor robot

T. Aussawarangkul · P. Wattanasinbumrung · W. Sawangsri (✉)
Department of Mechanical Engineering, Faculty of Engineering, Kasetsart University, Bangkok 10900, Thailand
e-mail: fengwps@ku.ac.th

S. A. Che Ghani
Faculty of Mechanical and Automotive Engineering Technology, Universiti Malaysia Pahang, 26600 Pekan, Pahang, Malaysia

C. N. Rosyidi et al. (eds.), *Proceedings of the 6th Asia Pacific Conference on Manufacturing Systems and 4th International Manufacturing Engineering Conference*, Lecture Notes in Mechanical Engineering,
https://doi.org/10.1007/978-981-99-1245-2_58

1 Introduction

Currently, part and spare parts are increasing due to the high demand market. Manufacturers have stocked those parts in warehouses to protect uncertain manufacturing processes of goods as well as sufficient supply to mass production. It is necessary to sort them out by type to make it easy to find and pick them up and save space for store workpieces. Lots of warehouses solve these problems by making vertical stores [1]. However, whenever storage is much higher from the floor, it is difficult to retrieve and store workpieces. Automatic robots and currently high technological devices are able to assist in every situation which makes tasks possibly handy and reduces accidents from high falls [2].

Thus, the 3-axis Cartesian robot is used to conduct the automatic warehouse system [3, 4]. This robot has been selected in this study due to not having a revolute joint, thus do not be concerned about the orientation of products. The system is performed with two comparative control methods namely included the PLC (programmable logic controller) and the Arduino. Therefore, the system performances have been evaluated and discussed on whether advantages or disadvantages.

PLC is one device that is easy to understand and common use in industry and daily life, e.g., traffic signals in every intersection that we pass every day using a PLC controller. Nowadays, traffic signals can detect emergency vehicles and dynamically manage the traffic light to facilitate and let them pass first [5]. This presents an effective, reliable, and stable PLC exchanging with a relatively high price.

Arduino is especially popular in education on automatic control or robotics [6] because of its cost-effectiveness and ease of programming with a lot of Arduino libraries [7]. The Arduino dataset contains more features than the PLC data set such as Functions, Comments, Lines of code, etc. [8]. In addition, the operated time comparison is appropriate for considering robot performance assessments. For example, Stephanie Glass et al. [9] presented the compared operative time between conventional laparoscopy and robotic-assisted Sacrocolpopexy. Therefore, a comparative study to investigate the performances of control algorithms between the PLC and the Arduino by considering the operating time as the primary parameter to compare between two control methods is conducted in this paper.

2 Experiment Implementations

2.1 Demo Warehouse

The demonstrated warehouse is constructed by aluminum profile 2020 V-slot and aluminum profile 2040 V-slot for robot guideway and steel angle bar for storage structure. Storage structure is built with up to three floors and four slots per floor that can place workpieces as shown in Fig. 2. Dimensions for each slot are 100 mm

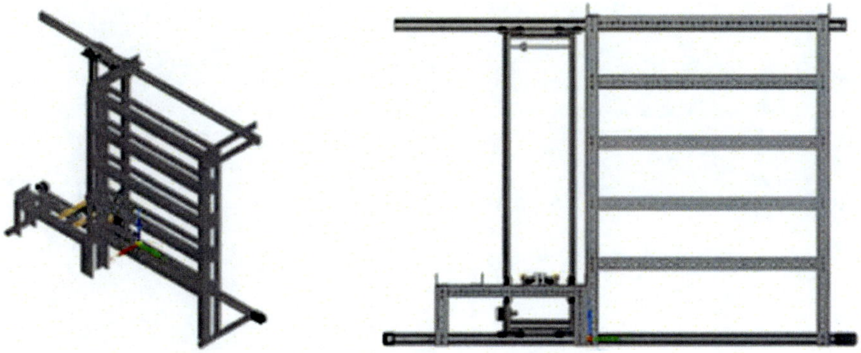

Fig. 1 Warehouse demonstration structural design

		4	3	2	1
		8	7	6	5
		12	11	10	9
Store Slot	Retrieve Slot				

Fig. 2 Diagram shows location of each slot

height × 180 mm width and palate size 800 mm × 90 mm, respectively (Figs. 1, 2 and 3).

The slider is designed shapely like a forklift using a lead screw to retrieve or store pallets at desired slot, presented in Fig. 4.

2.2 PLC Ladder Diagram Programming

The IEC-1131-3 is commonly use as a standard programming language for PLC [10, 11]. It consists of two textual languages, instruction list, and structured text, as well as two graphical languages, Ladder Diagram and Function Block Diagram, respectively. The PLC Delta AS332T from Delta Electronic is applied for control robot and build ladder diagram by the ISP Soft software. This is Delta's freeware that only works with Delta's device. Ethernet cables are used for communication between a PLC device and a computer. The main functions that control motors used in the ladder diagram are Absolute position control (DDRVA), Relative position control (DDRVI), and zero return (DZRN), respectively (Fig. 5).

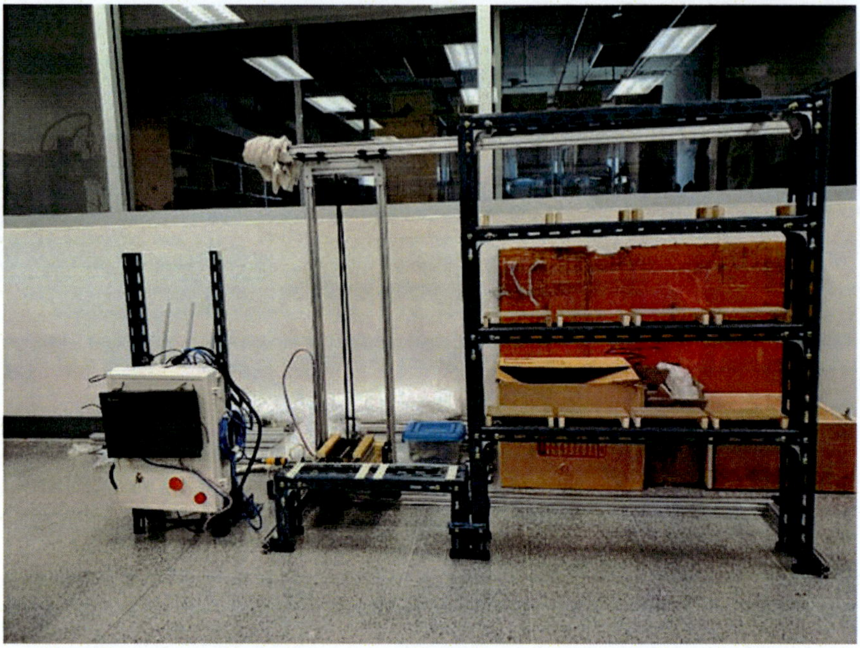

Fig. 3 Cartesian robot structure

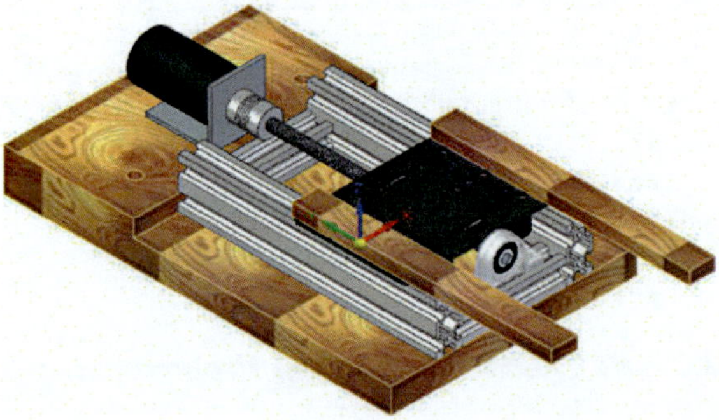

Fig. 4 The slider design with linear actuator

2.3 HMI Programming

The User Interface is created for HMI (human machine interface) by DOPSoft Delta's freeware program. HMI is commonly used for input from human and sent to PLC with IP address configuration (Figs. 6, 7, 8 and 9).

The developed system has three control modes can be selected. Full Auto, Semi-Auto and Manual. Full Auto mode, users only select the desired slot for placing a workpiece, then the system will automatically operate every step. Where the system

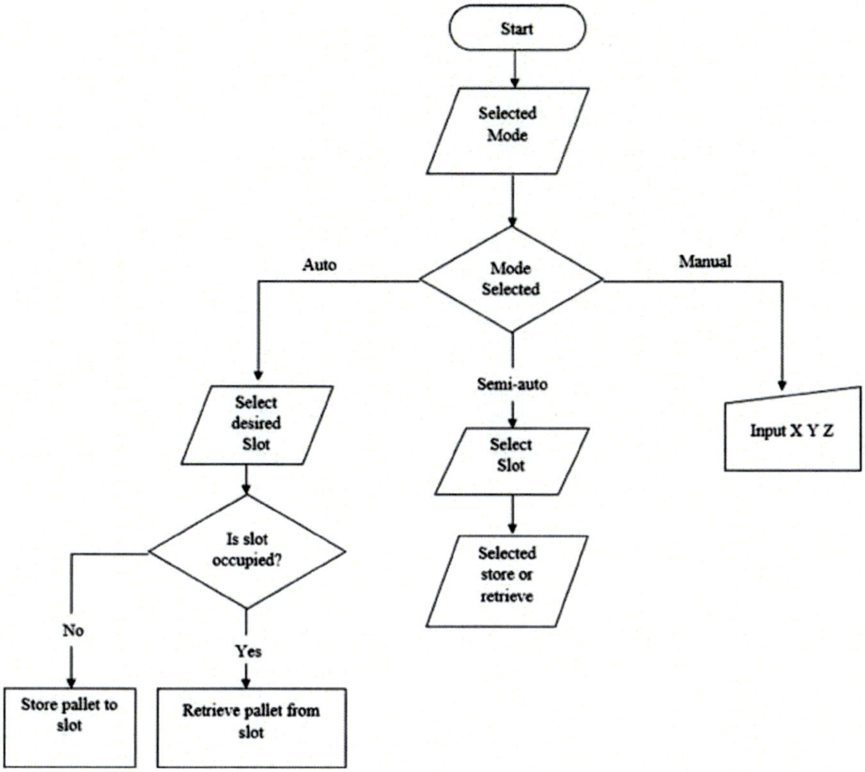

Fig. 5 PLC programming workflow diagram of 3 modes

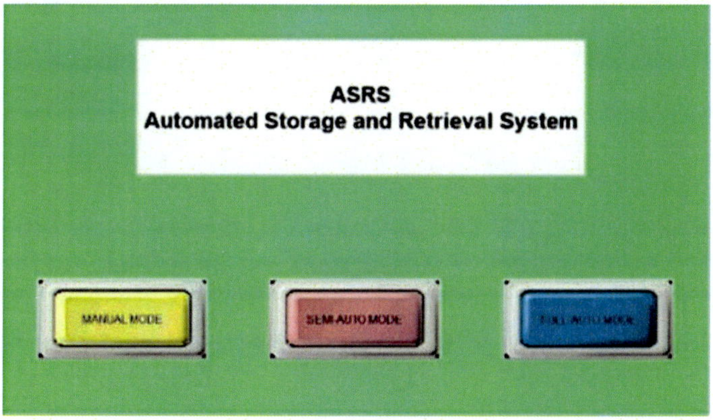

Fig. 6 Display screen of HMI, users have to select desired modes to work manual, semi-auto, or full auto

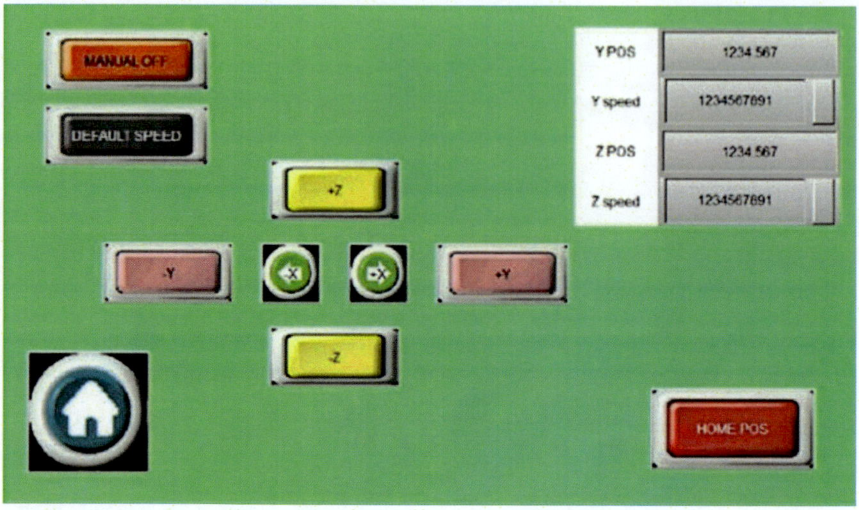

Fig. 7 Manual mode, users can manually control position of X, Y, Z and speed of robot via HMI screen

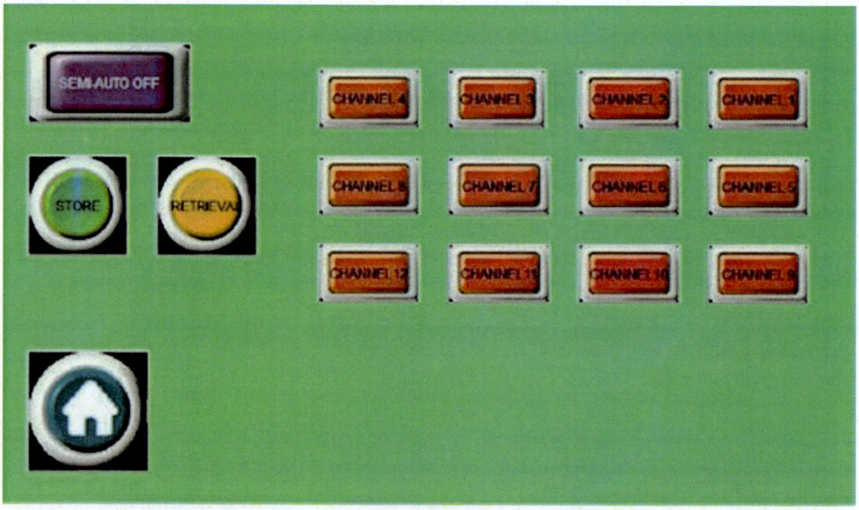

Fig. 8 Semi-auto mode, users can select the desired slot and choose to store or retrieve pallet

is recognizing that in that slot have any workpiece. If not, it will store new one. But if there is, it will be pick up that workpiece at that slot. Semi-Auto mode is for specific control of picking or storing and selecting desired slots. Furthermore, if users need to control any other movements of the robot, use the manual mode.

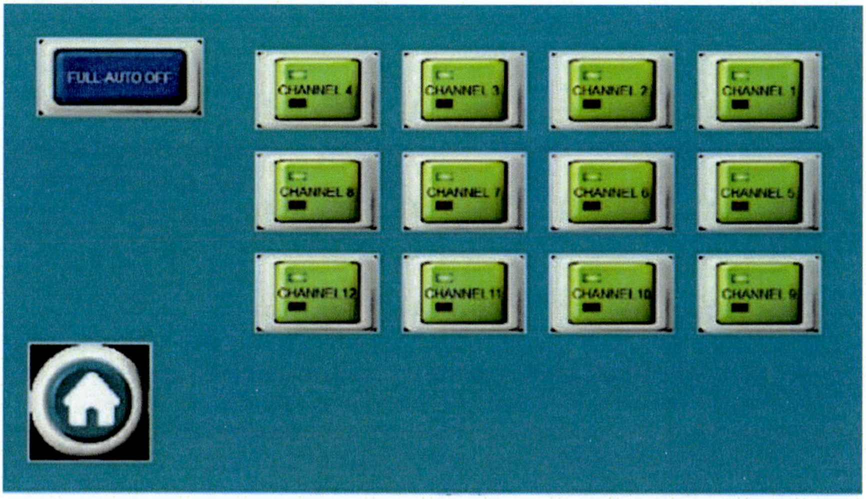

Fig. 9 Full auto mode, users only selected the desired slot and the system will operate every step

2.4 Arduino Programming

Besides the PLC, the Arduino board is a microcontroller for robot control. The Arduino Mega2560 has been selected for system control purposes. It has 54 Digital IO and PWM (Pulse Width Modulation) output mode that can control motor. C language and libraries are known as Arduino IDE which open-source programing is implemented for the system control algorithms. It contains various codes in libraries and also has a community helping with coding developments to support a variety of devices. Therefore, it is another option for deciding to choose Arduino as part of this research (Fig. 10).

3 Results

In experimental results. Time intervals from the initial station to each slot are collected and compared between two different control methods presented in Fig. 11. The average time spent on both control algorithms is not much different (7% difference) where Arduino is faster than PLC.

After collection of time intervals moving the object from the station to storing slot and vice versa as well as the average time of two modes. Thus, the power consumption of both methods is monitored.

From Fig. 12 shows the energy consumption of PLC with HMI consumes more power than Arduino MEGA 2560 around 173.33% without connecting any device to PLC or Arduino (Fig. 13).

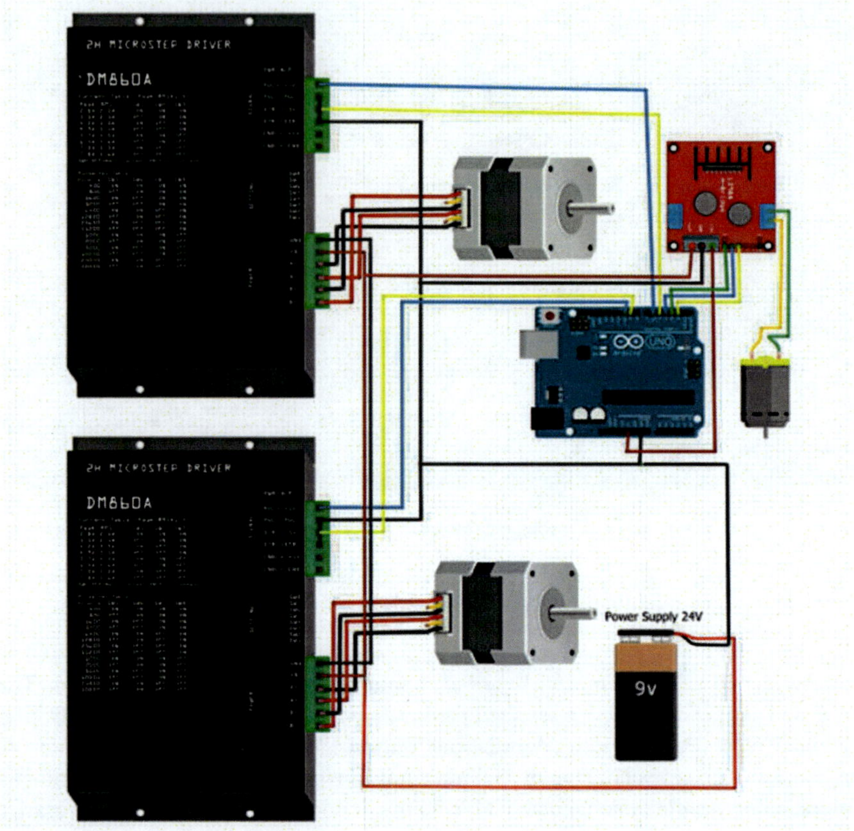

Fig. 10 Schematic wiring of Arduino

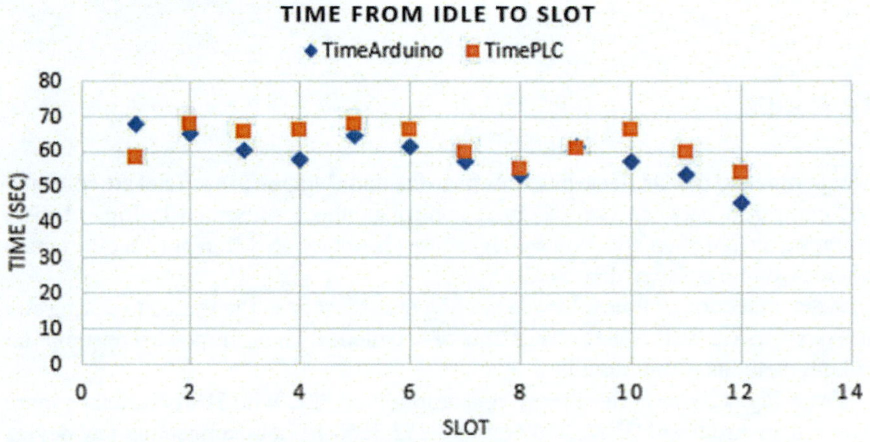

Fig. 11 Time intervals from the initial station to each slot

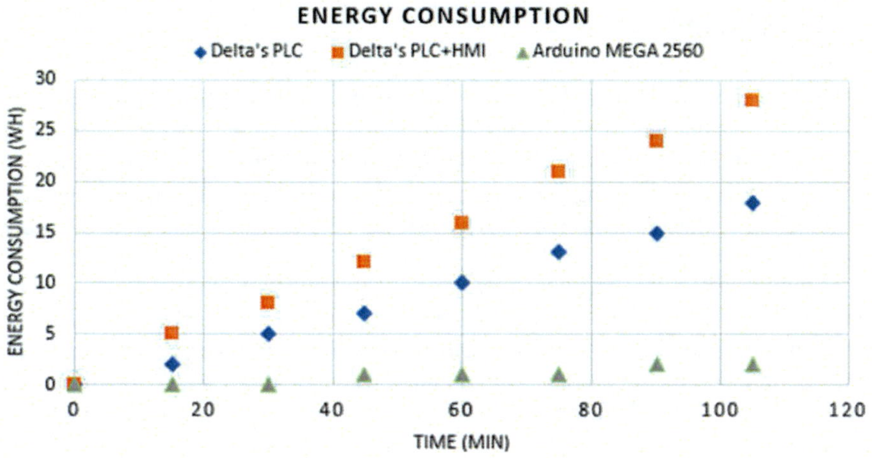

Fig. 12 Energy consumption of **a** Delta's PLC **b** Delta's PLC with HMI and **c** Arduino MEGA 2560

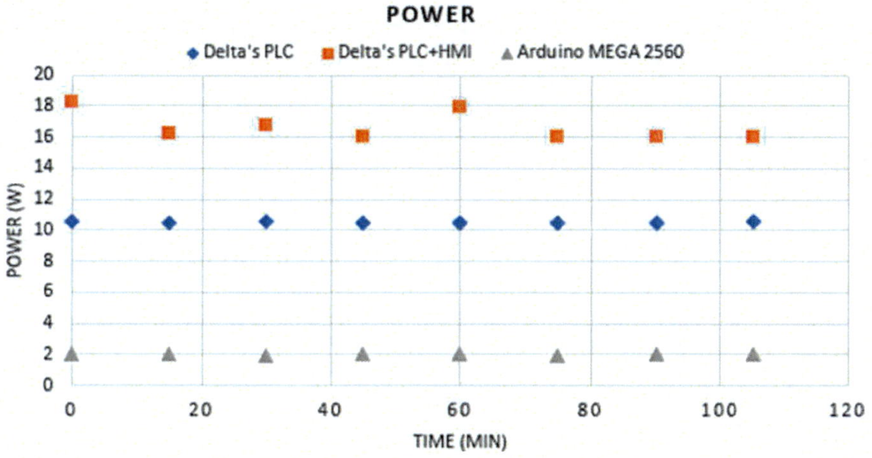

Fig. 13 Power of each device **a** Delta's PLC **b** Delta's PLC with HMI and **c** Arduino MEGA 2560

The results presented regarding energy consumption, Arduino consumed power of around 2 W while PLC with HMI used the power of around 18 W, and standalone PLC is around 10 W. This huge difference obviously illustrates why the PLC needs 24 V power supply while the Arduino only uses 5 V.

4 Conclusion

The comparative control methods to manipulate the demonstration of the automatic warehouse system are implemented. This comparative study presented results of PLC and Arduino are very close and going in the similar direction in overall aspects. However, programming language and structure including sequences of both methods might affect system running speed, if written in a concise structure, it might make the system faster.

Arduino has an advantage in less power consumption and the I/O pin can be program flexible to connect a variety of sensors but it requires expert programmers in c or c# languages. Meanwhile, PLC is often implemented at industrial level rather than Arduino due to reliability and has more industrial application connection protocols, and is rugged. Also, HMI is required if user want to manually control directly to PLC.

In experimental implementations, it can be seen that whatever is microprocessor can be applied to an automated robot system. It depends on which systems will be picked up and integrated into the devices. Recently, the ROS system is quite interesting and suitable for programming to work with automated robot systems. Thus, an automatic warehouse system has currently been conducted by using ROS, and implemented results will be reported in the near future.

References

1. Nicolas L, Yannick F, Ramzi H (2018) Order batching in an automated warehouse with several vertical lift modules: optimization and experiments with real data. Eur J Oper Res 267(3):958–976
2. Ortiz F et al (2005) An implementation of a teleoperated robot control architecture on a PLC and field-bus based platform. IFAC Proc Volumes 38(1):349–354
3. Fenercioğlu A, Soyaslan M, Közkurt C (2011) Automatic storage and retrieval system (AS/RS) based on Cartesian robot for liquid food industry
4. Morrison D et al (2018) Cartman: the low-cost Cartesian manipulator that won the amazon robotics challenge. In: 2018 IEEE international conference on robotics and automation (ICRA)
5. Amir S et al (2017) PLC based traffic control system with emergency vehicle detection and management
6. Candelas FA et al (2015) Experiences on using Arduino for laboratory experiments of automatic control and robotics. IFAC-PapersOnLine 48(29):105–110
7. Purdum J (2015) Arduino libraries. In: Purdum J (ed) Beginning C for Arduino: learn C programming for the Arduino. Apress, Berkeley, CA, pp 277–298
8. Canedo A et al (2021) ArduCode: predictive framework for automation engineering. IEEE Trans Autom Sci Eng 18(3):1417–1428
9. Glass Clark S et al (2022) Operative time for minimally invasive sacrocolpopexy: comparison of conventional laparoscopy versus robotic platform. J Minim Invasive Gynecol 29(9):1063–1067
10. Patel DM, Shah AK (2021) FPGA-PLC-based multi-channel position measurement system. ISA Trans 115:234–249
11. Öhman M, Johansson S, Årzén K-E (1997) Implementation aspects of the PLC standard IEC 1131-3. IFAC Proc Volumes 30(4):27–32

Materials

Study on Plasticity of Cu–Ni Alloy at Atomic Scale: Insight from Molecular Dynamics Simulation

M. Zaenudin, Ujiburrahman, Ade Sunardi, Yasya Khalif Perdana Saleh, and Adhes Gamayel

Abstract The development of alloys or various applications shows an extensive study, from the novel alloying elements to the novel techniques in processing the material. This effort is conducted to find the best suit of an alloy for a specific application. Most applications of metallic material consider its strength and plasticity; thus, an extensive study of these problems is required. This study presents an investigation of the plasticity of Cu–Ni alloy at an atomic scale employing molecular dynamics simulation. A cube-shaped model with a dimension of 20 nm for x, y, and z planes with a percentage of 50% Cu and 50% Ni. The system is subjected to heat treatment at various temperatures before the tensile test at a strain rate of $2.64e9/s$. The result shows that the heat treatment shows to have a slight effect on the strength of the Cu–Ni alloy. The stress that is evenly distributed along the z plane induces the response of the system with a slippage along the line defects between the crystal and generates the dislocation. This study has successfully given an insight into the plasticity of Cu–Ni alloy and its detailed features at the atomic scale.

Keywords Cu–Ni alloy · Molecular dynamics simulation · Plasticity

1 Introduction

Alloy is a material where two elements or more are mixed in such a way with different techniques to generate a new class of material with advantageous properties [1]. It has the potential to be used in various applications ranging from binary systems to multicomponent elements with different properties from its parent element [2]. Various alloys were developed to produce a material that can accommodate multiple needs of everyday human life, from simple applications like a chair to more complex

M. Zaenudin (✉) · Ujiburrahman · A. Sunardi · Y. K. P. Saleh · A. Gamayel
Department of Mechanical Engineering, Universitas Global Jakarta, Depok 16412, West Java, Indonesia
e-mail: mzaenudin@jgu.ac.id

C. N. Rosyidi et al. (eds.), *Proceedings of the 6th Asia Pacific Conference on Manufacturing Systems and 4th International Manufacturing Engineering Conference*, Lecture Notes in Mechanical Engineering,
https://doi.org/10.1007/978-981-99-1245-2_59

applications such as aircraft and nuclear reactors [3, 4]. Therefore, the development of alloy material, especially a metallic alloy, shows extensive research with a few factors that it is not only related to the properties of the alloy but also consider the possibility of its application, where the economic value becomes no less important.

One of the alloys that have many advantages is Cu–Ni alloy. Cu–Ni alloy is a binary alloy with various applications, such as metal foam due to the high corrosion resistance [5–7], a catalyst for hydrogenation from CO to methanol [8] and other nanoparticles [9], for thin films [10], and chemical vapour deposition of graphene [11]. Those applications require a deep understanding of thephysical, mechanical, chemical, or even electrical properties of the materials being discussed; thus, proper use of the material could be expected anytime soon.

In the case of mechanical properties, such as structural plasticity, that could be investigated from the relationship between stress and strain during a tensile test, the material should have excellent strength with sufficient plasticity. A study on the structural plasticity of a material with a certain ductility, including metallic material, is not far from the analysis of plastic deformation, where the deformation of a material is permanent and irreversible [12]. The application for this deformation is usually related to the forging proses of a metal sheet for a certain part of a tool or machine that tends to be lighter with various forms. Meanwhile, in another context, this deformation is described as a form of effect from damage produced during an impact, where the deformation process may be neither static nor dynamic. Generally, this kind of deformation is related closely to the concept of stress and strain, where the two will give a sign of a material with a particular property or another, and the material can be processed with a specific technique or another [13].

With that in mind, this paper tries to address the structural plasticity of Cu–Ni alloy at the atomic scale by means of molecular dynamics simulation. The effect of heat treatment on such alloy is relatively high [14]; thus, various heat treatment is applied. The model and simulation procedure is described in detail in the method section, and the detailed result and discussion are also addressed in the result and discussion section.

2 Method

Three software are used in this study, they are: (1) for the modelling part, the AtomSK software is employed [15], (2) for the simulation part, the LAMMPS software package is utilized [16], and (3) for post-processing part the OVITO software is used for visual and analysis purposes [17]. AtomSK [15] is software for manipulating and converting atomic data files. It could create a system of an atom consisting of nanograins of various sizes. In this study, the model created from AtomSK software is depicted in Fig. 1. The grain size of the model is about 8 nm, while the system itself has a size of 20 nm × 20 nm × 20 nm in x, y, and z direction, respectively, with FCC crystal at [100], [010], and [001] plane, respectively. It has a 50:50% percentage of each Cu and Ni element.

Fig. 1 The model of Cu–Ni alloy system with 50:50 percentage

At the beginning of the simulation, the system is initialized with minimum energy for equilibration purposes. This step is essential in molecular dynamics simulation because it helps to minimize the "jiggling and wiggling" phenomena associated with the initial system [18]. The simulation then continued by increasing the temperature to meet the heat treatment temperature; they varied from 700 to 1200 K with an increase of 100 K for each designed temperature. The temperature then dropped from the designed heat treatment temperature to 300 K for relaxation purposes before the tensile test step. The tensile test step is performed at 300 K with a strain rate of $2.64e9/s$ in z direction, while the x and y direction are set to be at a pressure of 0 MPa; thus the stress generated during the tensile test is focused on the z direction. Temperature profile during simulation process is shown in Fig. 2.

Interatomic potential adopted for molecular dynamics simulation is an essential part because it is one of the critical aspects that define the quality of the result. The interatomic potential of a molecular dynamics simulation is a mathematical function that calculates the potential energy of a given system of an atom relative to its position in space [19, 20]. It provides the particles in a system a value of the potential energy, thus allowing it to behave as an integral part of the system for various conditions of temperature and pressure. There are so many interatomic potentials developed nowadays for many elements. In this study, the interatomic potential of Cu-Ni was developed by Onat et al. [21]. The development of this interatomic potential is following the embedded atom method (EAM), which was first developed in 1984 for various metals [22, 23] by following the EAM formalism for defining the total energy from a given system with the equation as follows:

$$E_{tot} = \frac{1}{2}\sum_{ij} \Phi_{a_i-a_j}(r_{ij}) + \sum_{i} F_{a_i}(\overline{\rho}_i) \quad (1)$$

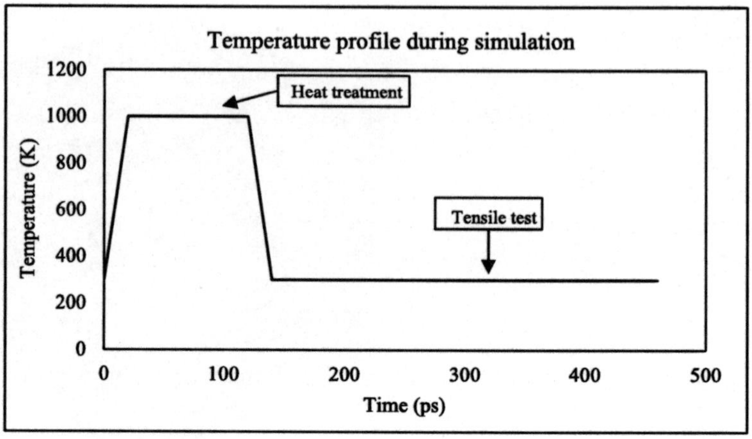

Fig. 2 Temperature profile of the simulation process

here, the $\Phi_{a_i-a_j}$ is the interaction of pair potential as a function of distance r_{ij} between atom i and j, namely a_i and a_j ($= A$ *or* B), while F_{a_i} is the embedding energy of an atom, namely a_i as a function of host electron density $\overline{\rho}_i$ induced at site i by all the atoms surrounding.

3 Result and Discussion

The plasticity response of metallic material mainly depends on the presence or absence of dislocations before straining. Thus, the strain rate during the tensile test is an essential factor. In this study, the value of $2.64e9/s$ which have been used in several studies [24–27]. Figure 3 shows the time evolution of the Cu-Ni alloy during the final stage of heat treatment and tensile test. The figure is analyzed with the displacement vector, and the magnitude of displacement is coloured with several levels. Blue is the lowest magnitude, while red-coloured atoms show the highest magnitude. The dislocations generated by the reaction of the Cu–Ni system, which resulted in material slippage along lattice planes, are generated from the slip between the crystal at the grain boundary. However, the evenly distributed stress induced by the straining reduces the focused deformation, thus allowing it to have a distributed deformation at the grain boundary of every crystal. The strain-induced plasticity during the tensile test shows that the difference between heat treatment temperatures insignificantly affects the tensile strength. This phenomenon is confirmed by the stress–strain curve associated with time evolution, as depicted in Fig. 4. The slight difference in the tensile strength shown in Fig. 4 is only observed for the system with or without heat treatment. The sample without heat treatment, i.e., remains at 300 K has the lowest tensile strength. Generally, the stress–strain curve has shown that the Cu–Ni alloy is a material with brittle properties when they are cold. Therefore, if we are about to process this material to meet the necessary complex shape, it can only be achieved with a relatively high temperature.

The deformation of nanograins induced by the straining shows to diffuses one cluster to another. Figure 5 shows the Cu-Ni system with the highest value of heat treatment temperature, i.e., 1200 K, for the standard view and the view after being analyzed by the displacement vector. Although this phenomenon seems to weaken the nanograins' structure, the introduction of twinning and slipping between the grains

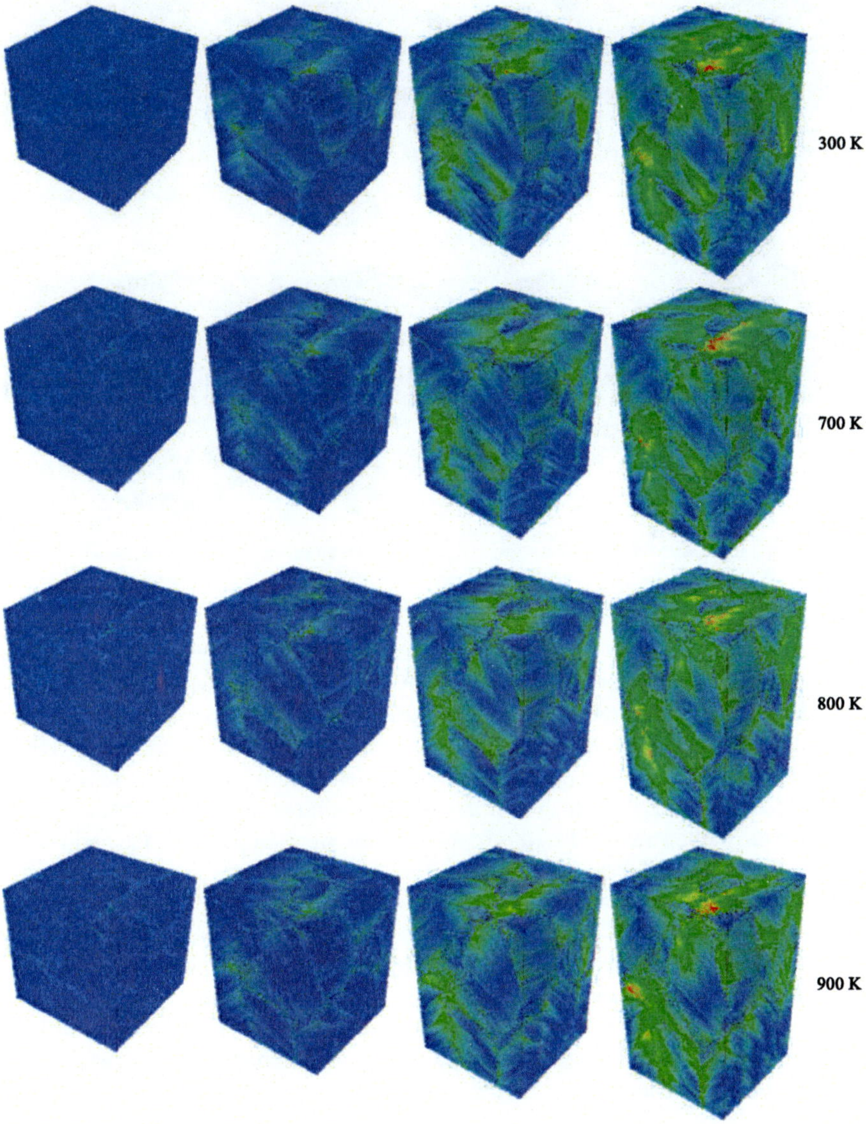

Fig. 3 Time evolution of the Cu–Ni alloy system during heat treatment and tensile test

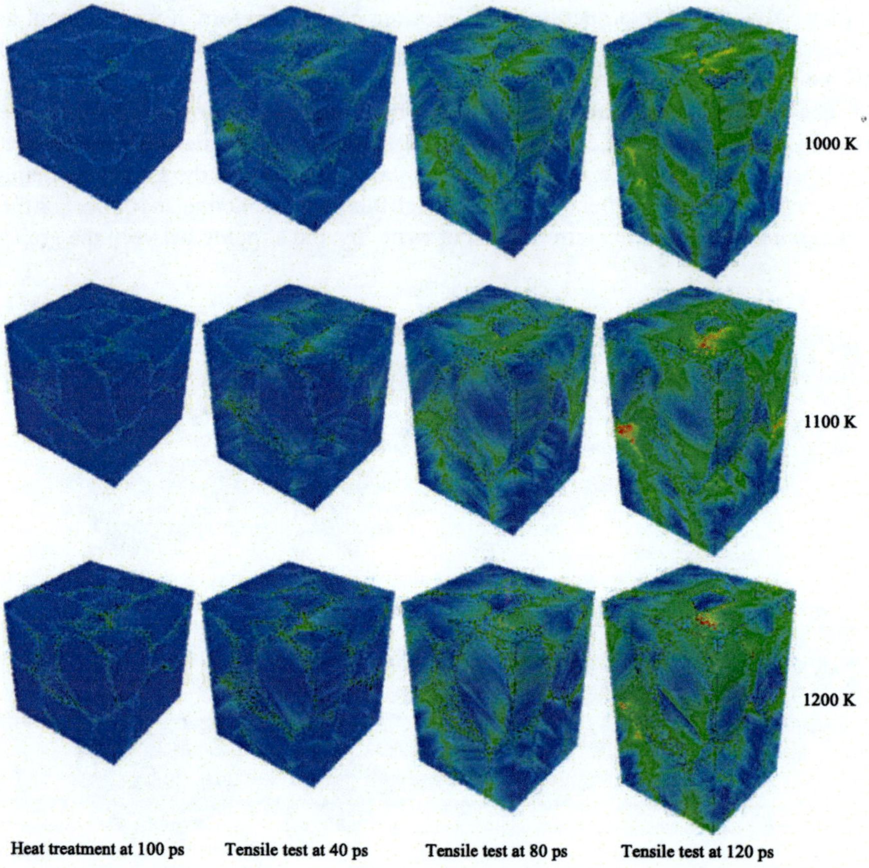

Fig. 3 (continued)

accommodate the plasticity well indicated by the low grains' displacement. Moreover, the stress-induced during straining, which will typically weaken the weakest part of the sample during the tensile test, in this case, is not employed. In molecular dynamics simulation, where the model could be designed with a perfect sample, the stress-induced during the tensile test is evenly distributed along the z plan. As the timesteps grow, the stress-induced is focused on the weakest part, i.e., the part where the grain boundary is located. The displacement vector analysis shows that the most probably crack will occur from the grain boundary as it shows the highest magnitude of displacement indicated by the difference of colour from the lower group in Fig. 5. In another case, the higher displacement magnitude is not necessarily showing the most probable site of crack propagation. Still, instead, it indicates a better bonding structure since, from the first step, the heat treatment process is performed. Furthermore, from the simulation, the distributed stress could help prevent the occurrence of cracks in the system.

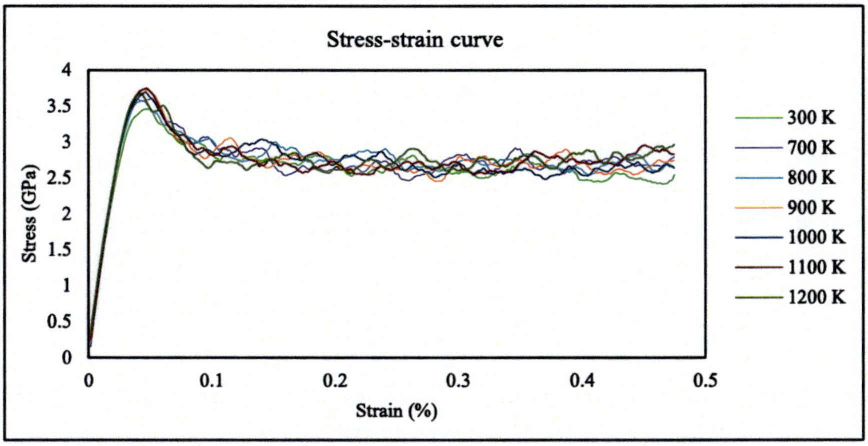

Fig. 4 Stress–strain curve of different temperature of heat treatment of polycrystal Cu–Ni

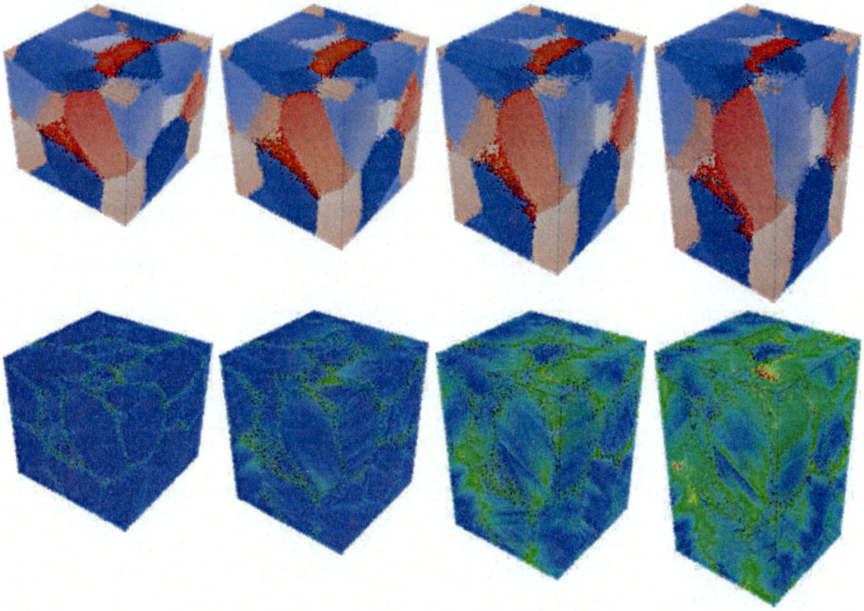

Heat treatment at 100 ps Tensile test at 40 ps Tensile test at 80 ps Tensile test at 120 ps

Fig. 5 Time evolution of Cu–Ni system with heat treatment temperature of 1200 K during the last stage of heat treatment and the stage of tensile test for a standard view (upper group) and displacement magnitude level view (lower group)

4 Conclusion

This study has successfully captured the detailed features of the plasticity of Cu–Ni alloy at the atomic scale. The heat treatment shows an insignificant effect on the mechanical properties of Cu–Ni alloy, but all the heat-treated alloy shows an increase in tensile strength. This is due to the reorientation and strengthening mechanism of the structure during heat treatment indicated by the diffusion between the crystal at the grain boundary. The deformation induced by the straining is mainly accommodated by the slip between the crystal. Since the stress induced during straining is distributed evenly, the Cu–Ni system shows to deform evenly, but the dislocation of that defines the plasticity is mainly generated by the dislocation slip along lattice planes.

Acknowledgements The authors would like to thank Jakarta Global University for its support throughout this project. This project is funded by the Ministry of Education, Culture, Research and Technology of Indonesia under the scheme of Penelitian Dosen Pemula (PDP) grant no. 0357/E5/AK.04/2022.

References

1. Faulkner JS (1982) The modern of theory of alloys. Prog Mater Sci (United Kingdom) 27
2. Luborsky FE (1983) Amorphous metallic alloys 1
3. Starke EA Jr, Staley JT (1996) Application of modern aluminum alloys to aircraft. Prog Aerosp Sci 32(2–3):131–172
4. Ukai S, Fujiwara M (2002) Perspective of ODS alloys application in nuclear environments. J Nucl Mater 1(307):749–757
5. Fabrication and electromagnetic interference shielding performance of open-cell foam of a Cu–Ni alloy integrated with CNTs
6. Ji K, Zhao H, Zhang J, Chen J, Dai Z (2014) Fabrication and electromagnetic interference shielding performance of open-cell foam of a Cu–Ni alloy integrated with CNTs. Appl Surf Sci 30(311):351–356
7. Mizushima I, Chikazawa M, Watanabe T (1996) Microstructure of electrodeposited Cu–Ni binary alloy films. J Electrochem Soc 143(6):1978
8. Studt F, Abild-Pedersen F, Wu Q, Jensen AD, Temel B, Grunwaldt JD, Nørskov JK (2012) CO hydrogenation to methanol on Cu–Ni catalysts: theory and experiment. J Catal 1(293):51–60
9. Ahmed J, Ramanujachary KV, Lofland SE, Furiato A, Gupta G, Shivaprasad SM, Ganguli AK (2008) Bimetallic Cu–Ni nanoparticles of varying composition (CuNi3, CuNi, Cu3Ni). Colloids Surf A 331(3):206–212
10. Pellicer E, Varea A, Pané S, Nelson BJ, Menéndez E, Estrader M, Surinach S, Baró MD, Nogués J, Sort J (2010) Nanocrystalline electroplated Cu–Ni: metallic thin films with enhanced mechanical properties and tunable magnetic behavior. Adv Func Mater 20(6):983–991
11. Bakharev PV, Huang M, Saxena M, Lee SW, Joo SH, Park SO, Dong J, Camacho-Mojica DC, Jin S, Kwon Y, Biswal M (2020) Chemically induced transformation of chemical vapour deposition grown bilayer graphene into fluorinated single-layer diamond. Nat Nanotechnol 15(1):59–66
12. Stronge WJ, Yu T (2012) Dynamic models for structural plasticity. Springer Science & Business Media
13. Yu MH, Ma GW, Li JC (2009) Structural plasticity: limit, shakedown and dynamic plastic analyses of structures. Springer Science & Business Media

14. Christofidou KA, Robinson KJ, Mignanelli PM, Pickering EJ, Jones NG, Stone HJ (2017) The effect of heat treatment on precipitation in the Cu–Ni–Al alloy Hiduron® 130. Mater Sci Eng A 24(692):192–198
15. Hirel P (2015) Atomsk: a tool for manipulating and converting atomic data files. Comput Phys Commun 1(197):212–219
16. Thompson AP, Aktulga HM, Berger R, Bolintineanu DS, Brown WM, Crozier PS, in't Veld PJ, Kohlmeyer A, Moore SG, Nguyen TD, Shan R (2022) LAMMPS-a flexible simulation tool for particle-based materials modeling at the atomic, meso, and continuum scales. Comput Phys Commun 271:108171
17. Stukowski A (2009) Visualization and analysis of atomistic simulation data with OVITO–the open visualization tool. Modell Simul Mater Sci Eng 18(1):015012
18. Stukowski A, Bulatov VV, Oppelstrup T, Zepeda-Ruiz LA (2018) Cross-scale molecular dynamics simulations of single crystal plasticity. In: NIC Symposium 2018 (No. FZJ-2018-02962). John von Neumann-Institut für Computing
19. Torrens I (2012) Interatomic potentials. Elsevier
20. O'connor DJ, Biersack JP (1986) Comparison of theoretical and empirical interatomic potentials. Nucl Instrum Methods Phys Res Sect B 15(1–6):14–19
21. Onat B, Durukanoğlu S (2013) An optimized interatomic potential for Cu–Ni alloys with the embedded-atom method. J Phys Condens Matter 26(3):035404
22. Daw MS, Baskes MI (1984) Embedded-atom method: derivation and application to impurities, surfaces, and other defects in metals. Phys Rev B 29(12):6443
23. Daw MS, Foiles SM, Baskes MI (1993) The embedded-atom method: a review of theory and applications. Mater Sci Rep 9(7–8):251–310
24. Zaenudin M, Abdulrazaq MN, Al-Zubaidi SS (2022) A review on molecular dynamics simulation of joining carbon-nanotubes and nanowires: joining and properties. Int J Integr Eng 14(4):137–159
25. Kasum K, Mulyana F, Zaenudin M, Gamayel A, Mohammed MN (2021) Molecular dynamics simulation on creep mechanism of nanocrystalline Cu–Ni alloy. Jurnal Fisika Flux: Jurnal Ilmiah Fisika FMIPA Universitas Lambung Mangkurat 18(1):67–74
26. Zaenudin M, Abdulrazaq MN, Al-Zubaidi SS (2020) Atomistic investigation on the effect of temperature on mechanical properties of diffusion-welded Aluminium–Nickel. Int J Integr Eng 12(5):62–69
27. Zaenudin M, Abdulrazaq MN, Al-Zubaidi S, Sunardi A (2020) Atomistic investigation on the role of temperature and pressure in diffusion welding of Al–Ni. J Eng Technol Sci 52(2)

Deformation Behaviour of a High Carbon Co–Cr–Mo Medium-Entropy Alloy via Thermal Oxidation

Rosdi Daud, H. Mas Ayu, R. Norizah, and A. K. Nasution

Abstract Thermal oxidation is widely used in substrate surface modification before applying bioceramic coating to enhance implant resistant against corrosion and biocompatibility. In this work, mechanical and microstructural characteristics of a medium-entropy alloy $Co_{61.9}Cr_{29.6}Mo_{6.5}C_{0.24}$ (in wt.%) was thermally oxidize under controlled furnace atmosphere, at 450 °C, 650 °C, 850 °C, 1050 °C and 1250 °C for 3 h. It has been demonstrated that by oxidizing high carbon Co–Cr–Mo alloy can create a reasonably thick, stable and hard Cr_2O_3 layer on the substrate at 1050 °C. The exchange of substrate colour from silver to brownish and lastly dark green of oxide layer was observed and is evident from the variations of the surface morphology which is also a cause response to the increment of temperature. Within the interval of extremely high temperature considered (1250 °C), the alloy exhibits several unusual features, such as presence of massive porosity with clumpy blended of distinct composition of oxide layer like $CoCr_2O_4$ and $Co_{0.8}Cr_{0.2}$. The results also demonstrate that Cr_2O_3 layer recede in thickness as the temperature reaches 1250 °C and experienced oxide layer spalling due to bounteous chemosorption of O_2 atoms react with dominant elements like cobalt and chromium in the substrate.

Keywords HC Co–Cr–Mo · Thermal oxidation · Medium-entropy alloy

R. Daud
Faculty of Mechanical and Automotive Engineering Technology (FTKMA), Universiti Malaysia Pahang, 26600 Pekan, Pahang, Malaysia

H. Mas Ayu (✉)
Faculty of Manufacturing and Mechatronic Engineering Technology (FTKPM), Universiti Malaysia Pahang, 26600 Pekan, Pahang, Malaysia
e-mail: masszee@ump.edu.my

R. Norizah
Department of Materials, Manufacturing and Industrial Engineering, School of Mechanical Engineering, Universiti Teknologi Malaysia, 81310 Skudai, Johor, Malaysia

A. K. Nasution
Department of Mechanical Engineering, Faculty of Engineering, Muhammadiyah University of Riau, Pekan Baru, Indonesia

C. N. Rosyidi et al. (eds.), *Proceedings of the 6th Asia Pacific Conference on Manufacturing Systems and 4th International Manufacturing Engineering Conference*, Lecture Notes in Mechanical Engineering,
https://doi.org/10.1007/978-981-99-1245-2_60

1 Introduction

The need for a new generation of metallic biomaterials with superior biocompatibility and mechanical properties to meet future demands of the medical field is appealing nowadays [1, 2]. Recently, many studies on high-entropy alloy (HEA) have gained unprecedented attention from researchers because of their promising properties [3, 4]. More recently, studies on medium-entropy alloy (MEA) which consisting only three equiatomic elements compared to five or more elements in HEAs, such as Co–Cr–Ni, has become more apparent since it exceeds high-entropy alloys in terms of strength, ductility and toughness [5]. Although Co–Cr–Ni MEAs possess potential in use for outer-space applications, such alloys have adverse effect upon application as biomedical implant. Early reports of seizing and loosening metallic implant were associated with metal staining due to wear debris and metal ions [6]. All these problems are mainly facilitated by the release of Cr and Ni ions into human body fluids [2].

A substitute material to the CoCr alloy such as Co–Cr–Mo MEA, have lately made it also popular in the making of dental and orthopedic implants [1]. Scientific literature reports that passive oxide layers form on the Co–Cr–Mo MEA surfaces are more stable than the metallic elements themselves, resulting in a barrier between the implant surface and the harsh aqueous environment, which effectively suppresses dissolving of the toxicity metallic elements in human body [7]. Even though, several other studies on this topic have been published, but their conclusions remain ambiguous. Specifically, the oxidation mechanism of HC Co–Cr–Mo MEA at high temperature needs serious investigations. Most importantly, because of the medium-entropy configuration of the alloy, the oxidation mechanism can be completely different from that of pure metals or binary alloys.

In the light of the background provided herein, the authors carried out an investigation on deformation behaviour of an equiatomic HC Co–Cr–Mo MEA when undergone various temperatures of thermal oxidation process in order to enhance surface morphologies which strongly influence in reducing metal ions release to human body. The surface morphology, elemental distributions, and oxide compositions during oxidation were analyzed to understand the evolution of the oxides. A wide temperature range (450–1250 °C) was considered to ensure the continuity and reliability of the study. Details of the study are elucidated as follow.

2 Materials and Methods

Round bar (14 mm in diameter) of HC Co–Cr–Mo MEA biomedical grade (ASTM F1537) were the material used in this study. The material was supplied with the following chemical compositions (in wt.%): C: 0.24; Cr: 29.6; Mo: 6.5; Si: 0.7; Ni: 0.1; Fe: 0.12; Mn: 0.7. N: 0.16 and Co: balance. The bar was cut using a precision cutter (Buehler, Isomet 4000) into disc size of 2 mm thick. Cylinder faces were wet ground using 500 Mesh and ultrasonically cleaned with acetone. The initial average of substrate surface roughness, Ra = 0.1 ± 0.02 μm. Thermal oxidation was carried out in muffle furnace at 450 °C, 650 °C, 850 °C, 1050 °C and 1250 °C for 3 h, which is below and above the annealing temperature (1121 °C) of HC Co–Cr–Mo MEA substrate [8, 9]. The surface morphology of the oxide HC Co–Cr–Mo MEA was observed using a Field Emission Scanning Electron Microscopy (FESEM, JEOL JSM-7800F, JEOL, Japan) set at an accelerating voltage of 5 kV. Elemental analysis was performed using dispersive energy X-ray (EDX LSM880-FESEM, Carl Zeiss AG, Germany). The crystal structure of the samples was characterized using X-ray diffraction (XRD) (Bruker D8 Model 2016, Berlin, Germany) with Cu Kα radiation (λ = 1.54060 Å), operated at 45 kV and 40 mA in the angular range 20–70° of 2θ. All the patterns were collected at an incident angle of 2° to minimise any interference from the substrate.

3 Results and Discussion

Morphological and structural properties. The FESEM images of the as-produced oxide HC Co–Cr–Mo MEA substrates are depicted in Fig. 1. It is assumed that the variation in the produced shapes, sizes and surface texture depends upon the thermal oxidation process and metallic element that reactant at particular temperature. The apparent grounded surface of HC Co–Cr–Mo MEA as indicate in Fig. 1a reveal that at 450 °C the microstructure does not change at all. However, the change of substrate colour from silver to brownish was observed. This behaviour is best explained due to formation of manganese oxide (MnO_2) on the substrate surface as provided in XRD results (Fig. 2). Even though the grounded surface still visible on oxidized substrate at 650 °C (Fig. 1b), closer look at the surfaces showed that there are notable nodules scattered on the substrate. This phenomenon can be explained by the possibility of Cr_2O_3 oxide particle begin to grow. The suspected oxide layer formation on this particular substrate was confirmed as XRD spectrum presented in Fig. 2 has identified Cr_2O_3 peak.

As the temperature increases to 850 °C (Fig. 1c), the surface morphology has deformed into denser and sharp peak like flake shape, which reflects more towards the deformation structure of Cr_2O_3 layer. This clearly shows that the growth of Cr_2O_3 layer by facile thermal oxidation route does affect the morphology to a certain extent but not the crystal structure which aligns well with the corresponding XRD pattern (Fig. 2). On the other hand, the substrate colour has change from brownish to dark green at this temperature.

Further inspection on the oxidized substrate at 1050 °C (Fig. 1d), the formation of cubic shape particle was observed. XRD analysis confirmed the rhombohedral structure of the material with lattice parameter, a = b = 4.958 Å; c = 13.594 Å. However, at this stage the substrate colour still remain the same as mentioned earlier. As seen in the Fig. 1e, the aggregation occurs only on substrate that encounter highest temperature (1250 °C) of oxidation. The micrographs indicate that the substrate has

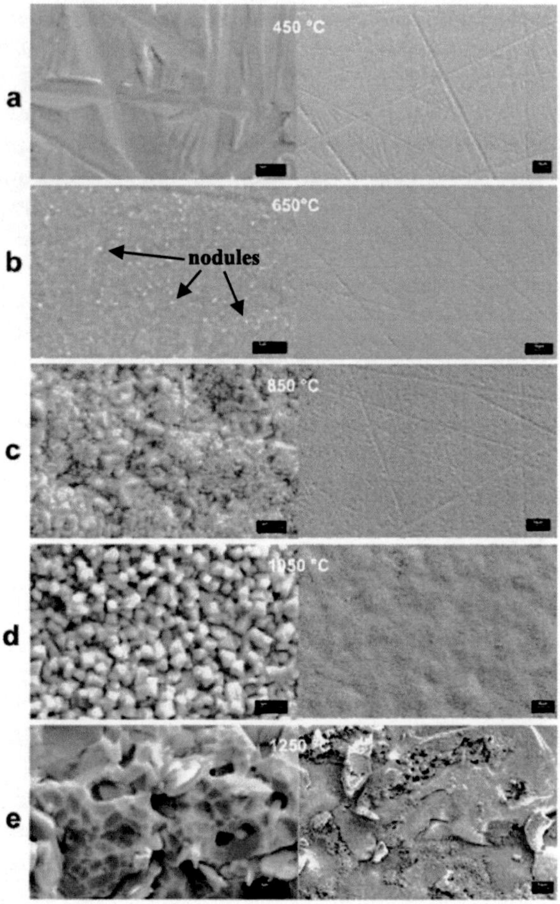

Fig. 1 The corresponding FESEM micrographs for oxidized HC Co–Cr–Mo MEA prepared at **a** 450, **b** 650, **c** 850, **d** 1050 and **e** 1250 °C for 3 h

protrusions and a rough heterogeneous oxide surface. Indeed, as evident from the FESEM images, there is also a presence of porosity and delamination of oxide layer which is not visible at lower oxidation temperature. It is fascinating findings that formation of Cr_2O_3 layer on the substrate at different oxidation temperature possess unchanged rhombohedral structure even though the substrate morphology varied as discernible in FESEM micrographs.

The corresponding EDX in the selected sites for the oxidized substrates at 450, 650, 850, 1050 and 1250 °C are shown in Fig. 3. The mapping images show a significant difference in the distribution of the chemical elements for each substrate. The spectrum displays the corresponding peaks dominant by cobalt (49.38 at.%) and chromium (26.61 at.%) which clearly reflect that oxidation at lowest temperature (450 °C) not much attain on the substrate (Fig. 3a). Although high amount detects in EDX, but absence of peaks compound related to chromium in XRD analysis which is believed at this stage chromium element have not reactant with any element yet due

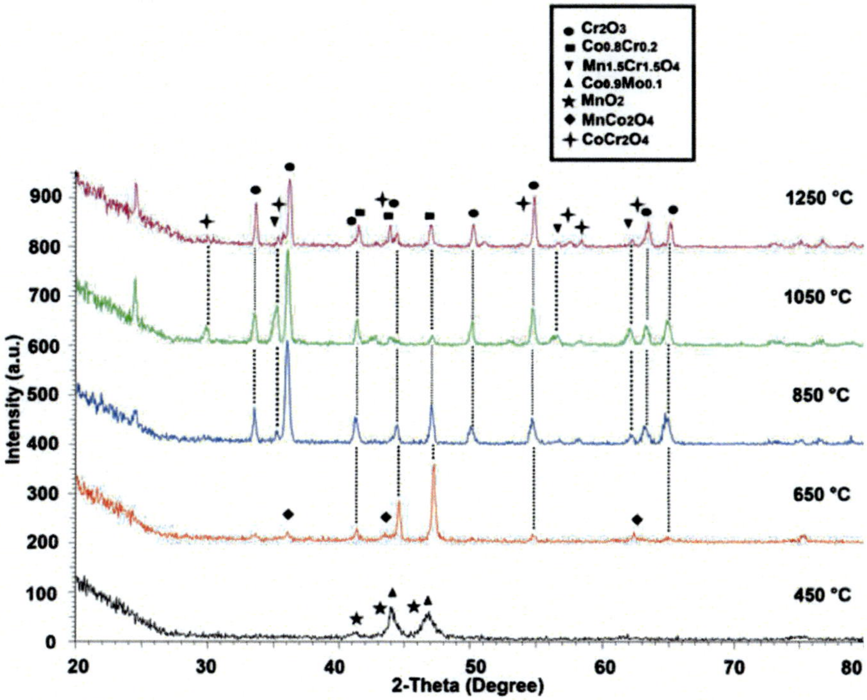

Fig. 2 The XRD pattern of oxidized HC Co–Cr–Mo MEA prepared at **a** 450, **b** 650, **c** 850, **d** 1050 and **e** 1250 °C for 3 h

to energy supplied during oxidation not sufficient to excite the chromium element [9]. The element detection probably due to substrate itself content high amount of chromium element as indicated in material composition earlier.

Further inspection on EDX results, detect that high atomic percentage of oxygen (33.68 at.% versus 7.6 at.%) in the oxidized surface at 650 °C (Fig. 3b). This scenario indicates that bounteous chemosorption of O_2 atoms has suddenly react with dominant elements such as cobalt and chromium and form into Cr_2O_3, $Co_{0.8}Cr_{0.2}$ and $MnCo_2O_4$. In contrast with the XRD result for 650 °C substrate, none manganese element was detected in the EDX pattern. This discrepancy can be explained by the possibility of oxide layer formation on the substrate surface during oxidation, obscuring features of manganese element to be detected. As temperature increased from 850 °C to 1050 °C, both substrates indicate similar elements with distinct amount of atomic percentage as indicate in Fig. 3c and d. In addition, a slight reduction of element chromium and oxygen while, almost double amount of manganese as temperature rise to 1050 °C. At this point, the grains have varied in cubic shapes that span over on the oxidized substrate indicating that the grains grow from the underlying solidified grains into an epitaxial trend (Fig. 1c and d) [2]. Furthermore, as per the EDX analysis results, substrate oxidized at highest temperature (1250 °C)

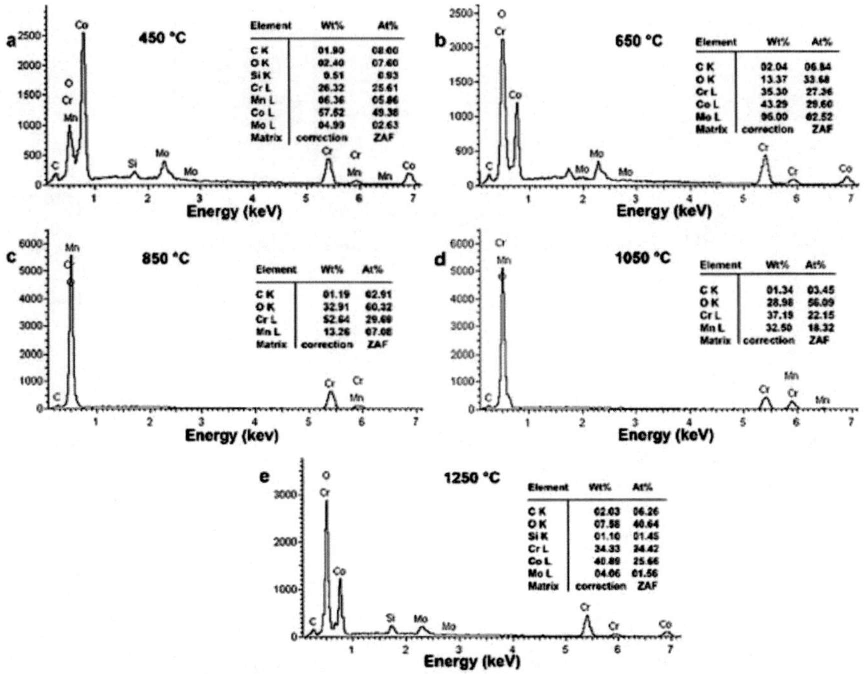

Fig. 3 The corresponding EDX analyses for oxidized HC Co–Cr–Mo MEA prepared at **a** 450, **b** 650, **c** 850, **d** 1050 and **e** 1250 °C for 3 h

shown to have almost all elements were identified (Fig. 3e). At this point, again manganese element varnish from the peaks. This finding is in good agreement with corresponding XRD pattern as minor spectra of $Mn_{1.5}Cr_{1.5}O_4$ oxide was seen. One possible reason might be ascribed to the $Mn_{1.5}Cr_{1.5}O_4$ oxide might dissolve and spall off from the substrate during oxidation process [10].

As previously noted, the oxidation temperature influences the size/morphology of the Cr_2O_3 oxide as evinced by the FESEM micrographs. With the changes in the temperatures, the establishment of new morphologies and different dispersibility states manifest. It is worth mentioning that the temperatures alter the crystalline phase of oxide layers.

4 Conclusions

In this study, the deformation behaviour of HC Co–Cr–Mo MEA via thermal oxidation process reveals distinct oxides formed at particular temperature. The presence of Cr_2O_3 layer was observed in all oxidized substrates except at the lowest oxidation temperature. However, it is worth noting that most of the oxide layer was spalled off and this scenario was only found on the substrate at highest oxidation temperature. It has been suggested that excessive temperature can cause adhesive failure in the case of an oxide layer. The post mortem of surface morphology, phase purity and elemental composition demonstrate that the pre-eminent Cr_2O_3 layer was performed at 1050 °C as stability of oxide layer significantly affecting the rate of ion release from the alloy in the long term. These results enable additional extensive investigations to validate further the correlation for these oxide layers with the release of toxicity ions in in-vitro testing. Overall, the reported outputs can provide relevant insight of the HC Co–Cr–Mo MEA to be utilized as a safe implant material in biomedical application.

Acknowledgements The authors would like to acknowledge the support by the Ministry of Higher Education Malaysia under the Fundamental Research Grant Scheme (FRGS) Ref: FRGS/1/2018/TK03/UMP/02/5 (University reference number RDU190114) and Collaborative Research Grant Scheme (CRGS) RDU192309 provided by Universiti Malaysia Pahang (UMP). Special thanks to Universiti Teknologi Malaysia (UTM) Skudai for providing additional financial assistance under CRG 26.0 grant program.

References

1. Gurel S, Nazarahari A, Canadinc D, Gerstein G, Maier HJ, Cabuk H, Bukulmez T, Cananoglu M, Yagci MB, Toker SM, Gunes S, Soykan MN (2022) From corrosion behavior to radiation response: a comprehensive biocompatibility assessment of a CoCrMo medium entropy alloy for utility in orthopedic and dental implants. Intermetallics 149:107680
2. Mace A, Khullar P, Bouknight C, Gilbert JL (2022) Corrosion properties of low carbon CoCrMo and additively manufactured CoCr alloys for dental applications. Dent Mater 38(7):1184–1193
3. Wetzel A, von der Au M, Dietrich PM, Radnik J, Ozcan O, Witt J (2022) The comparison of the corrosion behavior of the CrCoNi medium entropy alloy and CrMnFeCoNi high entropy alloy. Appl Surf Sci 601:154171
4. Günen A, Lindner T, Karakaş MS, Kanca E, Töberling G, Vogt S, Gök MS, Lampke T (2022) Effect of the boriding environment on the wear response of laser-clad AlCoCrFeNi high entropy alloy coatings. Surf Coat Technol 447:128830
5. Cao F, Munroe P, Zhou Z, Xie Z (2018) Medium entropy alloy CoCrNi coatings: enhancing hardness and damage-tolerance through a nanotwinned structuring. Surf Coat Technol 335:257–264
6. Wang Q, Eltit F, Feng R, Garbuz D, Duncan C, Masri BA, Greidanus N, Cox ME, Wang R (2022) Nature of fretting corrosion products in CoCrMo hip implants from in vivo study to in vitro simulation. Materialia 22:101433
7. Mas AH, Daud R, Kurniawan T, Alias J, Izman S, Shah A, Anwar M (2019) Improving biocompatibility of cobalt based alloy using chemical etching and mechanical treatment. Materialwiss Werkstofftech 50(3):254–259
8. Baddar BI, Shah A, Mas AH, Daud R, Dambatta MS (2020) Hydroxyapatite and thermal oxidation as intermediate layer on metallic biomaterial for medical implant: a review. J Adv Res Fluid Mech Thermal Sci 62(1):138–150
9. Hassan MA, Izman S, Daud R, Abdullah AS, Anwar M, Genasan K, Zaman T (2017) In-vitro biocompatibility study of hydroxyapatite coated on Co–Cr–Mo with oxide interlayer. Jurnal Teknologi 80

10. Li H, Wang M, Lou D, Xia W, Fang X (2020) Microstructural features of biomedical cobalt–chromium–molybdenum (CoCrMo) alloy from powder bed fusion to aging heat treatment. J Mater Sci Technol 45:146–156

Mechanical Performance Evaluation of Al_2O_3–ZrO_2–Ti Composite Materials

Nur Aziera, Mebrahitom Asmelash, Azmir Azhari, and Kushendarsyah Saptaji

Abstract The applications of ceramic based materials is becoming a trend due to their important mechanical and microstructural properties. This work evaluates the mechanical properties of ceramics composite of Al_2O_3–ZrO_2 reinforced with Titanium (Ti) particles. The material is known for the application of biomedical devices due to the optimized functional properties. The mechanical properties of Al_2O_3–ZrO_2 composites can be enhanced by incorporating Ti. To measure the mechanical properties, standard size samples of the ceramic composites were fabricated through powder processing method to consolidate the powders. A varying composition of Ti ranging from 5–45% by volume was used. Then, the measurements were conducted on each sample to determine the physical and mechanical performance of the composite. As a result, the newly produced composite materials with 75 vol.% Al_2O_3 + 10 vol.% ZrO_2 + 15 vol.% Ti showed better physical and mechanical properties than the other compositions based on the requirements in the biomedical applications.

Keywords Alumina · Zirconia · Titanium · Powder metallurgy · Flexural strength · Density · Ceramics

N. Aziera
Faculty of Mechanical and Automotive Engineering Technology, Universiti Malaysia Pahang, Pahang, Malaysia

M. Asmelash (✉) · A. Azhari
Faculty of Manufacturing and Mechatronics Engineering Technology, Universiti Malaysia, Pahang, Malaysia
e-mail: mebrahitoma@ump.edu.my

K. Saptaji
Department of Mechanical Engineering, Faculty of Engineering and Technology, Sampoerna University, Jakarta, Indonesia

C. N. Rosyidi et al. (eds.), *Proceedings of the 6th Asia Pacific Conference on Manufacturing Systems and 4th International Manufacturing Engineering Conference*, Lecture Notes in Mechanical Engineering,
https://doi.org/10.1007/978-981-99-1245-2_61

1 Introduction

Ceramic composite is known as biomaterial or ceramic biomaterial when it is used as a material to substitute the original part of living beings. Biomaterials have played a vital role in the medical field as its functionality to be an implant for replacing the damage or missing parts in human's body [1, 2].

Nowadays, the development of ceramic composites with the advantageous properties of high hardness, strength, able to withstand prolonged corrosion, high pressure and high temperature environments, excellent fracture toughness with high resistance to brittleness is the main issue in materials science [3]. However, the brittleness property of ceramic is the main problem that can be the cause of failure during the medical procedure of a living organism. Hence, there are several studies [4] to improve the mechanical properties of the ceramic composites. Kong et al. [5] used Hydroxyapatite (HA) to improve the mechanical properties of the ZrO_2–Al_2O_3 composites which was found compatible for biomedical applications. Moreover, Agac et al. [6] used a uniaxial pressure compact with a pressureless sintering od temperature 1450 °C and managed to develop a ZrO_2–Al_2O_3–TiO_2 composites.

This research mainly focuses on the effect of Ti reinforcement on the Alumina (Al_2O_3), Zirconia (ZrO_2) and Titanium (Ti) (Al_2O_3–ZrO_2–Ti) composite on the mechanical and physical properties for the application of biomedical devices.

2 Methodology

A standard die and punch with dimension 75 mm × 10 mm based on ASTMB925 standard was fabricated to produce the samples as shown in Fig. 1.

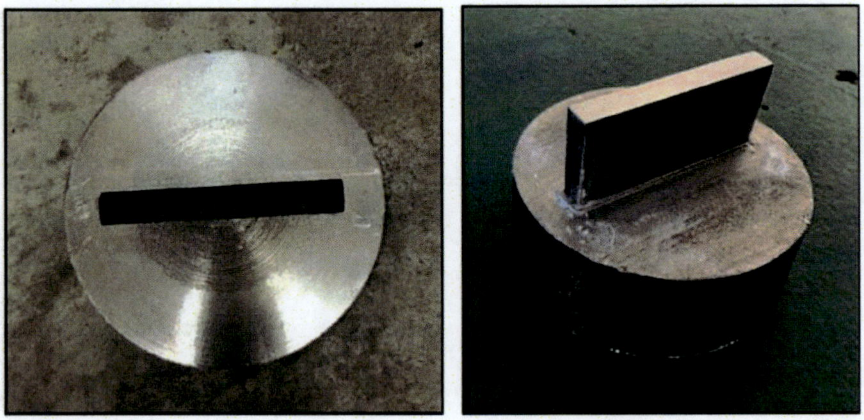

Fig. 1 The die and punch

Table 1 Table of the ratio for materials in composites

In composites			
Samples	Percentage of content		
	Al_2O_3, %	ZrO_2, %	Ti, %
S1	85	10	5
S2	80	10	10
S3	75	10	15
S4	70	10	20
S5	65	10	25
S6	60	10	30
S7	55	10	35
S8	50	10	40
S9	45	10	45

A powder metallurgy processing such as powder preparation, mixing and blending, compacting, sintering, and then finishing operations was followed to create ceramic composites. Powders Al_2O_3 with the size of 176 μm, ZrO_2 with the size of 150 μm and Ti with the size of 149 μm were used in the sample preparation. The ratio of Ti particles used for reinforcement is shown in Table 1. The total mass of the composites is roughly around 30 g for each sample by considering the average density of each material and assumption of thickness wanted for the sample with die's area. The three powder materials were mixed with the binders and water. After drying, grinding and mixing the powder, a mesh size 200 μm was used to make the sizes uniform. A uniaxial hydraulic compaction was used to form the green compact as shown in Fig. 2. The samples were sintered at 1450 °C with dwell time of 2 h and 10 °C increment per minute. The grinding operation was done by using sandpaper with grit size from 120 to 1200 for further microstructural investigation. The microstructure of the sintered samples was examined using the SEM-EDX to investigate the phases and compositions of the composite materials. Furthermore, the density and flexural strength evaluation was performed using Archimedes principle and 3-point bending test methods respectively.

3 Results and Discussion

Figure 3 shows the phase and elemental analysis results indicating the three phases namely, Al_2O_3, ZrO_2 and Ti as particles and oxide form as TiO_2 distributed evenly in the structure. The observation also demonstrates that the shapes of the phases are irregular with multiple cavities. The change in morphology and size of the powders has a direct impact on the phases developed in ceramic matrix [7].

The graph in Fig. 4 displays a decreasing density trend from S1 to S9, while both S3 and S8 were fluctuating and therefore not following the trend. The fluctuation may

Fig. 2 Rectangular-shaped samples after sintering sintering

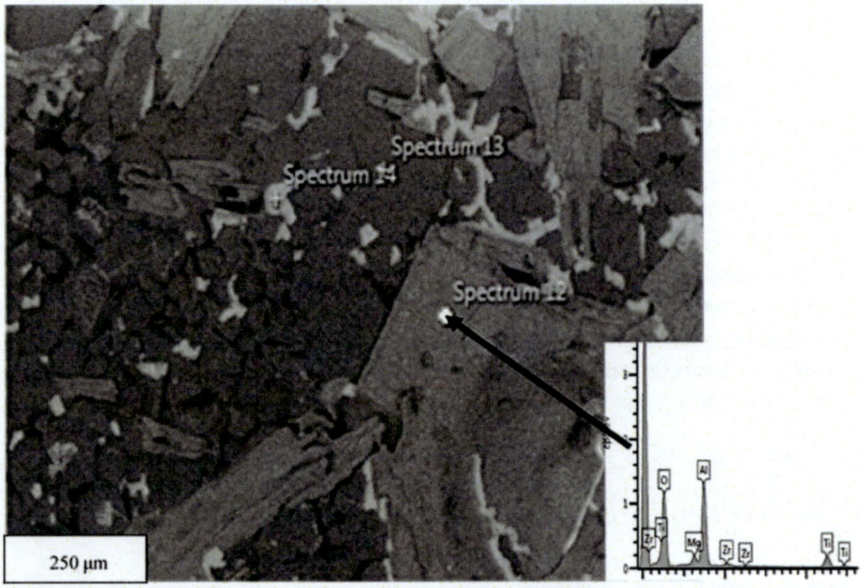

Fig. 3 Microstructure of sample 3 using electron microscope (SEM)

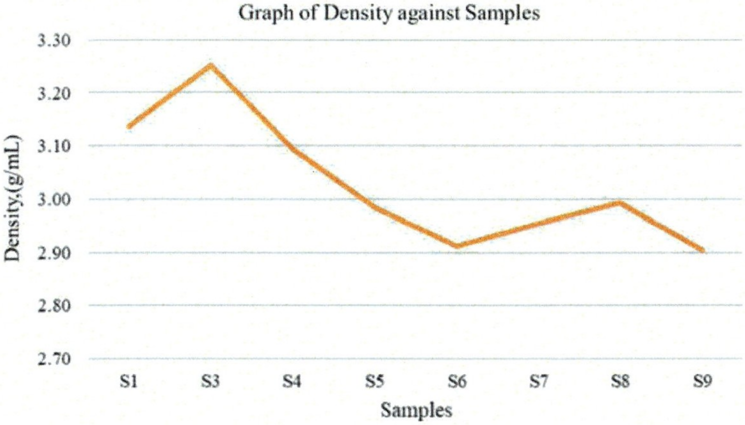

Fig. 4 Graph of density against Al_2O_3–ZrO_2–Ti composites

happen due to inconsistency of mass between each sample where S3 and S8 were having excess mass compared to another samples. According to studies, combining ceramic powders and metal powders with significant particle size differences may result in an increase in density in the mixed composites [8]. The reason for this is that smaller ceramic particles will be able to pack into vacant spaces between bigger metal particles. The density of composites is directly influenced by the morphology and size of the particles. In this work, the density is roughly decreasing by the increasing ratio of Ti particles. It could be due to the fact that Al_2O_3, ZrO_2, and Ti have different particle sizes. In this situation, the greater particle size of Ti may make the densification process more difficult, despite the possibility of smaller particle size ceramic powders being located within Ti particles. The lower the density of Al_2O_3–ZrO_2–Ti composites will lead to lower hardness. Ceramics material is classified to have high hardness when comparing to metal material, hence the decreasing hardness is equivalent to the increasing Ti particles from S1 to S9 samples.

Figure 5 shows the graph of force against displacement obtained from S1 to S9. The composition of S1 of the red colour curve denoted as 1 in the graph, are 85 vol.% Al_2O_3 + 10 vol.% ZrO_2 + 5 vol. % Ti. S1 has shown the obvious characteristic of ceramics where it needs the highest force to displace the sample when comparing to other samples. This is due to majority of compositions for S1 are made from 95 vol.% ZTA of ceramics. Ceramics material is familiar to its property of high hardness; hence it requires a lot of force to break the sample. The force needed to displace samples are reducing from S1 to S9 as shown on the graph above. It was concluded that the hardness of all the samples had declined significantly [9, 10].

Figure 6 shows the graph of flexure stress against flexure strain obtained from S1 to S9. S1 has portray the similar characteristic of stress–strain curve for pure ceramics where it doesn't have significant plastic phase. S1 which 95% of its composites were made from ceramic was not having high ductility that cause the curve to show the brittleness criteria where it went rupture drastically after elastic phase. As the ratio

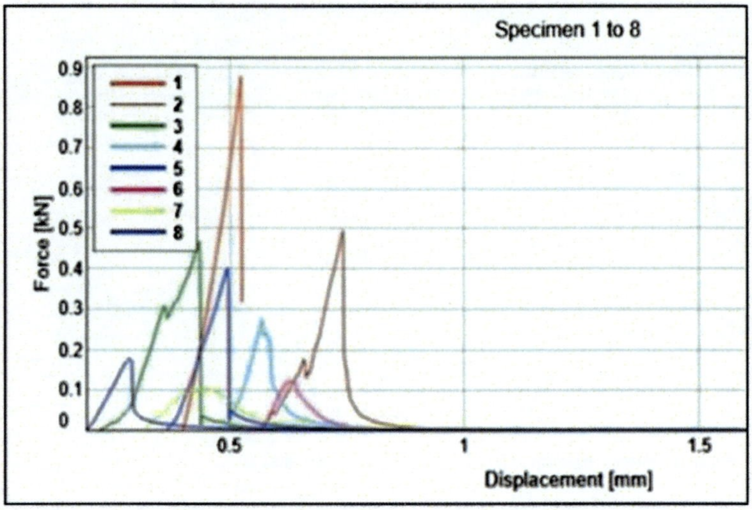

Fig. 5 Graph of force against displacement

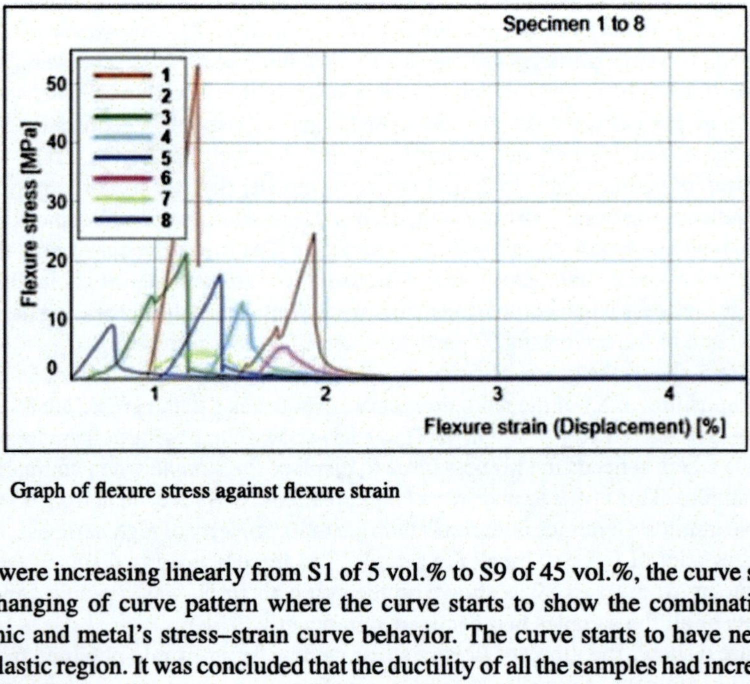

Fig. 6 Graph of flexure stress against flexure strain

of Ti were increasing linearly from S1 of 5 vol.% to S9 of 45 vol.%, the curve shows the changing of curve pattern where the curve starts to show the combination of ceramic and metal's stress–strain curve behavior. The curve starts to have necking and plastic region. It was concluded that the ductility of all the samples had increasing significantly.

4 Conclusion

In conclusion, the Al_2O_3–ZrO_2–Ti ceramic composites were fabricated to investigate the effect of Ti composition on the mechanical properties and its density. The increase of Titanium's ratio from 5 vol.% to 45 vol.% in the ceramic's composites of Al_2O_3–ZrO_2 was able to reduce the brittleness of the ceramics. The newly developed compositions have shown a superior characteristic such as has high hardness with ductility, lower brittleness that increase toughness and high in strength. The composite with 75 vol.% Al_2O_3 + 10 vol.% ZrO_2 + 15 vol.% Ti composition shows optimum mechanical and physical properties which is important for the applications in the biomedical devices.

Acknowledgements Authors would like to gratefully acknowledge the financial support from the Ministry of Higher Education Malaysia through FRGS/1/2019/TK03/UMP/02/25 (RDU1901161).

References

1. Huang J, Best SM (2022) Ceramic biomaterials for tissue engineering. Tissue Eng Ceram Polym 3–40. https://doi.org/10.1016/B978-0-12-820508-2.00007-6
2. Rahmani R, Kamboj N, Brojan M, Antonov M, Prashanth KG (2022) Hybrid metal-ceramic biomaterials fabricated through powder bed fusion and powder metallurgy for improved impact resistance of craniofacial implants. Materialia 24:101465. https://doi.org/10.1016/J.MTLA.2022.101465
3. Mebrahitom Asmelash G, Mamat O (2012) Processing and characterisation of $Al_2O_3SiO_2ZrO_2$ composite material. Int J Microstruct Mater Prop 7:64–76
4. Mebrahitom Asmelash G, Mamat O, Ahmad F (2015) Thermal shock and fatigue behavior of pressureless sintered Al_2O_3–SiO_2–ZrO_2 composites. J Adv Ceram 4:190–198. https://doi.org/10.1007/s40145-015-0146-0. Accessed 7 Sept 2022
5. Kong YM, Bae CJ, Lee SH, Kim HW, Kim HE (2005) Improvement in biocompatibility of ZrO_2–Al_2O_3 nano-composite by addition of HA. Biomaterials 26:509–517. https://doi.org/10.1016/J.BIOMATERIALS.2004.02.061
6. Agac O, Gozutok M, Sasmazel HT, Ozturk A, Park J (2017) Mechanical and biological properties of Al_2O_3 and TiO_2 co-doped zirconia ceramics. Ceram Int 43:10434–10441. https://doi.org/10.1016/J.CERAMINT.2017.05.080
7. Sabuan NA, Asmelash M, Azhari A, Mulubrhan F, Alemu T (2020) Investigation on the electrical conductivity of Al_2O_3–Ti ceramic composites using a pressureless sintering process. Mater Today Proc. https://doi.org/10.1016/j.matpr.2020.07.520
8. Sankhla AM, Patel KM, Makhesana MA, Giasin K, Pimenov DY, Wojciechowski S, Khanna N (2022) Effect of mixing method and particle size on hardness and compressive strength of aluminium based metal matrix composite prepared through powder metallurgy route. J Mater Res Technol 18:282–292. https://doi.org/10.1016/J.JMRT.2022.02.094
9. Asmelash AF, Mebrahitom G, Mamat O (2012) Toughening mechanisms of Al_2O_3–SiO_2–ZrO_2 composite materials. Ceramics-Silikáty 56:360–366
10. Giordano RA, Pelletier L, Campbell S, Pober R (1995) Flexural strength of an infused ceramic, glass ceramic, and feldspathic porcelain. J Prosthet Dent 73:411–418. https://doi.org/10.1016/S0022-3913(05)80067-8

Author Index

A
Abdillah, Irsyad Thalib, 447
Achmad, Fandi, 3, 205, 285
Achmad, T. P., 87
Afraah, Sayyidah Maulidatul, 365
Ajidarma, Praditya, 517
Alaudin, G. M. A., 87
Alghofari, Ahmad Kholid, 383
Alifin, Fakhri I., 457
Ali, Irfansyah, 509
Ali, M. Y., 21
Amalia, Tatinia A. R., 183
Amelia, Mia, 151, 447
Amran, Muhammad Zulhilmi Bin Shamsul, 401
Andini, Nurul Fitri Dwi, 447
Andini, Syania Hilda, 391
Andriyani, Hanin Tsimar, 247
Anisa, 607
Ardiyanto, Agus, 339
Aribowo, Wisnu, 51, 509
Ariqoh, Amanda Syifa, 495
Ariviani, Setyaningrum, 301
Asmelash, Mebrahitom, 59, 657
Astuti, Rahmaniyah Dwi, 67, 161
Atthariq, Muhammad Junus, 67
Aussawarangkul, Thitisak, 627
Awalia, Lili, 293
Azhari, Azmir, 59, 657
Azharudin, Muhammad Zuhairi, 115
Aziera, Nur, 657
Azizah, Dana, 447
Aziz, Fazilah Abdul, 425
Azlan, Nur Syahirah Mohamad, 539
Azmi, Nur Iffah Mohamed, 559

B
Basirisa, Muhammad Syahrain, 293

C
Cahyono, R. T., 87
Cakravastia, Andi, 457
Che Ghani, S. A., 627

D
Damayanti, Retno Wulan, 375, 495
Darmawi, Ahmad, 339
Daryanto, Yosef, 437
Daud, Rosdi, 649
Dwiastuti, Anik, 97

F
Fa'rifah, Riska Yanu, 285
Fauza, Gusti, 301, 527

G
Gamayel, Adhes, 639
Goesasi, Rachmat Zulkarnain, 485
Gultom, Shinta Uli Agnes Monica, 195

H
Halim, Abdul Hakim, 325
Hananta, Shafa Keysa Rinjani, 271
Handojo, Andreas, 331
Hasby, Fariz Muharram, 51
Hasian, Ferdinanda Pascha, 495
Hayati, Enty Nur, 375

C. N. Rosyidi et al. (eds.), *Proceedings of the 6th Asia Pacific Conference on Manufacturing Systems and 4th International Manufacturing Engineering Conference*, Lecture Notes in Mechanical Engineering,
https://doi.org/10.1007/978-981-99-1245-2

Hidayat, Yosi Agustina, 411
Hisjam, Muhammad, 131, 365, 593
Husniah, H., 569
Hutami, Audira, 517

I
Irianto, Dradjad, 139, 183, 485, 517
Iskandar, Bermawi P., 457, 569
Isnaini, Mohammad Mi'radj, 3, 139, 183
Izzati, Inaya, 577

J
Jauhari, Wakhid Ahmad, 97, 271, 375, 383
Jundullah, Muhammad Nihri, 607
Juniani, Anda Iviana, 97

K
Khairunnisa, Hasna, 339
Koeshardono, Fachri, 293
Kurnianto, Dicka Korintus, 495
Kurniawati, Amelia, 447

L
Laksono, Pringgo Widyo, 67, 213, 223, 233, 247, 257, 375, 383, 607
Liquiddanu, Eko, 257

M
Maharani, Dewi, 107
Ma'ruf, Anas, 195, 325
Mas Ayu, H., 649
Millenia, Aisyah Rizky, 301
Mohamed, Nik Mohd Zuki Nik, 425
Mokhtar, Mohd Saifuldin Mohd, 11
Mokhtar, Nazrin, 59
Muchjayadi, Mugo, 293

N
Nafiah, Latifu Itsnaini Khoirun, 607
Nandyasari, Tiara Priscilla, 437
Nasution, A. K., 649
Nisar, Muhammad Usman, 325
Nisfullaili, Josyika, 495
Norizah, R., 649
Nugroho, Imanuella Indah, 495
Nursanto, Djoko, 617
Nursiwi, Asri, 301
Nurtriandari, Erza, 485

P
Padmaningrum, Dwiningtyas, 301
Pambudi, Hardian Kokoh, 577
Permatasari, Hani Aninda Intan, 233
Pradana, Wali Hakim, 447
Prady, A. D. P., 87
Prasetya, Sonki, 617
Prasetyo, Hari, 301, 527
Prasiwi, Rosifa Sahara, 223
Pratama, Yoshan Ardhi, 257
Pujawan, Nyoman, 331
Pujiyanto, Eko, 593
Putri, Farhana Mazaya, 349

R
Rahman, M. M., 21
Ramadhandyna, Annisa Puja, 213
Razali, Noraini Mohd, 539
Ridwan, Ari Yanuar, 577
Rizaldi, Artamevia S., 317
Rizana, Afrin Fauzya, 3, 317, 447
Rose, Ahmad Nasser Mohd, 425
Rosida, Amirul Wika, 67
Rosyidi, Cucuk Nur, 213, 223, 233, 247, 271, 375, 383, 391
Rumanti, Augustina Asih, 151, 205

S
Safiei, W., 21
Saleh, Yasya Khalif Perdana, 639
Salma, Melati, 549
Santosa, Budi, 331
Saptaji, Kushendarsyah, 59, 657
Saraswati, Cindy, 527
Sari, Loviana Ajeng, 607
Sari, Santika, 107
Satriawan, Ardianto, 485
Sawangsri, Worapong, 627
Sazali, Norazlianie, 115, 123
Septiansyah, Fadlika, 617
Septiningrum, Lutfia, 3, 151, 205, 285
Setianingsih, Casi, 485
Shaharudin, M. A. H., 33
Singgih, Moses Laksono, 331, 473
Siswanto, N., 473
Soesanto, Rayinda P., 3, 317
Srikandi, Dessy, 593
Suhardi, Bambang, 67, 161, 349
Sukoyo, 51, 509
Sulistyowati, Wiwik, 97
Sunardi, Ade, 639
Suryadi, Aris, 293

Sutopo, Wahyudi, 131, 365
Sutrimo, 293
Suwignjo, P., 473
Syahliantina, Annisa, 161

T
Tengtarto, M. A. K., 473
Toat, Annas, 339
Trusaji, Wildan, 485
Tyastuti, Niken Utami, 161

U
Ujiburrahman, 639

W
Wangsaputra, R., 569
Wardiani, Gina Rahayu, 139
Wattanasinbumrung, Pakanun, 627
Wicaksono, Hendro, 131
Widyanti, Ari, 549
William, 51
Wiratmadja, Iwan Inrawan, 205, 285

Y
Yahya, Nafrizuan Mat, 559
Yuniaristanto, 131, 365
Yusoff, Ahmad Razlan, 11, 21, 33

Z
Zaenudin, M., 639
Zahid, Muhammed Nafis Bin Osman, 401
Zaid, Nik Nurjihan Nisrina Nik, 123